Lecture Notes in Physics

The Lecture Notes in Physics

The series Lecture Notes in Physics (LNP), founded in 1969, reports new developments in physics research and teaching – quickly and informally, but with a high quality and the explicit aim to summarize and communicate current knowledge in an accessible way. Books published in this series are conceived as bridging material between advanced graduate textbooks and the forefront of research and to serve three purposes:

- to be a compact and modern up-to-date source of reference on a well-defined topic

- to serve as an accessible introduction to the field to postgraduate students and nonspecialist researchers from related areas

- to be a source of advanced teaching material for specialized seminars, courses and schools

Both monographs and multi-author volumes will be considered for publication. Edited volumes should, however, consist of a very limited number of contributions only. Proceedings will not be considered for LNP.

Volumes published in LNP are disseminated both in print and in electronic formats, the electronic archive being available at springerlink.com. The series content is indexed, abstracted and referenced by many abstracting and information services, bibliographic networks, subscription agencies, library networks, and consortia.

Proposals should be sent to a member of the Editorial Board, or directly to the managing editor at Springer:

Christian Caron
Springer Heidelberg
Physics Editorial Department I
Tiergartenstrasse 17
69121 Heidelberg / Germany
christian.caron@springer.com

J.G. Muga
A. Ruschhaupt
A. del Campo (Eds.)

Time in Quantum
Mechanics - Vol. 2

 Springer

Editors
J. Gonzalo Muga
Universidad Pais Vasco
Depto. Quimica Fisica EHU
Apartado, 644
48080 Bilbao
Spain
jg.muga@ehu.es

Andreas Ruschhaupt
TU Braunschweig
Inst. Mathematische Physik
Mendelssohnstr. 3
38106 Braunschweig
Germany
a.ruschhaupt@tu-bs.de

Adolfo del Campo
Imperial College London
Inst. Mathematical Sciences
53 Prince's Gate
London
United Kingdom SW7 2PG
a.del-campo@imperial.ac.uk

Muga J.G., Ruschhaupt A., del Campo A. (Eds.), *Time in Quantum Mechanics - Vol. 2*,
Lect. Notes Phys. 789 (Springer, Berlin Heidelberg 2009),
DOI 10.1007/978-3-642-03174-8

Lecture Notes in Physics ISSN 0075-8450
ISBN 978-3-642-26193-0   e-ISBN 978-3-642-03174-8
DOI 10.1007/978-3-642-03174-8
Springer Heidelberg Dordrecht London New York

Cover design: Integra Software Services Pvt. Ltd., Pondicherry

Printed on acid-free paper

Springer is part of Springer Science+Business Media (www.springer.com)

Einstein changed dramatically our concept of time and thus of the world. By contrast, quantum mechanics, the other great twentieth century physical theory, has paid to time a much more modest and secondary attention, and most practitioners have even refused with stubborn determination to deal with some of its evident aspects, the "time observables," in our opinion without a good or sufficient reason. Less controversial but not at all less interesting and much influential have been the fundamental contribution of quantum mechanics to improve time measurement with atomic clocks, as well as the development of techniques to study quantum dynamics and characteristic timescales, both at theoretical and experimental levels, complementary to the knowledge on the structure and properties of matter derived from time-independent methods.

The aim of a workshop series at La Laguna, Spain, since the first edition in 1994, and of this book series is to promote and contribute to a more intense interplay between time and the quantum world. This volume fills some of the gaps left by the first one, recently re-edited. It begins with a historical review in Chap. 1. Most chapters orbit around fundamental concepts and time observables (Chaps. 2–6), or quantum dynamical effects and characteristic times (Chaps. 7–12). The book ends with a review on atomic clocks in Chap. 13. Several authors have participated in "Time in Quantum Mechanics" workshops at La Laguna or Bilbao, but we have not imposed this as a necessary condition. As in the first volume, our recommendation to all authors has been to write reviews that may serve both as an introductory guide for the noninitiated and a useful tool for the expert, leaving them full freedom for the choice of emphasis and presentation.

We would like to acknowledge the work, patience, and discipline of all contributors, as well as the support of the University of the Basque Country (UPV-EHU), Ministerio de Ciencia e Innovación (Spain), EU Integrated Project QAP, EPSRC QIP-IRC, German Research Foundation (DFG), and the Max Planck Institute for Complex Systems at Dresden, where much of our work was completed within the "Advanced Study Group" "Time: quantum and statistical mechanics aspects" organized by L. S. Schulman during the summer of 2008.

Bilbao, Braunschweig, London,   *J.G. Muga, A. Ruschhaupt, and A. del Campo*
                                                                    January 2009

Preface

> *But all the clocks in the city*
> *Began to whirr and chime:*
> *'O let not Time deceive you,*
> *You cannot conquer Time.*

> W. H. Auden

It is hard to think of a subject as rich, complex, and important as time. From the practical point of view it governs and organizes our lives (most of us are after all attached to a wrist watch) or it helps us to wonderfully find our way in unknown territory with the global positioning system (GPS). More generally it constitutes the heartbeat of modern technology. Time is the most precisely measured quantity, so the second defines the meter or the volt and yet, nobody knows for sure what it is, puzzling philosophers, artists, priests, and scientists for centuries as one of the enduring enigmas of all cultures. Indeed time is full of contrasts: taken for granted in daily life, it requires sophisticated experimental and theoretical treatments to be accurately "produced." We are trapped in its web, and it actually kills us all, but it also constitutes the stuff we need to progress and realize our objectives. There is nothing more boring and monotonous than the tick-tock of a clock, but how many fascinating challenges have physicists met to realize that monotony: Quite a number of Nobel Prize winners have been directly motivated by them or have contributed significantly to time measurement.[1] We feel that time flows, we feel it as an ever evolving, restless "now", and yet, from the perspective of relativity this unfolding of events at an always renewing present instant would in fact be "an illusion." Also, while the future awaits us and the past is gone, there is no time arrow making such a fundamental distinction in the microscopic equations of physics.

Physics does not capture time in its domain without residue, but it has of course much to say about time, an essential element of its theories and of our rationalization of nature. In the case of relativity, time plays a prominent, starring role:

---

[1] Here is a nonexhaustive list including award years: Isidor I. Rabi (1944), Charles H. Townes (1964), Alfred Kastler (1966), Norman F. Ramsey, Hans G. Dehmelt and Wolfgang Paul (1989), Steven Chu, Claude Cohen Tannoudji, and William D. Phillips (1997), John L. Hall and Theodor W. Hänsch (2005).

Contents

Chapter 1
Memories of Old Times: Schlick and Reichenbach on Time in Quantum Mechanics

José M. Sánchez-Ron

Space and time are the basic entities in physics; they provide the framework for any description of natural processes. As such, both have been throughout history the subject of many philosophical and scientific analyses (remember Newton's reflections and use of absolute space and time). The 20th century was specially fruitful in this regard. It could hardly have been otherwise, in as much as the first physics revolution that took place then – special (1905) and general relativity (1915) – was deeply dependent on the concepts of space and time. The fact that relativity appeared on the physics scenario before quantum mechanics and that space and time played such an important role in it meant that during most of the century the great majority of philosophical analyses of both concepts were based on Einstein's theory, while much less attention was dedicated to the implications that quantum physics had on them. Moritz Schlick (1882–1936), the leader of the Vienna Circle (the philosophical group that began its activities in 1924), and Hans Reichenbach (1891–1953), the main protagonists of the present chapter, are good examples of this, although they finally turned their attention also to the philosophy of quantum mechanics, the second being probably the most active of the philosophers of his time on this activity.

1.1 Introduction: The New Physics, via Relativity, Attracts the Philosophers

Restricting ourselves to the German-speaking world (in which, as a matter of fact, those philosophical interests first appeared), we have that Moritz Schlick was one of the earliest and more active "missionaries" of Einstein's relativity in the philosophical arena. A student of Max Planck, under whom he got his Ph.D. in physics in 1904, with a thesis on the reflection of light in inhomogeneous media, Schlick turned afterward his academic activity to philosophy and was soon attracted by the

J.M. Sánchez-Ron (✉)
Departamento de Física Teórica, Universidad Autónoma de Madrid, Spain,
josem.sanchez@uam.es

Sánchez-Ron, J.M.: *Memories of Old Times: Schlick and Reichenbach on Time in Quantum Mechanics*. Lect. Notes Phys. **789**, 1–13 (2009)
DOI 10.1007/978-3-642-03174-8_1   © Springer-Verlag Berlin Heidelberg 2009

many wonders of relativity, as can be seen, for example, in his 1915 article on "Die philosophische Bedeutung des Relativitätsprinzips" [39], or in his rather general exposition of Einstein's relativity theories that appeared in 1917, in two parts, in the scientific weekly journal *Die Naturwissenschaften*, as well as, expanded, in book format [40, 41].

Einstein was particularly attracted by Schlick's ideas. Thus, on April 19, 1920, he wrote to him[1]: "Your epistemology has made many friends. Even Cassirer had some works of acknowledgment for it . . . Young Reichenbach has written a very interesting paper about Kant & general relativity, in which he also gives your comparison with a calculating machine."

We find in this excerpt the names of two German-speaking philosophers who, together with Schlick, wrote extensively about relativity: Ernst Cassirer (1874–1945) and Hans Reichenbach. In due time, by the way, both left Germany and made the United States their country (both as professors of philosophy: Cassirer at Yale University since 1932 and Reichenbach at the University of California, Los Angeles, since 1938).

Cassirer, who had grown up philosophically as a member of the neo-Kantian school of Marburg, recognized that he had to revise his philosophical views so as to see whether they were consistent or not with Einstein's relativity theories. He was particularly interested in finding out whether the philosophical worldview that he had presented in his *Substanzbegriff und Funktionsbegriff* (1910), which was dominated by the Newtonian conceptions of space and time, was consistent with the new relativity world. Thus, after producing a manuscript about this subject and having sent it to Einstein, he wrote to him on May 10, 1920, thanking for his help[2]: "Please accept my cordial thanks for your kind willingness to glance briefly through my manuscript now . . . As far as the content of my text is concerned, it evidently does not propose to list all philosophical problems contained in the theory of relativity, let alone to solve them. I just wanted to try to stimulate general philosophical discussion and to open the flow of arguments and, if possible, to define a specific methodological direction. Above all, I would wish, as it were, to confront physicists and philosophers with the problems of relativity theory and bring about agreement between them. . . ."

The manuscript in question was published next year, in 1921, as a book entitled *Zur Einstein'schen Relativitätstheorie (Einstein's Theory of Relativity* [9]).

As to Reichenbach, he was also an early follower of the new relativity theories, not been in this sense one of the many who only became interested in them after the news of the eclipse expedition made Einstein and his theories world-famous in November 1919. "Due to my work in the army radio troops [signal corps]," he wrote in an autobiographical sketch for academic purposes [38, p. 2], "I became involved with radio technology and during the last year of the war [World War I], after I was transferred from active duty because of a severe illness I had con-

---

[1] [23, p. 510], [21, p. 317].
[2] [24, p. 255], [22, p. 158].

tracted at the Russian front, I began to work as an engineer for a Berlin firm specializing in radio technology (from 1917 until 1920). During this period, and in my capacity as physicist, I directed the loud-speaker laboratory of this firm... Soon thereafter, my father died and for the time being I could not give up my engineering position because I had to earn a salary in order to provide for my wife and myself. Nevertheless, in my spare time I studied the theory of relativity; I attended Einstein's lectures at the University of Berlin; at that time, his audience was very small because Einstein's name had not yet become known to a wider public. The theory of relativity impressed me immensely and led me into a conflict with Kant's philosophy. Einstein's critique of the space-time-problem made me realize that Kant's *a priori* concept was indeed untenable. I recorded the result of this profound inner change in a small book entitled *Relativitätstheorie und Erkenntnis Apriori* [1920]."

It gives an idea of Reichenbach's intentions what he wrote to Einstein (June 15, 1920) when asking permission to dedicate him his book *Relativitätstheorie und Erkenntnis Apriori*[3]: "You know that with this work my intention was to frame the philosophical consequences of your theory and to expose what great discoveries your physical theory have brought to epistemology... I know very well that very few among tenured philosophers have the faintest idea that your theory is a philosophical feat and that your physical conceptions contain more philosophy than all the multivolume works by the epigones of the great Kant. Do, therefore, please allow me to express these thanks to you with this attempt to free the profound insights of Kantian philosophy from its contemporary trappings and to combine it with your discoveries within a single system." To this letter, Einstein replied (June 30, 1920)[4]: "The value of the th. of rel. for philosophy seems to me to be that it exposed the dubiousness of certain concepts that even in philosophy were recognized as small change. Concepts are simply empty when they stop being firmly linked to experience."

Relativitätstheorie und Erkenntnis Apriori was not the only book Reichenbach dedicated to those matters. He also published *Axiomatik der relativistischen Raum-Zeit-Lehre* (1924) and *Philosophie der Raum-Zeit-Lehre* (1928). In them he developed a causal theory of time "according to which the concept of time is reduced to the concept of causality; since, on the other hand, measurement of space is also reduced to the measurement of time, space and time are therefore shown to be the 'causal structure of the world'" [38, p. 5].

1.2 Time in Quantum Physics: The Time–Energy Uncertainty Relation

So, we have seen that Einstein's relativity theories attracted the attention of the philosophers, first of the German-speaking ones. We can consider this as a sort of

---

[3] [24, p. 313–314], [22, p. 195].
[4] [24, p. 323], [22, p. 201].

entrance door of philosophers to the new physics that the new century was producing. But after relativity came quantum physics; therefore, we should ask ourselves if quantum physics, quantum mechanics in particular, attracted so much and so early, philosophical attention as relativity.[5]

"During the first decades of the development of quantum physics it was often stated that the concepts of space and time are intrinsically inapplicable at the quantum level, even when no doubt was implied as to the validity of these concepts in the domain of classical physics, both relativistic and pre-relativistic," wrote Henry Mehlberg [30, p. 235], a member of the great inter-war generation of teachers and students in physics, logic, and philosophy of science. What did he mean?

When faced with the problem of sustaining such assertion (Mehlberg did not offer any reference), I thought immediately of Niels Bohr, the great patron of quantum physics, and, indeed, I found soon a pertinent reference in a paper he wrote in 1935 to oppose Einstein–Podolsky–Rosen's 1935 famous critique of the quantum mechanical description of physical reality. There Bohr [6, p. 700] wrote,

> It is true that we have freely made use of such words as 'before' and 'after' implying time-relationships; but in each case allowance must be made for a certain inaccuracy, which is of no importance, however, so long as the time intervals concerned are sufficiently large compared with the proper periods entering in the closer analysis of the phenomena under investigation. As soon as we attempt a more accurate time description of quantum phenomena, we meet with the well-known paradoxes, for the elucidation of which further features of the interaction between the objects and the measuring instruments must be taken into account.

And he added [6, pp. 700–701],

> The decisive point as regards time measurements in quantum theory is now completely analogous to the argument concerning measurements of positions... Just as the transfer of momentum to the separate parts of the apparatus, - the knowledge of the relative positions of which is required for the description of the phenomenon -, has been seen to be entirely uncontrollable, so the exchange of energy between the object and the various bodies, whose relative motion must be known for the intended use of the apparatus, will defy any closer analysis. Indeed, it is *excluded in principle to control the energy which goes into the clocks without interfering essentially with their use as time indicators.*

And he then concluded,

> Just as in the question discussed above of the mutually exclusive character of any unambiguous use in quantum theory of the concepts of position and momentum, it is in the last resort this circumstance which entails the complementary relationship between any detailed time account of atomic phenomena on the one hand and the unclassical features of intrinsic stability of atoms, disclosed by the study of energy transfers in atomic reactions on the other hand.

---

[5] The content of the present chapter refers mainly to non-relativistic quantum mechanics; however, a relativistic theory will not introduce many fundamental differences in the topics I address here; only that, instead of just one time, we would have to consider as many local times as particles involved.

Bohr was referring, of course, to Heisenberg's uncertainty relations

$$\Delta x \cdot \Delta p \geq h/4\pi,$$
$$\Delta E \cdot \Delta t \geq h/4\pi,$$

where x represents the position, p the linear momentum, E the energy, t the time, and h Planck's constant.

The force and pertinence of Bohr's arguments seem obvious though not trivial – but, as far as I know, very few scholars addressed them explicitly. In an early paper, in which they tried to extend the uncertainty principle to relativistic quantum theory, Landau and Peierls [26] did. There, and referring to the energy uncertainty relation, they wrote [26, 27, p. 467],

> Clearly it does *not* signify that the energy can not be known exactly at a given time (for in that case the concept of energy would have no meaning), nor does it mean that the energy can not be measured with arbitrary accuracy within a short time. We must take into account the change caused by the process of measurement even in the case of a predictable measurement, i.e. of the difference between the result of the measurement and the state after the measurement. The relation then signifies that this difference causes an energy uncertainty of the order of $h/\Delta t$, so that on time Δt no measurement can be performed for which the energy uncertainty in *both* states is less that $h/\Delta t$.

However, it is legitimate to ask about the ideas on such questions by Heisenberg, the discoverer of the uncertainty principle. Well, neither in the 1927 paper in which he introduced the uncertainty relations nor in the lectures he delivered in Chicago in the spring of 1929 on "The physical principles of quantum theory" [18–20] did he pay special attention to the time–energy uncertainty relation nor, certainly, considered what it might imply for the meaning of time in the quantum domain. Similarly, when he introduced (beginning in the second edition) Heisenberg's principle of uncertainty in his influential *The Principles of Quantum Mechanics*, the always precise Paul Dirac [11] had nothing to say about the time–energy relation; actually, in the section dedicated to the uncertainty relations, he introduced only the position–momentum relation, a tactic that it is found also in the section that Landau and Lifshitz dedicated to the uncertainty relations in the volume dealing with non-relativistic quantum mechanics of his well-known course of theoretical physics. There, Landau and Lifshitz [25, p. 49] opted for writing $\Delta f \cdot \Delta g \approx hc$ and added that if one of the magnitudes, say f, is equal to the energy, E, and the other *operator* (g) does not depend explicitly on time, then $c = g$, and the uncertainty relation in the semiclassical case would be $\Delta E \cdot \Delta g \approx hg$.

Perhaps, Dirac and Landau and Lifshitz considered the non-commutativity of E and t (from which the uncertainty relation is derived) questionable if t is not an operator, but rather a c-number,[6] a circumstance that in his classic *Mathematische*

---

[6] C-numbers were introduced by Dirac [10, p. 562]: "The fact that the variables used for describing a dynamical system do not satisfy the commutative law means, of course, that they are not numbers in the sense of the word previously used in mathematics. To distinguish the two kinds of numbers,

Grundlagen der Quantenmechanik, John von Neumann [48, Chap. 5, Sect. 1] had already pointed out, although briefly and rather cryptically.

During the following decades there would be several attempts to prove rigorously the time–energy uncertainty relation, whose truth nobody seemed to doubt. Among those who made progress on this question figure Bohr and Rosenfeld [7], Mandel'shtam and Tamm [28], Fock [15], Aharonov and Bohm [1, 2], and Fujiwara [16]. The problem even made its way into a few textbooks, at least in two written by Russian scientists. The first one was the already mentioned text of Landau and Lifshitz. Section 44 of it is entitled "The uncertainty relation for the energy" [25, pp. 157–159] (note that no explicit reference is made to time, energy being the central physical concept in it). Reading it, it is obvious that time was the usual classical parameter, Δt the interval of time between two measurements, and ΔE "the difference between two values of energy measured exactly at two different instants of time."

The other Russian book is the fourth edition of Dmitrii Blokhintsev's [4, 5] quantum mechanics text, which had a whole section dedicated to "The law of conservation of energy and the special significance of time in quantum mechanics." There, Blokhintsev [5, p. 389] stated that "a relation between the uncertainty ΔE in the energy E at a given time t and the accuracy Δt with which the instant t is determined... does *not* exist in quantum mechanics, just as there is no relation $tH - Ht = ih/2\pi$ as distinct from the relation $xP_x - P_x x = ih/2\pi$." Recognizing, nevertheless, that that relation *was satisfied* in practice, he added, "We can, however, obtain the relation $[\Delta E \cdot \Delta t \geq h/4\pi]$ if the quantities ΔE and Δt are suitable interpreted" (his own option was to deal with a wave packet with group velocity v and having a dimension Δx, so that $\Delta t = \Delta x/v$, but he also referred, favorably, to Mandl'shtam and Tamm's paper [28]).

A good and concise statement of what the situation was at the beginning of the 1970s is the following, due to Aharonov and Petersen [3, p. 136]:

> As it is well known, the time-energy relation cannot be deduced from the commutation relations in the usual way, since the time is not a dynamical variable but a parameter. This has given rise to two different interpretations of the meaning of Δt. According to the first, Δt refers to the uncertainty in any dynamical 'time' defined by the system itself; for example, the position of the hand of a clock is such a dynamical variable. If the energy of the clock has been measured with an accuracy ΔE, then there must be an uncertainty in the position of the hand such that the corresponding $\Delta t \geq h/\Delta E$. According to the second interpretation, Δt refers to the period during which the energy measurement takes place. In other words, the uncertain time is not related to any dynamical variable belonging to the system itself but rather to the laboratory time which specifies when the energy is measured.

There would be, no doubt, much more to say on these questions.[7] However, I will not follow this route, because I am interested in Schlick and Reichenbach's reactions to quantum physics as regards time, specially in Reichenbach's, the most

---

we shall call the quantum variables q-numbers and the numbers of classical mathematics which satisfy the commutative law c-numbers."

[7] N. of E.: See Chap. 3 (first volume) by P. Busch on the time–energy uncertainty relation.

knowledgeable in quantum physics of those philosophers who first reacted to the relativity and quantum revolutions.[8] What I have already said proves, I think, that there were important – from the physical as well as from the philosophical point of view – problems related to the concept of time in quantum physics and that, although not always clear and abundant, there was enough material produced by physicists which a knowledgeable philosopher could, at least, mention.

1.3 Schlick on Quantum Theory

As mentioned before, Moritz Schlick, the former doctoral student of Max Planck and physicist turned philosopher, was one of the first German-speaking philosophers who paid attention to the implications that Einstein's relativity had on the space and time concepts considered from a philosophical point of view. Indeed, he published a large number of works on this subject. The question is: when quantum mechanics was formulated, and its philosophical implications became apparent, did he dedicate to the quantum as much attention and efforts as he had dedicated to relativity? The answer is a plain "no."

This does not mean, however, that the quantum did not make its way to some of his publications. Thus, in a paper dedicated to causality in contemporary physics, Schlick could not avoid referring to the novelties introduced by quantum mechanics [42, 44, p. 203]: "The most succinct description of the situation outlined is doubtless to say (as do the leading investigators of quantum problems), that the field of validity of the ordinary concepts of space and time is confined to the macroscopically observable; within atomic dimensions they are inapplicable." Such a drastic sentence certainly deserved a detailed justification, which, however, the paper does not include. Next year, during a lecture Schlick [43] delivered at the University of Berkeley in which he made use of the uncertainty relations, the argument was the traditional, that is, one in which only the position–momentum relation was considered. Nothing was said about the time–energy relation. With such theoretical baggage, Schlick could argue that the classical physics assertion that "a particle which at one moment has been observed at a definite particular place could be observed, after a definite interval of time, at another definite place" will cease to be true: if we take the value of the velocity of a particle and try to use it for an extrapolation to get a future position of the particle, "the Uncertainty Principle steps in to tell us that our attempt is in vain; our value of the velocity is no good for such a prediction, our

---

[8] To support the contention that Reichenbach was the most knowledgeable in quantum physics of the philosophers who first reacted to the relativity and quantum revolutions, I offer the following quotation from Carnap's autobiography in *The Library of Living Philosophers* [8, p. 14]: "After the Erlangen Conference [1923] I met Reichenbach frequently. Each of us, when hitting upon new ideas, regarded the other as the best critic. Since Reichenbach remained in close contact with physics through his teaching and research, whereas I concentrated more on other fields, I often asked him for explanations in recent developments, for example, in quantum-mechanics. His explanations were always excellent in bringing out the main points with great clarity."

own observation will have changed the velocity in an unknown way, therefore the particle will probably not be found in the predicted place and there is no possibility of knowing where it could be found" [43, 31, pp. 255–256].

Positions – that is, space – were, therefore, the subject of Schlick considerations, not time; "the particle will probably not be found in the predicted place," he wrote, but this "predicted place" will take place, as well as the previous one, at definite instants of time, not subject, apparently, to any uncertainty. This was made possible, obviously, by the use of the position–momentum uncertainty relation, as well as by ignoring the time–energy relation. Were it not ignored, could it be argued for time the same that was said about space? Naturally, the problem was (and still is) the special nature of time.[9]

1.4 Reichenbach on Time in Quantum Physics

Hans Reichenbach was more active in the philosophical analysis of quantum physics than Schlick (among other things because he lived more). His main contribution, an original one, was the introduction of a three-valued logic, in which a category called "indeterminate" stands between the truth values "true" and "false." The place where he gave a more detailed presentation of such ideas was his book *Philosophic Foundations of Quantum Mechanics* [33].[10]

In the preface of this work, Reichenbach [37, vi–vii], already installed in the Department of Philosophy of the University of California, Los Angeles, explained why he had become involved with quantum theory. Thus, and after referring to the first phases in the development of quantum mechanics, he stated that the time had arrived for attempting a serious philosophical study of the foundations of the theory. "Fully aware that philosophy should not try to construct physical results, nor try to prevent physicists from finding such results," he "nonetheless believed that a logical analysis of physics which did not use vague concepts and unfair excuses was possible." And he added,

> The philosophy of physics should be as neat and clear as physics itself; it should not take refuge in conceptions of speculative philosophy which must appear outmoded in the age of empiricism, nor use the operational form of empiricism as a way to evade problems of the logic of interpretations. Directed by this principle the author has tried in the present book to develop a philosophical interpretation of quantum physics which is free from metaphysics, and yet allow us to consider quantum mechanical results as statements about an atomic world as real as the ordinary physical world.

---

[9] Although not in the quantum realm, but in the relativistic one, Einstein pointed the specificity of time in his autobiographical notes when, after remembering the well-known mental experiment that he posed himself at the age of 16 (what would happen if he pursued a beam of light with the velocity of light), he added [12, p. 53]: "One sees that in this paradox the germ of the special relativity theory is already contained. Today everyone knows, of course, that all attempts to clarify this paradox satisfactorily were condemned to failure as long as the axiom of the absolute character of time, viz., of simultaneity, unrecognizedly was anchored in the unconscious."

[10] I will use the first paperback printing of this book [37]. An interesting review of the book was written by Mehlberg [29].

The purpose was, of course, sound and the results significant, but not so as regards the concept of time in quantum mechanics. Reichenbach, it is true, included the time–energy uncertainty relation alongside the position–momentum one, but his interpretation of them was not particularly interesting or new; he emphasized their implications with respect to causality, not with respect to time itself. And he said nothing about the time not being an operator but a mere parameter. However, we know that time was a concept in which he was specially interested. *The Direction of Time*, a posthumous work, assembled by his wife, Maria, from various manuscripts he left at the time of his death in April 1953 is proof of this.[11] However, the problem of the direction of time is part of several branches of classical physics (mechanics, electrodynamics, thermodynamics, statistical physics, cosmology), and we must not be surprised that the majority of the pages of *The Direction of Time* [35] were dedicated to what classical physics, thermodynamics, and statistical physics have to say concerning the observed asymmetry between past and future: 200 pages versus 63 dedicated to "The time in quantum physics." Besides, the question of the direction of time is not exactly the same as what is its nature, assuming such a thing, or expression, the "nature of time," makes sense.[12]

Early on the chapter of the book dedicated to time in quantum mechanics, Reichenbach considered the wave function of Schrödinger's equation, which occupies the central place in the theory. He pointed out that when the state changes in the course of time, the variable t enters as another argument into the function, which is then written in the form $\Psi(q, t)$, and that the differential equation which Schrödinger had constructed to express the fundamental law of change in quantum mechanics has the form

$$H_{op}\Psi(q, t) = c[\partial\Psi(q, t)/\partial t] , \qquad (1.1)$$

where $c = ih/2\pi$.

"The direction of time," wrote then Reichenbach [35, p. 209], "that is, the temporal direction in which the change occurs, manifests itself in the sign of the argument 't'." However, what happens if we change t by $-t$? The problem here is that contrary to what happens in classical physics, where the differential equations are of second order in time, with first derivatives absent, in quantum mechanics the latter are present. Therefore, one has that if $\Psi(q, t)$ is a solution of Schrödinger equation,

---

[11] Shortly before, the Institut Henry Poincaré published the text of a series of lectures Reichenbach [34] delivered at that Paris Institute on June 4, 6, and 7, 1952. Some of the themes of *The Direction of Time* were advanced there.

[12] I am aware that often the question of the "nature of time" is identified with "the direction of time." A splendid example of this is the collective book edited by Thomas Gold entitled *The Nature of Time* [17], in which, however, most contributions deal with the direction of time. Of course, with my comments I do not mean that the problem of the direction of time is not interesting or fundamental. I fully agree with what the theoretical astrophysicist Dennis Sciama [45, p. 6] wrote, "Time has always struck people as mysterious: mysterious, in fact, in a number of different ways. One thing that is mysterious about time is its directionality. What is it that underlies time's arrow? What, that is to say, is the source of the asymmetry between past and future, between earlier and later? Why, for example, can we remember the past but not the future?"

$\Psi(q, -t)$ is not, because the equation it satisfies is

$$H_{op}\Psi(q, -t) = -c \cdot [\partial\Psi(q, -t)/\partial t], \qquad (1.2)$$

which differs from the original in the minus sign on the right-hand side.[13] And here Reichenbach [35, pp. 209–210] wrote,

> There remains the problem of distinguishing between $\Psi(q, t)$ and $\Psi(q, -t)$. In order to discriminate between these two functions, we would first have to know whether (1.1) or (1.2) is the correct equation. But the sign of the term on the right in Schrödinger's equation can be tested observationally only if a direction of time has been previously defined. We use here the time direction of the macrocosmic systems by the help of which we compare the mathematical consequences of Schrödinger's equation with observations. Therefore, to attempt a definition of time direction through Schrödinger's equation would be reasoning on a circle; this equation merely presents us with the time direction which we introduced previously in terms of macrocosmic processes.

And he added,

> It may be recalled that even in classical physics the time direction of a molecule is not ascertainable from observations of the molecule, even if such direct observations could be made, but is determined only by comparison with macroprocesses, for which statistics define a time direction. In the same way, the time direction of a quantum-mechanical elementary process, like the movement of an electron, is determined only with reference to the time of macroprocesses.

> This consideration shows that the fundamental quantum-mechanical law governing the time development of physical systems does not distinguish one time direction from its opposite. Since the laws governing the observables of quantum physics are not causal laws, but probability laws, the reversibility of elementary processes assumes here the form of a symmetry of the relations connecting probability distributions. These connecting relations are strict laws expressible through a differential equation, namely, Schrödinger's equation.

"Time direction of a quantum-mechanical elementary process," he wrote, "is determined only with reference to the time of macroprocesses." Not a stimulating comment for anyone who would have thought that so radical physics revolution as quantum mechanics should affect also our ideas of what time is.

1.5 Reichenbach on Feynman's Theory of the Positron

One thing that strikes one when reading Reichenbach's book is how scarce the references to works of physicists who dealt with the quantum are, whether with quantum mechanics or with quantum electrodynamics. It seems as if it was more a philosophical inner discussion, illuminated mainly by quantum mechanics (mainly

---

[13] N. of E.: The standard "time-reversal invariance" argument is based on the commutation of the Hamiltonian with the antiunitary time-reversal operator, see, e.g., Chap. 12 in this volume. The time-reversed state of $\Psi(q, t)$ is $\Psi(q, t)^*$. If $\Psi(q, t)^*$ evolves for a time t, the resulting state is $\Psi(x, 0)^*$, which is the time-reversed state of $\Psi(x, 0)$. For a critical analysis see A.T. Holster, New J. Phys. **5**, 130 (2003).

Schrödinger's version), statistical mechanics, and the mathematical theory of probability. There is, however, an important exception: Reichenbach's reference to works of the Lausanne professor E. C. G. Stückelberg and the American Richard Feynman on the theory of positrons, where they considered positrons as electrons moving backward in time [46, 47, 13, 14].[14]

"Surprisingly enough," he wrote [35, pp. 263–264], "recent developments have demonstrated that the genidentity [that is, the physical identity of a thing] of material particles can be questioned more seriously than is done in Bose statistics. The difference between one and two, or even three, material particles can be shown to be a matter of interpretation; that is, this difference is not an objective fact, but depends on the language used for the description. The number of material particles, therefore, is contingent upon the extension rules of language. However, the interpretations thus admitted for the language of physics differ in one essential point from all others: they require an abandonment of the order of time."

Those "recent developments" were the "conceptions... developed by E. C. G. Stückelberg and R. P. Feynman. Their investigations showed that a positron – that is, a particle of the mass of an electron, but carrying a positive charge – can be regarded as an electron moving backward in time." To Reichenbach [35, pp. 266, 268] such interpretation "does not merely signify a reversal of time *direction*; it represents an abandonment of time *order*... This is the most serious blow the concept of time has ever received in physics. Classical mechanics cannot account for the direction of time; but it can at least define a temporal order. Statistical mechanics can define a temporal direction in terms of probabilities; but this definition presupposes time order for those atomic occurrences the statistical behavior of which supplies time direction. Quantum physics, it appears, cannot even speak of a unique time order of the processes, if further investigations confirm Feynman's interpretation, which is at present still under discussion."

Even without addressing the fundamental problem of putting in a sound theoretical basis for the time–energy uncertainty relation and deriving its consequences for the concept of time, Reichenbach had found that the realm of the quantum was a dangerous territory for the conception of time that classical physics had favored.

1.6 Epilogue

In his book *Philosophie der Raum-Zeit-Lehre*, and thinking in the case of the relativity theories, Reichenbach [32] [36, p. 109] stated that "philosophy of science has examined the problems of time much less than the problems of space. Time has generally been considered as an ordering schema similar to, but simpler than, that of space, simpler because it has only one dimension. Some philosophers have

---

[14] Feynman's work here was influenced by his previous collaboration with John Wheeler on an action-at-distance electrodynamics, in which they used retarded as well as advanced potentials; that is, electromagnetic waves moving forward and backward in time [49, 50].

believed that a philosophical clarification of space also provided a solution of the problem of time." It seems to me, after having reviewed here what Moritz Schlick and Hans Reichenbach – just two, certainly, German-speaking philosophers of science, but, nevertheless, very prominent ones – had to say about time in quantum mechanics, that the same comment can be applied to the analysis of time in the quantum domain. How different, and much less frequent, were the comments that the two uncertainty relations aroused apropos the time concept is a good example of such assertion. Of course, it is true – obviously true – that scientifically time is a much more problematic and difficult concept to define and study than space, but it is so fundamental! Without it, there would be nothing, just "something" (I resist calling it "world") unknowledgeable. With it, we have science, but also mystery, the mystery of a concept perhaps too difficult for us to fully understand.

References

1. Y. Aharonov, D. Bohm, Phys. Rev. **122**, 1649 (1961)
2. Y. Aharonov, D. Bohm, Phys. Rev. **134**, B1417 (1964)
3. Y. Aharonov, A. Petersen, in *Quantum Theory and Beyond*, T. Bastin (ed.) (Cambridge University Press, Cambridge, 1971), p. 135
4. D.I. Blokhintsev, *Osnovy Kvantoloi Mekhaniki* (Moscú, 1963)
5. D.I. Blokhintsev, *Quantum Mechanics* (Reidel, Dordrecht, 1964), English translation of *Osnovy Kvantoloi Mekhaniki*
6. N. Bohr, Phys. Rev. **48**, 696 (1935)
7. N. Bohr, L. Rosenfeld, Mathematisk-fysiske Meddeleser **XII**, 8 (1933), English version in *Selected Papers of Léon Rosenfeld*, R.S. Cohen, J. Stachel (eds.) (Reidel, Dordrecht, 1979), p. 357
8. R. Carnap, in *The Philosophy of Rudolf Carnap*, P.A. Schilpp (ed.) (Open Court, La Salle Illinois, 1963), p. 3
9. E. Cassirer, *Substance and Function & Einstein's Theory of Relativity* (Dover, New York, 1953)
10. P.A.M. Dirac, Proc. Roy. Soc. (London) A **110**, 561 (1926)
11. P.A.M. Dirac, *The Principles of Quantum Mechanics*, 2nd edn (Clarendon Press, Oxford, 1935)
12. A. Einstein, in *Albert Einstein: Philosopher-Scientist*, P.A. Schilpp (ed.) (Open Court, La Salle, IL, 1949), p. 3
13. R.P. Feynman, Phys. Rev. **74**, 1430 (1948)
14. R.P. Feynman, Phys. Rev. **76**, 749 (1949)
15. V. Fock, Zhurnal Eksperimental'noi i Teoretischeskoi Fiziki **42**, 1135 (1962)
16. I. Fujiwara, Prog. Theor. Phys. **44**, 1701 (1970)
17. T. Gold (ed.), *The Nature of Time* (Cornell University Press, Ithaca, 1967)
18. W. Heisenberg, Zeitschrift für Physik **43**, 172 (1927)
19. W. Heisenberg, *Die physikalischen Prinzipien der Quantentheorie* (S. Hirzel, Leipzig, 1930)
20. W. Heisenberg, *The Physical Principles of the Quantum Theory* (Dover, New York, 1930), English translation of *Die physikalischen Prinzipien der Quantentheorie*
21. A. Hentschel, transl., *The Collected Papers of Albert Einstein, vol. 9 (The Berlin Years: Correspondence, January 1919–April 1920)* (Princeton University Press, Princeton, 2004)
22. A. Hentschel, transl., *The Collected Papers of Albert Einstein, vol. 10 (The Berlin Years: Correspondence, May–December 1920, and Supplementary Correspondence, 1909–1920)* (Princeton University Press, Princeton, 2006)

23. D. Kormos Buchwald, R. Schulmann, J. Illy, D.J. Kennefick, T. Sauer (eds.), *The Collected Papers of Albert Einstein, vol. 9 (The Berlin Years: Correspondence, January 1919–April 1920)* (Princeton University Press, Princeton, 2004)

24. D. Kormos Buchwald, T. Sauer, Z. Rosenkranz, J. Illy, V.I. Holmes (eds.), *The Collected Papers of Albert Einstein, vol. 10 (The Berlin Years: Correspondence, May–December 1920, and Supplementary Correspondence, 1919–1920)* (Princeton University Press, Princeton, 2006)

25. L. Landau, E. Lifschitz, *Quantum Mechanics (Non-Relativistic Theory)*, 3rd edn (Butterworth-Heinemann, Oxford, 1977)

26. L. Landau, R. Peierls, Zeitschrift für Physik **69**, 56 (1931)

27. L. Landau, R. Peierls, English translation of Zeitschrift für Physik **69**, 56 (1931), in *Quantum Theory and Measurement*, J.A. Wheeler, W.H. Zurck (eds.) (Princeton University Press, Princeton, 1983), p. 465

28. L. Mandel'shtam, I.E. Tamm, Izvestiya Akademii nauk SSSR Seriya fizicheskaya **9**, 122 (1945), English translation: J. Phys. URSS **9**, 249

29. H. Mehlberg, Phil. Rev. **71**, 99 (1962), reprinted in *Time, Causality, and the Quantum Theory, vol. II (Time in a Quantized Universe)* (Reidel, Dordrecht, 1980), p. 203

30. H. Mehlberg, *Time, Causality, and the Quantum Theory, vol. II (Time in a Quantized Universe)* (Reidel, Dordrecht, 1980)

31. H. Mulder, B.F.B. van de Velde-Schlick (eds.), *Moritz Schlick: Philosophical Papers (1925–1936)*, vol. II (Reidel, Dordrecht, 1979)

32. H. Reichenbach, *Philosophie der Raum-Zeit-Lehre* (Walter de Gruyter, Berlin, 1928)

33. H. Reichenbach, *Philosophical Foundations of Quantum Mechanics* (University of California Press, Berkeley, 1944)

34. H. Reichenbach, Annales de l'Institut Henri Poincaré **13**, part 2, 109 (1952/1953), English translation in *Moritz Schlick: Philosophical Papers (1925–1936)*, vol. II, H. Mulder, B.F.B. van de Velde-Schlick (eds.) (Reidel, Dordrecht, 1979), p. 237

35. H. Reichenbach, *The Direction of Time* (University of California Press, Berkeley, 1956)

36. H. Reichenbach, *The Philosophy of Space & Time* (Dover, New York, 1957), English translation of *Philosophie der Raum-Zeit-Lehre* (Walter de Gruyter, Berlin, 1928)

37. H. Reichenbach, *Philosophical Foundations of Quantum Mechanics* (University of California Press, Berkeley, 1964), first California paperback printing

38. H. Reichenbach, in *Selected Writings, 1909–1953*, vol. I, M. Reichenbach, R.S. Cohen (eds.) (Reidel, Dordrecht, 1978)

39. M. Schlick, Zeitschrift für Philosophie und philosophische Kritik **159**, 129 (1915), English translation in *Moritz Schlick: Philosophical Papers (1909–1922)*, vol. I, H. Mulder, B.F.B. van de Velde-Schlick (eds.) (Reidel, Dordrecht, 1979), p. 153

40. M. Schlick, Die Naturwissenschaften **5**, 161, 177 (1917)

41. M. Schlick, *Raum und Zeit in der gegenwärtigen Physik* (Julius Springer Verlag, Berlin, 1917)

42. M. Schlick, Die Naturwissenschaften **19**, 145 (1931), English translation [48]

43. M. Schlick, University of California Publications in Philosophy **XV**, 99 (1932), reprinted in *Moritz Schlick: Philosophical Papers (1925–1936)*, vol. II, H. Mulder, B.F.B. van de Velde-Schlick (eds.) (Reidel, Dordrecht, 1979), p. 238

44. M. Schlick, in *Moritz Schlick: Philosophical Papers (1925–1936)*, vol. II, H. Mulder, B.F.B. van de Velde-Schlick (eds.) (Reidel, Dordrecht, 1979), p. 176

45. D. Sciama, in *The Nature of Time*, R. Flood, M. Lockwood (eds.) (Basil Blackwell, Oxford, 1986), p. 6

46. E.C.G. Stückelberg, Helvetica Physica Acta **14**, 588 (1941)

47. E.C.G. Stückelberg, Helvetica Physica Acta **15**, 23 (1942)

48. J. von Neumann, *Mathematische Grundlagen der Quantenmechanik* (Springer, Berlin, 1932)

49. J.A. Wheeler, R.P. Feynman, Rev. Mod. Phys. **17**, 157 (1945)

50. J.A. Wheeler, R.P. Feynman, Rev. Mod. Phys. **21**, 425 (1949)

Chapter 2
The Time-Dependent Schrödinger Equation Revisited: Quantum Optical and Classical Maxwell Routes to Schrödinger's Wave Equation[1]

Marlan O. Scully

2.1 Introduction

In a previous paper [1, 2] we presented quantum field theoretical and classical (Hamilton–Jacobi) routes to the time-dependent Schrödinger equation (TDSE) in which the time t and position $\mathbf{r}$ are regarded as parameters, not operators. From this perspective, the time in quantum mechanics is argued as being the same as the time in Newtonian mechanics. We here provide a parallel argument, based on the photon wave function, showing that the time in quantum mechanics is the same as the time in Maxwell equations.

The next section is devoted to a review of the photon wave function which is based on the premise that a photon is what a photodetector detects. In particular, we show that the time-dependent Maxwell equations for the photon are to be viewed in the same way we look at the time-dependent Dirac–Schrödinger equation for the (massive) π meson particle or (massless) neutrino.

In Sect. 2.3 we then recall previous work which casts the classical Maxwell equations into a form which is very similar to the Dirac equation for the neutrino. Thus, we are following de Broglie more closely than did Schrödinger, who followed a Hamilton–Jacobi approach to the quantum mechanical wave equation. In this way, with nearly a century of hindsight, we arrive naturally at the time-dependent Schrödinger equation without operator baggage. Figures 2.1 and 2.2 summarize the physics of the present chapter.

M.O. Scully (✉)
Texas A&M University, College Station, TX 77843; Princeton University, Princeton, NJ 08544, USA, mscully@Princeton.edu

[1] It is a pleasure to dedicate this chapter to David Woodling who has enriched our lives through his engineering and mechanical gifts and his insightful and gentle ways.

Scully, M.O.: *The Time-Dependent Schrödinger Equation Revisited: Quantum Optical and Classical Maxwell Routes to Schrödinger's Wave Equation.* Lect. Notes Phys. **789**, 15–24 (2009)
DOI 10.1007/978-3-642-03174-8_2   © Springer-Verlag Berlin Heidelberg 2009

	Photon (spin 1)	Neutrino (spin ½)	π Meson (spin 0)
Quantum Field Theory	$\left|\dot{\Psi}_\gamma\right\rangle = -\dfrac{1}{\hbar}H_\gamma\left|\Psi_\gamma\right\rangle$	$\left|\dot{\Psi}_\nu\right\rangle = -\dfrac{1}{\hbar}H_\nu\left|\Psi_\nu\right\rangle$	$\left|\dot{\Psi}_\pi\right\rangle = -\dfrac{1}{\hbar}H_\pi\left|\Psi_\pi\right\rangle$
"Wave Mechanics"	$\Psi_\gamma = \begin{bmatrix}\varphi_\gamma \\ \chi_\gamma\end{bmatrix}$ $\dot{\Psi}_\gamma = -\dfrac{i}{\hbar}\begin{bmatrix}0 & -cs\cdot p \\ cs\cdot p & 0\end{bmatrix}\Psi_\gamma$	$\Psi_\nu = \begin{bmatrix}\varphi_\nu \\ \chi_\nu\end{bmatrix}$ $\dot{\Psi}_\nu = -\dfrac{i}{\hbar}\begin{bmatrix}0 & -c\sigma\cdot p \\ c\sigma\cdot p & 0\end{bmatrix}\Psi_\nu$	$\dot{\Psi}_\pi = -\dfrac{i}{\hbar}\sqrt{p^2c^2 + m_0^2c^4}\,\Psi_\pi$ $\Rightarrow \dfrac{i\hbar}{2m_0}\nabla^2\Psi_\pi$
Classical limit: Eikonal physics	 $n(r)$ Ray optics $\delta\int n\,ds = 0$ Fermat's principle		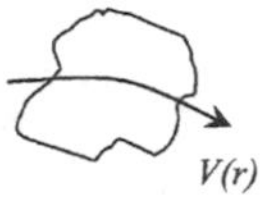 $V(r)$ Classical Mechanics $\delta\int L\,dt = 0$ Hamilton's principle

Fig. 2.1 Comparison of the quantum field, wave mechanical, and classical descriptions of the spin 1 photon, spin $\frac{1}{2}$ neutrino, and spin 0 meson; adapted from Scully and Zubairy "Quantum Optics" [3]

$$\boxed{\langle 0|\hat{E}(r)|\Psi(t)\rangle}$$

⇩

$$\boxed{i\hbar\frac{\partial\Psi(r,t)}{\partial t} = -\frac{\hbar^2}{2m}\nabla^2\Psi(r,t)}$$

⇧

$$\boxed{\Psi_m(r,t) = E(r,t) + iH(r,t)}$$

Fig. 2.2 *Top Down*: The time-dependent Schrödinger wave equation follows from the quantum optical "a photon is what a photodetector detects" definition. This is in accord with the usual wave function definition $\Psi(\mathbf{r},t) = \langle\mathbf{r}|\Psi(t)\rangle$ since $|\mathbf{r}\rangle = \hat{\Psi}^+(r)|0\rangle$. *Bottom Up*: The time-dependent Schrödinger wave follows nicely from the classical Maxwell equations by, for example, working with a combination of electric and magnetic fields

2.2 The Quantum Optical Route to the Time-Dependent Schrödinger Equation

Quantum optics is an offshoot of quantum field theory in which we are often interested in intense light beams such as provided by the laser. However the issue of the photon concept, and how we should think of the "photon," is a topic of current and reoccurring discussion.

Perhaps the most logical, at least the most operational, approach is to say that the photon is what a photodetector detects. In this spirit we consider the excitation of a single atom at point $\mathbf{r}$ at time t to be our photodetector and, following [3], write the probability of exciting the atom as

$$P_\Psi(\mathbf{r}, t) = \eta\langle\Psi\left|\hat{E}^\dagger(\mathbf{r}, t)\hat{E}(\mathbf{r}, t)\right|\Psi\rangle . \tag{2.1}$$

Several points should be made:

1. We consider the state $|\Psi\rangle$ to be a single photon state. For example, the state generated by the emission of a single photon (see [3], Eq. 6.3.18)

$$|\psi_\gamma\rangle = \sum_\mathbf{k} c_\mathbf{k}\,|\mathbf{k}\rangle , \tag{2.2}$$

where the state $|\mathbf{k}\rangle$ is expressed in terms of the radiation creation operator $\hat{a}^\dagger_\mathbf{k}$ as $|\mathbf{k}\rangle = \hat{a}^\dagger_\mathbf{k}\,|0\rangle$ and in the simple case of a scalar photon, we find

$$c_\mathbf{k} = g_\mathbf{k}\frac{e^{-i\mathbf{k}\cdot\mathbf{r}_0}}{(\nu_k - \omega_0) + i\Gamma/2} , \tag{2.3}$$

where $g_\mathbf{k}$ is the atom-field coupling constant, $\mathbf{r}_0$ is the atomic position vector, ν_k and ω_0 are the photon and atomic frequencies, and Γ is the atomic decay rate.
2. The uninteresting photodetection efficiency constant η will be ignored in the following.
3. $\hat{E}^\dagger(\mathbf{r}, t)$ and $\hat{E}(\mathbf{r}, t)$ are the creation and annihilation operators defined by

$$\hat{E}^\dagger(\mathbf{r}, t) = \sum_{\mathbf{k},\lambda}\boldsymbol{\varepsilon}^{(\lambda)}_\mathbf{k}\mathcal{E}_k\hat{a}^\dagger_{\mathbf{k},\lambda}e^{-i\nu_k t + i\mathbf{k}\cdot\mathbf{r}} , \tag{2.4}$$

where $\boldsymbol{\varepsilon}^\lambda_\mathbf{k}$ is the unit vector for light having polarization λ and wave vector $\mathbf{k}$, $\nu_k = ck = c|\mathbf{k}|$ and the electric field "per photon" $\mathcal{E}_k = \sqrt{\hbar\nu_k/2\varepsilon_0 V}$, where we use MKS units so that $\varepsilon_0\mu_0 = 1/c^2$ and V is the quantization volume.

Next we insert a sum over a complete set of states, $\sum_n |n\rangle\langle n| = 1$ in Eq. (2.1) and note that since there is only one photon in ψ_γ (and $\hat{E}(\mathbf{r}, t)$ annihilates it), only the vacuum term $|0\rangle\langle 0|$ will contribute. Hence we have

$$P_{\psi_\gamma}(\mathbf{r}, t) = \langle\psi_\gamma|\hat{E}^\dagger(\mathbf{r}, t)|0\rangle\langle 0|\hat{E}(\mathbf{r}, t)|\psi_\gamma\rangle , \tag{2.5}$$

and we are therefore led to define the single photon detection amplitude as

$$\Psi_\mathcal{E}(\mathbf{r}, t) = \langle 0|\hat{E}(\mathbf{r}, t)|\psi_\gamma\rangle . \tag{2.6}$$

As shown in detail in Sect. 2.4, the one photon state $|\psi_\gamma\rangle$ yields

$$\Psi_\mathcal{E}(\mathbf{r}, t) = \frac{\mathcal{E}}{\Delta r}\Theta\left(t - \frac{\Delta r}{c}\right)e^{-i(t-\Delta r/c)(\omega - i\Gamma/2)} , \tag{2.7}$$

where $\mathcal{E}$ is a constant, Δr is the distance from the atom to the detector, and $\Theta(x)$ is the usual step function. More generally we have

$$\Psi_{\mathcal{E}}(\mathbf{r}, t) = \langle 0|\hat{E}(\mathbf{r}, t)|\psi_\gamma\rangle = \left\langle 0\left|\sum_{\mathbf{k},\lambda} \hat{\varepsilon}_{\mathbf{k}}^\lambda \sqrt{\frac{\hbar v_k}{2\varepsilon_0 V}} \hat{a}_{\mathbf{k},\lambda} \mathrm{e}^{-iv_k t + i\mathbf{k}\cdot\mathbf{r}}\right|\psi_\gamma\right\rangle . \tag{2.8}$$

The field is sharply peaked about the frequency ω so that we may replace the frequency v_k as it appears in the square root factor by ω and write

$$\Psi_{\mathcal{E}}(\mathbf{r}, t) = \sqrt{\frac{\hbar\omega}{2\varepsilon_0}} \varphi_\gamma(\mathbf{r}, t) , \tag{2.9}$$

where

$$\varphi_\gamma(\mathbf{r}, t) = \sum_{\mathbf{k},\lambda} \hat{\varepsilon}_{\mathbf{k}}^{(\lambda)} \left\langle 0\left|\hat{a}_{\mathbf{k},\lambda}\frac{\mathrm{e}^{-iv_k t + i\mathbf{k}\cdot\mathbf{r}}}{\sqrt{V}}\right|\psi_\gamma\right\rangle . \tag{2.10}$$

The complete "photon wave function" also involves the magnetic analog of the proceeding. To that end we write

$$\Psi_{\mathcal{H}}(\mathbf{r}, t) = \langle 0|\hat{H}(\mathbf{r}, t)|\psi_\gamma\rangle , \tag{2.11}$$

where $\hat{H}(\mathbf{r}, t)$ is the annihilation operator for the magnetic field which is given by

$$\hat{H}(\mathbf{r}, t) = \sum_{\mathbf{r},\lambda} \frac{\mathbf{k}}{k} \times \hat{\varepsilon}_{\mathbf{k}}^{(\lambda)} \sqrt{\frac{\hbar v_k}{2\mu_0}} \hat{a}_{\mathbf{k},\lambda}\frac{\mathrm{e}^{-iv_k t + i\mathbf{k}\cdot\mathbf{r}}}{\sqrt{V}} , \tag{2.12}$$

and we introduce the notation

$$\Psi_{\mathcal{H}}(\mathbf{r}, t) = \sqrt{\frac{\hbar\omega}{2\mu_0}} \chi_\gamma(\mathbf{r}, t) , \tag{2.13}$$

where

$$\chi_\gamma(\mathbf{r}, t) = \left\langle 0\left|\sum_{\mathbf{k},\lambda} \frac{\mathbf{k}}{k} \times \hat{\varepsilon}_{\mathbf{k}}^{(\lambda)} a_{\mathbf{k},\lambda}\frac{\mathrm{e}^{-iv_k t + i\mathbf{k}\cdot\mathbf{r}}}{\sqrt{V}}\right|\psi_\gamma\right\rangle . \tag{2.14}$$

Finally, we write $\varphi_\gamma(\mathbf{r}, t)$ and $\chi_\gamma(\mathbf{r}, t)$ in matrix form as

$$\varphi_\gamma = \begin{bmatrix} \varphi_x \\ \varphi_y \\ \varphi_z \end{bmatrix} , \quad \chi_\gamma = \begin{bmatrix} \chi_x \\ \chi_y \\ \chi_z \end{bmatrix} , \tag{2.15}$$

in terms of which Maxwell equations may be written as

$$i\hbar\frac{\partial}{\partial t}\begin{bmatrix}\varphi_\gamma\\\chi_\gamma\end{bmatrix}=\begin{bmatrix}0 & -c\mathbf{s}\cdot\mathbf{p}\\c\mathbf{s}\cdot\mathbf{p} & 0\end{bmatrix}\begin{bmatrix}\varphi_\gamma\\\chi_\gamma\end{bmatrix},\qquad(2.16)$$

where $\mathbf{p}=\frac{\hbar}{i}\nabla$ and

$$s_x=\begin{bmatrix}0 & 0 & 0\\0 & 0 & -1\\0 & 1 & 0\end{bmatrix},\quad s_y=\begin{bmatrix}0 & 0 & 1\\0 & 0 & 0\\-1 & 0 & 0\end{bmatrix},\quad s_z=\begin{bmatrix}0 & -1 & 0\\1 & 0 & 0\\0 & 0 & 0\end{bmatrix}\qquad(2.17)$$

are the 3×3 matrices for the (spin 1) photon.

Finally, we note the close correspondence with the two-component (spin $\frac{1}{2}$) neutrino,

$$\begin{bmatrix}\varphi_{\text{photon}}\\\chi_{\text{photon}}\end{bmatrix}\longleftrightarrow\begin{bmatrix}\varphi_{\text{neutrino}}\\\chi_{\text{neutirno}}\end{bmatrix},\qquad(2.18)$$

and the Dirac equation for the neutrino

$$i\hbar\frac{\partial}{\partial t}\begin{bmatrix}\varphi_\nu\\\chi_\nu\end{bmatrix}=\begin{bmatrix}0 & -c\sigma\cdot\mathbf{p}\\c\sigma\cdot\mathbf{p} & 0\end{bmatrix}\begin{bmatrix}\varphi_\nu\\\chi_\nu\end{bmatrix},\qquad(2.19)$$

where σ is given in terms of the 2×2 Pauli matrices and $\mathbf{p}=\frac{\hbar}{i}\nabla$.

We conclude by noting that, just as in the quantum field theory [4, 5] route to the Schrödinger equation, the appearance of $\frac{\partial}{\partial t}$ and ∇ in Eq. (2.16) has not arisen from operator arguments. In the next sections, we follow a de Broglie wave–particle duality path to the Schrödinger equation.

2.3 The Classical Maxwell Route to the Schrödinger Equation

In the previous section, we followed a top-down quantum field route to the Schrödinger equation, see Fig. 2.2. In particular, we saw that the quantum optical analysis of the single photon wave equation provided an interesting connection between the Schrödinger (Dirac) equations for photons and neutrinos.

In the present section, we start with the classical Maxwell equations and obtain a Schrödinger equation for the combination $\mathbf{E}+i\mathbf{H}$ which previous workers [6, 7] call the photon wave function. It is then natural to follow de Broglie and associate a wave function with matter waves. This provides another (operator-free) route to the Schrödinger equation.

Thus, we define the "classical" photon wave function as

$$\Psi_m(\mathbf{r}, t) = \mathbf{E}(\mathbf{r}, t) + i\mathbf{H}(\mathbf{r}, t) = \begin{bmatrix} \psi_x(\mathbf{r}, t) \\ \psi_y(\mathbf{r}, t) \\ \psi_z(\mathbf{r}, t) \end{bmatrix}, \tag{2.20}$$

where the subscript m stands for Maxwell. Along the lines of the discussion in Sect. 2.2, we may write the Maxwell equations as

$$i\hbar\dot{\Psi}_m(\mathbf{r}, t) = -c\mathbf{s} \cdot \mathbf{p}\Psi_m(\mathbf{r}, t), \tag{2.21}$$

where $\mathbf{p} = \frac{\hbar}{i}\nabla$, as before, but now

$$s_x = \begin{bmatrix} 0 & 0 & 0 \\ 0 & 0 & -i \\ 0 & i & 0 \end{bmatrix}, \quad s_y = \begin{bmatrix} 0 & 0 & i \\ 0 & 0 & 0 \\ -i & 0 & 0 \end{bmatrix}, \quad s_z = \begin{bmatrix} 0 & -i & 0 \\ i & 0 & 0 \\ 0 & 0 & 0 \end{bmatrix}. \tag{2.22}$$

The present s matrix is related to the s of Sect. 2.2 by the factor i. It also should be noted that the present photon wave function ψ_m is a 1×3 matrix whereas that of 2.2 is a 1×6 matrix. That is, the quantum optical analysis involves a two-component wave function in Ψ_ε and $\Psi_{\mathcal{H}}$; in the present analysis we find it convenient to combine the electric and magnetic contributions at the outset.

Since the energy per photon is $\hbar\omega = \hbar ck = cp$, we write

$$i\hbar\dot{\Psi}_m(\mathbf{r}, t) = H\Psi_m(\mathbf{r}, t), \tag{2.23}$$

where the Hamiltonian is given by

$$H = -c\mathbf{s} \cdot \mathbf{p}. \tag{2.24}$$

The natural extension of this Schrödinger equation for the spin one massless photon to the case of a spin zero particle of mass m is clear. That is, since $E = \sqrt{m_0^2c^4 + p^2c^2}$ is the finite mass extension of $E = pc$, we follow the lead of de Broglie and write

$$i\hbar\dot{\Psi}(\mathbf{r}, t) = \sqrt{m_0^2c^4 + p^2c^2}\,\Psi(\mathbf{r}, t), \tag{2.25}$$

where $\mathbf{p} = \frac{\hbar}{i}\nabla$, just as it is for the photon.

Hence when $m_0c^2 \gg pc$ we may write $\sqrt{m_0^2c^4 + p^2c^2} \cong \frac{p^2}{2m} + m_0c^2$, and we have

$$i\hbar\dot{\Psi}(\mathbf{r}, t) = \frac{-\hbar^2}{2m_0}\nabla^2\Psi(\mathbf{r}, t), \tag{2.26}$$

which is the non-relativistic wave equation, again obtained without introducing operator-valued time or momentum.

2.4 The Single Photon and Two Photon Wave Functions

The photon wave function concept really comes into its own when solving problems involving photon–photon correlations. Then, as is explained in [8], the two photon wave function

$$\psi^{(2)}(\mathbf{r}_1, t_1; \mathbf{r}_2, t_2) \equiv \langle 0 | \hat{E}(\mathbf{r}_2, t_2) \hat{E}(\mathbf{r}_1, t_1) | \Psi \rangle \qquad (2.27)$$

is the subject of interest. Under some conditions this may be written in terms of single photon wave functions, as in the case of two photon cascade discussed below. Some of the calculational details will be given since the physics (and the devil) is in the details.

Consider first the single photon wave function. From Eqs. (2.3) and (2.4) and ignoring polarization, we find

$$\langle 0 | \hat{E}(\mathbf{r}, t) | \psi_\gamma \rangle = \sqrt{\frac{\hbar}{2\varepsilon_0 V}} \sum_{\mathbf{k}} (\nu_k)^{1/2} \mathbf{g}_{\mathbf{k}} e^{-i\nu_k t} e^{i\mathbf{k}\cdot(\mathbf{r}-\mathbf{r}_0)} \frac{1}{(\nu_k - \omega) + i\Gamma/2} . \qquad (2.28)$$

We now evaluate this function by converting the sum into an integral. The ϕ- and θ-integrations can be carried out by choosing a coordinate system in which the vector $\mathbf{r}-\mathbf{r}_0$ points along the z-axis. We then carry out the integration over $|\mathbf{k}|$ by evaluating the density of states and matrix elements at resonance. We are left with the integral

$$\int_{-\infty}^{\infty} d\nu_k \frac{e^{-i\nu_k t + i\nu_k \Delta r/c}}{(\nu_k - \omega) + i\Gamma/2},$$

which is evaluated via contour methods and where $\Delta r = |\mathbf{r} - \mathbf{r}_0|$ is the distance from the atom located at position $\mathbf{r}_0$ to the detector. For $t < \Delta r/c$, the contour lies in the upper half-plane and if $t > \Delta r/c$, in the lower half-plane. On performing the integration, we find

$$\langle 0 | \hat{E}(\mathbf{r}, t) | \psi_\gamma \rangle = \frac{\mathcal{E}}{\Delta r} \Theta \left(t - \frac{\Delta r}{c} \right) e^{-i(t - \frac{\Delta r}{c})(\omega - i\Gamma/2)} , \qquad (2.29)$$

where Θ is a unit step function and $\mathcal{E}$ is an overall constant with the units of electric field.

Next we consider the problem of "interrupted" emission, see Fig. 2.3. The first photon, associated with the $a \leftrightarrow b$ transition, is described in the long time limit by our "old friend"

Fig. 2.3 Figure illustrating
decay of atom excited to state
a at a rate γ_a to non-decaying
level b. Upon detection of
$a \to b$ photon, population in
level b is transferred to b' by
means of an external field
indicated by *wavy line*. Level
b' decays to c at rate γ_b

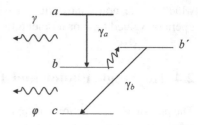

$$|\gamma\rangle = \sum_{\mathbf{k}} \frac{g_{a,\mathbf{k}} e^{-i\mathbf{k}\cdot\mathbf{r}}}{(\omega_{ab} - c|\mathbf{k}|) - i\gamma_a} |1_{\mathbf{k}}\rangle . \tag{2.30}$$

Likewise the second photon, associated with the $b' \to c$ transition, is given in the long time limit by

$$|\phi\rangle = \sum_{\mathbf{q}} \frac{g_{b,\mathbf{q}} e^{-i(\mathbf{q}\cdot\mathbf{r}-cqt_0)}}{(\omega_{ac} - c|\mathbf{q}|) - i\gamma_b} |1_{\mathbf{q}}\rangle , \tag{2.31}$$

where t_0 is the time of detection of γ photon and the transfer from $b \to b'$.

Using (2.30) and (2.31), it is easy to calculate the two photon wave function $\Psi^{(2)}(\mathbf{r}_1, t_1; \mathbf{r}_2, t_2)$ as defined by (2.27). We find

$$\psi^{(2)}(\mathbf{r}_1, t_1; \mathbf{r}_2, t_2) = \psi_\gamma(\mathbf{r}_1, t_1)\psi_\phi(\mathbf{r}_2, t_2) + \psi_\phi(\mathbf{r}_1, t_1)\psi_\gamma(\mathbf{r}_2, t_2) , \tag{2.32}$$

where

$$\psi_\gamma(\mathbf{r}_i, t_i) = \frac{\varepsilon_\gamma}{\Delta r_i} \Theta\left(t_i - \frac{\Delta r_i}{c}\right) e^{-\gamma_a(t_i - \frac{\Delta r_i}{c})} e^{-i\omega_{ab}(t_i - \frac{\Delta r_i}{c})}, \tag{2.33}$$

and

$$\psi_\phi(\mathbf{r}_i, t_i) = \frac{\varepsilon_\phi}{\Delta r_i} \Theta\left(t_i - t_0 - \frac{\Delta r_i}{c}\right) e^{-\gamma_a(t_i - t_0 - \frac{\Delta r_i}{c})} e^{-i\omega_{bc}(t_i - t_0 - \frac{\Delta r_i}{c})}, \tag{2.34}$$

where $i = 1, 2$ designates the detector positions.

2.5 Conclusions

One motive for this chapter is to show that the time appearing in the classical Maxwell equations is the same as the time parameter which appears in the TDSE. Thus, the times appearing in classical mechanics and electrodynamics and quantum mechanics are all the same.

Another motivation involves the definition of the photon wave function in terms of the electric and magnetic operators as

$$\Psi_{\mathcal{E}}(\mathbf{r}, t) = \langle 0 | \hat{E}(\mathbf{r}) | \Psi(t) \rangle, \tag{2.35}$$

and

$$\Psi_{\mathcal{H}}(\mathbf{r}, t) = \langle 0 | \hat{H}(\mathbf{r}) | \Psi(t) \rangle. \tag{2.36}$$

Equations (2.6) and (2.11) are the analog of the matter wave probability amplitudes

$$\Psi(\mathbf{r}, t) = \langle 0 | \hat{\psi}(\mathbf{r}) | \Psi(t) \rangle \tag{2.37}$$

discussed at length in Sect. 2.1.

As explained in [3], the discussion of the proceeding paragraph serves to put the nice question of Kramers [9] in perspective. Specifically, Kramers asks,

When in 1924 De Broglie suggested that material particles should show wave phenomena ... such a comparison was of great heuristic importance. Now that wave mechanics has become a consistent formalism one could ask whether it is possible to consider the Maxwell equations to be a kind of Schrödinger equation of light particles ...?

Kramers answers his question in the negative, he says,

Thus it is natural to ask what are the ϕ's for photons? Strictly speaking there are no such wave functions! One may not speak of particles in a radiation field in the same sense as in the elementary quantum mechanics of systems of particles as used in the last chapter. The reason is that the wave equation ... solutions of Schrödinger's time dependent wave function corresponding to an energy E_λ have a circular frequency $\omega_\lambda = +E_\lambda/\hbar$, while the monochromatic solutions of the wave equation have both $\pm\omega_\lambda$.

In other words, Kramers is saying that "the *real* electric wave has both $\exp(-i\nu_k t)$ and $\exp(i\nu_k t)$ parts while the matter wave has only $\exp(-i\nu_p t)$ type terms."

However, from the quantum optical perspective, we see that the photon wave functions (2.35) and (2.36) and the matter wave function (2.37) are identical in spirit. An earlier discussion of the importance of the analytical (positive frequency) signal in this context was given by Sudarshan [10].

The present measurement theory, "a-photon-is-what-a-photodetector-detects" point-of-view is discussed further in [3]. We have also included in Sect. 2.4 a detailed photon–photon correlation analysis [8] for the convenience of the reader.

Acknowledgments I would like to thank R. Arnowitt, C. Summerfield, and S. Weinberg for useful and stimulating discussions. This work was supported by the Robert A. Welch Foundation grant number A-126 and the ONR award number N00014-07-1-1084.

References

1. M. Scully, J. Phys. Conf. Ser. **99**, 012019 (2008); Wilhelm and Else Heraeus-Seminar (no. 395) in Honor of Prof. M. Kleber (Blaubeuren, Germany, Sept 2007)

2. K. Chapin, M. Scully, M.S. Zubairy, in *Frontiers of Quantum and Mesoscopic Thermodynamics Proceedings* (28 July–2 August 2008), Physica E, to be published
3. M. Scully, M.S. Zubiary, *Quantum Optics* (Cambridge Press, Cambridge, 1997)
4. S. Weinberg, *The Quantum Theory of Fields I* (Cambridge Press, Cambridge, 2005)
5. S. Schweber, *An Introduction to Relativistic Quantum Field Theory* (Harper and Row, New York, 1962)
6. R. Good, T. Nelson, *Classical Theory of Electric and Magnetic Fields* (Academic Press, New York, 1971)
7. R. Oppenheimer, Phys. Rev. **38**, 725 (1931)
8. M. Scully, in *Advances in Quantum Phenonema*, E. Beltrametti, J.-M. Lévy-Leblond (eds.) (Plenum Press, New York, 1995)
9. H. Kramers, *Quantum Mechanics* (North Holland, Amsterdam, 1958)
10. E.C.G. Sudarshan, Phys. Rev. Lett. **10**, 277 (1963)

Chapter 3
Post-Pauli's Theorem Emerging Perspective on Time in Quantum Mechanics

Eric A. Galapon

3.1 Introduction

In a Hilbert space setting, Pauli's well-known theorem asserts that no self-adjoint operator exists that is conjugate to a semibounded or discrete Hamiltonian [55]. Pauli's argument goes as follows. Assume that there exists a self-adjoint operator $\mathbf{T}$ conjugate to a given Hamiltonian $\mathbf{H}$, that is, $[\mathbf{T}, \mathbf{H}] = i\hbar\mathbf{I}$; such an operator conjugate to the Hamiltonian is known as a time operator. Since $\mathbf{T}$ is self-adjoint, the operator $\mathbf{U}_\varepsilon = \exp(-i\varepsilon\mathbf{T})$ is unitary for all real number ε. Now if φ_E is an eigenvector of $\mathbf{H}$ with the eigenvalue E, then, according to Pauli, the conjugacy relation $[\mathbf{T}, \mathbf{H}] = i\hbar\mathbf{I}$ implies that $\mathbf{T}$ is a generator of energy shifts so that $\mathbf{H}\mathbf{U}_\varepsilon\varphi_E = (E + \varepsilon)\varphi_{E+\varepsilon}$; this means that $\mathbf{H}$ has a continuous spectrum spanning the entire real line because ε is an arbitrary real number. Hence, the 'inevitable' conclusion that if the Hamiltonian is semibounded or discrete no self-adjoint time operator $\mathbf{T}$ will exist to satisfy $[\mathbf{T}, \mathbf{H}] = i\hbar\mathbf{I}$. A modern reading of Pauli's theorem is that the conjugacy relation $[\mathbf{T}, \mathbf{H}] = i\hbar\mathbf{I}$ implies that the pair $\mathbf{T}$ and $\mathbf{H}$ form a system of imprimitivities over the entire real line, so that when $\mathbf{H}$ is semibounded or discrete $\mathbf{T}$ cannot be self-adjoint [59].

It is Pauli's theorem that has distilled the idea that self-adjointness and conjugacy of a time operator for a semibounded or discrete Hamiltonian cannot be imposed simultaneously [29, 43, 56, 59, 53, 60, 44]. Since quantum observables are postulated to be self-adjoint operators in the earlier days of quantum mechanics [63], the non-existence of self-adjoint time operator has been interpreted to mean that time is not a dynamical observable but a mere parameter marking the evolution of a quantum system [58, 36, 54, 7, 2]. However, it is likewise widely recognized that time acquires dynamical significance in questions involving the occurrence of an event [59, 8, 52, 12] – when a nucleon decays [15] or when a particle arrives at a given spatial point [51, 39] or when a particle emerges from a potential barrier [48].

E.A. Galapon (✉)
Theoretical Physics Group, National Institute of Physics, University of the Philippines, Diliman, Quezon City, Philippines, eric.galapon@up.edu.ph

Galapon, E.A.: *Post-Pauli's Theorem Emerging Perspective on Time in Quantum Mechanics*. Lect. Notes Phys. **789**, 25–63 (2009)
DOI 10.1007/978-3-642-03174-8_3   © Springer-Verlag Berlin Heidelberg 2009

Moreover, there is the time–energy uncertainty principle, a reasonable interpretation of which requires more than a parametric treatment of time [8, 1, 3, 9, 16, 17, 40–42]. This opposing view on time in quantum mechanics precipitated to what is now known as the quantum time problem.

Pauli's theorem has been so ingrained into the physicist's psyche that it stifled serious sustained research on the quantum dynamical aspect of time until quite recently. The realization that quantum observables are not necessarily self-adjoint but may be non-self-adjoint as first moments of positive operator-valued measures (POVM) has brought a resurgence of interest on the quantum time problem. The introduction of POVM observables has opened up the possibility of entertaining non-self-adjoint, conjugate time operators as quantum observables, because such operators may be first moments of certain POVMs [59, 2, 10]; the quantized free time of arrival operator is an example of such a non-self-adjoint operator conjugate to the free Hamiltonian [14]. Since Pauli's theorem has been understood to mean that **T** and **H** are each other's generator of translations in their respective spectral measures, it has been the belief that a time operator must inevitably be non-self-adjoint for semibounded Hamiltonians and must cardinally be covariant and a POVM observable [29, 59, 60, 2, 8, 9, 14, 4]. Now covariance requires that a time operator must at least have a completely continuous spectrum. This altogether denies the possibility of constructing self-adjoint time operators that are bounded and compact for semibounded Hamiltonians.

However, while a sustained development in the dynamical aspect of time under the motivation of POVM observables is in progress, an unexpected development has emerged: a counter example to Pauli's theorem in the Hilbert space formulation of quantum mechanics is constructed, exposing the subtle assumptions that go into Pauli's arguments that cannot be sustained. In [19] we have shown the consistency of a bounded and self-adjoint operator conjugate to a discrete and semibounded Hamiltonian, contrary to Pauli's claim. There we have explicitly shown that the quantized classical free time of arrival for a spatially confined particle is self-adjoint, compact, and conjugate to the Hamiltonian in a non-dense subspace of the Hilbert space. This in effect has demonstrated that the non-self-adjointness of the same formal quantized operator for a particle in the real line has nothing to do with the semiboundedness of the free Hamiltonian, again, contrary to expectations due to Pauli's theorem. The existence of such self-adjoint time operators has opened up a new window through which the quantum time problem can be viewed from a different perspective.

In this chapter, we synthesize the progress that we have made since the appearance of [19], in particular, to our solution to the quantum time of arrival problem in the interacting case [50, 4, 49, 21]. Our solution consists of generalizing the time of arrival for a spatially confined particle in [19] under more general boundary conditions and in the presence of an interaction potential. This generalization led to the introduction of the confined quantum time of arrival (CTOA) operators, which are both conjugate and self-adjoint [24, 25, 22]. The dynamical behaviors of the eigenfunctions of the CTOA operators lead to a coherent theory of quantum arrival in one dimension that can yield both time of arrival distributions and at the same give a mechanism for the appearance of particle at the moment of its arrival [26, 23]. The resulting theory of quantum arrival invites us to reconsider our beliefs on time operators and on the role of time in quantum measurement theory.

3.2 Quantum Canonical Pairs

3.2.1 Canonical Pairs in Hilbert Spaces

We cannot start to appreciate the significance of the counter example to Pauli's theorem without a clear understanding of the properties of a canonical pair in a Hilbert space. To the physicist, a canonical pair is a pair of operators $(\mathbf{Q}, \mathbf{P})$ satisfying the canonical commutation relation, $[\mathbf{Q}, \mathbf{P}] = i\hbar\mathbf{I}$, (CCR), but a quantum canonical pair is much more elaborate than that. Failure to recognize its ramifications can lead to unwarranted claims and conclusions regarding the properties of such a pair [11]. Let $\mathcal{H}$ be the system Hilbert space, which we assume to be infinite dimensional. If we seek a pair of operators in $\mathcal{H}$, $\mathbf{Q}$ and $\mathbf{P}$, with respective domains $\mathcal{D}_{\mathbf{Q}}$ and $\mathcal{D}_{\mathbf{P}}$, satisfying the CCR, then two facts must be recognized:

1. No pair $(\mathbf{Q}, \mathbf{P})$ exists to satisfy the CCR in the entire Hilbert space $\mathcal{H}$.
   That is, there are no $\mathbf{Q}$ and $\mathbf{P}$ such that $[\mathbf{Q}, \mathbf{P}]\varphi = i\hbar\varphi$ for all φ in $\mathcal{H}$, or $[\mathbf{Q}, \mathbf{P}] = i\hbar\mathbf{I}_{\mathcal{H}}$, where $\mathbf{I}_{\mathcal{H}}$ is the identity in $\mathcal{H}$. A pair $(\mathbf{Q}, \mathbf{P})$ can at most satisfy the CCR in a proper subspace, $\mathcal{D}_c$, of $\mathcal{H}$; that is, the relation $[\mathbf{Q}, \mathbf{P}]\varphi = i\hbar\varphi$ holds only for all those φ in $\mathcal{D}_c$, where $\mathcal{D}_c$ is always smaller than $\mathcal{H}$. Thus a canonical pair in a Hilbert space is a triple $\mathcal{C}(Q, P; \mathcal{D}_c)$ – a pair of Hilbert space operators, $\mathbf{Q}$ and $\mathbf{P}$, together with a non-trivial, proper subspace $\mathcal{D}_c$ of $\mathcal{H}$, which we refer to as the canonical domain. The canonical domain may or may not be dense;[1] it may not even be invariant under either $\mathbf{Q}$ or $\mathbf{P}$. These subtle properties of the canonical domain generally forbid us from acting arbitrarily with $\mathbf{Q}$ and $\mathbf{P}$ on $\mathcal{D}_c$. Failure to pay attention to these small details can lead to erroneous generalizations; for example, the conclusion that the spectra of $\mathbf{Q}$ and $\mathbf{P}$ are the entire real line because they satisfy the CCR (see, for example, [11]) requires, at least, the canonical domain be dense and invariant under $\mathbf{Q}$ and $\mathbf{P}$. When even just one of these conditions is not satisfied, the conclusion no longer holds. In general the commutator domain, $\mathcal{D}_{com} = \mathcal{D}_{\mathbf{QP}} \cap \mathcal{D}_{\mathbf{PQ}}$, the domain in which $(\mathbf{QP} - \mathbf{PQ})$ is defined in the Hilbert space, does not coincide with the canonical domain. That is, the canonical commutation relation $[\mathbf{Q}, \mathbf{P}]\varphi = i\hbar\varphi$ does not hold in general for arbitrary elements of $\mathcal{D}_{com}$ but only for certain elements of a subset $\mathcal{D}_c$ of $\mathcal{D}_{com}$.
2. There are canonical pairs in the *same* Hilbert space that do not share the same properties.

---

[1] A subspace $\mathcal{D}$ of $\mathcal{H}$ is dense if the only vector orthogonal to all elements of $\mathcal{D}$ is the zero vector.

This means that, for a given Hilbert space $\mathcal{H}$, we can find different pairs of operators $(\mathbf{Q}_j, \mathbf{P}_j)$ acting in $\mathcal{H}$, together with corresponding subspaces $\mathcal{D}_j$, such that we have the canonical pairs $\mathcal{C}(\mathbf{Q}_j, \mathbf{P}_j; \mathcal{D}_j)$. The pairs $(\mathbf{Q}_j, \mathbf{P}_j)$ and $(\mathbf{Q}_{j'}, \mathbf{P}_{j'})$ may be different in the sense that there is *no* unitary operator $\mathbf{U}$ such that $\mathbf{Q}_{j'} = \mathbf{U}\mathbf{Q}_j\mathbf{U}^{-1}$ and $\mathbf{P}_{j'} = \mathbf{U}\mathbf{P}_j\mathbf{U}^{-1}$. For such cases, the pairs $\mathcal{C}(\mathbf{Q}_j, \mathbf{P}_j; \mathcal{D}_j)$ will have different spectral properties, e.g., one pair may be self-adjoint, another non-self-adjoint. Also it is possible that for a given operator $\mathbf{Q}$ there may be several distinct $\mathbf{P}_k$'s with corresponding subspaces $\mathcal{D}_k$ – that is, $\mathbf{P}_k \neq \mathbf{P}_{k'}$ and $\mathcal{D}_k \neq \mathcal{D}_{k'}$ for $k \neq k'$ – such that for every k we have the canonical pair $\mathcal{D}(\mathbf{Q}, \mathbf{P}_k; \mathcal{D}_k)$. Of course for a given $\mathbf{P}$ and $\mathcal{D}_c$, such that we have the canonical pair $\mathcal{C}(Q, P; \mathcal{D}_c)$, we can also find another operator $\mathbf{P}' = \mathbf{P} + \mathbf{F}$ with $[\mathbf{Q}, \mathbf{F}]\varphi = 0$ for all φ in $\mathcal{D}_c$ such that we have another canonical pair $\mathcal{C}'(\mathbf{Q}, \mathbf{P}'; \mathcal{D}_c)$. But we mean more than that: For $\mathbf{P}_k \neq \mathbf{P}_{k'}$ there may not be an $\mathbf{F}$ such that $\mathbf{P}_{k'} = \mathbf{P}_k + \mathbf{F}$. We will illustrate later how these different cases may arise in certain physical systems.

3.2.2 Classification of Hilbert Space Solutions to the CCR

For a given Hilbert space $\mathcal{H}$, we refer to a canonical pair $\mathcal{C}(\mathbf{Q}, \mathbf{P}; \mathcal{D}_c)$, with $\mathbf{Q}$ and $\mathbf{P}$ both operators in $\mathcal{H}$, as a solution to the CCR.[2] Solutions split into two major *categories*, according to whether the canonical domain $\mathcal{D}_c$ is dense or not. We shall say that a canonical pair is of *dense-category* if the corresponding canonical domain is dense; otherwise, it is of *closed category*. Solutions under these categories further split into distinct *classes* of unitary equivalent pairs, and each class will have its own set of properties. Under such categorization of solutions, the CCR in a given Hilbert space $\mathcal{H}$ assumes the form $[\mathbf{Q}, \mathbf{P}] \subset i\hbar\mathbf{P}_{\bar{\mathcal{D}}_c}$, where $\mathbf{P}_{\bar{\mathcal{D}}_c}$ is the projection operator onto the closure $\bar{\mathcal{D}}_c$ of the canonical domain $\mathcal{D}_c$. If the pair $\mathcal{C}(\mathbf{Q}, \mathbf{P}; \mathcal{D}_c)$ is of dense category, then the closure of $\mathcal{D}_c$ is just the entire $\mathcal{H}$, so that $\mathbf{P}_{\bar{\mathcal{D}}_c}$ is the identity $\mathbf{I}_{\mathcal{H}}$ of $\mathcal{H}$. In fact, we are considering a more general solution set to the CCR than has been considered so far. The traditional reading of the CCR in $\mathcal{H}$ is the form $[\mathbf{Q}, \mathbf{P}] \subset i\hbar\mathbf{I}_{\mathcal{H}}$, which is just the dense category.

It can be shown that the canonical and commutator domains coincide for dense category canonical pairs, that is, $\mathcal{D}_c = \mathcal{D}_{\text{com}}$; on the other hand, the canonical domain is smaller than and contained in the commutator domain for closed category canonical pairs, that is, $\mathcal{D}_c \subset \mathcal{D}_{\text{com}}$ [27]. Since only the dense category solutions have been the subject of investigations so far, we have gotten used to dealing with canonical pairs in the entire commutator domain and may feel suspicious with the

---

[2] We avoided to use the more mathematically accurate term *representation* in favor of the term *solution*. The reason is that *representation* carries an extra connotation in physics in which it usually implies *equivalence*. For example, we have position and momentum representations, and we know that these two representations are equivalent so that it does not matter which one we use in describing our system. In fact, the use of phrases such as *representations of the Heisenberg pair* in physics literature has added to the confusion on the exact nature of quantum canonical pairs, in particular, giving the impression that different canonical pairs have similar properties.

closed category solutions. However, the confined time of arrival operators, together
with their Hamiltonians, form such a class of canonical pairs, and they, as we will
see later, have an unambiguous physical origin.

3.2.3 Is There a Preferred Solution to the CCR?

We discussed above that for a given Hilbert space $\mathcal{H}$ there are numerous solutions
to the canonical commutation relation that do not necessarily share the same properties. So is there a preferred solution to the CCR? Should we accept only solutions
of dense or closed category of a specific class? Let us see how different solutions
may arise in a given Hilbert space and see how each solution may represent different
systems.

Let us consider the well-known position and momentum operators in three different configuration spaces: The entire real line, $\Omega_1 = (-\infty, \infty)$; the bounded segment
of the real line, $\Omega_2 = (0, 1)$; and the half line $\Omega_3 = (0, \infty)$. Quantum mechanics
in each of these happens in the Hilbert spaces $\mathcal{H}_1 = L^2(\Omega_1)$, $\mathcal{H}_2 = L^2(\Omega_2)$, and
$\mathcal{H}_3 = L^2(\Omega_3)$, respectively. The position operators, $\mathbf{Q}_j$, in $\mathcal{H}_j$, for all $j = 1, 2, 3$,
arise from the fundamental axiom of quantum mechanics that the propositions for
the location of an elementary particle in different volume elements of Ω_j are compatible (see Jauch [45] for a detailed discussion for Ω_1, which can be extended to
Ω_2 and Ω_3). They are self-adjoint and are given by the operators $(\mathbf{Q}_j\varphi)(q) = q\varphi(q)$
for all φ in the domain $\mathcal{D}_{\mathbf{Q}_j} = \{\varphi \in \mathcal{H}_j : \mathbf{Q}_j\varphi \in \mathcal{H}_j\}$. Note that $\mathbf{Q}_1$ and $\mathbf{Q}_3$ are
both unbounded, while $\mathbf{Q}_2$ is bounded.

Now each of the configuration spaces, Ω_1, Ω_2, and Ω_3, has an identifying property. Ω_1 is fundamentally homogeneous – points there are physically indistinguishable. On the other hand, Ω_2 and Ω_3 are not homogeneous, the boundaries being the
distinguishing factor. However, their inhomogeneities are not the same, their number
of boundaries being different. These properties can be expressed mathematically in
terms of the respective representation of translation in each of these configuration
spaces. Translation in Ω_1 is isomorphic to the additive group of real numbers; in
Ω_2, to the group of rotations of the circle; in Ω_3, to the semigroup of additive positive numbers. Thus in $\mathcal{H}_1$ and $\mathcal{H}_2$ there are one-parameter unitary operators $\mathcal{U}_1(s)$,
$\mathcal{U}_2(s)$ representing translations in $\mathcal{H}_1$ and $\mathcal{H}_2$, respectively. And in $\mathcal{H}_3$ there is a
completely one-parameter semigroup $\mathcal{U}_3(s)$ representing translations. If we define
the momentum operator as the generator of translation in the configuration space,
then the momentum operator in $\mathcal{H}_j$ is the operator $\mathbf{P}_j$ defined on all vectors φ for
which the limit $\hbar \lim_{s \to 0}(is)^{-1}(\mathcal{U}_j(s) - \mathbf{I}_{\mathcal{H}})\varphi = \mathbf{P}_j\varphi$ exists. Explicitly, it is given by
$(\mathbf{P}_j\varphi)(q) = -i\hbar\varphi'(q)$.

In each $\mathcal{H}_j$, there exists a dense common subspace $\mathcal{D}_j$ of $\mathbf{Q}_j$ and $\mathbf{P}_j$, which is
invariant under $\mathbf{Q}_j$ and $\mathbf{P}_j$, for which we have the canonical pair $\mathcal{C}_j(\mathbf{Q}_j, \mathbf{P}_j; \mathcal{D}_j)$.
The $\mathcal{C}_j$'s are of the same dense category, but they belong to different classes: $\mathbf{Q}_1$ and
$\mathbf{P}_1$ are both self-adjoint, having absolutely continuous spectra spanning the entire
real line Re and forming a system of imprimitivities in Re, and their restrictions

in $\mathcal{D}_1$ are essentially self-adjoint. $\mathbf{Q}_2$ is self-adjoint with an absolutely continuous spectra in (a, b), and its restriction in $\mathcal{D}_2$ is essentially self-adjoint; $\mathbf{P}_2$ is self-adjoint with a pure point spectrum, but its restriction in $\mathcal{D}_2$ is not essentially self-adjoint. $\mathbf{Q}_3$ is self-adjoint with an absolutely continuous spectra in $(0, \infty)$, and its restriction in $\mathcal{D}_3$ is essentially self-adjoint; $\mathbf{P}_3$ is maximally symmetric and non-self-adjoint, thus without any self-adjoint extension. These varied properties of the position and momentum canonical pairs are obviously the consequences of the underlying properties of their respective configuration spaces.

So is there a preferred solution to the CCR? Recall that there is only one separable Hilbert space; that is, all separable Hilbert spaces are isomorphically equivalent to each other, so that there are unitary operations transforming one Hilbert space to another. The three Hilbert spaces, $\mathcal{H}_1$, $\mathcal{H}_2$, and $\mathcal{H}_3$, are separable, and hence can be transformed to a common Hilbert space $\mathcal{H}_C$, together with all the operators in them, including their respective position and momentum operators. The canonical pairs, $\{\mathcal{C}_1, \mathcal{C}_2, \mathcal{C}_3\}$, are then solutions of the CCR in the same Hilbert space $\mathcal{H}_C$. And we have seen that they are of dense category solutions, but of different classes – and, most important, they represent different physical systems. If we look at the diverse properties of the above $\mathcal{C}_j$'s, we can see that these properties are reflections of the fundamental properties of the underlying configuration spaces of their respective physical systems.

It is then misguided to prefer one solution of the CCR over the rest or to require a priori a particular category of a specific class of a solution without a proper consideration of the physical context against which the solution is sought. For example, if we insist that only canonical pairs forming a system of imprimitivities over the real line are acceptable, then, within the context of position–momentum pairs, we are imposing homogeneity in all configuration spaces. But why impose the homogeneity of, say, Ω_1 in intrinsically inhomogeneous configuration spaces like Ω_2 and Ω_3?

From the position–momentum example, it can be concluded that the set of properties of a specific solution to the CCR is consequent to a set of underlying fundamental properties of the system under consideration, or to the basic definitions of the operators involved, or to some fundamental axioms of the theory, or to some postulated properties of the physical universe. That is, a specific solution to the CCR is canonical in some *sense*, i.e., of a particular category and of a particular class. It is conceivable to impose that a given pair be canonical as a priori requirement based, say, from its classical counterpart, but not without a deeper insight into the underlying properties of the system. In other words, we do not impose in what *sense* a pair is canonical if we do not know much, we derive in what *sense* instead. Furthermore, if a given pair is known to be canonical in some *sense*, then we can learn more about the system or the pair by studying the structure of the *sense* the pair is canonical [19].

We can appreciate this statement further by noting that finding solutions to the CCR in a Hilbert space is akin to solving a differential equation in which there is no preferred solution until appropriate boundary or initial condition is imposed. Also as in differential equations where the imposed conditions may not admit a solution, the CCR may, too, not admit a solution for a given required set of properties of the sought-after canonical pair. For example, we may require a solution in $L^2(0, \infty)$ for the position and momentum operator pair that are both self-adjoint. But, while the position operator is self-adjoint, the momentum operator cannot be self-adjoint. Hence, no solution exists for the sought pair. It is clear though that the reason for the non-existence of solution lies in the inhomogeneous property of the underlying configuration space; and the non-self-adjointness of the momentum operator, being the generator of translation in the configuration space, is a statement of this fact in the system Hilbert space. It is important to bear in mind that, while the set of properties we require of a canonical pair may be physically motivated, the mathematical structure of Hilbert spaces is under no obligation to submit to our wishes. If no solution exists, it may be because our required properties are inconsistent or not physically possible in the first place. And if we pay attention to why no solution exists or why a certain class of solutions exists, we may have a better understanding of the physical underpinnings of the system under consideration.

3.2.4 Example: Dense and Closed Category Solutions to the Time–Energy Canonical Commutation Relation

Consider a particle in a force-free interval $[-l, l]$. The system Hilbert space is $\mathcal{H} = L^2[-l, l]$. Let the Hamiltonian be $(\mathbf{H}\varphi)(q) = -\hbar^2\varphi''(q)/2\mu$ subject to the boundary conditions $\varphi(-l) = e^{-2i\gamma}\varphi(l)$ and $\varphi'(-l) = e^{-2i\gamma}\varphi'(l)$, for some fixed γ with $0 < |\gamma| < \pi/2$. We demonstrate that there exist at least two self-adjoint time operators conjugate to $\mathbf{H}$, forming dense and closed category solutions to the time–energy canonical commutation relation $[\mathbf{T}, \mathbf{H}]\varphi = i\hbar\varphi$. Moreover, we find these time operators to be both compact and hence non-covariant.

Now there exists a compact and self-adjoint operator $\mathbf{T}_1$ such that $\mathbf{T}_1$ and $\mathbf{H}$ form a canonical pair of dense category [20]. This operator has the integral representation

$$(\mathbf{T}_1\varphi)(q) = \int_{-l}^{l} \left[i\hbar \sum_{k,k'}' \frac{\varphi_k^{(\gamma)}(q)\varphi_{k'}^{(\gamma)}(q')^*}{E_k - E_{k'}} \right] \varphi(q')\,dq' , \tag{3.1}$$

where the $\phi_k^{(\gamma)}(q)$'s are the eigenfunctions of $\mathbf{H}$ and the E_k's are the corresponding eigenvalue (see Eq. 3.42), in which the primed sum indicates that $k = k'$ is excluded. That is, the pair $\mathbf{T}_1$ and $\mathbf{H}$ satisfy the canonical commutation relation in some dense subspace $\mathcal{D}_c^{(1)}$ of $\mathcal{H}$,

$$([\mathbf{T}_1, \mathbf{H}]\varphi)(q) = i\hbar\varphi(q), \text{ for all } \varphi(q) \in \mathcal{D}_c^{(1)} , \tag{3.2}$$

$$\mathcal{D}_c^{(1)} = \left\{ \varphi(q) = \sum_k a_k\varphi_k^{(\gamma)}(q), \sum_k |a_k|^2 < \infty, \sum_k a_k = 0 \right\} . \tag{3.3}$$

Since the canonical domain is dense, i.e., orthogonal only to the zero vector, the canonical pair $\mathcal{C}(\mathbf{T}_1, \mathbf{H}; \mathcal{D}_c^{(1)})$ is of dense category. $\mathbf{T}_1$ is compact because its kernel is square integrable.

There exists a compact and self-adjoint operator $\mathbf{T}_2$ such that $\mathbf{T}_2$ and $\mathbf{H}$ form a canonical pair of closed category. This operator has the integral representation

$$(\mathbf{T}_2\varphi)(q) = \int_{-l}^{l} \left[\frac{\mu}{4\hbar \sin \gamma} (q + q') \left(e^{i\gamma} H(q - q') + e^{-i\gamma} H(q' - q) \right) \right] \varphi(q')\, dq', \quad (3.4)$$

where $H(q - q')$ is the Heaviside function. That is, the pair $\mathbf{H}$ and $\mathbf{T}_2$ satisfies the canonical commutation relation in a non-dense subspace of $\mathcal{H}$,

$$([\mathbf{T}_2, \mathbf{H}]\varphi)(q) = i\,\hbar\varphi(q) \quad \text{for all} \quad \varphi(q) \in \mathcal{D}_c^{(2)} , \qquad (3.5)$$

$$\mathcal{D}_c^{(2)} = \left\{ \int_{-l}^{l} \varphi(q)\, dq = 0, \ \varphi(-l) = \varphi(l) = 0, \ \varphi'(-l) = \varphi'(l) = 0 \right\} . \quad (3.6)$$

Since the canonical domain $\mathcal{D}_c^{(2)}$ is not dense, i.e., it is orthogonal to any vector $\phi(q) = constant \neq 0$, the canonical pair $\mathcal{C}(\mathbf{T}_2, \mathbf{H}; \mathcal{D}_c^{(2)})$ is of closed category. $\mathbf{T}_2$ is likewise compact because its kernel is square integrable.

This example shows that it is possible for a given Hamiltonian to have numerous distinct associated time operators. So which time operator? Learning from our example with the position and momentum operators, we cannot know which one until we knew the origins of these operators or studied their properties. Remember that the quantum time problem has many aspects – quantum arrival, quantum traversal, quantum tunneling – so it is possible that one time operator is more appropriate than the other for a certain aspect of the quantum time problem. While both operators are compact, it is not necessary that they will exhibit the same dynamical behaviors, so that they do not necessarily represent the same aspect of time.

The rest of the chapter is devoted to solving the time of arrival problem by finding the appropriate time of arrival operator solution to the time–energy canonical commutation relation. From this we will uncover the physical origin of the time operator $\mathbf{T}_2$ and find the operator $-\mathbf{T}_2$ as a time of arrival operator for a spatially confined particle under certain boundary condition. More importantly, we will find that the solution set for such a system in the presence of a continuous interaction potential consists of self-adjoint, compact, and conjugate time operators – the confined time of arrival (CTOA) operators. It will become clear why the CTOA operators are appropriately referred to as time of arrival operators; and, in the process, it will be made evident why compact and non-covariant time operators can be physically meaningful.

3.3 Time of Arrival Operators

3.3.1 The Quantum Time of Arrival Problem

The quantum time of arrival problem seeks to find the time of arrival distribution of a quantum particle at a given arrival point in the configuration space, for a given initial state [50]. A solution to the problem within standard quantum mechanics constitutes finding a time of arrival operator $\mathbf{T}$ conjugate to the system Hamiltonian $\mathbf{H}$, $[\mathbf{H}, \mathbf{T}]\varphi = i\hbar\varphi$, that admits a spectral resolution, not necessarily projection valued, from which the time of arrival distribution can be computed in the standard way.

However, the consensus is that no time of arrival operator can be constructed in the most general case of arbitrary arrival point and of arbitrary interaction potential. In the early days of quantum mechanics, the reason for this consensus is that no self-adjoint time of arrival operator can be constructed, in accordance with Pauli's theorem. This belief has been bolstered by the observation that the quantization of the free classical time of arrival expression is non-self-adjoint and admits no self-adjoint extension. But even with the acceptance of POVMs as observables, so that the non-self-adjoint free time of arrival operator can be interpreted as a POVM observable [14], the belief is still there that no time of arrival operator can be constructed.

In one dimension, the most quoted reason is that the classical time of arrival at some point x, which is given by

$$T_x(q, p) = -\text{sgn}(p)\sqrt{\frac{\mu}{2}} \int_x^q \frac{dq'}{\sqrt{H(q, p) - V(q')}} \,, \qquad (3.7)$$

where H is the Hamiltonian and V is the interaction potential and (q, p) are the position and momentum at $t = 0$, does not admit a sensible quantization because Eq. (3.7) is generally not everywhere real and single valued in the entire phase space [50, 57]. Moreover, the known existence of obstruction to quantization in Euclidean space forbids the existence of quantization that satisfies the Dirac Poisson-bracket-commutator correspondence; this implies that even if somehow we can quantize Eq. (3.7), then in general the quantized time of arrival operator is not conjugate with the Hamiltonian. For these reasons it is believed that if a theory of quantum arrival existed it could not rest on the spectral resolution of a time of arrival operator.

However, it is now clear that these objections to the construction of time of arrival operators can be overcome. In this section, we describe how a time of arrival operator conjugate to a given Hamiltonian can be constructed. And in later sections we will describe how self-adjoint, compact, and conjugate time of arrival operators can be obtained from this time of arrival operator. We will see how these self-adjoint operators can address not only the quantum time of arrival problem but also the question of appearance of particles in quantum mechanics.

3.3.2 The Idea of Supraquantization

Quantization seeks to derive the quantum counterpart of a classical observable $f(q, p)$ by some associative mapping $\mathcal{Q}$ of the real-valued function $f(q, p)$ to a maximally symmetric operator $\mathbf{F}$ in the system Hilbert space $\mathcal{H}$, i.e., $\mathcal{Q}(f) \mapsto \mathbf{F}$. A paramount requirement of quantization is that the Poisson bracket of two (classical) observables quantizes into the commutator of the separately quantized observables, in particular, $\mathcal{Q}(\{f, g\}) = (i\hbar)^{-1}[\mathcal{Q}(f), \mathcal{Q}(g)]$. However, there is a well-known obstruction to quantization in Euclidean space (and other spaces) which says that no quantization exists that satisfies the Poisson-bracket-commutator correspondence requirement for all observables [32, 33, 31, 34, 30, 35, 38, 61]. This is unsatisfactory because the said correspondence is necessary, for example, in ensuring that required evolution properties of a certain class of observables are satisfied.

Thus in [18, 21] we addressed the problem of obstruction to quantization by proposing the method of supraquantization – the construction of quantum observables without quantization and the subsequent quantum mechanical derivation of its classical counterpart. The central idea of supraquantization is that quantum observables can be grouped meaningfully into distinct classes of observables, with each class possessing a set of properties that distinguishes the observables of the class from other observables not belonging to the class. It is the central problem of supraquantization to determine this set of properties shared by the class of observables. Once these properties are known, the observables of the class are determined by imposing the axioms of quantum mechanics in conjunction with other principles of physics and by requiring that the quantum observables of the class reduce to their classical counterparts in the classical limit.

For a specific class of classical observables, the required supraquantization may be accomplished by referring to one of the members of the class and employing a transfer principle to the rest. The transfer principle can be expressed as follows: Each element of a class of observables shares a common set of properties with the rest of its class such that when a particular property is identified for a specific element of the class that property can be transferred to the rest of the class without discrimination. This, together with the axioms of quantum mechanics and the correspondence principle, allows us to infer a general property of the observables of the class by solving a particular observable of the class and then abstracting from that particular observable the sought after property.

The idea of supraquantization is employed in [18, 21] in constructing time of arrival operators without quantization and is described in the rest of this section. We describe below how solving for the free quantum time of arrival operator without quantization leads to solving the time of arrival operator in the presence of interaction potential using the transfer principle. This consequently leads to the derivation of the classical time of arrival from pure quantum mechanical consideration.

3.3.3 Construction of Time of Arrival Operators Without Quantization

Before we proceed, observe that by changing variables from (q, p) to $(\tilde{q} = q - x, \tilde{p} = p)$ in Eq. (3.7), we find that Eq. (3.7) becomes the time of arrival at the origin for the potential $\tilde{V}(\tilde{q}) = V(\tilde{q} + x)$. Hence it is sufficient for us to consider the time of arrival at the origin in the development to follow, for when the arrival point is different from the origin we only have to appropriately change variables. Our problem now is to find the appropriate time of arrival operator $\mathbf{T}$ at the origin for a given Hamiltonian $\mathbf{H}$ for a quantum particle in one dimension.

While our ultimate goal is to construct self-adjoint and conjugate Hilbert space time of arrival operators for a given interaction potential, we pose the construction problem in the rigged Hilbert space $\Phi^{\times} \supset \mathcal{H}_{\infty} \supset \Phi$ as an intermediate step, where Φ is the fundamental space of infinitely differentiable functions in the real line with compact supports, $\mathcal{H}_{\infty} = L^2(-\infty, \infty)$ is the Hilbert space closure of Φ under the usual metric of quantum mechanics and $\Phi^{\times}$ is the space of functionals on Φ. The operator $\mathbf{T}$ that we seek generally maps Φ into $\Phi^{\times}$ and has the integral representation

$$(\mathbf{T}\varphi)(q) = \int_{-\infty}^{\infty} \langle q | \mathbf{T} | q' \rangle \, \varphi(q') \mathrm{d}q', \tag{3.8}$$

for all $\varphi(q)$ in Φ. In this form, solving for $\mathbf{T}$ is finding for the kernel $\langle q | \mathbf{T} | q' \rangle$. But how do we determine the kernel $\langle q | \mathbf{T} | q' \rangle$ without resorting to quantization?

We accomplish this in two steps. First, it is by identifying the set of properties of the time of arrival operator $\mathbf{T}$ and the set of appropriate physical principles that will uniquely identify $\mathbf{T}$. Second, it is by implementing the transfer principle where we solve the construction of the free time of arrival operator using the results of first step to abstract out the general form of the time of arrival operator kernel.

Let us enumerate the properties of the time of arrival $\mathbf{T}$ that we require:

1. Being a time of arrival operator, it must at least evolve according to $d\mathbf{T}/dt = -\mathbf{I}_{\Phi}$, where $\mathbf{I}_{\Phi}$ is the identity in Φ, at least in the neighborhood of $t = 0$. This condition translates to the generalized canonical commutation relation

$$\left([\mathbf{H}^{\times}, \mathbf{T}]\varphi\right)(q) = i\hbar\varphi(q) \tag{3.9}$$

for all $\varphi(q)$ in Φ, where $\mathbf{H}^{\times}$ is the rigged Hilbert space extension of $\mathbf{H}$ in $\Phi^{\times}$. This is a property arising from the axioms of quantum mechanics.

2. For the operator $\mathbf{T}$ to be identifiable as a time of arrival operator, it must reduce to the classical time of arrival in the classical limit. That is, it must satisfy the condition

$$T_0(q, p) = \lim_{\hbar \to 0} T_{\hbar}(q, p), \tag{3.10}$$

where $T_0(q, p)$ is the classical time of arrival at the origin and $T_\hbar(q, p)$ is the Wigner transform of the kernel

$$T_\hbar(q, p) = \int_{-\infty}^{\infty} \left\langle q + \frac{v}{2} \middle| \mathbf{T} \middle| q - \frac{v}{2} \right\rangle \exp\left(-i\frac{v\,p}{\hbar}\right) dv . \tag{3.11}$$

This is a property arising from the correspondence principle.

3. The operator $\mathbf{T}$ must satisfy the time reversal symmetry $\Theta \mathbf{T} \Theta^{-1} = -\mathbf{T}$, where Θ is the time reversal operator; this translates to the condition

$$\langle q| \mathbf{T} |q'\rangle^* = - \langle q| \mathbf{T} |q'\rangle . \tag{3.12}$$

4. In keeping with the requirement that quantum observables must yield real expectation values, we require $\mathbf{T}$ to be Hermitian; this translates into the condition

$$\langle q'| \mathbf{T} |q\rangle^* = \langle q| \mathbf{T} |q'\rangle . \tag{3.13}$$

The time reversal and hermicity conditions already restrict the functional form of the kernel. Equation (3.12) implies that $\langle q| \mathbf{T} |q'\rangle$ is purely complex, so that $\langle q| \mathbf{T} |q'\rangle = i\tau(q, q')$, where $\tau(q, q')$ is real valued. On the other hand, Eq. (3.13) implies that $\tau(q, q') = -\tau(q', q)$, which allows us to write $\tau(q, q') = T(q, q')S(q, q')$, where $T(q, q') = T(q', q)$ and $S(q, q') = -S(q', q)$. The time reversal and hermicity conditions then dictate that the kernel is of the form $\langle q| \mathbf{T} |q'\rangle = iT(q, q')S(q, q')$, with $T(q, q')$ and $S(q, q')$ to be determined. The conjugacy (3.9) and the correspondence principle (3.10) conditions will further restrict the forms of $T(q, q')$ and $S(q, q')$, but we will find the above-enumerated conditions are not sufficient to uniquely identify $\mathbf{T}$. However, supplementing them with the condition that only parameters of the system enter into the construction will identify $\mathbf{T}$ uniquely.

Recognize that solving for the time of arrival operator $\mathbf{T}$ for a given Hamiltonian $\mathbf{H}$ is essentially solving for the generalized canonical commutation relation (3.9) under certain conditions specified by Eqs. (3.10), (3.12), and (3.13). This is an example of what we have discussed earlier that canonical pairs may arise as solutions to specific problems. We will see below that there are in fact numerous solutions to Eq. (3.9), but only a particular solution will be found acceptable.

3.3.4 Non-interacting Case

We now solve the free particle kernel without quantization.[3] The problem is to determine the unknown $T(q, q')$ and $S(q, q')$ for the free particle. Substituting the free Hamiltonian and the time of arrival operator back into Eq. (3.9) and after performing

---

[3] Here we give a more transparent solution than the one provided in [21].

two integrations by parts, we arrive at

$$([\mathbf{H}, \mathbf{T}]\varphi)(q) = \int_\Sigma i \left[-\frac{\hbar^2}{2\mu} \frac{\partial^2 T(q, q')}{\partial q^2} + \frac{\hbar^2}{2\mu} \frac{\partial^2 T(q, q')}{\partial q'^2} \right] S(q, q')\varphi(q')\,dq'$$
$$+ \int_\Sigma i \frac{\hbar^2}{2\mu} \left[-\left\{ 2\frac{\partial T}{\partial q}\frac{\partial S}{\partial q} + T(q, q')\frac{\partial^2 S}{\partial q^2} \right\} \right.$$
$$+ \left. \left\{ 2\frac{\partial T}{\partial q'}\frac{\partial S}{\partial q'} + T(q, q')\frac{\partial^2 S}{\partial q'^2} \right\} \right] \varphi(q')\,dq' , \tag{3.14}$$

where Σ is the support of $\varphi(q)$. We now have to choose $T(q, q')$ and $S(q, q')$ such that $\mathbf{T}$ is a solution to the generalized CCR. The first term can be made to vanish without specifying $S(q, q')$ by imposing that $T(q, q')$ is a continuous solution to the partial differential equation

$$-\frac{\hbar^2}{2\mu} \frac{\partial^2 T(q, q')}{\partial q^2} + \frac{\hbar^2}{2\mu} \frac{\partial^2 T(q, q')}{\partial q'^2} = 0 . \tag{3.15}$$

This has the general solution $T(q, q') = f(q + q') + g(q - q')$, where f and g are arbitrary; however, since $T(q, q') = T(q', q)$, g must be an even function.

Moreover, functions $T(q, q')$ and $S(q, q')$ must be chosen such that the second term equals the right-hand side of Eq. (3.9). This is only possible if the bracketed quantity in the second term involves the Dirac delta function and, perhaps, its derivatives. Since $T(q, q')$ has already been chosen to satisfy a second-order partial differential equation, so that it has continuous second-order partial derivatives, the delta function must only come from $S(q, q')$. The general form of $S(q, q')$ that leads to the required Dirac delta function is $S(q, q') = C\,H(q - q') + B\,H(q' - q)$, where C and B are constants and $H(x)$ is the Heaviside step function. But since $S(q, q') = -S(q', q)$, we must have $B = -C$ or $S(q, q') = C\,\text{sgn}(q - q')$, where $\text{sgn}(x) = H(x) - H(-x)$ is the sign function.

Substituting $S(q, q') = C\,\text{sgn}(q - q')$ back into the second term leads to

$$([\mathbf{H}, \mathbf{T}]\varphi)(q) = \left\{ -2C\frac{\hbar}{\mu} \left[\frac{\partial T}{\partial q} + \frac{\partial T}{\partial q'} \right]_{q'=q} \right\} i\hbar\varphi(q) . \tag{3.16}$$

Hence $\mathbf{T}$ is a solution to the CCR if the bracketed quantity is unity. If we choose $C = -\mu/\hbar$, we must have

$$\left[\frac{\partial T}{\partial q} + \frac{\partial T}{\partial q'} \right]_{q'=q} = \frac{1}{2} . \tag{3.17}$$

This implies that $T(q, q')$ has finite first derivatives along the line $q = q'$. For a solution of the form $T(q, q') = f(q + q')$, the only possible solution satisfying (3.17) is given by $T(q, q') = (q + q')/4$; however, the general solution satisfying

(3.17) is given by $T'(q, q') = (q + q')/4 + h((q - q')^2)$, where $h(x)$ continuous second derivative everywhere.

We now require that the correspondence principle be satisfied. That is, the classical time of arrival at the origin, given by $T = -\mu q/p$, must be derived from $\mathbf{T}$. We first consider the solution $T(q, q') = (q + q')/4$ so that we have the kernel

$$\langle q | \mathbf{T} | q' \rangle = \frac{\mu}{i\hbar} \frac{(q + q')}{4} \operatorname{sgn}(q - q') . \qquad (3.18)$$

This is an acceptable solution if the corresponding operator $\mathbf{T}$ reduces to the classical free time of arrival. Indeed we have its Wigner transform

$$\int_{-\infty}^{\infty} \langle q + \frac{v}{2} | \mathbf{T} | q - \frac{v}{2} \rangle \exp\left(-i\frac{v\,p}{\hbar}\right) dv = \frac{\mu}{2i\hbar} q \int_{-\infty}^{\infty} \operatorname{sgn}(v) e^{-ivp/\hbar} \, dv = -\mu \frac{q}{p} . \quad (3.19)$$

Thus Eq. (3.18) solves the kernel problem. However, as we have noted above, the function $T(q, q')$ is not, so far, uniquely determined because we can have, say, the solution $T'(q, q') = (q + q')/4 + a(q - q')^2$ for some real constant a. This gives another solution $\mathbf{T}'$ whose Wigner transform is given by

$$\int_{-\infty}^{\infty} \langle q + \frac{v}{2} | \mathbf{T}' | q - \frac{v}{2} \rangle \exp\left(-i\frac{v\,p}{\hbar}\right) dv = -\mu \frac{q}{p} - 4a\mu\hbar^2 \frac{1}{p^3} , \qquad (3.20)$$

which reduces to the classical free time of arrival as $\hbar \to 0$. For a fixed we find that the two operators $\mathbf{T}$ and $\mathbf{T}'$ satisfy the correspondence principle in the classical limit.

Hence the enumerated properties of the time of arrival operator that we seek is not sufficient to uniquely identify the solution. They must somehow be supplemented. Notice that $\mathbf{T}'$ involves a constant a with unit of inverse length that cannot be generated from the available parameters of the free particle, which are $\hbar$ and μ. Hence to consider $\mathbf{T}'$ we have to introduce another constant of nature which is questionable because quantum mechanics does well without the need for such another physical constant. Thus restricting ourselves to solutions involving parameters of the system alone, we are left with the solution $T(q, q') = (q+q')/4$, and the free time of arrival operator kernel is given by Eq. (3.18).

3.3.5 Interacting Case

Having solved the free particle kernel, we proceed in implementing the transfer principle. We hypothesize that all time of arrival operator kernels assume the same form. Thus, from Eq. (3.18), the kernel takes on the form

$$\langle q | \mathbf{T} | q' \rangle = \frac{\mu}{i\hbar} T(q, q') \operatorname{sgn}(q - q') , \qquad (3.21)$$

where $T(q, q')$ depends on the interaction potential and which we refer to as the kernel factor. From the time reversal symmetry and hermicity requirement, we require that $T(q, q')$ be real valued and symmetric, $T(q, q') = T(q', q)$. We determine $T(q, q')$ by imposing condition (3.9) on **T**. Substituting Eq. (3.21) back into the left-hand side of Eq. (3.9) and performing two successive integration by parts, we arrive at

$$([\mathbf{H}^\times, \mathbf{T}]\varphi)(q) = i\hbar \left(\frac{dT(q, q)}{dq} + \frac{\partial T(q', q')}{\partial q} + \frac{\partial T(q, q)}{\partial q'} \right) \varphi(q) \, dq$$

$$-i\frac{\mu}{\hbar} \int_\Sigma \left[\left(-\frac{\hbar^2}{2\mu} \frac{\partial^2}{\partial q^2} + V(q) \right) T(q, q') - \left(-\frac{\hbar^2}{2\mu} \frac{\partial^2}{\partial q'^2} + V(q') \right) T(q, q') \right]$$

$$\times \mathrm{sgn}(q - q')\varphi(q') \, dq', \tag{3.22}$$

where Σ is the support of $\varphi(q)$. We point out that our ability to arrive at this expression has been made possible by extending the formulation in a rigged Hilbert space.

If $\mathbf{H}^\times$ and $\mathbf{T}$ are to be conjugate in Φ, then the right-hand side of Eq. (3.22) must reduce to Eq. (3.9). Following the lead of the free particle case, Eq. (3.22) reduces to Eq. (3.9) if we impose that $T(q, q')$ solves the partial differential equation

$$-\frac{\hbar^2}{2\mu} \frac{\partial^2 T(q, q')}{\partial q^2} + \frac{\hbar^2}{2\mu} \frac{\partial^2 T(q, q')}{\partial q'^2} + (V(q) - V(q')) T(q, q') = 0 \tag{3.23}$$

and satisfies the condition

$$\frac{dT(q, q)}{dq} + \frac{\partial T(q, q')}{\partial q} \bigg|_{q=q'} + \frac{\partial T(q, q')}{\partial q'} \bigg|_{q'=q} = 1. \tag{3.24}$$

Equation (3.24) defines a family of operators conjugate to the Hamiltonian in the sense required by Eq. (3.9).

The conditions guaranteeing uniqueness of $T(q, q')$ is found by appealing to the transfer principle. We determine the set of conditions satisfied by the free particle kernel factor that ensures it uniqueness and then impose the same conditions in the interacting case. By inspection, the free particle kernel factor $T(q, q') = (q + q')/4$ satisfies both (3.23) and (3.24), and it satisfies the conditions

$$T(q, q) = \frac{q}{2}, \quad T(q, -q) = 0. \tag{3.25}$$

These two conditions uniquely identify the free particle solution. By virtue of the transfer principle, we impose the same conditions (3.25) on the solution to Eq. (3.23). It can be established that the boundary conditions (3.25) guarantee that boundary condition (3.24) is satisfied; moreover, Eqs. (3.25) already imposes the condition $T(q, q') = T(q', q)$ [21]. Likewise, it can be shown that for continuous potentials Eq. (3.23) has a unique solution in the entire qq'-plane [22, 47].

In general, the solution to the time kernel equation falls into whether the system under consideration is linear or non-linear. For linear systems, systems with the interacting potential $V(q) = aq^2 + bq + c$ for some constants a, b, and c, the solution is given by $T_{lin}(q, q') = T_0(q, q')$, where

$$T_0(q, q') = \frac{1}{2} \int_0^\eta {}_0F_1\left(; 1; \frac{\mu}{2\hbar^2} [V(\eta) - V(s)]\right) ds \Big|_{\eta = \frac{q+q'}{2}}, \qquad (3.26)$$

in which ${}_0F_1$ is a specific hypergeometric function. On the other hand, for non-linear systems with everywhere analytic potentials, the solution to the time kernel equation is given by $T_{nli}(q, q') = T_0(q, q') + T_1(q, q') + T_2(q, q') + \cdots$, where $T_1(q, q')$, $T_2(q, q')$, ... are determined by the potential and are found by solving explicitly the time kernel equation for the given potential.

The correspondence principle can now be established via the Wigner transform of the kernel of **T**. For linear systems, we have $T_\hbar(q, p) = t_0(q, p)$; while for non-linear systems, $T_\hbar(q, p) = t_0(q, p) + \hbar^2 t_1(q, p) + \hbar^4 t_2(q, p) + \cdots$, where $t_0(q, p)$ is given by

$$t_0(q, p) = -\sum_{k=0}^\infty (-1)^k \frac{(2k-1)!!}{k!} \frac{\mu^{k+1}}{p^{2k+1}} \int_0^q [V(q) - V(q')]^k \, dq', \qquad (3.27)$$

and the $\hbar^{2j} t_j(q, p)$'s arise from the $T_j(q, q')$ in the solution for $T_{nli}(q, q')$. For potentials continuous in the neighborhood of the origin, Eq. (3.27) can be summed to yield the classical time of arrival $T_0(q, p)$ at the origin. That is, $t_0(q, p)$ is an expansion of $T_0(q, p)$ in the neighborhood of the arrival point; for this reason, we referred to $t_0(q, p)$ as the local time of arrival. Hence we have our operator **T** reducing to the classical time of arrival in the classical limit. And we have solved the supraquantization of the time of arrival operator, at least, for continuous potentials.

3.3.6 Time of Arrival Operators in Representation-Free Form

In the construction of the confined time of arrival operators in the next section, we will need to explicitly write our time of arrival operator **T** in terms of the position and momentum operators, **q** and **p**, that is, in representation-free form. **T** must be written in them such that, in coordinate representation, the kernel of **T** is given by Eq. (3.21). Moreover, **T** must be in the form such that the generalized time–energy CCR, $([\mathbf{H}^\times, \mathbf{T}]\varphi)(q) = i\hbar\varphi(q)$, assumes the formal relation $[\mathbf{H}, \mathbf{T}] = i\hbar\mathbf{I}$, where **H** is the formal Hamiltonian $\mathbf{H} = \mathbf{p}^2/2\mu + V(\mathbf{q})$ [28].

This can be accomplished by Weyl quantizing the transform $T_\hbar(q, p)$ of the kernel. For everywhere analytic potentials, $T_\hbar(q, p)$ is an expansion in $q^n p^{-m}$ for positive integers n and m. The explicit operator form of **T** is then obtained with the formal replacement scheme

$$\frac{q^n}{p^m} \mapsto \mathbf{T}_{-m,n} = \frac{1}{2^n} \sum_{j=0}^{n} \binom{n}{j} \mathbf{q}^j \mathbf{p}^{-m} \mathbf{q}^{n-j} \qquad (3.28)$$

in $\mathcal{T}_\hbar(q, p)$. In this form, the canonical commutation relation formally reads $[\mathbf{H}, \mathbf{T}] = i\hbar\mathbf{I}$; this can be shown by using the known algebra of the formal operators $\mathbf{T}_{-m,n}$ due to Bender and Dunne [28, 5, 6].

For linear systems, it is clear that the formal time of arrival operator $\mathbf{T}$ is obtained by Weyl quantizing the classical local time of arrival since $\mathcal{T}_\hbar(q, p) = t_0(q, p)$. On the other hand, for non-linear systems, $\mathbf{T}$ is the quantization of $t_0(q, p)$ plus quantum corrections required to satisfy the time–energy commutation relation; due to the existence of obstruction to quantization in Euclidean space, $\mathbf{T}$ (for non-linear systems) cannot be constructed, except for linear systems, via direct quantization of the classical time of arrival.

3.3.7 Example: The Harmonic Oscillator

Let us consider the harmonic oscillator whose interaction potential is given by $V(q) = \mu\omega^2 q^2/2$. Substituting the potential back into the general expression for the classical time of arrival (3.7) yields the time of arrival at the origin $T_0(q, p) = -\tan^{-1}(\mu\omega q/p)/\omega$, for position q and momentum p at $t = 0$. On the other hand, the local time of arrival at the origin is given by

$$t_0(q, p) = -\sum_{k=0}^{\infty} \frac{(-1)^k}{2k+1} \mu^{2k+1} \omega^{2k} \frac{q^{2k+1}}{p^{2k+1}}, \qquad (3.29)$$

obtained by either expanding $T_0(q, p)$ or using Eq. (3.27). We demonstrate how Eq. (3.29) can be obtained using supraquantization.

Substituting the harmonic oscillator potential back into Eq. (3.23), the kernel factor can be solved to yield [21]

$$\begin{aligned}
T(q, q') &= \frac{\hbar}{2\mu\omega} \sum_{k=0}^{\infty} \frac{1}{(2k+1)!} \left(\frac{\mu\omega}{2\hbar}\right)^{2k+1} (q+q')^{2k+1} (q-q')^{2k} \\
&= \frac{\hbar}{2\mu\omega} \frac{\sinh\left(\frac{\mu\omega}{2\hbar}(q^2 - q'^2)\right)}{(q-q')}.
\end{aligned} \qquad (3.30)$$

This solution is unique and analytic in the entire qq'-plane. Using the identity

$$\int_{-\infty}^{\infty} y^{m-1} \operatorname{sgn}(y) e^{-ixy} dy = 2(m-1)!/(i^m x^m), \qquad (3.31)$$

the Wigner transform of the kernel can be calculated to yield

$$T_\hbar(q,p) = \frac{1}{2i\,\omega} \sum_{k=0}^{\infty} \frac{1}{(2k+1)!} \left(\frac{\mu\omega}{2\hbar}\right)^{2k+1} (2q)^{2k+1} \int_{-\infty}^{\infty} v^{2k} \mathrm{sgn}(v) \exp\left(-i\frac{v\,p}{\hbar}\right) dv$$

$$= -\sum_{k=0}^{\infty} \frac{(-1)^k}{2k+1} \mu^{2k+1} \omega^{2k} \frac{q^{2k+1}}{p^{2k+1}} \,. \tag{3.32}$$

We find that $T_\hbar(q,p)$ coincides exactly with the local time of arrival in the neighborhood of the origin.

Now quantizing the local time of arrival yields the harmonic oscillator time of arrival operator in formal representation-free form

$$\mathbf{T} = -\frac{1}{\omega} \sum_{k=0}^{\infty} \frac{(-1)^k}{2k+1} \mu^{2k+1} \omega^{2k+1} \mathbf{T}_{-2k-1,2k+1} \,. \tag{3.33}$$

This can shown to be formally conjugate with the formal harmonic oscillator Hamiltonian

$$\mathbf{H} = \frac{1}{2\mu}\mathbf{p}^2 + \frac{\mu\omega^2}{2}\mathbf{q}^2 \,. \tag{3.34}$$

That is, $[\mathbf{H}, \mathbf{T}] = i\hbar\mathbf{I}$ on using the formal canonical commutation relation $[\mathbf{q}, \mathbf{p}] = i\hbar\mathbf{I}$ [28]. Here we see that the result of supraquantization coincides with the Weyl quantization of the local time of arrival.

3.3.8 Example: The Quartic Oscillator

Let us consider the quartic oscillator whose interaction potential is given by $V = \lambda q^4/4$. The local time of arrival can be calculated from Eq. (3.27) to give

$$t_0(q,p) = -\sum_{j=0}^{\infty} (-1)^j \frac{\Gamma(5/4)\Gamma(j+1/2)}{\sqrt{\pi}\, 2^j \Gamma(j+5/4)} \mu^{j+1} \lambda^j \frac{q^{4j+1}}{p^{2j+1}} \,. \tag{3.35}$$

In the following, we demonstrate that $T_\hbar(q,p) = t_0(q,p) + \mathcal{O}(\hbar^2)$. And this is just a special case of our general result on the equality of $T_\hbar(q,p)$ and $t_0(q,p)$ only in the limit of vanishing or infinitesimal $\hbar$ for non-linear systems.

Substituting the quartic oscillator back into Eq. (3.23), the kernel factor can be solved to yield [21]

$$T(q,q') = \frac{1}{4} \sum_{k=0}^{\infty} \sum_{j=2k}^{\infty} \sigma_{k,j} \left(\frac{\mu\lambda}{16\hbar^2}\right)^{j-k} (q+q')^{4j+1-6k} (q-q')^{2j} \,, \tag{3.36}$$

where the $\sigma_{k,j}$'s are constants given by

$$\sigma_{0,j} = \frac{\Gamma(5/4)}{8^j \, j! \, \Gamma(j+5/4)} , \tag{3.37}$$

$$\sigma_{k,j} = \sum_{m=0}^{j-2k} \frac{\sigma_{k-1,2(k-1)+m}}{8^{j-2k+1-m} \, (-j)_{j-2k+1-m} \left(-\frac{1}{4} - j + \frac{3k}{2}\right)_{j-2k+1-m}} , \tag{3.38}$$

for $k = 1, 2, \ldots$ and $j \geq 2k$, in which $(a)_n$ is the Pochammer symbol. This solution is likewise unique.

Using the same identity, we can similarly calculate the Wigner transform of the kernel to yield

$$T_\hbar(q, p) = -\sum_{j=0}^{\infty} (-1)^j \frac{\Gamma(5/4)\Gamma(j+1/2)}{\sqrt{\pi} 2^j \Gamma(j+5/4)} \mu^{j+1} \lambda^j \frac{q^{4j+1}}{p^{2j+1}}$$

$$-\hbar^2 \cdot \frac{1}{4} \sum_{j=2}^{\infty} (-1)^j \sigma_{1,j}(2j)! \mu^j \lambda^{j-1} \frac{q^{4j-5}}{p^{2k+1}} - \cdots$$

$$= \tau_0(q, p) + \hbar^2 \tau_1(q, p) + \hbar^4 \tau_2(q, p) + \cdots , \tag{3.39}$$

where the $\tau_j(q, p)$'s are independent of $\hbar$. Comparing the leading term, which is independent of $\hbar$, we find that it is just the local time of arrival at the origin. Thus the classical time of arrival is recovered from the time of arrival operator.

Quantizing the local time of arrival yields the quartic oscillator time of arrival operator in formal representation-free form

$$\mathbf{T} = -\sum_{j=0}^{\infty} (-1)^j \frac{\Gamma(5/4)\Gamma(j+1/2)}{\sqrt{\pi} 2^j \Gamma(j+5/4)} \mu^{j+1} \lambda^j \mathbf{T}_{-4j-1,2k+1}$$

$$-\hbar^2 \cdot \frac{1}{4} \sum_{j=2}^{\infty} (-1)^j \sigma_{1,j}(2j)! \mu^j \lambda^{j-1} \mathbf{T}_{-4j+5,2k+1} - \cdots . \tag{3.40}$$

This can be shown to be formally conjugate with the formal quartic oscillator Hamiltonian

$$\mathbf{H} = \frac{1}{2\mu} \mathbf{p}^2 + \lambda \mathbf{q}^4 . \tag{3.41}$$

That is, $[\mathbf{H}, \mathbf{T}] = i\hbar\mathbf{I}$ on using the formal canonical commutation relation $[\mathbf{q}, \mathbf{p}] = i\hbar\mathbf{I}$ [28]. We find that the leading term is just the Weyl quantization of the local time of arrival. However, the appearance of terms in powers of $\hbar$ shows that quantization fails to yield a formal time of arrival operator that is conjugate with the Hamiltonian. This is a manifestation of the obstruction to quantization in Euclidean space.

3.4 Confined Time of Arrival Operators

Technically the confined time of arrival (CTOA) operators are the projections of the formal operator $\mathbf{T}$ in the Hilbert space $\mathcal{H}_l = L^2[-l, l]$ [24, 25, 22, 28]. Physically they are the time of arrival operators for a confined particle in the interval $[-l, l]$ under certain boundary conditions imposed on the momentum operator of the confined particle. Since the time of arrival operator $\mathbf{T}$ is explicitly in terms of the position and momentum operators, a CTOA operator is the operator $\mathbf{T}$ written in terms of the position and momentum operator of the confined particle. The CTOA operator is then obtained by specifying the position and momentum operator. The position operator is unique and is given by the bounded operator $\mathbf{q}$, $(\mathbf{q}\varphi)(q) = q\varphi(q)$ for all $\varphi(q)$ in $\mathcal{H}_l$. On the other hand, the momentum operator is not unique and has to be considered carefully. We assume the system to be conservative and we require that the evolution of the system be generated by a purely kinetic Hamiltonian in the *absence* of interaction potential. The former requires a self-adjoint Hamiltonian to ensure that time evolution is unitary. The latter requires a self-adjoint momentum operator commuting with the kinetic energy operator.

These two requirements are only satisfied by the following choice of the momentum operator. For every $|\gamma| < \pi/2$, define the self-adjoint momentum operator $(\mathbf{p}_\gamma \phi)(q) = -i\hbar\phi'(q)$, with domain $\mathcal{D}_{p_\gamma}$ consisting of those vectors $\phi(q)$ in $\mathcal{H}_l$ with square integrable first derivatives and that satisfy the boundary condition $\phi(-l) = e^{-2i\gamma}\phi(l)$. With $\mathbf{p}_\gamma$ self-adjoint, the Hamiltonian is purely kinetic in the non-interacting case, i.e., $\mathbf{H}_\gamma = (2\mu)^{-1}\mathbf{p}_\gamma^2$. The momentum and the Hamiltonian then commute and have the common set of eigenvectors

$$\phi_k^{(\gamma)}(q) = \frac{1}{2l} \exp\left[i(\gamma + k\pi)\frac{q}{l}\right] , \tag{3.42}$$

with respective eigenvalues $p_{k,\gamma} = \hbar(\gamma + k\pi)l^{-1}$, $E_k = p_{k,\gamma}^2(2\mu)^{-1}$, for all $k = 0$, $\pm 1, \pm 2, \ldots$. Since $\mathbf{T}$ depends on the momentum operator, the projection of $\mathbf{T}$ in $\mathcal{H}_l$ is a family of operators $\{\mathbf{T}_\gamma\}$, with each $\mathbf{T}_\gamma$ corresponding to the momentum $\mathbf{p}_\gamma$. The operators $\mathbf{T}_\gamma$ are the confined time of arrival operators.

To find the $\mathbf{T}_\gamma$'s explicitly we need to have the explicit forms of the operators $\mathbf{T}_{-m,n}$ in $\mathcal{H}_l$ for every γ and positive integer m and n, in particular their kernels in coordinate representation. There are two cases to consider: for non-periodic boundary conditions, $\gamma \neq 0$; and for periodic boundary condition, $\gamma = 0$. Both give the compact and self-adjoint operators

$$\mathbf{T}_{-m,n}^{\gamma \neq 0} = \frac{1}{2} \sum_{j=0}^{n} \binom{n}{j} \mathbf{q}^j \mathbf{p}_\gamma^{-m} \mathbf{q}^{n-j}, \quad \mathbf{T}_{-m,n}^{\gamma=0} = \frac{1}{2} \sum_{j=0}^{n} \binom{n}{j} \mathbf{q}^j \mathbf{P}_0^{-m} \mathbf{q}^{n-j} , \tag{3.43}$$

where $\mathbf{P}_0^{-1} = \mathbf{E}\mathbf{p}_0^{-1}\mathbf{E}$ in which $\mathbf{E}$ is the projector onto the subspace orthogonal to the null space of $\mathbf{p}_0$. The physical motivation and mathematical justification of using $\mathbf{P}_0^{-1}$ is elaborated in [19].

For continuous potentials, the index m is of the form $m = 2s + 1$ for some positive integer s. Now in coordinate representation, the operators $\mathbf{T}^{\gamma}_{-2s-1,n}$ in the Hilbert space $\mathcal{H}_l$ assume the integral form $\int_{-l}^{l} \langle q | \mathbf{T}^{\gamma}_{-2s-1,n} | q' \rangle \varphi(q') dq'$, where the kernels are given by [22]

$$\langle q | \mathbf{T}_{-2s-1,n} | q' \rangle = \left(\frac{q + q'}{2} \right)^n \langle q | \mathbf{p}^{-2s-1} | q' \rangle , \qquad (3.44)$$

where

$$\langle q | \mathbf{p}^{-2s-1}_{\gamma \neq 0} | q' \rangle = \frac{1}{2 \sin \gamma} \frac{(-1)^s (q - q')^{2s}}{\hbar^{2s+1}(2s)!} \left(e^{i\gamma} H(q - q') + e^{-i\gamma} H(q' - q) \right), (3.45)$$

$$\langle q | \mathbf{P}^{-2s-1}_{0} | q' \rangle = \frac{i(-1)^s}{2\hbar^{2s+1}} \frac{(q - q')^{2s}}{(2s)!} \operatorname{sgn}(q - q') - \frac{i}{l} \frac{(-1)^s}{2\hbar^{2s+1}} \frac{(q - q')^{2s+1}}{(2s + 1)!} . \qquad (3.46)$$

Notice that the second term in Eq. (3.46) can be obtained by integrating the factor of $\operatorname{sgn}(q - q')$ in the first term with respect to the variable $v = (q - q')$. This observation will be important below.

Because the $\mathbf{T}^{\gamma}_{-2s-1,n}$'s are Fredholm integral operators in coordinate representation, the projection of the formal operator $\mathbf{T}$ in $\mathcal{H}_l = L^2[-l, l]$ is the family of Fredholm integral operators

$$\left\{ (\mathbf{T}_{\gamma} \varphi)(q) = \int_{-l}^{l} \langle q | \mathbf{T}_{\gamma} | q' \rangle \varphi(q') dq', \quad |\gamma| \leq \pi/2 \right\} . \qquad (3.47)$$

Using the coordinate representation of the operators $\mathbf{T}_{-2s-1,n}$ in $\mathcal{H}_l$ for a given γ, the kernel of $\mathbf{T}_{\gamma}$ can be shown to be given by [22]

$$\langle q | \mathbf{T}_{\gamma \neq 0} | q' \rangle = -\mu \frac{T(q, q')}{\hbar \sin \gamma} \left(e^{i\gamma} H(q - q') + e^{-i\gamma} H(q' - q) \right) , \qquad (3.48)$$

$$\langle q | \mathbf{T}_{\gamma = 0} | q' \rangle = \frac{\mu}{i\hbar} T(q, q') \operatorname{sgn}(q - q') - \frac{\mu}{il\hbar} B(q, q') , \qquad (3.49)$$

where $T(q, q')$ is the kernel factor and $B(q, q')$ is given by

$$B(q, q') = \int_{0}^{(q-q')} T(u, v) dv \Bigg|_{u = q + q'} , \qquad (3.50)$$

in which $T(u, v) = T(q, q')$ with $u = (q + q')$ and $v = (q - q')$. Equation (3.50) follows from the observation made above concerning Eq. (3.46).

For continuous potentials the kernel $\langle q | \mathbf{T}_\gamma | q' \rangle$ is square integrable; and since it is symmetric, the confined time of arrival operator $\mathbf{T}_\gamma$ is compact and self-adjoint.[4] The compactness of $\mathbf{T}_\gamma$ implies that $\mathbf{T}_\gamma$ has a complete set of square integrable eigenfunctions with a pure discrete spectrum. This further implies that the confined time of arrival operators cannot be covariant as their eigenvalues do not constitute the real line. We will see later on that even though the $\mathbf{T}_\gamma$'s are non-covariant they are physically meaningful. Moreover, we will provide for a numerical evidence that covariance may be recovered in the limit as the confining length l increases indefinitely.

With the kernel factor $T(q, q') = (q + q')/4$ for the free particle, the origin of the operator $-\mathbf{T}_2$ in Sect. 3.2.4 is now clear: It is just the confined time of arrival operator for the given system for the given boundary conditions on the Hamiltonian.

3.5 Conjugacy of the Confined Time of Arrival Operators

We now address the conjugacy of the CTOA operators $\mathbf{T}_\gamma$ with their respective Hamiltonians $\mathbf{H}_\gamma$. We will show that the commutator domain and the canonical domain do not coincide and that the confined time of arrival and Hamiltonian pair is a canonical pair of closed category. The proof of the non-triviality of the canonical domain is given elsewhere [27].

3.5.1 Non-periodic Boundary Conditions

We first determine the commutator domain $\mathcal{D}_{\mathrm{com}}$ of $\mathbf{T}_\gamma$ and the Hamiltonian $\mathbf{H}_\gamma$ for $\gamma \neq 0$; for continuous potentials, the potential energy operator is defined everywhere in $\mathcal{H}_l$ so that the domain of $\mathbf{H}_\gamma$ equals the domain of the kinetic energy operator. Now $\mathcal{D}_{\mathrm{com}}$ consists of all vectors common to the domains of the operators $\mathbf{T}_\gamma \mathbf{H}_\gamma$ and $\mathbf{H}_\gamma \mathbf{T}_\gamma$. Since $\mathbf{T}_\gamma$ is defined everywhere in the Hilbert space, the domain of $\mathbf{T}_\gamma \mathbf{H}_\gamma$ is just the domain of $\mathbf{H}_\gamma$, $\mathcal{D}_{\mathbf{T}_\gamma \mathbf{H}_\gamma} = \mathcal{D}_{\mathbf{H}_\gamma}$. Since $\mathcal{D}_{\mathbf{H}_\gamma}$ is dense, $\mathcal{D}_{\mathbf{T}_\gamma \mathbf{H}_\gamma}$ is dense. On the other hand, the domain of $\mathbf{H}_\gamma \mathbf{T}_\gamma$ consists of those vectors φ in $\mathcal{H}$ such that $\mathbf{T}_\gamma \varphi$ falls into the domain $\mathcal{D}_{\mathbf{H}_\gamma}$ of $\mathbf{H}_\gamma$. In particular, it consists of those $\varphi(q)$ such that $(\mathbf{T}_\gamma \varphi)(-l) = e^{-2\gamma}(\mathbf{T}_\gamma \varphi)(l)$ and $(\mathbf{T}_\gamma \varphi)'(-l) = e^{-2\gamma}(\mathbf{T}_\gamma \varphi)'(l)$.

One can explicitly work out the implementation of these boundary conditions to yield

---

[4] In [22] two confined time of arrival operators are introduced: The algebra preserving CTOA operators, which are just the operators we refer to here as the confined time of arrival operators; and the quantized CTOA operators, which are the projections of the Weyl-quantized local time of arrival in the Hilbert space $\mathcal{H}_l$. The two are the same for linear systems, while the later is the leading term of the former for non-linear systems.

$$\mathcal{D}_{\mathbf{H}_\gamma \mathbf{T}_\gamma} = \left\{ \varphi(q) \in L^2[-l,l] : \ \varphi'(q) \in L^2[-l,l], \ \int_{-l}^{l} \phi_T(q')\varphi(q')\mathrm{d}q' = 0, \right.$$

$$\left. \int_{-l}^{l} \varphi_T(q')\varphi(q')\mathrm{d}q' + il \sin\gamma \left(e^{-i\gamma}\varphi(l) + e^{i\gamma}\phi(-l) \right) = 0 \right\} , \quad (3.51)$$

where

$$\phi_T(q') = \int_{-l}^{l} \frac{\partial T(q,q')}{\partial q}\mathrm{d}q, \quad \varphi_T(q') = \int_{-l}^{l} \frac{\partial^2 T(q,q')}{\partial q^2}\mathrm{d}q . \quad (3.52)$$

Since $\mathcal{D}_{\mathbf{T}_\gamma \mathbf{H}_\gamma} = \mathcal{D}_{\mathbf{H}_\gamma}$ the domain of the commutator is found by restricting the domain of $\mathbf{H}_\gamma \mathbf{T}_\gamma$ in the domain of $\mathbf{H}_\gamma$. Explicitly the commutator domain is given by

$$\mathcal{D}_{[\mathbf{H}_\gamma, \mathbf{T}_\gamma]} = \left\{ \varphi(q) \in \mathcal{D}_{\mathbf{H}_\gamma} : \ \varphi'(q) \in L^2[-l,l], \ \int_{-l}^{l} \phi_T(q')\varphi(q')\mathrm{d}q' = 0, \right.$$

$$\left. \int_{-l}^{l} \varphi_T(q')\varphi(q')\mathrm{d}q' + 2il \sin\gamma \, e^{-i\gamma}\varphi(l) = 0 \right\} . \quad (3.53)$$

Since $T(q,q')$ has continuous first and second partial derivatives with respect to both variables, the functions $\phi_T(q')$ and $\varphi_T(q')$ are both square integrable in $[-l,l]$ so that they belong to $\mathcal{H}_l$. This implies that the condition $\int_{-l}^{l} \phi_T(q')\varphi(q')\mathrm{d}q' = 0$ requires that the vector ϕ_T in $\mathcal{H}_l$ be orthogonal to the domain $\mathcal{D}_{\mathbf{H}_\gamma \mathbf{T}_\gamma}$. The commutator domain is then not dense.

For all φ in the commutator domain, the commutator can be explicitly evaluated to yield

$$([\mathbf{H}_\gamma, \mathbf{T}_\gamma]\varphi)(q) = i\hbar \left(\frac{\mathrm{d}T(q,q)}{\mathrm{d}q} + \left. \frac{\partial T(q,q')}{\partial q} \right|_{q'=q} + \left. \frac{\partial T(q,q')}{\partial q'} \right|_{q'=q} \right) \varphi(q)$$

$$- \frac{\mu}{\hbar \sin\gamma} \int_{-l}^{l} \left(-\frac{\hbar^2}{2\mu} \frac{\partial T(q,q')}{\partial q^2} + \frac{\hbar^2}{2\mu} \frac{\partial T(q,q')}{\partial q'^2} (V(q) - V(q'))T(q,q') \right)$$

$$\times \left(e^{i\gamma} H(q-q') + e^{-i\gamma} H(q'-q) \right) \varphi(q')\mathrm{d}q'$$

$$+ \frac{\hbar}{2\sin\gamma} e^{-i\gamma}\varphi'(l) \left(\left. T(q,q')\right|_{q'=-l} - \left. T(q,q')\right|_{q'=l} \right)$$

$$- \frac{\hbar}{2\sin\gamma} e^{-i\gamma}\varphi(l) \left(\left. \frac{\partial T(q,q')}{\partial q'} \right|_{q'=-l} - \left. \frac{\partial T(q,q')}{\partial q'} \right|_{q'=l} \right) . \quad (3.54)$$

We arrive at this expression by performing two successive integration by parts in evaluating the action of the operator $\mathbf{T}_\gamma \mathbf{H}_\gamma$.

The bracketed quantity in the first term equals one, by virtue of the boundary conditions on the kernel factor Eq. (3.24); and the second term vanishes, since $T(q,q')$ solves Eq. (3.23). Then the required commutator is given by

$$\left([\mathbf{H}_\gamma, \mathbf{T}_\gamma]\varphi\right)(q) = i\hbar\varphi(q) + \frac{\hbar}{2\sin\gamma}e^{-i\gamma}\varphi'(l)\left(T(q,q')\big|_{q'=-l} - T(q,q')\big|_{q'=l}\right)$$

$$-\frac{\hbar}{2\sin\gamma}e^{-i\gamma}\varphi(l)\left(\frac{\partial T(q,q')}{\partial q'}\bigg|_{q'=-l} - \frac{\partial T(q,q')}{\partial q'}\bigg|_{q'=l}\right). \quad (3.55)$$

If the second and third terms vanish, then $\mathbf{H}_\gamma$ and $\mathbf{T}_\gamma$ are conjugate in their entire commutator domain. However, they do not generally vanish because the vectors $\varphi(q)$ in $\mathcal{D}_{[\mathbf{H}_\gamma, \mathbf{T}_\gamma]}$, together with their first derivatives, do not necessarily vanish at the boundaries.

But if the commutator is further restricted to those elements of the commutator domain that satisfies $\varphi(\partial) = 0 \left(\text{or } \varphi(-l) = \varphi(l) = 0\right)$ and $\varphi'(\partial) = 0 \left(\text{or } \varphi'(-l) = \varphi'(l) = 0\right)$, the pair forms a canonical pair,

$$\left([\mathbf{H}_\gamma, \mathbf{T}_\gamma]\varphi\right)(q) = i\hbar\varphi(q) \text{ for all } \varphi(q) \in \mathcal{D}_{\text{can}}(\mathbf{H}_\gamma, \mathbf{T}_\gamma), \quad (3.56)$$

$$\mathcal{D}_{\text{can}}(\mathbf{H}_\gamma, \mathbf{T}_\gamma) = \left\{\varphi(q) \in \mathcal{D}_{\mathbf{H}_\gamma} : \varphi(\partial) = 0, \ \varphi'(\partial) = 0, \int_{-l}^{l}\phi_T(q')\varphi(q')\mathrm{d}q' = 0,\right.$$

$$\left.\int_{-l}^{l}\varphi_T(q')\varphi(q')\mathrm{d}q' = 0\right\}. \quad (3.57)$$

The canonical domain is orthogonal to the vectors ϕ_T and φ_T, hence not dense. The pair $\mathbf{H}_\gamma$ and $\mathbf{T}_\gamma$ then forms a canonical pair of closed category.

3.5.2 Periodic Boundary Condition

The commutator domain for $\mathbf{H}_0$ and $\mathbf{T}_0$ can be similarly derived. It is given by

$$\mathcal{D}_{[\mathbf{H}_0, \mathbf{T}_0]} = \left\{\varphi(q) \in \mathcal{D}_{\mathbf{H}_0} : \int_{-l}^{l}\phi_{T,B}(q')\varphi(q')\mathrm{d}q' = 0, \right.$$

$$\left.\int_{-l}^{l}\varphi_{T,B}(q')\varphi(q')\mathrm{d}q' + 4l\varphi(l) = 0\right\}, \quad (3.58)$$

where

$$\phi_{T,B}(q') = \left(T(q,q')\big|_{q=l} + T(q,q')\big|_{q=-l}\right) - \frac{1}{l}\left(B(q,q')\big|_{q=l} - B(q,q')\big|_{q=-l}\right),$$

$$(3.59)$$

$$\varphi_{T,B}(q') = \left(\frac{\partial T(q,q')}{\partial q}\bigg|_{q=l} + \frac{\partial T(q,q')}{\partial q}\bigg|_{q=-l} \right)$$
$$- \frac{1}{l} \left(\frac{\partial B(q,q')}{\partial q}\bigg|_{q=l} - \frac{\partial B(q,q')}{\partial q}\bigg|_{q=-l} \right). \tag{3.60}$$

Since $T(q,q')$ and $B(q,q')$ are both continuous, together with their first derivatives, $\phi_{TB}(q')$ and $\varphi_{TB}(q')$ both belong to the Hilbert space. Then because of the condition $\int_{-l}^{l} \phi_{TB}(q')\varphi(q')dq' = 0$, the commutator domain is not dense.

For all φ in the commutator domain, the commutator can be explicitly evaluated to yield

$$([\mathbf{H}_0, \mathbf{T}_0]\varphi)(q) = i\hbar\varphi(q)$$
$$- \frac{\mu}{i\hbar l} \int_{-l}^{l} \left[-\frac{\hbar^2}{2\mu} \frac{\partial^2 B(q,q')}{\partial q^2} + \frac{\hbar^2}{2\mu} \frac{\partial^2 B(q,q')}{\partial q'^2} + (V(q) - V(q'))B(q,q') \right] \varphi(q')dq'$$
$$- \frac{i\hbar}{2} \varphi'(l) \left[\left(T(q,q')\big|_{q'=l} + T(q,q')\big|_{q'=-l} \right) + \frac{1}{l} \left(B(q,q')\big|_{q'=l} - B(q,q')\big|_{q'=-l} \right) \right]$$
$$+ \frac{i\hbar}{2} \varphi(l) \left[\left(\frac{\partial T(q,q')}{\partial q'}\bigg|_{q'=l} + \frac{\partial T(q,q')}{\partial q'}\bigg|_{q'=-l} \right) \right.$$
$$\left. + \frac{1}{l} \left(\frac{\partial B(q,q')}{\partial q'}\bigg|_{q'=l} - \frac{\partial B(q,q')}{\partial q'}\bigg|_{q'=-l} \right) \right], \tag{3.61}$$

where we have performed the same simplifications done above to arrive at the first term. Again the pair is not conjugate in the entire commutator domain. However, we see that further restriction of the commutator domain yields a canonical domain. In contrast to the non-periodic case, vanishing of the $\varphi(q)$'s and their derivatives at the boundaries is not sufficient to enforce conjugacy of the pair.

Let us define the integral operator $(\varphi\mathbf{B})(q) = \int_{-l}^{l} \langle q| \mathbf{B} |q'\rangle \varphi(q')dq'$ with kernel given by

$$\langle q| \mathbf{B} |q'\rangle = -\frac{\hbar^2}{2\mu} \frac{\partial^2 B(q,q')}{\partial q^2} + \frac{\hbar^2}{2\mu} \frac{\partial^2 B(q,q')}{\partial q'^2} + (V(q) - V(q'))B(q,q'). \tag{3.62}$$

This does not vanish because $B(q,q')$ is not a solution to Eq. (3.23). Thus restricting the commutator domain to those vectors that belong to the null space of $\mathbf{B}$ and to those vectors, together with their first derivative, that vanish at the boundaries yields a canonical domain. Then we have the canonical commutation relation

$$([\mathbf{H}_0, \mathbf{T}_0]\varphi)(q) = i\hbar\varphi(q) \text{ for all } \varphi(q) \in \mathcal{D}_{\text{can}}(\mathbf{H}_0, \mathbf{T}_0), \tag{3.63}$$

$$\mathcal{D}_{\mathrm{can}}(\mathbf{H}_0, \mathbf{T}_0) = \left\{ \varphi(q) \in \mathcal{D}_{\mathbf{H}_0} : \ \varphi(\partial) = 0, \ \varphi'(\partial) = 0, \ \int_{-l}^{l} \phi_{TB}(q')\varphi(q')\mathrm{d}q' = 0, \right.$$

$$\left. \int_{-l}^{l} \varphi_{TB}(q')\varphi(q')\mathrm{d}q' = 0, (\mathbf{B}\varphi)(q) = 0 \right\} . \qquad (3.64)$$

The canonical domain is again not dense because there are non-zero vectors orthogonal to the canonical domain; moreover, it is contained in the null space of $\mathbf{B}$, which is not dense. The pair $\mathbf{H}_0$ and $\mathbf{T}_0$ then forms a canonical pair of closed category.

The above results are meaningful only if the canonical domain is not trivial. In the following, we will consider the non-interacting case to demonstrate the non-triviality of the canonical domain of the confined time of arrival operators. In the process, we will be introducing a method of constructing explicitly the vectors of the canonical domain.

3.5.3 Example: Non-interacting Case

3.5.3.1 Non-periodic Boundary Conditions

For the non-interacting case, the kernel factor is given by $T(q, q') = (q + q')/4$. From this, the vectors ϕ_T and φ_T orthogonal to the canonical domain can be determined to yield $\phi_T(q') = l/2$ and $\varphi_T(q') = 0$. The canonical domain reduces to

$$\mathcal{D}_{\mathrm{can}}(\mathbf{H}_\gamma, \mathbf{T}_\gamma) = \left\{ \varphi(q) \in \mathcal{D}_{\mathbf{H}_0} : \ \varphi(\partial) = 0, \ \varphi'(\partial) = 0, \ \int_{-l}^{l} \varphi(q)\mathrm{d}q = 0 \right\} . \quad (3.65)$$

Clearly $\mathcal{D}_{\mathrm{can}}$ is not dense because it is orthogonal to the non-zero vector $\phi(q) = constant$. We now show that this subspace is not trivial and in fact it contains an infinite number of elements.

We do so by constructing explicit elements of the canonical domain. Every element of $\mathcal{D}_{\mathbf{H}_\gamma}$ can be expanded in terms of the complete eigenfunctions of the momentum operator, $\varphi(q) = \sum_k \varphi_k \, \phi_k^{(\gamma)}(q)$, where $\varphi_k = \int_{-l}^{l} \phi_k^{(\gamma)*}(q) \, \varphi(q)\mathrm{d}q$. For $\varphi(q)$ to belong to the domain of the Hamiltonian, it must have a square integrable second derivative; this imposes the condition $\sum_k k^4 |\varphi_k|^2 < \infty$ on the φ_k's. Now $\varphi(q)$ belongs to the canonical domain if it satisfies all the required conditions, which translates to conditions on the coefficients φ_k's. The conditions $\varphi(\partial) = 0$, $\varphi'(\partial) = 0$, and $\int_{-l}^{l} \varphi(q)\mathrm{d}q$ translate respectively into

$$\sum_{k=-\infty}^{\infty} (-1)^k \varphi_k = 0, \quad \sum_{k=-\infty}^{\infty} (-1)^k k \, \varphi_k = 0, \quad \sum_{k=-\infty}^{\infty} \frac{(-1)^k}{k\pi + \gamma} \varphi_k = 0 . \quad (3.66)$$

For vectors in the domain of the Hamiltonian, Eqs. (3.66) determine a system of three equations determining three of the φ_k's in terms of the rest of the coefficients. This allows us to construct the basic vectors that belong to the canonical domain, that is, the vectors with the smallest number of expansion coefficients. Let k_0, k_1, k_2, and k_3 be any integer, with $k_i \neq k_j$ for $i \neq j$. Then the vector

$$\varphi(q) = \varphi_{k_0} \phi_{k_0}^{(\gamma)}(q) + \varphi_{k_1} \phi_{k_1}^{(\gamma)}(q) + \varphi_{k_2} \phi_{k_2}^{(\gamma)}(q) + \varphi_{k_3} \phi_{k_3}^{(\gamma)}(q) \tag{3.67}$$

belongs to the domain of the Hamiltonian. We can fix φ_{k_0} and determine the three other coefficients for $\varphi(q)$ to belong to the canonical domain. Equations (3.66) then yield the system of equations in matrix form

$$\begin{pmatrix} (-1)^{k_1} & (-1)^{k_2} & (-1)^{k_3} \\ (-1)^{k_1} k_1 & (-1)^{k_2} k_2 & (-1)^{k_3} k_3 \\ \frac{(-1)^{k_1}}{k_1 \pi + \gamma} & \frac{(-1)^{k_2}}{k_2 \pi + \gamma} & \frac{(-1)^{k_3}}{k_3 \pi + \gamma} \end{pmatrix} \begin{pmatrix} \varphi_{k_1} \\ \varphi_{k_2} \\ \varphi_{k_3} \end{pmatrix} = (-1)^{k_0+1} \varphi_{k_0} \begin{pmatrix} 1 \\ k_0 \\ \frac{1}{\pi k_0 + \gamma} \end{pmatrix} \tag{3.68}$$

that determines the coefficients φ_{k_1}, φ_{k_2}, and φ_{k_3} in terms of φ_{k_0}.

For a given φ_{k_0}, the coefficients φ_{k_1}, φ_{k_2}, and φ_{k_3} are determined if the determinant of the matrix

$$\mathbf{C} = \begin{pmatrix} (-1)^{k_1} & (-1)^{k_2} & (-1)^{k_3} \\ (-1)^{k_1} k_1 & (-1)^{k_2} k_2 & (-1)^{k_3} k_3 \\ \frac{(-1)^{k_1}}{k_1 \pi + \gamma} & \frac{(-1)^{k_2}}{k_2 \pi + \gamma} & \frac{(-1)^{k_3}}{k_3 \pi + \gamma} \end{pmatrix} \tag{3.69}$$

does not vanish. The determinant is given by

$$\det \mathbf{C} = (-1)^{k_1 + k_2 + k_3} \pi^2 \frac{(k_3 - k_2)(k_1 - k_3)(k_1 - k_2)}{(\pi k_3 + g)(\pi k_2 + g)(\pi k_1 + g)}, \tag{3.70}$$

which does not vanish for different positive integers k_1, k_2, and k_3. Hence the canonical domain of the pair $\mathbf{H}_\gamma$ and $\mathbf{T}_\gamma$ is non-trivial and is in fact infinite dimensional.

The above method of constructing the basic vectors of the canonical domain holds for an arbitrary confined time of arrival operator. In showing for the non-triviality of the canonical domain, one needs only to construct the matrix $\mathbf{C}$ from the boundary conditions and show that the determinant does not vanish. We will use this procedure in what follows.

3.5.3.2 Periodic Boundary Condition

For the periodic case, the canonical domain is not only determined by the free kernel factor but also by the kernel $B(q, q')$, which is obtained from Eq. (3.50) and is given by $B(q, q') = (q^2 - q'^2)/4$. Given $T(q, q')$ and $B(q, q')$, we have the relevant vectors and kernel $\phi_{TB}(q') = q'/2$, $\varphi_{TB}(q') = 0$, and $\langle q | \mathbf{B} | q' \rangle = -\hbar^2/2\mu$, respectively. Then the canonical domain reduces to

$$\mathcal{D}_{\text{can}}([\mathbf{H}_0, \mathbf{T}_0]) = \left\{ \varphi(q) \in \mathcal{D}_{\mathbf{H}_0} : \varphi(\partial) = 0, \; \varphi'(\partial) = 0, \; \int_{-l}^{l} q'\varphi(q')\mathrm{d}q' = 0, \right.$$

$$\left. \int_{-l}^{l} \varphi(q')\mathrm{d}q' = 0 \right\} . \tag{3.71}$$

The condition $\int_{-l}^{l} \varphi(q')\mathrm{d}q' = 0$, which follows from the requirement that $\varphi(q)$ must belong to the null space of $\mathbf{B}$, restricts the canonical domain to those $\varphi(q)$ with coefficient $\varphi_0 = 0$, $\varphi(q) = \sum_{k \neq 0} \varphi_k \varphi_k^{(\gamma)}(q)$, with $\varphi_k = \int_{-l}^{l} \varphi_k^{(0)*}(q)\varphi(q)\mathrm{d}q$; and since $\varphi(q)$ must belong to the domain of the Hamiltonian, the coefficients must also satisfy the requirement $\sum_k k^4 |\varphi_k|^2$. The rest of the conditions on the canonical domain read

$$\sum_{k \neq 0}(-1)\varphi_k = 0, \quad \sum_{k \neq 0}(-1)^k k\varphi_k = 0, \quad \sum_{k \neq 0} \frac{(-1)^k}{k}\varphi_k = 0 . \tag{3.72}$$

The conditions (3.72) again determine three of the coefficients in terms of the rest of the coefficients. For three different integers, k_1, k_2, and k_3, the matrix of the coefficients is given by

$$\mathbf{C} = \begin{pmatrix} (-1)^{k_1} & (-1)^{k_2} & (-1)^{k_3} \\ (-1)^{k_1}k_1 & (-1)^{k_2}k_3 & (-1)^{k_3}k_3 \\ \frac{(-1)^{k_1}}{k_1} & \frac{(-1)^{k_2}}{k_2} & \frac{(-1)^{k_3}}{k_3} \end{pmatrix} . \tag{3.73}$$

The determinant of $\mathbf{C}$ can be calculated to yield

$$\det \mathbf{C} = (-1)^{k_1+k_2+k_3} \frac{(k_2 - k_1)(k_1 - k_3)(k_2 - k_3)}{k_1 k_2 k_3} , \tag{3.74}$$

which does not vanish for different non-zero k_j's. Hence the canonical domain is not trivial and is infinite dimensional.

3.6 Dynamics of the Eigenfunction of the Confined Time of Arrival Operators

3.6.1 CTOA Operator Eigenvalue Problem

Let us first summarize the properties of the eigenfunctions and eigenvalues of the CTOA operators. The immediate properties of the CTOA operators follow from their being compact self-adjoint operators. This implies that they have a complete orthogonal set of (square integrable) eigenfunctions with corresponding discrete eigenvalues. Moreover, their compactness implies that their eigenvalues are bounded and they get denser as they approach the value zero. The boundary condition

$T(q, q) = q/2$ dictates that the trace of $\mathbf{T}_\gamma$ vanishes, $\text{Tr}[\mathbf{T}_\gamma] = \int_{-l}^{l} \langle q | \mathbf{T}_\gamma | q \rangle \, dq \propto \int_{-l}^{l} T(q, q) \, dq = 0$. This implies that the sum of the eigenvalues of $\mathbf{T}_\gamma$ is zero, which means that $\mathbf{T}_\gamma$ has positive and negative eigenvalues.

For a fixed γ, the relationship between eigenfunctions corresponding to positive and negative eigenvalues can be established from the following symmetry of the kernels:

$$\langle q | \mathbf{T}_{\gamma'} | q' \rangle = -\langle -q | \mathbf{T}_{\gamma'} | -q' \rangle^* , \quad \langle q | \mathbf{T}_{\gamma''} | q' \rangle = -\langle q | \mathbf{T}_{\gamma''} | q' \rangle^* , \quad (3.75)$$

where $0 < |\gamma'| < \pi/2$ and $|\gamma''| = 0, \pi/2$. These symmetries correspond to the respective symmetries of the confined time of arrival operators

$$\Pi^{-1} \Theta^{-1} \mathbf{T}_{\gamma'} \Theta \Pi = -\mathbf{T}_{\gamma'}, \quad \Theta^{-1} \mathbf{T}_{\gamma''} \Theta = -\mathbf{T}_{\gamma''}. \quad (3.76)$$

If $\varphi_{\tau,\gamma'}$ is an eigenfunction of $\mathbf{T}_{\gamma'}$ with the eigenvalue τ, then the first symmetry dictates that the vector $\Theta \Pi \varphi_{\tau,\gamma'}$ is an eigenfunction of $\mathbf{T}_{\gamma'}$ with the corresponding eigenvalue $-\tau$. Also if $\varphi_{\tau,\gamma''}$ is an eigenfunction of $\mathbf{T}_{\gamma''}$ with the eigenvalue τ, then the second symmetry dictates that $\Theta \varphi_{\tau,\gamma''}$ is an eigenfunction of $\mathbf{T}_{\gamma''}$ with the corresponding eigenvalue $-\tau$. This is consistent with the above result that the eigenvalues sum to zero.

Analytical calculations for the non-interacting case [24, 25] and numerical computations for the interacting case [22][5] show that the confined time of arrival operators are non-degenerate, and they come in pairs of positive and negative eigenvalues, in accordance with the above results. In particular, the eigenfunctions can be written in the form $\varphi_{\gamma,n}^\pm$, $n = 1, 2, 3, \ldots$, where $\mathbf{T}_\gamma \varphi_{\gamma,n}^\pm = \pm \tau_{\gamma,n} \varphi_{\gamma,n}^\pm$, with $\tau_{\gamma,n} > 0$. The eigenvalues are ordered according to $\tau_{\gamma,1} > \tau_{\gamma,2} > \tau_{\gamma,3} > \ldots$. From the above symmetries, we have the relationships $\varphi_{\gamma',n}^-(q) = [\varphi_{\gamma',n}^+(-q)]^*$ and $\varphi_{\gamma'',n}^-(q) = [\varphi_{\gamma'',n}^+(q)]^*$. The eigenfunctions are classifiable into nodal (vanishing at some point in $[-l, l]$) or non-nodal (non vanishing anywhere in $[-l, l]$). The eigenfunctions further come in pairs of nodal and non-nodal ones, i.e.,

$$P_k = \left\{ \varphi_{\gamma,2k+1}^\pm, \varphi_{\gamma,2k+2}^\pm \right\}, \quad (3.77)$$

for $k = 0, 1, 2, \ldots$, with corresponding pair of eigenvalues $\tau_k^\pm = \{\tau_{2k+1}^\pm, \tau_{2k+2}^\pm\}$ (except perhaps for the first two largest eigenvalue–eigenfunctions), such that $\tau_{2k+1}^\pm$ and $\tau_{2k+2}^\pm$ have neighboring values. The significance of this pairing is best

---

[5] Due to the intractability of solving the kernel factor for non-linear systems, the CTOA operator eigenvalue problem for these systems has been investigated using the leading term of the kernel factor, which is known for any potential. We expect that the results are already representative of the exact results because the corrections to the leading term differ from the largest correction by a factor of $\hbar^2$ in the classical limit. On the other hand, the full solution for linear systems have been used in the numerical investigation. The reader is referred to [22] for a detailed discussion of the numerical computations.

appreciated in the limit for arbitrarily large l, which we will consider later. Nodal and non-nodal eigenfunctions alternate.

3.6.2 Dynamics of the CTOA Operator Eigenfunctions

Using symmetry arguments, it can be established that the negative eigenvalue–eigenfunctions have exactly the same dynamics as those of the positive eigenvalue–eigenfunctions in the time-reversed direction. It is then sufficient for us to consider in detail the dynamical behaviors of the positive eigenvalue–eigenfunctions. Our analysis is based on the numerical evaluation of $\varphi_{\gamma,n}^+(q, t) = (e^{-i\mathbf{H}_\gamma t/\hbar} \varphi_{\gamma,n}^+)(q)$.

Our numerical simulation shows remarkable similarity of the time evolutions of the eigenfunctions across the set of potentials considered, which included $V(q) = \{0, q, q^2/2, q^2/2 + q\}$ for linear systems and $V(q) = \{q^4, \sin q, e^{-q^2/2}\}$ for non-linear systems [22]. Foremost, the uncertainty of the position operator with respect to the evolved eigenfunction $\varphi_{\gamma,n}^+(q, t)$, which is given by

$$(\Delta\mathbf{q})(t) = \sqrt{\langle \varphi_{\gamma,n}^+(t) | \mathbf{q}^2 | \varphi_{\gamma,n}^+(t) \rangle - \langle \varphi_{\gamma,n}^+(t) | \mathbf{q} | \varphi_{\gamma,n}^+(t) \rangle^2}, \qquad (3.78)$$

obtains its minimum value at the time equal to the eigenvalue corresponding to the eigenfunction $\varphi_{\gamma,n}^+(q)$. Moreover, the minimum uncertainty decreases with increasing n so that the eigenfunctions become increasingly localized at the origin at their eigenvalues with n. For eigenfunctions that are not eigenfunctions of the parity operator, the expectation value of the position operator evolves such that it crosses the origin at their respective eigenvalues; these eigenfunctions arise in general for $\gamma \neq 0, \pi/2$. The non-nodal eigenfunctions evolve such that the probability density is maximum at the origin at their eigenvalues. On the other hand, the nodal eigenfunctions evolve with the probability density having two peaks approaching the origin, with the time of closest approach at their eigenvalues.

In short, the CTOA-operator eigenfunctions (within numerical accuracy) evolve according to Schrödinger's equation such that the *event of the position expectation value assuming zero* and the *event of the position uncertainty being minimum* occur at the same instant of time equal to their corresponding eigenvalues. Figures (3.1) and (3.2) show these behaviors for non-interacting and harmonic oscillator cases, respectively.

We have referred to this dynamical property of the eigenfunctions as their unitary arrival at the arrival point at their respective eigenvalues, unitary arrival because they exhibit the dynamical behavior during their unitary time evolution according to Schrödinger equation. The unitary arrival of the eigenfunctions support the interpretation that the confined time of arrival operators are first time of arrival operators, the corresponding eigenvalues being the first unitary time of arrival of the eigenfunctions. This dynamical behavior of the eigenfunctions is important because it gives an instance in which the time of occurrence of a quantum event – in our case unitary arrival at a certain point in the configuration space – is solved in the form of

(a)   (b)

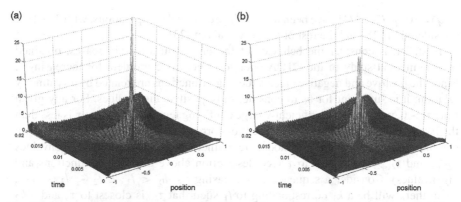

Fig. 3.1 (Color online) Non-interacting case: (a) $n = 20$ and (b) $n = 21$ evolved eigenprobability densities, $|\varphi_n(q, t)|^2$, for $\gamma = 0.01$, with $\hbar = l = m = 1$. Both symmetrically collapse at the origin at their respective eigenvalues, 0.0081 and 0.0079. (Reprinted from [24])

(a)   (b)

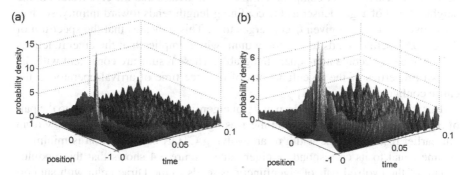

Fig. 3.2 (Color online) Harmonic oscillator: The evolved eigenprobability densities, $|\varphi_n(q, t)|^2$, for the (a) 9 th non-nodal ($\gamma = 0$) and (b) 8 th nodal ($\gamma = \pi/2$) positive eigenvalue–eigenfunctions of the harmonic oscillator CTOA operator for $l = 1$. The eigendensities take their maximum values at the arrival point at their respective eigenvalues. The eigendensities become more localized at the arrival point at the eigenvalue as n increases. (Reprinted from [22])

an eigenvalue problem of a self-adjoint operator, which is conjugate to the system Hamiltonian, with the eigenfunctions as the states exhibiting the event at a later time equal to their respective eigenvalues.

3.7 Dynamical Behaviors in the Limit of Large Confining Lengths and the Appearance of Particle

3.7.1 Properties of the Eigenfunctions and Eigenvalues for Large Confining Lengths

We now describe the dynamical behavior of the CTOA-operator eigenfunctions as the confining length l gets arbitrarily large [26, 23]. The representative potentials

$V(q) = \{0, q^2/2, q^4/4\}$ have been used in the numerical computations, with $V(q) = 0$ considered in [26] and the rest of the potentials in [23].

First, let us consider the behavior of the eigenvalues as l increases, obtained by a numerical study of the CTOA-operator eigenvalue problem for increasing l. Since the positive and negative eigenvalue–eigenfunctions are related by a symmetry relation, it is sufficient for us to consider only positive eigenvalues. For any given time $\tau > 0$, we can find a sufficiently large confining length l_0 such that τ is less than the largest eigenvalue of the CTOA operator corresponding to l_0. For any such l_0, we can find a pair of nodal and non-nodal eigenfunctions P_k such that their eigenvalues τ_{2k+1} and τ_{2k+2} are closest to τ (see description above of the eigenfunctions and eigenvalues). Consider a sequence of increasing l's, $l_0 < l_1 < l_2 < l_3 < \cdots$. Then there will be a k_1 corresponding to l_1 such that τ_{k_1} is closest to τ; and a k_2 corresponding to l_2 such that τ_{k_2} is closest to τ; and so on. Our computation indicates that as l gets larger, τ_{2k+1} approaches τ_{2k+2}. That is, the pair of eigenfunctions P_k becomes degenerate. Our computation likewise indicates that the eigenvalues in the neighborhood of τ get denser as the confining length tends toward infinity, so that, for a given τ, τ_{k_r} for a given l_r converges to τ. This indicates that the spectrum of the CTOA operator tends to the continuum, which implies that the eigenfunctions will eventually become non-square integrable. These results are consistent with the known properties of the eigenfunctions of the free time of arrival operator in the entire configuration space [14].

But what is the behavior of the pair P_k of eigenfunctions corresponding to the pair of eigenvalues closest to the given time τ as l gets arbitrarily large? We know from our earlier results that the width of an evolving CTOA eigenfunction is minimum at time equal to its corresponding eigenvalue. Figure 3.4 shows that the modulus square of the evolved pair of eigenfunctions tends to the Dirac delta with support at the arrival point as l_r tends to infinity. Then the CTOA eigenfunctions evolve to a singular support at the arrival point at their respective eigenvalues in the limit; in particular $|\varphi_{n_r}(q, \tau_{k_r})|^2$ tends to the position eigenfunction at the arrival point (see Figs. 3.3 and 3.4). These numerical observations are consistent for all the potentials considered, suggesting that the behaviors of the CTOA operators in the limit of arbitrarily large l are the same.

3.7.2 The Appearance of Particle

The limiting dynamical behavior described above gives us the picture of quantum arrival within standard quantum mechanics as that of unitarily evolving CTOA eigenfunction to a localized support at the arrival point.[6] If we accept the dogma that particles are wave packets of singular support, then the CTOA eigenfunctions for arbitrarily large l's are particles at their respective eigenvalues. This endorses a mechanism for the localization of the wave function in space and in time at the

---

[6] This and the following sections are excerpts from [23].

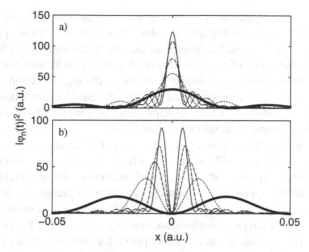

Fig. 3.3 Non-interacting case: Probability density $|\varphi_n(\tau)|^2$ versus position at the corresponding closest eigenvalue τ_n to $t = 0.01$, for the even (*upper figure*) and odd (*lower figure*) eigenfunctions. The different lines are associated with $l = 1$ (*thick solid line*), $l = 2$ (*dashed line*), $l = 3$ (*long-dashed line*), $l = 4$ (*dotted-dashed line*), and $l = 5$ (*thin solid line*). (Reprinted from [26])

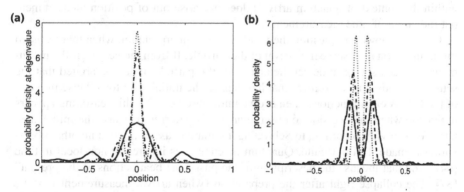

Fig. 3.4 Harmonic oscillator: (**a**) The evolved eigenprobability densities at their respective eigenvalues corresponding to the non-nodal eigenfunctions with eigenvalues nearest to $\tau = 0.11$ for lengths $l = 1$ (*solid line*), $l = 2$ (*dashed line*), and $l = 3$ (*dotted line*), for the harmonic oscillator CTOA operator T_0. (**b**) The corresponding nodal eigenfunctions for lengths $l = 2$ (*solid line*), $l = 3$ (*dashed line*), and $l = 4$ (*dotted line*). Peak increases with l. (Reprinted from [23])

registration of the particle: Consider a quantum particle prepared in some initial state ψ_0. Without loss of generality, we can assume that ψ_0 has a compact support. We can then enclose the system by a box of very large length (for all practical purposes infinite), with the support of ψ_0 lying completely in the box. Since the CTOA eigenfunctions are complete we can decompose ψ_0 in terms of these eigenfunctions. Now if we presuppose that particle detectors somehow respond only to a localized wave packet or localized energy, then registration or arrival of the particle at the arrival point at time τ can be interpreted as the detection of the component

eigenfunction whose eigenvalue is τ, that is, unitarily arriving (essentially collapsing for arbitrarily large l) at the arrival point. This implies that the appearance or
arrival of the particle is a combination of a collapse of the initial wavefunction into
one of the eigenfunctions of the time of arrival operator right after the preparation
of the initial state followed by a unitary evolution of the eigenfunction.

The emergent interpretation contrasts with the standard interpretation of the collapse of the spatial wavefunction on the appearance of particle. In standard quantum
mechanics, when a quantum object is prepared in some initial state ψ_0 and when an
observable of the object is measured at a later time T, then the state at the moment
of measurement, which is $\psi(T) = U_T \psi_0$, where U_t is the time evolution operator, collapses randomly into one of the eigenfunctions of the observable. Now the
consensus is that the appearance of particle is a position measurement, so that the
appearance at point q_0 at some time T is the projection of the evolved wavefunction $\psi(q, T) = U_T \psi_0(q)$ to the eigenfunction $\delta(q - q_0)$ of the position operator.
But according to our quantum arrival description, the collapse occurs much earlier
than the appearance of the particle, with the initial state collapsing to one of the
eigenfunctions of the time of arrival operator right after the preparation and with
the particle appearing later at the moment the eigenfunction has evolved to a state
of localized support at the arrival point. The appearance of particles then (at least
within the context of quantum arrival) does not arise out of position measurement
but rather out of time measurement.

One may, however, question the validity of our interpretation when there was no
initial intention to observe the arrival of the particle. If from the very start the instru-
ment has been set up to detect the arrival of the particle, it can be argued that the
setup is already a measurement that has caused the initial state to collapse into one
of the CTOA eigenfunctions, then evolve until observed. But this reasoning appears
untenable when the decision to observe arrival is deferred, because the initial state
has been evolving according to Schrödinger equation, assuming that no other observation is made. But not quite. Quantum mechanics is inherently non-local in time
[64]. And that means "the description of the past must bear actions of the present"
[37]. The collapse right after the preparation (when arrival measurement is to be
made) and the Schrödinger evolution right after the preparation (when some other
measurement is to be made) are two potentialities that are simultaneously true for
the system, which one is realized depends on the decision what to do with the system
at the moment. Temporal non-locality then replaces the spatial non-locality inherent
in the spontaneous localization of the wavefunction in the standard interpretation.

3.8 Quantum Time of Arrival Distribution

3.8.1 The Formalism

In [26] it is shown that the well-known time of arrival distribution for a quantum-
free particle due to Kijowski [46] can be extracted from the confined time of arrival
operators for the non-interacting case in the limit of infinite confining length.

This indicates that the time of arrival distribution in the entire real line can be approximated by the time of arrival distribution from the compact confined time of arrival operators [23]. This is important as solving for the eigenvalue problem for the time of arrival operator $\mathbf{T}$ (in the entire real line) is intractable, at least, at the moment. Once the kernel factor for a given interaction is already known, the eigenvalue problem for a confined time of arrival operator can be readily computed numerically [13, 62]. For linear systems, the kernel factor $T_{lin}(q, q')$ is readily available because it equals $T_0(q, q')$ which is expressible only in terms of the interaction potential. On the other hand, for non-linear systems, the kernel factor is difficult to solve exactly. However, the kernel factor for such systems has the expansion $T_{nli}(q, q') = T_0(q, q') + T_1(q, q') + \cdots$. Since the succeeding terms contribute to the time of arrival operator in powers of $\hbar^2$, the leading term can approximate the kernel factor, $T(q, q') \approx T_0(q, q')$, in the non-linear case.

Let us for the moment assume that the initial state of the quantum particle, $\psi_0(q)$, has a compact support. We now enclose the support of $\psi_0(q)$ by a very large box centered at the arrival point. Since the CTOA operator $\mathbf{T}_{\gamma=0}$ satisfies the same time reversal symmetry as that of the time of arrival operator $\mathbf{T}$ in the unbounded space, we use $\mathbf{T}_{\gamma=0}$ to approximate $\mathbf{T}$ or we take $\mathbf{T}_{\gamma=0}$ as coarse-grained version of $\mathbf{T}$ [23]. We then construct the confined time of arrival operator $\mathbf{T}_{\gamma=0}$ for the given box. Once $\mathbf{T}_0$ is constructed, compute

$$P_{\psi_0}(x, t) = \sum_{\tau_{0,s}^l \leq t} \left| \langle \varphi_{0,s}^l | \psi_0 \rangle \right|^2 , \qquad (3.79)$$

where $\varphi_{0,s}^l$ and $\tau_{0,s}^l$ are the eigenfunctions and eigenvalues of $\mathbf{T}_0$. The overlap $\langle \varphi_{0,s}^l | \psi_0 \rangle$ is the probability amplitude that the initial state will collapse into the sth eigenfunction right after the preparation. With this interpretation of the overlap, $P_{\psi_0}(x, t)$, in the limit of infinite l, can be naturally interpreted as the probability that one of the components of the eigenfunctions with corresponding eigenvalues less than or equal to t shall have unitarily evolved to a localized wavefunction at the arrival point x. If our detector is what we have presupposed above, then $P_{\psi_0}(x, t)$ is the probability of detection or arrival at x after some time t. Given $P_{\psi_0}(x, t)$ the time of arrival probability density is found by differentiation with respect to time, $\Pi_{\psi_0}(x, t) = \partial_t P_{\psi_0}(x, t)$. The peaks of $\Pi_{\psi_0}(x, t)$ determine the most likely times of arrival at the given arrival point.

If the initial state has infinite tails, we can always approximate it with arbitrary accuracy by a function ψ_l whose support lies entirely in the interval $[-l, l]$ such that $\psi_l \to \psi_0$ as $l \to \infty$. Then $P_{\psi_0}(x, t)$ is computed as above. The whole process can be implemented numerically by choosing the confining length to be large enough. The probability density can then be obtained by numerical interpolation and differentiation. Caveat – this procedure is meaningful only when the time of arrival operator $\mathbf{T}$ has completely continuous spectrum in the Hilbert space $\mathcal{H}_\infty$.

3.8.2 Example: The Harmonic Oscillator Time of Arrival Problem

Let us consider the harmonic oscillator time of arrival problem [23] at the origin. We choose our initial states to be particular Gaussians of the form $\varphi(q) = N e^{-(q-q_0)^2/(4\delta q^2)+i p_0 q/\hbar}$. We compare the time of arrival distribution $\Pi_\varphi(x = 0, t)$ computed using our algorithm above with the classical time of arrival $T(q_0, p_0) = -\omega^{-1} \tan^{-1}(q_0/\omega p_0)$. Figure 3.5a shows the computed time of arrival distribution for a fixed average position q_0 and for varying average momentum p_0. Evidently the time of arrival distribution becomes localized with increasing average momentum. In fact the most likely time of arrival approaches the classical time of arrival as average momentum increases; that is, the quantum time of arrival distribution becomes increasingly localized at the classical time of arrival. This implies that the quantum first time of arrival distribution approaches the classical distribution for arbitrarily large momenta or for high-energy oscillators. For small momenta, the most likely time of arrival is shorter than the classical time of arrival, so that quantum oscillators are, on the average, faster than their classical counterparts.

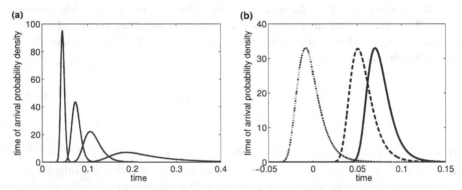

Fig. 3.5 (a) The probability density for $q_0 = 2.25$ and $p_0 = 10, 20, 30, 50$. (b) Evidence for covariance in the limit of infinite l. The probability densities corresponding to times $t = 0$ (*solid line*), $t = 0.02$ (*dashed line*), $t = 0.08$ (*dotted line*) for $p_0 = -30$, $q_0 = 2.25$, $\hbar = m = \sigma = 1$. (Reprinted from [23])

One desirable property of the time of arrival distribution is covariance; that is, translation in time should not affect the distribution. Covariance has been a primary requirement on time operators, and it is the lack of covariance of the CTOA operators (they being compact) that their introduction has been initially doubted upon. Figure 3.5b gives evidence of covariance of the distribution for times smaller than the period of the harmonic oscillator time evolution operator. The given initial state is evolved through different times. These evolved states are used as the initial states in the computation for the TOA distribution. If the distribution is covariant, the resulting distributions must be translations of each other. This is evident in the figure. Thus while the CTOA operators are non-covariant, covariance may naturally emerge in the limit of infinite confining length.

3.9 Conclusion

Since the clarification of Pauli's theorem and the introduction of the self-adjoint and conjugate quantization of the classical free time of arrival for a spatially confined particle in [19], we have already made progress toward a better understanding of the quantum time problem. Our solution to the quantum time of arrival problem via the self-adjoint and conjugate confined time of arrival operator represents such a progress. However, our progress has opened a cache of issues confronting standard quantum mechanics at the foundational level.

Our investigations on the confined quantum time of arrival operators show that the spectral properties of a time operator can acquire unambiguous interpretation independent of the postulates of standard quantum measurement theory. In particular, the spectral properties may be tied with the dynamics of the system, such as the eigenfunctions of the confined quantum time of arrival operators unitarily arriving at some point at their respective eigenvalues. This calls for a certain modification of the standard quantum measurement theory, which makes no exceptions on the interpretation of the spectral properties of observables. How the modified quantum theory of measurement will eventually look like is not yet clear. But we foresee that the modified theory will require classification of quantum observables into, which we can provisionally call, temporal and non-temporal observables, with the former falling into the standard quantum measurement theory and the latter falling into the more general measurement theory that accommodates time as a dynamical observable. But how can this classification be made? We suspect that temporal observables comprise the Hilbert space solutions to the time–energy canonical commutation relation for a given Hamiltonian.

The most surprising aspect of the confined time of arrival operators is that it connects the quantum time of arrival problem and the problem of the appearance of particle in quantum mechanics. In standard quantum mechanics, the appearance of particle arises out of the collapse of the wave function brought about by position measurement, and the collapse occurs not before but at the moment the particle appears on our detectors. However, the dynamics of the confined time of arrival operators suggests that the appearance of particle arises as a combination of the collapse of the initial wave function into one of the eigenfunctions of the time of arrival operator, followed by the unitary Schrödinger evolution of the eigenfunction. This implies that particles, at least within quantum arrival setting, do not arise out of position measurements but out of time of arrival measurements and that the collapse of the wave function on the appearance of particle is not fundamental but decomposable into a series of casually separated processes. This invites us to reconsider our beliefs on quantum measurement theory.

Finally, the clarification of "Pauli's theorem" has brought us the realization that the time–energy canonical commutation relation has a large class of physical solutions, contrary to expectations precipitated by Pauli's theorem. We have already demonstrated that physically meaningful Hilbert space solutions to the canonical commutation relation split into categories. But that is as far as we know. There is a need to pursue the investigation of the whole class of solutions to the canonical commutation relation in general and the relationship of these solutions to the quantum time problem. We note that only one class of solutions, the dense category canonical pairs that form a system of imprimitivities, has been given attention to in physics literatures, and the rest of the solutions are ignored or are assumed not to exist. The quantum time problem is multifaceted. With our solution to the quantum time arrival problem in the form of a solution to the time–energy CCR, it is not unreasonable to entertain the idea that the different solutions to the time–energy canonical commutation relation may be identified with the different facets of the quantum time problem.

Acknowledgments The works reported here were funded by the National Research Council of the Philippines through Grant No. I-81-NRCP, the UP Office of the Vice Chancellor for Research and Development through UP-OVCRD Outright Research Grants PNSE-050509 and PNSE-070703, and by the Office of the Vice Chancellor for Academic Affairs through UP System Grants 2004 and 2007. The theory of confined time of arrival operators was developed with the help of R. Caballar, R. Bahague, R. Vitancol, and A. Villanueva. The idea of taking the limit of the confining length to infinity arose out of a collaboration with J.G. Muga, I. Egusquiza, and F. Delgado. The author especially acknowledges J.G. Muga for his encouragements that have become a major motivation for the author's personal crusade to understand time in quantum mechanics.

References

1. Y. Aharonov, D. Bohm, Phys. Rev. **10**, 1127 (1969)
2. H. Atmanspacher, A. Amann, Int. J. Theor. Phys. **37**, 629 (1998)
3. M. Bauer, P.A. Mello, Ann. Phys. **111**, 38 (1978)
4. A. Baute, R. Sala, J.P. Palao, J.G. Muga, I.L. Egusquiza, Phys. Rev. A **61**, 022118 (2000)
5. C. Bender, G. Dunne, Phys. Rev. D **40**, 2739 (1989)
6. C. Bender, G. Dunne, Phys. Rev. D **40**, 3504 (1989)
7. Ph. Blanchard, A. Jadczyk, Helv. Phys. Acta **69**, 613 (1996)
8. P. Busch et al., Phys. Lett. A **191**, 357 (1994)
9. P. Busch et al., Annals of Phys. **237**, 1 (1995)
10. P. Busch, M. Grabowski, P. Lahti, *Operational Quantum Physics* (Springer, Berlin, 1995)
11. C. Cohen-Tannoudji, *Quantum Mechanics*, vol. 1 (John Willey and Sons, New York, 1977)
12. V. Delgado, J.G. Muga, Phys. Rev. A **56**, 3425 (1997)
13. L.M. Delves, J.L. Mohamed, *Computational Methods for Integral Equations* (Cambridge University Press, Cambridge, 1985)
14. I.L. Egusquiza, J.G. Muga, Phys. Rev A **61**, 012104 (2000)
15. E. Eisenberg, L.P. Horwitz, Ad. Chem. Phys. **XCIX**, 245 (1997)
16. H. Fick, F. Engelmann, Zeitschrift fur Physik **178**, 551 (1964)
17. V. Fock Kirilov, J. Phys. (USSR) **11**, 112 (1974)
18. E.A. Galapon, Opts. Specs. **91**, 399 (2001)
19. E.A. Galapon, Proc. R. Soc. Lond. A **487**, 451 (2002)
20. E.A. Galapon, Proc. R. Soc. Lond. A **487**, 2671 (2002)
21. E.A. Galapon, J. Math. Phys. **45**, 3180 (2004); quant-ph/0207044
22. E.A. Galapon, Int. J. Mod. Phys. A **21**, 6351 (2006)
23. E.A. Galapon, Proc. Roy. Lond. A, doi: 10.1098/rspa.2008.0278
24. E.A. Galapon, R. Caballar, R.T. Bahague, Phys. Rev. Lett. **93**, 180406 (2004)
25. E.A. Galapon, R. Caballar, R.T. Bahague, Phys. Rev. A **72**, 062107 (2005)
26. E.A. Galapon, F. Delgado, J.G. Muga, I. Egusquiza, Phys. Rev. A **72**, 042107 (2005)

27. E.A. Galapon, J. Nable, *Conjugacy of the Confined Time of Arrival Operators*, in preparation
28. E.A. Galapon, A. Villanueva, J. Phys. A Math. Theor. **41**, 455302 (2008)
29. R. Giannitrapani, Int. J. Theor. Phys. **36**, 1575 (1997)
30. M.J. Gotay, J. Nonlinear Sci. **6**, 469 (1996)
31. M.J. Gotay, J. Math. Phys. **40**, 2107 (1999)
32. M.J. Gotay, J. Grabowski, Can. Math. Bull. **22**, 140 (2001)
33. M.J. Gotay, J. Grabowski, H.B. Grundling, Proc. Am. Soc. **128**, 237 (2000)
34. M.J. Gotay, H.B. Grundling, Rep. Math. Phys. **40**, 107 (1997)
35. M.J. Gotay, H. Grundling, C.A. Hurst, Trans. A.M.S. **348**, 1579 (1996)
36. K. Gottfried, *Quantum Mechanics*, vol. 1 (Benjamin/Cummings, Reading, 1966), p. 248
37. B. Greene, *The Fabric of the Cosmos* (Vintage Books, New York, 2004)
38. H.J. Groenwold, Physica **12**, 405 (1946)
39. N. Grot, C. Rovelli, R.S. Tate, Phys. Rev. A **54**, 4676 (1996)
40. W. Heisenberg, Z. Phys. **43**, 172 (1927)
41. J. Hilgevoord, Am. J. Phys. **64**, 1451 (1996)
42. J. Hilgevoord, Am. J. Phys. **66**, 396 (1998)
43. A.S. Holevo, Rep. Math. Phys. **13**, 379 (1978)
44. P.R. Holland, *The Quantum Theory of Motion* (Press Syndicate, Cambridge University Press, 1993)
45. J.M. Jauch, *Foundations of Quantum Mechanics* (Addison-Wesley, Reading, MA, 1968)
46. J. Kijowski, Rep. Math. Phys. **6**, 362 (1974)
47. M. Krzyzanski, *Partial Differential Equation of Second Order*, vol. II (Polish Scientific Publishers, Warza, 1971)
48. R. Landauer, Rev. Mod. Phys. **66**, 217 (1994)
49. J. León, J. Julve, P. Pitanga, F.J. de Urríes, Phys. Rev. A **61**, 062101 (2000)
50. J.G. Muga, C.R. Leavens, Phys. Rep. **338**, 353 (2000)
51. J.G. Muga, J. Palao, P. Sala, Superlat. Microstruct. **24**, 23 (1998)
52. J.G. Muga, R. Sala Mayato, I.L. Egusquiza (eds.), *Time in Quantum Mechanics*, vol. 1, Lect. Notes Phys. **734** (Springer, Berlin, 2008)
53. V.S. Olhovsky, E. Recami, Nuovo Cimento **22**, 263 (1974)
54. R. Omnes, in *The Interpretation of Quantum Mechanics*, P.W. Anderson et al. (eds.) (Princeton University Press, Harvard, 1994)
55. W. Pauli, in *Handbuch der Physik*, vol. V/1, S. Flugge (ed.) (Springer-Verlag, Berlin, 1926), p. 60
56. D. Park, in *Fundamental Questions in Quantum Mechanics*, L. Roth, A. Inomata (eds.) (Gordon and Breach Science Publishers, New York, 1984)
57. A. Peres, *Quantum Theory: Concepts and Methods* (Kluwer Academic Publisher, Dordrecht, 1995)
58. J.J. Sakurai, *Modern Quantum Mechanics* (Addison-Wesley Publishing Company, Reading, MA, 1985)
59. M.D. Srinivas, R. Vijayalakshmi, Pramana **16**, 173 (1981)
60. M. Toller, Phys. Rev. A **59**, 960 (1999)
61. L. van Hove, Proc. Roy. Acad. Sci. Belgium **26**, 1 (1951)
62. R. Vitancol, E.A. Galapon, Int. J. Mod. Phys. C **19**, 821 (2008)
63. J. von Neumann, *Mathematische Grundlagen der Quantenmechanik* (Springer, Berlin, 1955).
64. J. Wheeler, in *Quantum Theory and Measurement*, J. Wheeler, W.H. Zurek (eds.) (Princeton University Press, Princeton, 1984), p. 9

Chapter 4
Detector Models for the Quantum Time of Arrival

Andreas Ruschhaupt, J. Gonzalo Muga, and Gerhard C. Hegerfeldt

> *There is so much in that moment!*
> John Asherby

4.1 The Time of Arrival in Quantum Mechanics

Quantum particles are characterized, for a given preparation, by a fundamental stochasticity of their observable features, such as positions, momenta, energies, or times, e.g., times of arrival at a detector in time-of-flight experiments. In quantum theory the preparation stage is encoded in a wave function,[1] whereas averages or statistical moments of the observables are calculated by a well-known prescription (the expectation value integral) from self-adjoint operators and their powers: this is at least the case for position, momentum, or energy. In fact, the entire statistical distributions are given by the square modulus of the overlap of the wave function with the corresponding eigenstates. It may be argued that, for a general observable, the choice of operator is as much of an art as a science, since none of the "rules of quantization" known is of universal validity; there is not necessarily a one-to-one relation between classical and quantum observables, so that a single classical quantity may be related to several (observable!) quantum quantities; and, there is frequently no obvious trace of the apparatus or experimental procedure in the formalism. In spite of these difficulties, the validity of the chosen operator is eventually confirmed or

A. Ruschhaupt (✉)
Institut für Mathematische Physik, TU Braunschweig, Mendelssohnstr. 3, 38106 Braunschweig, Germany, a.ruschhaupt@tu-bs.de

J.G. Muga
Departamento de Química Física, UPV-EHU, Apdo 644, 48080 Bilbao, Spain, jg.muga@ehu.es

G.C. Hegerfeldt
Institut für Theoretische Physik, Universität Göttingen, Friedrich–Hund–Platz 1, D-37077 Göttingen, Germany, hegerf@theorie.physik.uni-goettingen.de

[1]More generally in a density operator, but such a generalization is not essential to us here.

Ruschhaupt, A. et al.: *Detector Models for the Quantum Time of Arrival*. Lect. Notes Phys. **789**, 65–96 (2009)
DOI 10.1007/978-3-642-03174-8_4   © Springer-Verlag Berlin Heidelberg 2009

denied by comparison with experiment, and by consistency arguments, as for position, momentum, or energy. The fate of time as a random variable, realized, for instance, by the instantaneous clicks of a detector, has been, however, much more problematic. In fact, amazingly, the theoretical study of time observables has been in practice abandoned, if not banned, for decades, except for a handful of disperse contributions (e.g., by Wigner, Aharonov and Bohm, Allcock, Kijowski, Srinivas, and Werner) [69]. The main reason for this strange state of affairs has been the dominance of the paradigm imposed by Pauli's argument against the existence of self-adjoint time operators, according to which "time is only a parameter" in quantum mechanics. Instead of doing something about it, given the appalling evidence that time is also a random variable in the laboratories, most physicists have instead preferred to repeat the above mantra.

There are also complementary reasons that explain, at least partially, the neglect of time observables. One is that assuming classical translation of the particles has been an excellent approximation in practice, but in modern atom optics experiments with ultra-cold atoms this is not necessarily the case. Other reason is that in many experiments the signal is actually proportional to the (unproblematic) position probability density $\rho(x, t)$, rather than to a proper time-of-arrival density $\Pi(x, t)$. There is unfortunately a widespread confusion about this distinction, see Sect. 4.5.3 below for more details, and the expression "time of flight" is nowadays applied quite standardly in the cold-atom community to experiments in which essentially the position probability density, and not a time-of-arrival (TOA) density, is monitored. We are thus far from being free from constraining prejudices on time observables and much remains to be done on experiments and theory, but progress has been made. The fundamental theoretical lacuna for time observables has in more recent times struck the attention of several researchers[2] and led us, in particular, to embark on an ongoing research program.

The arrival time has been investigated at a fundamental level without making the detector explicit, i.e., idealized arrival time distributions – based on positive operator-valued measures (POVM) instead of a self-adjoint time operator – have been derived. These ideal theories use symmetry or consistency arguments (e.g., a correct classical limit), but need the support of experiments and the relation to auxiliary operational models which capture the essential aspects of these experiments.

In reverse, it is important to know what a particular experiment is really measuring and, again, the operational models play a useful intermediate role in this regard. An early attempt to incorporate the apparatus in an explicit, though certainly simplified manner is due to Allcock, by means of a complex, localized imaginary potential that absorbs the quantum wave packet [4–6]. The rationale for its use is that the incident particle in the measurement process is subjected to an irreversible transformation from the incident channel into some other final channels. It could, for example, be ionized or emit a photon after having been excited by a laser. The

---

[2]For reviews on the arrival time see [63, 69] and for more recent work [31, 83, 7, 47, 39, 81, 22, 40, 58, 8, 49, 12, 50, 3, 88, 43].

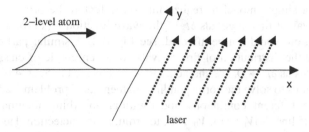

Fig. 4.1 Scheme of the atom-laser model

wave function component in the incident channel would thus lose norm at a rate proportional to the detection signal, a process mimicked by a complex absorbing potential.

The subject of this chapter is a more detailed model which has been developed since 2000 [64, 31, 32, 47, 48, 81, 82, 70, 33]: in essence it consists of a two-level atom wavepacket in the ground state sent toward a laser-illuminated region where the atom is excited, see Fig. 4.1; the detection time of the first spontaneous photon provides an operational definition of the arrival time that can be compared to more abstract theories and ideal quantities for the particle. We will show that the complex potential detector model can be justified in some limit from the atom-laser model. Moreover, we shall see that by means of different limiting operations of the laser field shape or intensity and manipulations of the time-dependent signal, such as "deconvolutions" and normalizations at the level of expectation values, or of the initial state, via "operator normalization," a wealth of information on a moving atom wavepacket is available: not only TOA distributions but also probability densities, kinetic energy densities, current densities, and other local densities.

The structure of the chapter is the following: in the rest of this section Allcock's papers [4–6] and the local density concept are reviewed; in Sect. 4.2 the basic atom-laser model is examined in more detail; the complex potential detector models and their connection to the atom-laser model will be revisited in Sect. 4.3; Sect. 4.4 deals with the concept of operator normalization and its application to arrival times; and Sect. 4.5 will show how other local densities, in particular the kinetic energy density, can be measured by the atom-laser model in certain limits; finally, pulsed measurement of the quantum time of arrival will be examined in Sect. 4.6, while Sect. 4.7 provides a summary.

4.1.1 The Papers of Allcock

Modern research on the quantum TOA owes much to Allcock's seminal work [4–6]. Looking for an ideal quantum arrival time concept, he considered, for a particle moving in one dimension, that arrival time measuring devices should rapidly transfer any probability that appears at $x > 0$ ($x = 0$ being the "arrival position") from the incident channel into various orthogonal and time-labeled measurement channels. As a simple model to realize this basic feature he proposed a pulsed, periodic removal, at time intervals δt, of the wave function $\psi(x)$ for $x > 0$, while the $x < 0$ region would not be affected, see Fig. 4.2. A similar particle removal would provide the distribution of first arrivals for an ensemble of classical, freely moving particles as $\delta t \to 0$. We shall elaborate on this idea in Sect. 4.6.

The difficulties to solve the corresponding mathematical problem lead Allcock to study instead a different, continuous model with an absorbing imaginary potential in the right half line, $-iV_0\Theta(x)$, $V_0 > 0$, to simulate the detection. He argued that the two models should lead to similar conclusions with a time resolution of the order δt in the chopping model or $\hbar/2V_0$ in the complex potential model. He then solved the Schrödinger equation with the complex potential and noticed that for $V_0 \gg |E|_{max}$, where E_{max} is a maximal relevant energy in the wave packet, the apparatus response vanishes, $-\delta N(t)/\delta t \to 0$, with $N(t) = \langle \psi(t)|\psi(t)\rangle$, because of quantum mechanical reflection. This is one of the first discussions of the quantum Zeno effect, not yet known by this name at that time.[3]

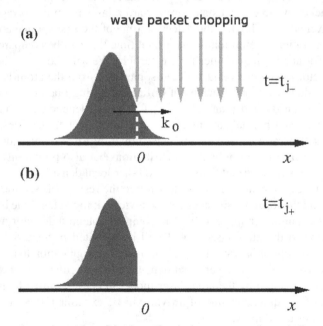

Fig. 4.2 Schematic representation of the time-of-arrival measurement by periodic projection of the wave function onto the subspace $x < 0$ at times $t_j = j\delta t$, $j = \ldots, -1, 0, 1, \ldots$. Figures (**a**) and (**b**) represent two instants immediately before and after the elimination of norm at $x > 0$

---

[3]Eight years later Misra and Sudarshan [62] generalized this result studying the passage of a system from one predetermined subspace to its orthogonal subspace: the periodic projection method in the limit $\delta t \to 0$ was presented as a natural way to model a continuous measurement, but it did not lead to a time distribution of the passage but to its suppression [62], imposing Dirichlet boundary conditions [35]. This lack of a "trustworthy algorithm" to compute TOA and related distributions has been much debated since then, see [54, 34, 57] for a review.

Allcock thus disregarded the strong absorption limit (equivalently the short δt limit) because of reflection, and only considered the weak absorption limit, in which the detection takes a long time but all particles are eventually detected. Under the assumption that the measured arrival time distribution Π was a convolution of an ideal distribution Π_{id} and an apparatus function R (the distribution for the particle survival at rest in the imaginary potential, $R(t) = 2V_0\Theta(t)e^{-2V_0t/\hbar}/\hbar$),

$$\Pi = \Pi_{id} * R \, ,$$

he suggested as an *approximate* solution for the unknown ideal distribution the positive expression[4]

$$\Pi_K(t) = \frac{\hbar}{m} \langle \psi_f(t)|\widehat{k}^{1/2}\delta(\widehat{x})\widehat{k}^{1/2}|\psi_f(t)\rangle \, , \tag{4.1}$$

where $\widehat{k}$ and $\widehat{x}$ are the wave number and position operators, m is the mass, and the averages are computed with the freely moving wave function ψ_f, without the complex potential (unless stated otherwise the initial states have only positive momenta). He also obtained the (not positive semidefinite) current density as the exact solution for Π_{id} in the $V_0 \to 0$ limit,

$$J(t) = \frac{\hbar}{2m} \langle \psi_f(t)|[\widehat{k}\delta(\widehat{x}) + \delta(\widehat{x})\widehat{k}]|\psi_f(t)\rangle \, . \tag{4.2}$$

This last result makes sense classically, but in quantum mechanics $J(t)$ is not positive semidefinite even for states composed only of positive momenta [4–6, 63].

$\Pi_K(t)$ has been later rederived in different ways and is nowadays known as "Kijowski's distribution" [55, 63, 47]. Kijowski [55] inferred it from an axiomatic approach inspired by the arrival time of classical particles. Much more recently, the distribution Π_K has been related to the positive operator-valued measure (POVM) generated by the eigenstates of the Aharonov–Bohm (maximally symmetric) time-of-arrival operator [1, 63, 41]. This method emphasizes the fact that self-adjointness is not necessary to generate quantum probability distributions.

Werner generalized Kijowski's distribution characterizing all possible covariant arrival time observables [89] and also showed a connection to absorption processes described by a semigroup [90]. Kijowski's distribution has also been compared to other approaches and generalized for multi-particle systems and for systems subject to interaction potentials [67, 68, 13, 15, 14, 16, 17].

4.1.2 Local Densities

The quantum mechanical probability flux (4.2) and Kijowski's distribution (4.1) are two quantizations of the same classical local density. Other local densities will

---

[4]Operators will wear a hat when confusion is possible with corresponding functions or c-numbers.

appear along the chapter so a lightning review of this concept is in order. For a classical dynamical variable $A(q, p)$ of position and momentum, its local density, $\alpha_A(x)$, is

$$\alpha_A(x) = \int dp D(x, p) A(x, p) = \int dq dp D(q, p) \delta(x - q) A(q, p) ,$$

where $D(q, p)$ is the phase space density. To quantize this expression one can use

$$\delta(x - \widehat{x}) = |x\rangle\langle x|,$$

and consider, for a point x, the operator $\widehat{A}(x) = \widehat{A}|x\rangle\langle x|$, or rather one of its many symmetrizations, as a quantum density (operator) for the observable $\widehat{A}$. For a given state $|\psi\rangle$, the expectation value $\langle\psi|\widehat{A}(x)|\psi\rangle$ would then be a candidate for the value of the local density at the point x of the observable $\widehat{A}$. The simplest case is $\widehat{A} = 1$, and it corresponds to the particle density, $\rho(x) = |\langle x|\psi\rangle|^2$. If $\widehat{A}$ does not commute with $|x\rangle\langle x|$, there are infinitely many "combinations" (orderings) to construct a quantum density [26, 66], for example,

$$\widehat{A}^{1/2}|x\rangle\langle x|\widehat{A}^{1/2} , \qquad (4.3)$$

$$\frac{1}{2}\left(\widehat{A}|x\rangle\langle x| + |x\rangle\langle x|\widehat{A}\right) , \qquad (4.4)$$

$$\frac{1}{2}\left(\widehat{A}^{1/2}|x\rangle\langle x|\widehat{A}^{1/2}\right) + \frac{1}{4}\left(\widehat{A}|x\rangle\langle x| + |x\rangle\langle x|\widehat{A}\right) . \qquad (4.5)$$

Different symmetrizations may have a perfectly respectful status as physically observable and measurable quantities, and different orderings may be associated with latent properties realized via different experimental measurement procedures as we shall see. They may also be related more indirectly to observables and yet carry valuable physical information.

In addition to the case of the flux and Π_K already mentioned, in which $\widehat{A} = \widehat{p}/m$, a second important example of this quantum non-uniqueness for a single classical quantity is the kinetic energy density, [27, 28, 77], with $\widehat{A} = \widehat{p}^2/2m$ as the kinetic energy operator. Operational procedures devised to achieve a TOA distribution may actually provide a kinetic energy density, see Sects. 4.5.1 and 4.5.2. It may also be possible to find simple approximate relations between ρ, Π_K, J, and kinetic energy densities for wave packets with rather well-defined momentum, see Sect. 4.5.3, but, in general, they are fundamentally different functions.

4.2 The Basic Atom-Laser Model

Apart from Allcock's work, several "toy models" for arrival time measurements have been put forward [2, 63, 16, 21, 78, 44, 80]. Instead, the aim of Damborenea et al. in [31], following [64], was to work out a more realistic model describing a TOA measurement by quantum optical means. The basic idea is that a region of space may be illuminated by a laser and upon entering the region an atom will start emitting photons. The first-photon emission can be taken as a measure of the arrival time of the atom in that region, see Fig. 4.1.[5] The pumping of the atom to the excited state and the photon emission takes some time and therefore produce delays with respect to some "ideal" arrival time of the atom. Increasing the laser intensity, and considering shorter lifetimes, does not solve the problem because the atom tends to be reflected from the laser region without ever emitting a photon. This is further discussed in Sect. 4.2.1 where the details of the model will be given.

A way out of these difficulties is to "subtract" the delays from the first-photon probability density by means of a deconvolution with an atom at rest, see Sect. 4.2.2. This results in a distribution which, for shorter and shorter lifetime of the atomic level, converges to J, the quantum mechanical probability flux, and provides a way to measure it.

4.2.1 Effective One-Dimensional Equations

We assume that a region is illuminated by a laser sheet, treated semiclassically, on the $y-z$ plane with wave vector in y-direction. There will be no z-dependence in the full Hamiltonian so this coordinate may be ignored. We will use the quantum jumps approach [51–53, 76] which is essentially equivalent to quantum trajectories [25] and to the Monte Carlo wave function approach [30]. Using also the rotating-wave approximation, and in a laser adapted interaction picture to get rid of an explicit time dependence, the effective two-dimensional Hamiltonian for the undetected atom which results from the quantum jump approach after tracing over the photon part is

$$H_{2D} = \frac{\widehat{P}_x^2}{2m} + \frac{\widehat{P}_y^2}{2m} + \frac{\hbar}{2} \begin{pmatrix} 0 & \Omega(x)e^{-ik_L y} \\ \Omega(x)e^{ik_L y} & -2\Delta_L - i\gamma \end{pmatrix},$$

with $|1\rangle \equiv \begin{pmatrix} 1 \\ 0 \end{pmatrix}$, $|2\rangle \equiv \begin{pmatrix} 0 \\ 1 \end{pmatrix}$, and where $\widehat{p}_x = -i\hbar\frac{\partial}{\partial x}$, $\widehat{p}_y = -i\hbar\frac{\partial}{\partial y}$ are the momentum operator components in $x - y$ space, Ω is the Rabi frequency, k_L is the laser wave number, and γ is the Einstein coefficient of level 2, i.e., its decay rate or inverse lifetime. We will examine the stationary Schrödinger equation

$$E_{2D}\chi(x, y) = H_{2D}\chi(x, y) \tag{4.6}$$

for wave functions with the atom incident in the ground state with well-defined momentum and energy, $E_{2D} = \frac{\hbar^2}{2m}\left(k_x^2 + k_y^2\right)$.

---

[5]A way to realize this atom-laser model experimentally is by using a Lambda system in which the laser drives the transition between levels 1 and 2 whereas 2 decays very fast and preferentially to a third non-interacting sink state 3 [74].

The two-dimensional equation (4.6) can be transformed into an effective one-dimensional one by inserting the ansatz

$$\chi(x, y) = \begin{pmatrix} \phi^{(1)}(x)e^{ik_y y} \\ \phi^{(2)}(x)e^{i(k_y+k_L)y} \end{pmatrix} .$$

Then we get

$$\frac{\hbar^2 k_x^2}{2m} \begin{pmatrix} \phi^{(1)}(x) \\ \phi^{(2)}(x) \end{pmatrix} = \left[\frac{p_x^2}{2m} + \frac{\hbar}{2} \begin{pmatrix} 0 & \Omega(x) \\ \Omega(x) & -2\Delta - i\gamma \end{pmatrix} \right] \begin{pmatrix} \phi^{(1)}(x) \\ \phi^{(2)}(x) \end{pmatrix} ,$$

where

$$\Delta = \Delta_L - \frac{\hbar}{2m}(2k_y k_L + k_L^2)$$

is an effective detuning which includes Doppler and recoil terms. We have thus reduced the original problem to a one-dimensional equation with effective Hamiltonian

$$H_c = \widehat{p}^2/2m + \frac{\hbar}{2} \begin{pmatrix} 0 & 0 \\ 0 & -i\gamma \end{pmatrix} + \frac{\hbar}{2} \begin{pmatrix} 0 & \Omega(x) \\ \Omega(x) & -2\Delta \end{pmatrix} . \qquad (4.7)$$

(The x subindex is dropped in the momentum operator since only effective one-dimensional equations in the x-direction are considered hereafter.) We shall concentrate now on the on-resonance ($\Delta = 0$) case that may be achieved by normal incidence ($k_y = 0$) and by adjusting the laser detuning to cancel the recoil frequency $k_L^2 \hbar/(2m)$.

4.2.1.1 Stationary Scattering Eigenfunctions

Let us first solve the eigenvalue equation

$$H_c \Phi_k = E \Phi_k, \quad \text{where } \Phi_k(x) \equiv \begin{pmatrix} \phi_k^{(1)}(x) \\ \phi_k^{(2)}(x) \end{pmatrix} \text{ and } E = \hbar^2 k^2/2m , \qquad (4.8)$$

for waves incident from the left in the ground state (dropping the subindex x in k too). In the simplest solvable model $\Omega(x) = \Theta(x)\Omega$, which corresponds to a sharp laser beam boundary at $x = 0$ and a semi-infinite field extension. If the atoms are slow enough the result of the experiment should not depend on the beam width beyond a minimal length necessary for detection.

For $x < 0$, left-incoming states must be of the form

$$\Phi_k(x) = \frac{1}{\sqrt{2\pi}} \begin{pmatrix} e^{ikx} + R_1 e^{-ikx} \\ R_2 e^{-iqx} \end{pmatrix}, \quad x < 0, \quad k > 0 ,$$

where q satisfies

$$E + i\hbar\gamma/2 = \hbar^2 q^2/2m ,$$

with $\mathrm{Im}\, q > 0$ for boundedness, while R_1 and R_2 are the reflection amplitudes. Note that although E is real, the complete wave functions will not be orthogonal, in accordance with the non-hermiticity of H_c.

To obtain the form of $\boldsymbol{\Phi}_k(x)$ for $x \geq 0$, we denote by $|\lambda_+\rangle$ and $|\lambda_-\rangle$ the eigenstates of the matrix $\frac{1}{2}\begin{pmatrix} 0 & \Omega \\ \Omega & -i\gamma \end{pmatrix}$ corresponding to the eigenvalues $\lambda_\pm$,

$$\lambda_\pm = -\frac{i}{4}\gamma \pm \frac{i}{4}\sqrt{\gamma^2 - 4\Omega^2} ,$$

$$|\lambda_\pm\rangle = \begin{pmatrix} 1 \\ 2\lambda_\pm/\Omega \end{pmatrix} .$$

Note that $|\lambda_\pm\rangle$ are not orthogonal and have not been normalized. Nevertheless one can write $\boldsymbol{\Phi}_k$ as a superposition of the form

$$\sqrt{2\pi}\boldsymbol{\Phi}_k(x) = C_+|\lambda_+\rangle e^{ik_+x} + C_-|\lambda_-\rangle e^{ik_-x}, \qquad x \geq 0 .$$

From the eigenvalue equation (4.8), one obtains for $x \geq 0$

$$k_\pm^2 = k^2 - 2m\lambda_\pm/\hbar = k^2 + i\gamma m/2\hbar \mp im\sqrt{\gamma^2 - 4\Omega^2}/2\hbar ,$$

with $\mathrm{Im}\, k_\pm > 0$ for boundedness. From the continuity of $\boldsymbol{\Phi}_k(x)$ and its derivative at $x = 0$ one gets R_1, R_2, $C_\pm$ and the wave functions explicitly [31].

4.2.1.2 Wavepackets

Wavepackets $\boldsymbol{\Psi} = \begin{pmatrix} \psi^{(1)} \\ \psi^{(2)} \end{pmatrix}$ are treated as superpositions of the stationary eigenfunctions. This is easy for an initial ground state wave packet $\begin{pmatrix} \psi^{(1)}(x,t_0) \\ 0 \end{pmatrix}$ coming in from the far left side and with t_0 in the remote past. Indeed, if $\widetilde{\psi}(k)$ denotes the momentum amplitude the wave packet would have as a freely moving packet at $t = 0$, then

$$\boldsymbol{\Psi}(x, t) = \int_0^\infty dk\, \widetilde{\psi}(k)\, \boldsymbol{\Phi}_k(x)\, e^{-i\hbar k^2 t/2m} . \qquad (4.9)$$

The normalization is chosen such that $||\boldsymbol{\Psi}(t_0)||^2 = 1$, for t_0 in the remote past. The probability, $N(t)$, of no photon detection for a wave packet from t_0 up to time t is given by

$$N(t) = ||e^{-iH_c(t-t_0)/\hbar}\boldsymbol{\Psi}(t_0)||^2 ,$$

and the probability density, $\Pi(t)$, for the first-photon detection by

$$\Pi(t) = -\frac{\mathrm{d}N}{\mathrm{d}t} = \frac{\mathrm{i}}{\hbar}\langle\Psi(t)|H_{\mathrm{c}} - H_{\mathrm{c}}^{\dagger}|\Psi(t)\rangle \ .$$

Since $H_{\mathrm{c}} - H_{\mathrm{c}}^{\dagger} = -\mathrm{i}\gamma\hbar|2\rangle\langle2|$, the first-photon probability density is given by

$$\Pi(t) = \gamma \int_{-\infty}^{\infty} \mathrm{d}x \, |\psi^{(2)}(x,t)|^2 \ .$$

4.2.1.3 The Reflection Problem and the No-Detection Probability

The probability of no photon detection at all is $N(t = \infty)$. Only $\psi^{(1)}$ contributes to this and, because $\Omega(x) = \Omega\Theta(x)$, only for $x < 0$, since, for $x > 0$, the ground state part will eventually be pumped by the laser to the excited state. So we get

$$N(t = \infty) = 1 - \int_{t_0}^{\infty} \mathrm{d}t' \, \Pi(t') = \int_0^{\infty} \mathrm{d}k \, |R_1(k)|^2 |\widetilde{\psi}(k)|^2 \ .$$

As a consequence, $\Pi(t)$ is, in general, not normalized to 1.

$|R_1|^2$ and therefore the probability for missing an atom increases with Ω, the strength of the laser driving. On the other hand, for $k \to \infty$ reflection becomes negligible. Hence for faster atoms reflection does not pose a problem. For later purposes we also consider increasingly large γ, the other parameters kept fixed. In this case the state vector for $x > 0$ becomes simply the plane wave with wave number k in the ground state. This means that for increasing γ there is less and less reflection, but also less and less absorption, i.e., photon detection, so that the effect of the laser on the atom decreases. Moreover, if both γ and Ω go to infinity with γ/Ω kept fixed, then $R_1 \to -1$ and all atoms are reflected without having been detected.[6] Depending on the parameters, the delay and reflection problem may be either very relevant or negligible [31].

4.2.2 The Connection to an Idealized Arrival Time Distribution

We have just seen that avoiding reflection by weak driving, $\Omega/\gamma \ll 1$, would cause a severe delay problem since the laser would take more time to pump the atom to the excited state, see Fig. 4.3.

Should it not be possible to somehow compensate the delay in $\Pi(t)$ by that of the atom at rest and thus arrive, in some limit, at a delay-free ideal distribution? To

---

[6]More precisely, if Ω and γ are of the same order of magnitude and $\gamma \gg \frac{\hbar k}{m}k$ or, alternatively, if $\Omega/\gamma \ll 1$ and $\Omega^2/\gamma \gg \frac{\hbar k}{m}k$, then $R_1 \approx -1$. Physically, both conditions mean that the distance traveled by the atom in the time it takes for a photon to be scattered is much less than the de Broglie wavelength.

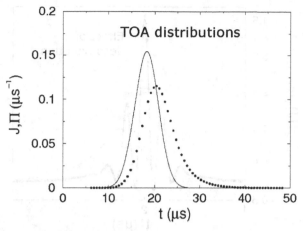

Fig. 4.3 Time-of-arrival distributions: Flux J (*solid line*, here indistinguishable from Kijowski's Π_K) and Π (first photon, *dots*). Note the delay in Π. The initial state is a minimum-uncertainty-product Gaussian for the center-of-mass motion of a single Cesium atom in the ground state with $\langle v \rangle = 9.0297$ cm/s, $\langle x \rangle = -1.85$ μm, and $\Delta x = 0.26$ μm; $\Omega = 0.0999\gamma$; all figures are for the transition $6^2 P_{3/2} - 6^2 S_{1/2}$ of Cesium with $\gamma = 33.3$ MHz

achieve this, we assume the "experimental" arrival time distribution $\Pi(t)$ to be the convolution of a hypothetical ideal distribution, Π_{id}, with the detection probability density $W(t)$ for an atom at rest [56], which is put in the laser field in the ground state,

$$\Pi = \Pi_{\mathrm{id}} * W,$$

$$W(t) = \frac{\gamma \Omega^2}{4|S|^2} e^{-\gamma t/2} |e^{St/2} - e^{-St/2}|^2 \Theta(t),$$

where

$$S = \frac{1}{2}(\gamma^2 - 4\Omega^2)^{1/2}.$$

The delay in Π is then contained in W and the hypothetical ideal distribution Π_{id} is obtained by deconvolution from $\tilde{\Pi}_{\mathrm{id}} = \tilde{\Pi}/\tilde{W}$, where the tildes denote the Fourier transforms, e.g., $\tilde{\Pi}(v) = \int \mathrm{d}t e^{-ivt} \Pi(t)$. One finally obtains, using (4.9) and taking the $\gamma \to \infty$ limit [31],

$$\Pi_{\mathrm{id}}(t) \to J(0, t),$$

which is the flux for the *free* wave function $\psi_f(x, t)$ at $x = 0$, i.e., the one evolving without laser, see also Fig. 4.4. This is an extremely interesting result since J is a

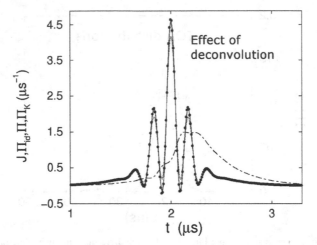

Fig. 4.4 Excellent agreement between Π_{id} (*filled circles*) and J (*solid line*); deviations from Π_K (*dotted line*) and Π (*dot-dashed line*). The initial wave packet is a coherent combination $\psi = 2^{-1/2}(\psi_1 + \psi_2)$ of two Gaussian states for the center-of-mass motion of a single Cesium atom that become separately minimal uncertainty packets (with $\Delta x_1 = \Delta x_2 = 0.021\,\mu$m and average velocities $\langle v \rangle_1 = 18.96$ cm/s, $\langle v \rangle_2 = 5.42$ cm/s) at $x = 0$ and $t = 2\,\mu$s; $\Omega = 0.37\gamma$

natural candidate for the arrival time distribution. We note that J is normalized to 1 for a wave function which has only positive-momentum components.

4.2.3 Generalizations of the Atom-Laser Model

Different generalizations of the atom-laser model have been worked out:

In [32] the effects of a finite laser width (in contrast to the semi-infinite laser described before) are examined. In that setting, the flux can again be obtained from the probability density for the arrival of the first photon in a given limit.

In [45] the three-dimensional formulation of the atom-laser model is investigated in detail. It is shown that within typical conditions for optical transitions the results of the simple one-dimensional version are generally valid. Differences that may occur are consequences of Doppler and momentum-transfer effects. Ways to minimize these are discussed.

In [82] the idea of the atom-laser model is transferred to an ionization model. A method to achieve perfect detection of ultra-cold atoms at fixed incident velocity is proposed.

4.3 Complex Potentials

Allcock [4–6] modeled the arrival detector by a local and energy-independent complex potential, i.e., he put forward a simplified operational approach for arrival times. The operational (unnormalized) time-of-arrival distribution Π is then identified as the absorption rate

$$\Pi = -dN/dt \ ,$$

where N is the decreasing norm of the quantum wave packet for the undetected particle. Later, Muga and coworkers used more complicated functional forms for the complex potential, including a real part, to show that the reflections induced by the step model could be partially avoided [65, 75]. Imaginary potentials have also been used in the phenomenological event-enhanced quantum theory (EEQT) of Blanchard and Jadczyk [18, 20, 19] to simulate time-of-arrival and traversal time measurements [21, 75, 78–80], and Halliwell [44] derived the imaginary step potential used by Allcock from an abstract two-level detector model.

The agreement between the deconvolution performed by Allcock with a weak imaginary step potential and the deconvolution of Sect. 4.2.2, in the short lifetime (weak driving) limit, suggests that there must be a connection. Indeed, the laser-atom model can be reduced to a local and energy-independent complex potential model in the weak driving regime. To see this first at a heuristic and intuitive level, let us write the stationary Schrödinger equation for the effective Hamiltonian (4.7) in components, putting $\Delta = 0$,

$$E\phi^{(1)}(x) = \frac{\widehat{p}^2}{2m}\phi^{(1)}(x) + \frac{\hbar}{2}\Omega(x)\phi^{(2)}(x) \ , \tag{4.10}$$

$$E\phi^{(2)}(x) = \frac{\widehat{p}^2}{2m}\phi^{(2)}(x) + \frac{\hbar}{2}\Omega(x)\phi^{(1)}(x) - i\hbar\gamma/2\,\phi^{(2)}(x) \ . \tag{4.11}$$

If $\hbar\gamma$ is much larger than the rest of energies in (4.11), we can approximate $\phi^{(2)}$ from this equation as

$$\phi^{(2)}(x) \approx \frac{\Omega(x)}{i\gamma}\phi^{(1)}(x) \ .$$

Putting this into Eq. (4.10) we get

$$E\phi^{(1)}(x) \approx \frac{\widehat{p}^2}{2m}\phi^{(1)}(x) - iV_0(x)\phi^{(1)}(x) \ , \tag{4.12}$$

where

$$V_0(x) = \frac{\hbar\Omega(x)^2}{2\gamma} \ , \tag{4.13}$$

i.e., there results a local, purely imaginary, and energy-independent potential for state 1.

For a wavepacket with stationary components satisfying (4.12), $\psi^{(1)}(t)$ satisfies the one-dimensional Schrödinger equation

$$i\hbar\frac{d}{dt}\psi^{(1)}(t) \approx \left[\widehat{p}^2/2m - iV_0(\widehat{x})\right]\psi^{(1)}(t) .\tag{4.14}$$

Since $\psi^{(2)}(t) \to 0$, one has

$$N(t) \approx ||\psi^{(1)}(t)||^2 ,$$

and so

$$\Pi(t) = -\frac{dN}{dt} \approx \frac{2}{\hbar}\int_{-\infty}^{\infty} dx\, V_0(x)\left|\psi^{(1)}(x, t)\right|^2 .$$

Therefore, in the simple semi-infinite laser case $\Omega(x) = \Omega\Theta(x)$, we get

$$\Pi(t) \approx \frac{2V_0}{\hbar}\int_0^{\infty} dx\, |\psi^{(1)}(x, t)|^2 ,\tag{4.15}$$

where $V_0 = \hbar\Omega^2/2\gamma$ is now a constant. Equations (4.14) and (4.15) relate the atom-laser model to the complex potential model of Allcock, see also [44].

An interesting case is the limit of the delta potential $V_0(x) = -iV_0\delta(x)$, for which the signal $\Pi(t)$ provides the probability density,

$$\Pi(t) \approx \frac{2V_0}{\hbar}\left|\psi^{(1)}(0, t)\right|^2 .$$

However, this is *not* the probability density for the *freely moving* wave unless the additional limit $V_0 \to 0$ is taken.

4.3.1 Complex Potential from Feshbach's Theory

A more rigorous and systematic derivation of the complex potential follows from using Feshbach's projector techniques [37, 38, 61]. They let us obtain an exact generalized, non-local and energy-dependent, complex potential treatment for laser and atom parameters that cannot be described with the simple local and energy-independent approach and also set its conditions of validity.

We shall now include detuning, as in (4.7),

$$H_c = K\mathbf{1} + V = \frac{1}{2m}\begin{pmatrix} \widehat{p}^2 & 0 \\ 0 & \widehat{p}^2 \end{pmatrix} + \frac{\hbar}{2}\begin{pmatrix} 0 & \Omega(x) \\ \Omega(x) & -(2\Delta + i\gamma) \end{pmatrix} ,$$

where $K = \widehat{p}^2/2m$ is the kinetic energy operator and $\mathbf{1} = |1\rangle\langle 1| + |2\rangle\langle 2|$ the unit operator in the internal state subspace. We want to solve the eigenvalue equation (4.8). Therefore, we may apply the complementary projectors

$$P = |1\rangle\langle 1|, \quad Q = |2\rangle\langle 2| ,$$

to write, using the standard manipulations of the partitioning technique [37, 38, 61], an exact integro-differential equation for the ground state amplitude $\phi_k^{(1)}$,

$$E\phi_k^{(1)}(x) = K\phi_k^{(1)}(x) + \int_{-\infty}^{\infty} dx' \langle x, 1|V_{\text{opt}}(E)|x', 1\rangle\phi_k^{(1)}(x') ,$$

where the exact complex non-local and energy-dependent "optical" potential is

$$V_{\text{opt}}(E) = PV_{\text{opt}}(E)P = PVP + PVQ(E + i0 - QHQ)^{-1}QVP ,$$

and its coordinate representation is given by

$$\langle x, 1|V_{\text{opt}}(E)|x', 1\rangle = -\frac{mi}{4}\frac{e^{i|x-x'|q}}{q}\Omega(x)\Omega(x') ,$$

with $q = \frac{\sqrt{2mE}}{\hbar}(1+\mu)^{1/2}$, $\text{Im}\, q \geq 0$, and $\mu = \frac{2\Delta+i\gamma}{2E/\hbar}$. The exact equation for the excited state is

$$\phi_k^{(2)}(x) = -\frac{im}{2\hbar}\int_{-\infty}^{\infty} dx' \frac{e^{i|x-x'|q}}{q}\Omega(x')\phi_k^{(1)}(x') .$$

If we set $\Omega(x) = \Omega_M\omega(x)$, where Ω_M is the maximum of $|\Omega(x)|$, we get by partial integration

$$\int_{-\infty}^{\infty} dx' \frac{e^{i|x-x'|q}}{q}\omega(x')f(x') = \frac{1}{\mu}\frac{i\hbar^2}{mE}\omega(x)f(x) + \mathcal{O}\left(\frac{1}{\mu^2}\right), \qquad (4.16)$$

for some function $f(x)$ and $2E/\hbar$ being constant.

Let us now examine the limit $\mu \to \infty$ keeping $2E/\hbar$ and Ω_M constant. To apply perturbation theory, we assume an asymptotic expansion of $\phi_k^{(1)}$ in powers of $1/\mu$ and compare equal powers of $1/\mu$ on both sides taking (4.16) into account. It follows that

$$E\phi_k^{(1)} = K\phi_k^{(1)}(x) + W(x)\phi_k^{(1)}(x) + \mathcal{O}\left(\frac{1}{\mu^2}\right) ,$$

where

$$W(x) = \frac{\Omega(x)^2\hbar}{2(2\Delta + i\gamma)} , \qquad (4.17)$$

i.e., there results a local and energy-independent approximation $W(x)$ to the exact optical potential. For the excited state, we get

$$\phi_k^{(2)}(x) = \frac{1}{2\Delta + i\gamma} \Omega(x)\phi_k^{(1)}(x) + \mathcal{O}\left(\frac{1}{\mu^2}\right), \tag{4.18}$$

i.e., the population of the excited state is very small compared to the ground state population when the local potential applies.

Note that (4.17) includes real and imaginary parts that can be modified by modulating the laser intensity and/or detuning. This may be used to control and optimize the atom dynamics and detection with quantum optical means [71].

Another interesting limit which yields the same effective potential is $\mu \to \infty$ while $2E/\hbar$ and Ω_M^2/μ remain constant [81].

The time-dependent two-component state vector for the undetected atom $\boldsymbol{\Psi}$ is obtained by solving the time-dependent Schrödinger equation corresponding to H. For the atom incident from the left in the ground state, $\boldsymbol{\Psi}$ is given by a linear superposition of the stationary waves. If the local and energy-independent potential $W(x)$ is a good approximation for all stationary waves composing the wavepacket, then the ground state component satisfies a closed time-dependent Schrödinger equation

$$i\hbar \frac{\partial}{\partial t} \psi^{(1)}(x, t) \approx \frac{\widehat{p}^2}{2m} \psi^{(1)}(x, t) + W(x)\psi^{(1)}(x, t),$$

whereas the excited state component becomes

$$\psi^{(2)}(x, t) \approx \frac{1}{2\Delta + i\gamma} \Omega(x)\psi^{(1)}(x, t).$$

As a consequence the photon detection rate, or operational arrival time distribution, is given by

$$\Pi = -\frac{dN}{dt} \approx -\frac{\partial}{\partial t} \int_{-\infty}^{\infty} dx \left(1 + \frac{\Omega^2(x)}{(2\Delta + i\gamma)^2}\right) |\psi^{(1)}(x, t)|^2$$

$$\approx -\frac{\partial}{\partial t} \int_{-\infty}^{\infty} dx |\psi^{(1)}(x, t)|^2.$$

For $\Delta = 0$ this result corresponds to the phenomenological one-channel models that make use of an imaginary potential [4–6, 21, 78–80], which can now be physically interpreted as $W_{\Delta=0}(x) = -iV_0(x)$ where $V_0(x)$ is given by Eq. (4.13). The physical significance of these approximations may also be understood by applying the Markovian limit in the time-dependent version of the partitioning technique due to Zwanzig [91].

4.3.2 Multiple Photon Case

In general the atom will emit many photons in consecutive fluorescence cycles and the distribution of the photon detection times may be of interest. The multiple photon case can also be studied by the quantum jumps approach. In this section we shall assume that the following conditions are satisfied: $\Delta = 0$, $\gamma \gg \Omega_M$ and the energy is assumed to be small, i.e., $E \ll \hbar\gamma/2$, but large with respect to the recoil energy. This simplifies the resetting of the wave function after a photon has been detected.

With the abbreviation $g(x) = \frac{\Omega(x)}{\sqrt{\gamma}}$ the effective ground state potential becomes $W(x) = -i\frac{\hbar}{2}g(x)^2$ and, using the above assumptions, the quantum jump algorithm takes the following form:

(1) Start with an atom in the ground state.

(2) In each time step ($\delta t \ll \gamma, \Omega_M$) evaluate the detection probability δP between t and $t + \delta t$,

$$\delta P = \gamma \left\| \psi^{(2)}(t) \right\|^2 \delta t = \left\| g \cdot \psi^{(1)}(t) \right\|^2 \delta t .$$

(3) Choose uniformly a random number $r \in [0, 1]$.

(4a) If $r > \delta P$ then no detection occurs and the wave function of the atom at time $t + \delta t$ is given by

$$\psi^{(1)}(x, t + \delta t) = \frac{1}{\sqrt{1 - \delta P}} \left[1 - \frac{i}{\hbar}\delta t \left(\frac{\widehat{p}^2}{2m} - i\frac{\hbar}{2}g(x)^2 \right) \right] \psi^{(1)}(x, t) ,$$

$$\psi^{(2)}(x, t + \delta t) = -\frac{i}{\sqrt{\gamma}}g(x)\psi^{(1)}(x, t + \delta t) .$$

(4b) If $r < \delta P$ then a photon is detected and the wave function collapses and is given at time $t + \delta t$ by

$$\psi^{(1)}(x, t + \delta t) = \frac{1}{\left\| \psi^{(2)}(t) \right\|} \psi^{(2)}(x, t) = -i \frac{1}{\left\| g \cdot \psi^{(1)}(t) \right\|} g(x)\psi^{(1)}(x, t) ,$$

$$\psi^{(2)}(x, t + \delta t) = 0 .$$

(For a more general treatment including the effect of recoil see [72].)

(5) Let $t + \delta t \to t$ and go to (2) for possible further detections until a certain maximum propagation time is reached. This would complete one "trajectory" with a record of photon emissions at a sequence of time steps. For obtaining averaged statistical results many trajectories starting at step (1) must be run.

It is remarkable that only the ground state $\psi^{(1)}$ has to be taken into account for this algorithm, i.e., for dealing with the multiple-photon case, under these approximations.

Note that this algorithm is also essentially coincident with the "PDP algorithm" of phenomenological "event-enhanced quantum theory" (EEQT) applied for arrival times using one complex potential detector (see, for example, [78, 20]). While the function $g(x)$ has only the abstract meaning of a "detector sensitivity" in EEQT, it has a precise physical content in the framework of the fluorescence model, namely $g(x) = \frac{\Omega(x)}{\sqrt{\gamma}}$, and we have also provided the conditions for the validity of the phenomenological model.

4.4 Quantum Arrival Times and Operator Normalization

This and the following two sections will deal with the cases disregarded by Allcock because of mathematical or physical difficulties, corresponding to strong interactions (driving) in the continuous case and to short pulses in the discrete measurement model. The good news is that these limits may indeed provide information on several local densities of the freely moving atom when harnessed with different treatments of the measured signal or the initial state. This section, in particular, provides an operational interpretation of Kijowski's distribution, which had remained elusive for a long time [63, 31], in the physically attractive limit of strong laser field and fast spontaneous emission. The trick is to compensate for the undetected atoms by modifying the initial state with the "operator normalization" technique proposed by Brunetti and Fredenhagen [24].

4.4.1 Operator-Normalized Arrival Times

For certain observables not all repetitions of the experiment give a measurement result. When detecting the atom's arrival at a laser region, there may be reflection or transmission without detection.

It is possible to normalize "by hand,"

$$\Pi_N(t) := \frac{\Pi(t)}{\int \Pi(t')\,\mathrm{d}t'} \,,$$

where

$$\Pi(t) = -\mathrm{d}N/\mathrm{d}t = \langle \psi_{\text{in}}(0)|\widehat{\Pi}_t|\psi_{\text{in}}(0)\rangle \,,$$

$$\widehat{\Pi}_t = -\frac{\mathrm{d}\widehat{N}}{\mathrm{d}t}, \quad \widehat{N} = U_c^{\dagger}(t, -\infty)U_c(t, -\infty), \tag{4.19}$$

and

$$U_c(t, t_0) = \mathrm{e}^{-\mathrm{i}H_c(t-t_0)/\hbar}\mathrm{e}^{-\mathrm{i}H_0 t_0/\hbar} \,. \tag{4.20}$$

Note that we have taken the average over incoming, asymptotic, and freely moving states.

The alternative is to normalize at the level of operators. For that end let us define a detection probability operator assuming for simplicity a complex potential treatment and a semi-infinite laser (the more general two-channel case or finite-width lasers may be handled similarly [47]),

$$
\widehat{B} = \int_{-\infty}^{\infty} dt \, \widehat{\Pi}_t = 1 - \widehat{N}_{t \to \infty}
$$
$$
= 1 - \widehat{S}^\dagger \widehat{S} .
$$

Here $\widehat{S}$ is the scattering operator which connects incoming and outgoing freely moving asymptotic states, $\psi_{\text{out}} = \widehat{S} \psi_{\text{in}}$. The expectation value for the incoming asymptote is

$$
P_{\text{detection}} = \langle \psi_{\text{in}} | \widehat{B} | \psi_{\text{in}} \rangle = 1 - \langle \psi_{\text{out}} | \psi_{\text{out}} \rangle .
$$

For a semi-infinite laser, and left incidence,

$$
P_{\text{detection}} = 1 - \int_0^\infty dp \, |\phi_{\text{in}}(p)|^2 |R(p)|^2 ,
$$
$$
\langle p | \widehat{B} | p' \rangle = (1 - |R(p)|^2) \delta(p - p') .
$$

Then we normalize at the "operator level" as follows:

$$
\widehat{\Pi}_t^{\text{ON}} = \widehat{B}^{-1/2} \widehat{\Pi}_t \widehat{B}^{-1/2} .
$$

The corresponding distribution $\Pi^{\text{ON}}(t) = \langle \psi_{\text{in}} | \widehat{\Pi}_t^{\text{ON}} | \psi_{\text{in}} \rangle$ is, by construction, normalized to 1. The physical meaning of this formal operation is to modify ψ_{in} by the operators $\widehat{B}^{-1/2}$ to compensate the detection losses. In the example of the semi-infinite laser, in p-representation,

$$
\psi_{\text{in}}(p) \to \psi'(p) = \psi_{\text{in}}(p)[1 - |R(p)|^2]^{-1/2}/C ,
$$

where C is a normalization constant. This preparatory "filtering" operation previous to the time measurement can be performed in principle by passing the initial cloud through a potential with the adequate scattering amplitude. However, the unbounded nature of the operator $\widehat{B}^{-1/2}$ restricts such a method to states with a low-energy cutoff or which vanish at $p = 0$ fast enough.

4.4.2 Kijowski Distribution as a Limiting Case

If the complex potential model is applicable, the distribution in Eq. (4.15) is again not normalized to 1, and it is, therefore, natural to employ an operator normalization. Now the operator-normalized distribution is obtained as

$$\Pi^{\text{ON}}(t) = \frac{2V_0}{\hbar} \int_0^\infty dx \int dk dk' \, \overline{\tilde{\psi}(k)} \tilde{\psi}(k')$$

$$\times (1 - |R(k)|^2)^{-\frac{1}{2}} (1 - |R(k')|^2)^{-\frac{1}{2}}$$

$$\times \overline{T(k)} T(k') \, e^{i\hbar(k^2 - k'^2)t/2m} e^{-i(\overline{\kappa} - \kappa')x} \, .$$

In the limit of strong interaction, $V_0 \to \infty$, this goes to Kijowski's distribution,

$$\Pi^{\text{ON}}(t) \to \Pi_K(t) \quad \text{for } V_0 \to \infty \, . \tag{4.21}$$

The advantage of the one-channel model is that it provides a simple calculational tool for further, more complicated arrival time problems and that, by simple limits and operator normalization, it is related to the operational fluorescence approach as well as to the axiomatic distribution of Kijowski.

Nevertheless, even in the general two-channel model, the operator-normalized distribution $\Pi^{\text{ON}}(t)$ approaches Kijowski's axiomatic distribution for large γ and Ω, with $\gamma^2/\Omega^2 = $ constant.

$$\Pi^{\text{ON}}(t) \to \Pi_K(t) \text{ for } \gamma \to \infty, \quad \gamma^2/\Omega^2 = \text{const} \, . \tag{4.22}$$

Experimentally, Ω is easier to adjust than γ.[7] Therefore we also consider the limit of large Ω, with γ held fixed. For $\gamma \to \infty$ one again obtains Kijowski's distribution, but for finite γ there is a delay in the arrival times. We can try to eliminate this, as before, by a deconvolution with the first-photon distribution, $W(t)$, of an atom at rest in the laser field, making the ansatz

$$\Pi^{\text{ON}}(t) = \Pi_{\text{id}}(t) * W(t) \tag{4.23}$$

for an ideal distribution $\Pi_{\text{id}}(t)$. For any value of γ and in the limit of strong driving [47],

$$\Pi_{\text{id}}(t) \to \Pi_K(t) \text{ for } \Omega \to \infty, \quad \gamma = \text{const} \, . \tag{4.24}$$

4.4.3 Operator-Normalized Quantum Arrival Times for Two-Sided Incidence and in the Presence of Interactions

An obvious limitation of the semi-infinite models (with the laser in explicit form or with a complex potential) is that one cannot study the arrival at a point, $x = 0$, say, for a state incident from both sides, $x > 0$ and $x < 0$ [59, 69]. Similarly, if one is interested in the arrivals within an interaction region, the semi-infinitely extended measurement will severely affect the dynamics of the unperturbed system

---

[7] γ may also be adjustable though, in particular when it is an effective decay rate resulting from some forced driving into a rapidly decaying state.

on one side. It is possible to apply, instead of the semi-infinite interaction, a weak
and narrow, minimally perturbing, absorbing potential combined with "operator normalization" to these two more elusive arrival time problems.

Let us consider an asymptotically free, moving wave packet impinging on a
potential $-iV_\varepsilon$ located in $-\varepsilon \leq x \leq \varepsilon$ and a real potential U localized between
a and b. The Hamiltonian is given by

$$H_c = \frac{\widehat{p}^2}{2m} - iV_\varepsilon \chi_{[-\varepsilon,\varepsilon]} + U\chi_{[a,b]}(\widehat{x}) , \qquad (4.25)$$

where $\chi_{[a,b]}$ is one in $[a, b]$ and zero elsewhere. In the atom-laser model we have
$V_\varepsilon = \hbar\Omega^2/2\gamma$. For the limit $\varepsilon \to 0$ we consider two cases:
  (a) $V_\varepsilon \sim \varepsilon^{-1}$,
  (b) $V_\varepsilon \sim \varepsilon^{-\alpha}$, $0 < \alpha < 1$.
In the limit $\varepsilon \to 0$, case (a) yields an imaginary delta-function potential whereas
case (b) implies a weaker and less perturbing measurement.

4.4.3.1 Free Wave Packet

The easiest case is a free incoming wave packet of positive momenta, see Fig. 4.5a.
Its Hamiltonian is given by Eq. (4.25) with $U = 0$. Imposing standard matching
conditions, in the limit $\varepsilon \overset{(a)}{\to} 0$ one obtains the same result as for an imaginary δ-like
potential. In the limit $\varepsilon \overset{(b)}{\to} 0$ the measurement region obviously has no effect on
the motion of the atoms. However, by operator normalization a finite distribution is
obtained in both limits,

$$\Pi^{ON} \to \Pi_K . \qquad (4.26)$$

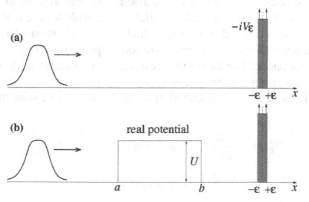

Fig. 4.5 Measurement scheme for the arrival time distribution at $x = 0$ of a wave packet coming
from the left. (**a**) Free wave packet. (**b**) Tunneling wave packet

Now let us look at the case of general states with $U = 0$. The Hamiltonian H_c in Eq. (4.25) commutes with the parity operator and, as a consequence, so does the operator $\widehat{\Pi}_t$. Therefore, its matrix elements between symmetric and antisymmetric states vanish, so that the operator normalization of $\widehat{\Pi}_t$ can be performed in the subspaces of symmetric and antisymmetric states independently by expanding in symmetric and antisymmetric eigenfunctions, respectively. Although for antisymmetric states the wave function vanishes at $x = 0$, the operator normalization preserves a finite arrival time distribution even when the width of the measurement region contracts to 0. A state ψ can be written in terms of its symmetric and antisymmetric part, and in the limit $\varepsilon \overset{(a)}{\to} 0$ and $\varepsilon \overset{(b)}{\to} 0$ one obtains

$$\Pi_\psi^{\mathrm{ON}}(t) = \frac{\hbar}{2\pi m} \sum_\pm \left| \int_0^\infty dk \, \tilde{\psi}(\pm k)\sqrt{k} \, e^{-i\hbar k^2 t/2m} \right|^2 . \qquad (4.27)$$

This expression has been proposed in Refs. [55, 69] on more heuristic grounds as a generalization of the distribution from left (or right) incoming states to general free states.

4.4.3.2 Effect of a Real Barrier

When the real barrier is present, $U \neq 0$, see Fig. 4.5b, the operator-normalized arrival time distribution of a tunneled particle takes, with $\varepsilon \overset{(b)}{\to} 0$, the form

$$\Pi_{\mathrm{pot}}^{\mathrm{ON}}(t) = \frac{\hbar}{2\pi m} \left| \int dk \, \tilde{\psi}(k)e^{-i\hbar k^2 t/2m} \sqrt{k} \, \frac{T(k)}{|T(k)|} \right|^2 .$$

The effect of the additional potential is the introduction of a phase factor which comes from the transmission amplitude $T(k)$ of the real potential. This means that the Hartman effect can be observed [46] in the arrival time distribution, i.e., the arrival time peak becomes independent of the barrier width.

In the above setup an incoming free particle was prepared far to the left and then interacted with an external real potential. If one includes the real potential as part of the preparation procedure through which the particle passes far away on the left, the incident state for operator normalization purposes would be the normalized transmitted wave packet. For the positive results, i.e., transmissions, the normalized incoming free state is then characterized by $T(k)\tilde{\psi}(k)/(\int dk \, |T(k)\tilde{\psi}(k)|^2)^{1/2}$ instead of $\tilde{\psi}(k)$. Applying Kijowski's distribution to the incoming free state thus prepared gives

$$\Pi_K^T(t) = \frac{\hbar}{2\pi m} \left| \int dk \, \tilde{\psi}(k)e^{-i\hbar k^2 t/2m} \sqrt{k} \, T(k) \right|^2$$

$$\times \left(\int dk \, |T(k)\tilde{\psi}(k)|^2 \right)^{-1} . \qquad (4.28)$$

This expression coincides with the proposal of Refs. [60, 69, 13, 15].

4.5 Kinetic Energy Densities

In this section we continue exploring the strong interaction limit, first via "normalization by hand" in Sect. 4.5.1 and then by deconvolution in Sect. 4.5.2. This will produce a kinetic energy density that is approximately related to the other local densities in Sect. 4.5.3 by expanding over the momentum width.

4.5.1 Kinetic Energy Density, Fluorescence, and Atomic Absorption Rate in an Imaginary Potential Barrier

In the complex potential model and for large V_0, $V_0 \gg \hbar^2 k^2 / 2m$, one has in leading order (the barrier is now located between 0 and L and the wavepacket comes from the left)

$$\Pi(t) \simeq \frac{\hbar^2}{\pi m \sqrt{m V_0}} \int_0^\infty dk \int_0^\infty dk' \, \widetilde{\psi}^*(k) \widetilde{\psi}(k') \, e^{i\hbar(k^2 - k'^2)t/2m} kk' \, .$$

This expression is independent of the barrier length L as a result of the large V_0 limit, so the same result is obtained with an imaginary step potential $-iV_0\Theta(\widehat{x})$ or with a very narrow barrier as in the previous section.

The normalization constant is given by $\int \Pi(t) dt \simeq 2\hbar k_0 (m V_0)^{-1/2}$, where the mean wave vector reads $k_0 = \int |\widetilde{\psi}(k)|^2 k \, dk$, and the normalized absorption rate ("normalization by hand") is

$$\Pi_N(t) \simeq \frac{\hbar}{2\pi m k_0} \int_0^\infty dk \int_0^\infty dk' \, \widetilde{\psi}^*(k) \widetilde{\psi}(k') \, e^{i\hbar(k^2 - k'^2)t/2m} kk' \, . \qquad (4.29)$$

With the freely moving wave packet $\psi_f(x, t)$, this can finally be rewritten in the form

$$\Pi_N(t) \simeq \frac{\hbar}{m k_0} \langle \psi_f(t) | \widehat{k} \delta(\widehat{x}) \widehat{k} | \psi_f(t) \rangle \, . \qquad (4.30)$$

Now, the right-hand side is just the expectation value at time t of the kinetic energy density operator corresponding to Eq. (4.3) (we shall call it $\widehat{\tau}^{(1)}$, and similarly $\widehat{\tau}^{(2,3)}$ correspond to Eqs. (4.4) and (4.5)[8]) evaluated at the origin! Thus, if $p_0 = k_0 \hbar$ is the initial average momentum we have, in the limit $V_0 \to \infty$,

$$\lim_{V_0 \to \infty} \Pi_N(t) = \frac{2}{p_0} \langle \widehat{\tau}^{(1)}(x = 0) \rangle_t \, . \qquad (4.31)$$

---

[8]These three operators lead to the three versions of the kinetic energy density commonly found in applications [9, 23, 27, 11, 10, 86, 87, 66].

Note that the averages are computed with the *freely moving* wave function.

The results are different from the ones obtained by operator normalization as applied in the previous section, i.e., by normalizing Π^{ON} to 1. It is possible, however, to get the same kinetic energy density in (4.31) by using $p_0/2$ as normalization constant [48].

4.5.2 Kinetic Energy Density from First-Photon Measurement and Deconvolution

If the atom-laser model is used to implement the complex potential V_0, then $V_0 = \hbar\Omega^2/2\gamma$, with $\hbar\gamma$ much larger than the kinetic energy. The limit $V_0 \to \infty$ implies a simultaneous change of Ω and γ but, experimentally, the Rabi frequency Ω is easier to adjust than γ. To overcome this problem, we describe now a procedure that allows to keep the value of γ fixed.

We again consider the atom-laser model but now for the limit $\Omega \to \infty$ and $\gamma = \text{const}$. In that case, the simplified description of the evolution of the wave function by means of the imaginary potential is not feasible, and one has to solve the full two-channel problem. This has been done in Refs. [31, 47] and "normalizing by hand" with a constant the resulting photon detection rate $\Pi_N(t)$ becomes

$$\Pi_N(t) \simeq \frac{\hbar}{2\pi m k_0} \int_0^\infty dk\,dk'\, \widetilde{\psi}^*(k)\widetilde{\psi}(k')\, e^{i\hbar(k^2 - k'^2)t/2m} \frac{\gamma kk'}{\gamma + i\hbar(k^2 - k'^2)/m}\,.$$

For $\hbar\gamma$ large compared to the kinetic energy of the incident atom, Eq. (4.29) is recovered, but for finite γ there is a delay in the detection rate. This can be eliminated by means of a deconvolution with the first-photon distribution $W(t)$ for an atom at rest [31, 47]. Using the ansatz

$$\Pi(t) = \Pi_{\text{id}}(t) * W(t)\,,$$

and deconvolution,

$$\Pi_{\text{id}}(t) \simeq \frac{2}{p_0} \langle \widehat{\tau}^{(1)}(x = 0)\rangle_t$$

holds as before.

4.5.3 Relations Between Different Local Densities

Here we briefly discuss a formal connection between the arrival time distribution of Kijowski (4.1) and the kinetic energy densities $\langle \widehat{\tau}^{(1)}(x = 0)\rangle_t$ and $\langle \widehat{\tau}^{(2)}(x = 0)\rangle_t$. For wave packets peaked around some k_0 in momentum space, the operator $\widehat{k}^{1/2}$ acting on ψ_f in Eq. (4.1) can be expanded in terms of $(\widehat{k} - k_0)$,

$$\widehat{k}^{1/2} = k_0^{1/2} + \frac{1}{2}k_0^{-1/2}(\widehat{k} - k_0) - \frac{1}{8}k_0^{-3/2}(\widehat{k} - k_0)^2 + \mathcal{O}\left((\widehat{k} - k_0)^3\right) . \qquad (4.32)$$

We take k_0 to be the first moment of the momentum distribution, $k_0 = \int |\widetilde{\psi}(k)|^2 k\,dk$. Inserting the expansion in Eq. (4.32) into Eq. (4.1) yields in zeroth order a very simple result,

$$\Pi_K(t) = v_0|\psi_f(0,t)|^2 + \mathcal{O}(\widehat{k} - k_0) ,$$

i.e., the particle density times the average velocity $v_0 = k_0\hbar/m$. In first order in $(\widehat{k} - k_0)$ one obtains the flux at $x = 0$,

$$\Pi_K(t) = J(0,t) + \mathcal{O}\left((\widehat{k} - k_0)^2\right) ,$$

and to second order the expression

$$\Pi_K(t) = J(0,t) + \frac{1}{2p_0}\Delta(0,t) + \mathcal{O}\left((\widehat{k} - k_0)^3\right) ,$$

where

$$\Delta(0,t) = \langle\widehat{\tau}^{(1)}(x=0)\rangle_t - \langle\widehat{\tau}^{(2)}(x=0)\rangle_t .$$

For states with positive momentum, which we are considering here, the first order, namely the flux, is correctly normalized to 1 and so is the second order since the time integral over Δ is easily shown to vanish. This difference only provides a local-in-time correction to J that averages out globally. Its quantum nature can be further appreciated by the more explicit expression

$$\frac{1}{2p_0}\Delta = \frac{\hbar^2}{8mp_0}\frac{\partial^2|\psi_f(0,t)|^2}{\partial x^2} .$$

4.6 Disclosing Hidden Information Behind the Quantum Zeno Effect: Pulsed Measurement of the Quantum Time of Arrival

This section completes our analysis of the strong interaction regime and deals with pulsed measurements simulated by instantaneous projections as in Fig. 4.2. When they are performed very frequently the wave packet is totally reflected. The first discussions of the Zeno effect, understood as the hindered passage of the system between orthogonal subspaces because of frequent instantaneous measurements, emphasized its problematic status and regarded it as a failure to simulate or define quantum passage-time distributions [4–6, 62]. We shall see, however, that in fact there is a "bright side" of the effect: by normalizing the little bits of norm removed at each projection step,

$$\Pi_{\text{Zeno}} = \lim_{\delta t \to 0} \frac{-\delta N/\delta t}{1 - N(\infty)} \,,$$

a physical time distribution defined for the freely moving system emerges,[9] the same kinetic energy density found in the previous section [33].

4.6.1 Zeno Time Distribution

To find Π_{Zeno} we shall put the parallelism hinted by Allcock between the pulsed measurement and the continuous measurement on a firmer, more quantitative basis. We shall define formally the pulsed and continuous measurement models as well as an intermediate auxiliary model [7] that will be a useful bridge between the two.

The "chopping process" amounts to a periodic projection of the wave function onto the $x < 0$ region at instants separated by a time interval δt. The wave functions immediately after and before the projection at the instant t_j are related by[10]
$\psi(x, t_{j_+}) = \psi(x, t_{j_-})\Theta(-x)$.

The wave at $x > 0$ may also be canceled with a "kicked" imaginary potential $\widehat{V}_k = \widehat{V}\delta t\, F_{\delta t}(t)$, where the subscript "k" stands for "kicked" and $F_{\delta t}(t) = \sum_{j=-\infty}^{\infty} \delta(t - j\delta t)$,

$$\widehat{V} = -i\widehat{V}_I = -iV_0\Theta(\widehat{x}) \,, \tag{4.33}$$

provided

$$V_0\delta t \gg \hbar \,. \tag{4.34}$$

The general (and exact) evolution operator is obtained by repetition of the basic unit

$$\widehat{U}_k(0, \delta t) = e^{-i\widehat{H}_0\delta t/\hbar}e^{-i\widehat{V}\delta t/\hbar} \,, \tag{4.35}$$

where $\widehat{H}_0 = -(\hbar^2/2m)\partial^2/\partial x^2$.

For the continuous model, the evolution under the imaginary potential (4.33) is given by

$$\widehat{U}(0, \delta t) = e^{-i(\widehat{H}_0+\widehat{V})\delta t/\hbar} = e^{-i\widehat{H}_0\delta t/\hbar}e^{-i\widehat{V}\delta t/\hbar} + \mathcal{O}(\delta t^2[\widehat{V}, \widehat{H}_0]/\hbar^2) \,.$$

Comparing with Eq. (4.35) we see that the kicked and continuous models agree when

---

[9]There are other "positive" uses of the Zeno effect, such as reduction of decoherence in quantum computing, see, e.g., [73, 36].

[10]Experimental realizations of repeated measurements will rely on projections with a finite frequency and pulse duration that provide approximations to the ideal result. Feasible schemes may be based on pulsed localized resonant laser excitation [85] or sweeping with a detuned laser [29].

$$\delta t^2 |\langle[\widehat{V}, \widehat{H}_0]\rangle|/\hbar^2 \ll 1 . \tag{4.36}$$

At first sight a large $V_0 \delta t/\hbar$, see Eq. (4.34), seems to be incompatible with this condition so that the three models would not agree [33]. In fact the numerical calculations show a better and better agreement between the continuous and pulsed models as $V_0 \to \infty$ when δt and V_0 are linked by some predetermined (large) constant α, $\delta t = \alpha \hbar/V_0$.

Figures 4.6 and 4.7 illustrate this agreement: in Fig. 4.6 the average absorption time, $\langle t \rangle = \int_0^\infty (-dN/dt) t dt/[1 - N(\infty)]$, is shown versus δt (chopping) and $\hbar/2V_0$ (complex potential) for a Gaussian wave packet sent from the left toward the origin. The lines bend at high coupling because of reflection. The normalized absorption distribution as $V_0 \to \infty$ has been derived in Sect. 4.5.1, see Eq. (4.30):

$$\Pi_N(t) = \frac{\hbar}{mk_0} \langle \psi_f(t) | \widehat{k} \delta(\widehat{x}) \widehat{k} | \psi_f(t) \rangle , \tag{4.37}$$

where $k_0 \hbar$ is the initial average momentum. Figure 4.7 shows for a more challenging state, a combination of two Gaussians, that this ideal distribution becomes indistinguishable from the normalized chopping distribution when δt is small enough. Even the minor details are reproduced and differ from J and Π_k, also shown.

To understand the compatibility "miracle" of the inequalities (4.34) and (4.36), we apply the Robertson–Schrödinger (generalized uncertainty principle) relation,

$$\left| \langle[\widehat{V}, \widehat{H}_0]\rangle \right| \leq 2|\Delta V_I| \Delta H_0 ,$$

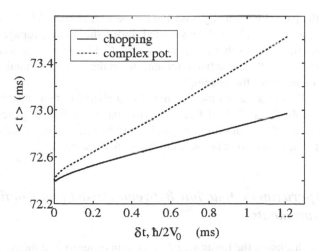

Fig. 4.6 Average absorption times evaluated from $-dN/dt$ (normalized) for the projection method and the continuous (complex potential) model. The initial state is a minimum uncertainty product Gaussian for ^{23}Na atoms centered at $x_0 = -500 \,\mu$m with $\Delta x = 23.5 \,\mu$m and average velocity 0.365 cm/s. In all numerical examples negative momentum components of the initial state are negligible

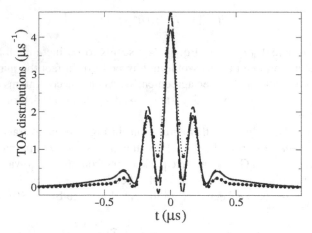

Fig. 4.7 Time-of-arrival distributions: Flux J (*dashed line*), Π_K (*solid line*), Π_{Zeno} (*big sparse dots*), and Π_{chopping} (for pulses separated by $\delta t = 0.266$ ns, *dotted line*). The initial wave packet is a combination $\psi = 2^{-1/2}(\psi_1 + \psi_2)$ of two Gaussian states for the center-of-mass motion of a single cesium atom that become separately minimal uncertainty packets (with $\Delta x_1 = \Delta x_2 = 0.021\,\mu$m and average velocities $\langle v \rangle_1 = 18.96$ cm/s, $\langle v \rangle_2 = 5.42$ cm/s) at $x = 0$ and $t = 0\,\mu$s

where Δ denotes here the standard deviation. Since $|\Delta V_I|$ is rigorously bounded at all times by $V_0/2$,[11] imposing $\delta t V_0 = \alpha \hbar$ with $\alpha \gg 1$, a sufficient condition for dynamical agreement among the models is

$$V_0 \gg \Delta H_0 . \tag{4.38}$$

For large V_0 the packet is basically reflected by the wall so that ΔH_0 tends to retain its initial value and Eq. (4.38) will be satisfied during the whole propagation.

This implies in summary that $\Pi_{\text{Zeno}} = \Pi_N$, Eq. (4.37), a very remarkable result, which again illustrates that an active intervention on the system dynamics may provide an ideal quantity for the system.

The proposed normalization method may be applied to other measurements as well, i.e., not only for a TOA of freely moving particles, but, in general, to first passages between orthogonal subspaces, and it will be interesting to find out in each case the ideal time distribution brought out by normalization.

4.6.2 An Approximate Relation Between Pulsed and Continuous Measurements

So far we have discussed the limits δt, $\hbar/V_0 \to 0$ in order to find the corresponding time distribution. We shall now relate the pulsed and continuous measurements

---

[11]If N_+ is the norm in $x > 0$, $|\Delta V| = V_0(N_+ - N_+^2)^{1/2}$ which is maximal at $N_+ = 1/2$.

approximately for finite, non-zero values of δt and $\hbar/V_0$, when they are sufficiently large to make reflection negligible: the average detection time is delayed with respect to the ideal limit corresponding to Π_{Zeno} as

$$\langle t \rangle \approx \langle t \rangle_{\text{Zeno}} + \delta t/2 \approx \langle t \rangle_{\text{Zeno}} + \frac{\hbar}{2V_0},$$

see Fig. 4.6, since, once a particle is in $x > 0$, $\frac{\hbar}{2V_0}$ and $\delta t/2$ are precisely the average lifetimes in the continuous and discrete measuring models, respectively.[12] This suggests an approximate agreement between projection and continuous dynamics when $\delta t \approx \hbar/V_0$ is satisfied. For large V_0, this is asymptotically not in contradiction with the requirement of a large α since $V_0^{-1} - (\alpha V_0)^{-1} \to 0$ as $V_0 \to \infty$; in any case quantum reflection breaks down the linear dependence, see Fig. 4.6.

A similar relation between pulsed and continuous measurements was described by Schulman [84] and has been tested experimentally [85]. The simplest model in [84] may be reinterpreted as a two-level atom in a resonant laser field, $\widehat{H} = \frac{\hbar}{2} \begin{pmatrix} 0 & \Omega \\ \Omega & -i\gamma \end{pmatrix}$. The relation between pulsed and continuous measurements follows by comparing the exponential decay for the effective 2-level Hamiltonian with Rabi frequency Ω and excited state lifetime $1/\gamma$, with the decay dynamics when $\gamma = 0$ and the system is projected every δt into the ground state. It takes the form [84] $\delta t = \frac{4}{\gamma}$ for $\gamma/\Omega \ll 1$ (weak driving). In our TOA model we have a different set of parameters. Using $\widehat{H} = \widehat{H}_0 + \frac{\hbar}{2} \begin{pmatrix} 0 & \Omega \Theta(\widehat{x}) \\ \Omega \Theta(\widehat{x}) & -i\gamma \end{pmatrix}$ gives $V_0 = \frac{\hbar \Omega^2}{2\gamma}$, so that $\delta t \approx \hbar/V_0$ becomes

$$\delta t \approx 2\frac{\gamma}{\hbar \Omega^2},$$

different from Schulman's relation, as it may be expected since the pulsed evolution depends on Ω in Schulman's model but not in our case, where it is only driven by the kinetic energy Hamiltonian $\widehat{H}_0$.

4.7 Summary

In this chapter we have discussed a model to measure the time of arrival of an atom. The basic idea is that upon entering a laser-illuminated region the atom will start emitting photons, and the first-photon detection can be taken as a measure of the arrival time of the atom in that region. We have shown the explicit connection between this atom-laser model and approaches where the detector is modeled in a

---

[12]The origin ordinate would be slightly above $\langle t \rangle_{\text{Zeno}}$ for optimized straight lines. Reflection at small δt (or high V_0) favors the detection of faster particles and bends the $\langle t \rangle$ lines toward shorter times, as in Fig. 4.6.

simplified manner by a local complex potential, with the arrival probability distribution given by the decrease rate of the norm of the quantum wave function.

By using deconvolution techniques, we can measure, according to this model, the quantum mechanical flux in a limit. The quantum mechanical flux is in many cases a good approximation to Kijowski's distribution, an ideal reference distribution for the arrival time of a particle. By applying operator normalization we can get also Kijowski's distribution directly from the fluorescence in some limits, which provides an operational interpretation of Kijowski's distribution.

The quantum mechanical flux and Kijowski's distribution are examples of local densities associated with a single classical local density. Other examples are the quantum kinetic energy densities that could also be measured in some limits.

Summarizing, we have shown that several local densities of an atomic wavepacket (quantum mechanical flux, Kijowski's time-of-arrival distributions, kinetic energy densities) can be measured for different limits of the laser shape or intensity and by means of different manipulations of the fluorescence signal or the initial state. One could in this manner observe quantum dynamical effects that have not been realized experimentally so far, for example, the backflow effect (negative fluxes for positive–momentum wavepackets), and quite generally distinguish quantum from classical dynamics and arrival times. Of course actual measurements will generally approximate these limits and operations only to some degree and this will determine the quantity which is really measured. We emphasize though that it is only through extreme operations and idealized limits that we get access to the properties of the bare, freely moving system.

There are still many open questions. For example, the atom-laser model has been examined until now on a single-particle level. An interesting task is to extend the model and examine many-particle effects. Another route of research is to apply the atom-laser model to other time quantities.

Acknowledgments This work could not have been completed without the aid of A. del Campo, J. A. Damborenea, J. Echanobe, I. L. Egusquiza, V. Hannstein, B. Navarro, and D. Seidel. We also acknowledge "Acciones Integradas" of the German Academic Exchange Service (DAAD) and Ministerio de Educación y Ciencia and additional support from the Max Planck Institute for the Physics of Complex Systems; MEC (FIS2006-10268-C03-01); and UPV-EHU (GIU07/40). AR acknowledges support by the German Research Foundation (DFG).

References

1. Y. Aharonov, D. Bohm, Phys. Rev. **122**, 1649 (1961)
2. Y. Aharonov, J. Oppenheim, S. Popescu, B. Reznik, W.G. Unruh, Phys. Rev. A **57**, 4130 (1998)
3. M.M. Ali, D. Home, A.S. Majumdar, A.K. Pan, Phys. Rev. A **75**, 042110 (2007)
4. G.R. Allcock, Ann. Phys. (N.Y.) **53**, 253 (1969)
5. G.R. Allcock, Ann. Phys. (N.Y.) **53**, 286 (1969)
6. G.R. Allcock, Ann. Phys. (N.Y.) **53**, 311 (1969)
7. D. Alonso, R.S. Mayato, C.R. Leavens, Phys. Rev. A **66**, 042108 (2002)
8. Ch. Anastopoulos, N. Savvidou, J. Math. Phys. **47**, 122106 (2006)
9. P.W. Ayers, R.G. Parr, A. Nagy, Int. J. Quant. Chem. **90**, 309 (2002)

10. R.F.W. Bader, *Atoms in Molecules: A Quantum Theory* (Clarendon, Oxford, 1990)
11. R.F.W. Bader, P.M. Beddall, J. Chem. Phys. **56**, 3320 (1972)
12. O. del Barco, M. Ortuño, V. Gasparian, Phys. Rev. A **74**, 032104 (2006)
13. A.D. Baute, R.S. Mayato, J.P. Palao, J.G. Muga, I.L. Egusquiza, Phys. Rev. A **61**, 022118 (2000)
14. A.D. Baute, I.L. Egusquiza, J.G. Muga, R. Sala Mayato, Phys. Rev. A **61**, 052111 (2000)
15. A.D. Baute, I.L. Egusquiza, J.G. Muga, Phys. Rev. A **64**, 012501 (2001)
16. A.D. Baute, I.L. Egusquiza, J.G. Muga, Phys. Rev. A **64**, 014101 (2001)
17. A.D. Baute, I.L. Egusquiza, J.G. Muga, Phys. Rev. A **65**, 032114 (2002)
18. Ph. Blanchard, A. Jadczyk, Phys. Lett. A **175**, 157 (1993)
19. Ph. Blanchard, A. Jadczyk, Ann. Phys. (Lpz.) **4**, 583 (1995)
20. Ph. Blanchard, A. Jadczyk, Phys. Lett. A **203**, 260 (1995)
21. Ph. Blanchard, A. Jadczyk, Helv. Phys. Acta **69**, 613 (1996)
22. R.S. Bondurant, Phys. Rev. A **69**, 062104 (2004)
23. M. Brack, B.P. van Zyl, Phys. Rev. Lett. **86**, 1574 (2001)
24. R. Brunetti, K. Fredenhagen, Phys. Rev. A **66**, 044101 (2002)
25. H. Carmichael, *An Open Systems Approach to Quantum Optics* (Springer, Berlin, 1993)
26. L. Cohen, P. Loughlin, J. Mod. Opt. **49**, 539 (2002)
27. L. Cohen, J. Chem. Phys. **70**, 788 (1979)
28. L. Cohen, J. Chem. Phys. **80**, 4277 (1984)
29. C.S. Chuu, F. Schreck, T.P. Meyrath, J.L. Hanssen, G.N. Price, M.G. Raizen, Phys. Rev. Lett. **95**, 260403 (2005)
30. J. Dalibard, Y. Castin, K. Mølmer, Phys. Rev. Lett. **68**, 580 (1992)
31. J.A. Damborenea, I.L. Egusquiza, G.C. Hegerfeldt, J.G. Muga, Phys. Rev. A **66**, 052104 (2002)
32. J.A. Damborenea, I.L. Egusquiza, G.C. Hegerfeldt, J.G. Muga, J. Phys. B: At. Mol. Opt. Phys. **36**, 2657 (2003)
33. J. Echanobe, A. del Campo, J.G. Muga, Phys. Rev. A **77**, 032112 (2008)
34. P. Facchi, S. Pascazio, Prog. Opt. **42**, 147 (2001)
35. P. Facchi, S. Pascazio, A. Scardicchio, L.S. Schulman, Phys. Rev. A **65**, 012108 (2001)
36. P. Facchi, S. Tasaki, S. Pascazio, H. Nakazato, A. Tokuse, D.A. Lidar, Phys. Rev. A **71**, 022302 (2005)
37. H. Feshbach, Ann. Phys. (N.Y.) **5**, 357 (1958)
38. H. Feshbach Ann. Phys. (N.Y.) **19**, 287 (1962)
39. E.A. Galapon, R.F. Caballar, R.T. Bahague, Phys. Rev. Lett. **93**, 180406 (2004)
40. E.A. Galapon, F. Delgado, J.G. Muga, I. Egusquiza, Phys. Rev. A **74**, 042107 (2005)
41. R. Giannitrapani, Int. J. Theor. Phys. **36**, 1575 (1997)
42. R. Golub, S. Felber, R. Gähler, E. Gutsmiedl, Phys. Lett. A **148**, 27 (1990)
43. A. Gozdz, M. Debicki, Phys. Atom. Nucl. **70**, 529 (2007)
44. J.J. Halliwell, Prog. Theor. Phys. **102**, 707 (1999)
45. V. Hannstein, G.C. Hegerfeldt, J.G. Muga, J. Phys. B: At. Mol. Opt. Phys. **38**, 409 (2005)
46. T.E. Hartman, J. Appl. Phys. **33**, 3427 (1962)
47. G.C. Hegerfeldt, D. Seidel, J.G. Muga, Phys. Rev. A **68**, 022111 (2003)
48. G.C. Hegerfeldt, D. Seidel, J.G. Muga, B. Navarro, Phys. Rev. A **70**, 012110 (2004)
49. G.C. Hegerfeldt, J.T. Neumann, L.S. Schulman, J. Phys. A **39**, 14447 (2006)
50. G.C. Hegerfeldt, J.T. Neumann, L.S. Schulman, Phys. Rev. A **75**, 012108 (2007)
51. G.C. Hegerfeldt, T.S. Wilser, in *Classical and Quantum Systems*. Proceedings of the Second International Wigner Symposium, July 1991, H.D. Doebner, W. Scherer, F. Schroeck (eds.) (World Scientific, Singapore, 1992), p. 104
52. G.C. Hegerfeldt, Phys. Rev. A **47**, 449 (1993)
53. G.C. Hegerfeldt, D.G. Sondermann, Quantum Semiclass. Opt. **8**, 121 (1996).
54. D. Home, A. Whitaker, Ann. Phys. **258**, 237 (1997)
55. J. Kijowski, Rep. Math. Phys. **6**, 361 (1974)

56. M.S. Kim, P.L. Knight, K. Wodkiewicz, Opt. Commun. **62**, 385 (1987)
57. K. Koshino, A. Shimizu, Phys. Rep. **412**, 191 (2005)
58. L. Lamata, J. León, Concepts Phys. **2**, 49 (2005)
59. C.R. Leavens, Phys. Rev. A **58**, 840 (1998)
60. J. León, J. Julve, P. Pitanga, F.J. de Urríes, Phys. Rev. A **61**, 062101 (2000)
61. R.D. Levine, *Quantum Mechanics of Molecular Rate Processes* (Oxford, London, 1969)
62. B. Misra, E.C.G. Sudarshan, J. Math. Phys. **18**, 756 (1977)
63. J.G. Muga, C.R. Leavens, Phys. Rep. **338**, 353 (2000)
64. J.G. Muga, A.D. Baute, J.A. Damborenea, I.L. Egusquiza, arXiv: *quant-ph/0009111*
65. J.G. Muga, S. Brouard, D. Macías, Ann. Phys. (N.Y.) **240**, 351 (1995)
66. J.G. Muga, J.P. Palao, R. Sala, Phys. Lett. A **238**, 90 (1998)
67. J.G. Muga, R. Leavens, J.P. Palao, Phys. Rev. A **58**, 4336 (1998)
68. J.G. Muga, J.P. Palao, C.R. Leavens, Phys. Lett. A **253**, 21 (1999)
69. J.G. Muga, R. Sala, I.L. Egusquiza (eds.), *Time in Quantum Mechanics*, vol. 1, Lect. Notes Phys. **734** (Springer, Berlin, 2008)
70. J.G. Muga, D. Seidel, G.C. Hegerfeldt, J. Chem. Phys. **122**, 154106 (2005)
71. B. Navarro, I.L. Egusquiza, J.G. Muga, G.C. Hegerfeldt, J. Phys. B: At. Mol. Opt. Phys. **36**, 3899 (2003)
72. B. Navarro, I.L. Egusquiza, J.G. Muga, G.C. Hegerfeldt, Phys. Rev. A **67**, 063819 (2003)
73. H. Nazakato, T. Takazawa, K. Yuasa, Phys. Rev. Lett. **90**, 060401 (2003)
74. M.K. Oberthaler, R. Abfalterer, S. Bernet, J. Schmiedmayer, A. Zeilinger, Phys. Rev. Lett. **77**, 4980 (1996)
75. J.P. Palao, J.G. Muga, S. Brouard, A. Jadczyk, Phys. Lett. A **233**, 227 (1997)
76. M.B. Plenio, P.L. Knight, Rev. Mod. Phys. **70**, 101 (1998)
77. R.W. Robinett, Am. J. Phys. **63**, 823 (1995)
78. A. Ruschhaupt, Phys. Lett. A **250**, 249 (1998)
79. A. Ruschhaupt, in *Decoherence: Theoretical, Experimental and Conceptual Problems*, Ph. Blanchard et al. (eds.), Lect. Notes Phys. **538** (Springer, Berlin, 2000), p. 259
80. A. Ruschhaupt, J. Phys. A: Math. Gen. **35**, 10429 (2002)
81. A. Ruschhaupt, J.A. Damborenea, B. Navarro, J.G. Muga, G.C. Hegerfeldt, Europhys. Lett. **67**, 1 (2004)
82. A. Ruschhaupt, B. Navarro, J.G. Muga, J. Phys. B: At. Mol. Opt. Phys. **37**, L313 (2004)
83. J. Ruseckas, B. Kaulakys, Phys. Rev. A **66**, 052106 (2002)
84. L.S. Schulman, Phys. Rev. A **57**, 1509 (1998)
85. E.W. Streed, J. Mun, M. Boyd, G.K. Campbell, P. Medley, W. Ketterle, D.E. Pritchard, Phys. Rev. Lett. **97**, 260402 (2006)
86. A. Tachibana, J. Chem. Phys. **115**, 3497 (2001)
87. M.W. Thomas, R.F. Snider, J. Stat. Phys. **2**, 61 (1970)
88. G. Torres-Vega, Phys. Rev. A **75**, 032112 (2007)
89. R. Werner, J. Math. Phys. **27**, 793 (1986)
90. R. Werner, Ann. Inst. Henri Poincaré **47**, 429 (1987)
91. M. Zwanzig, J. Chem. Phys. **33**, 1338 (1960)

Chapter 5
Dwell-Time Distributions in Quantum Mechanics

José Muñoz, Iñigo L. Egusquiza, Adolfo del Campo, Dirk Seidel,
and J. Gonzalo Muga

> *LOCATIONS and times—what is it in me that meets them all,*
> *whenever and wherever, and makes me at home?*
> Walt Whitman

5.1 Introduction

Time observables in quantum mechanics have a long and debated history [33]. In
spite of the fact that random time variables, measured after a system is prepared, are
common in laboratories, most often it has been argued that questions about time in
quantum mechanics should best be left alone, as illustrated by the frequent reference
to Pauli's theorem. Alternatively, the emphasis has been laid on characteristic times,
i.e., single time quantities characterizing a process such as tunneling or decay. This,
in many ways, runs counter to the usual procedure in quantum mechanics, where
additionally to the average value of a quantity we require prediction of higher order
moments of that quantity; in other words, the probability distribution.

J. Muñoz (✉)
Departamento de Química Física, UPV-EHU, Apdo. 644, 48080 Bilbao, Spain,
josemunoz@saitec.es

I.L. Egusquiza
Department of Theoretical Physics, The University of the Basque Country, Apdo. 644, 48080
Bilbao, Spain, inigo.egusquiza@ehu.es

A. del Campo
Institute for Mathematical Sciences, Imperial College London, SW7 2PG, UK;
QOLS, The Blackett Laboratory, Imperial College London, Prince Consort Rd., SW7 2BW, UK,
a.del-campo@imperial.ac.uk

D. Seidel
Departamento de Química Física, UPV-EHU, Apdo. 644, 48080 Bilbao, Spain,
dirk_x_seidel@yahoo.de

J.G. Muga
Departamento de Química Física, UPV-EHU, Apdo. 644, 48080 Bilbao, Spain, jg.muga@ehu.es

Muñoz, J. et al.: *Dwell-Time Distributions in Quantum Mechanics*. Lect. Notes Phys. **789**, 97–125
(2009)
DOI 10.1007/978-3-642-03174-8_5

With regard to the time-of-arrival observable several such distributions have been proposed and studied, see volume 1 of "Time in Quantum Mechanics" [12, 28], and Chaps. 3 and 4 in this volume. In this chapter we analyze the dwell time, which appears to be a much simpler time observable because the associated operator will indeed be self-adjoint (over the adequate domain). At first sight, it could be thought that this statement contradicts Pauli's theorem, which asserts that no self-adjoint time observable can exist with canonical commutation relations with a semi-bound Hamiltonian. However, the dwell time is an *interval* quantity, as opposed to the *instant* quantity that the time of arrival describes. Its associated operator, therefore, should *commute* with the Hamiltonian, as opposed to presenting canonical commutation with it.

The dwell time of a particle in a region of space and its close relative, the delay time [43], are rather fundamental quantities that characterize the duration of collision processes, the lifetime of unstable systems [13], the response to perturbations [19], ac-conductance in mesoscopic conductors [7], or the properties of chaotic scattering [29]. In addition, the importance of dwell and delay times is underlined by their relation to the density of states and to the virial expansion in statistical mechanics [8]. We could thus hardly fail to study and characterize in detail such a prominent quantity. For a sample of theoretical studies on the quantum dwell time, see [13, 19, 48, 5, 22, 8, 44, 11, 50, 45, 23, 49, 20, 46, 3, 27, 47]. A recurrent topic has been its role and decomposition in tunneling collisions. Instead, we shall focus here on different, so far overlooked but fundamental aspects, namely, the measurability and physical implications of its distribution and its second moment.

Despite the nice properties of the dwell-time operator, the relevance of the concept and average value in many different fields, or the apparent formal simplicity stated above, the dwell time is actually rather subtle and remains elusive and challenging in many ways. In particular, a direct and sufficiently noninvasive measurement, so that the statistical moments are produced by averaging over individual dwell-time values, is yet to be discovered. If the particle is detected (and thus localized) at the entrance of the region of interest, its wavefunction is severely modified ("collapsed"), so that the times elapsed until a further detection when it leaves the region do not reproduce the ideal dwell-time operator distribution and depend on the details of the localization method. Proposals for operational, measurement-based approaches to traversal times which model the detectors and study the effect of localization have been discussed by Palao et al. [37] and by Ruschhaupt [39]. For attempts to measure the dwell time with continuous or kicked "clocks" coupled to the particle's presence in the region, see [47, 1]. All operational approaches to the quantum dwell time known so far have provided only its average, and indirectly, by deducing it from its theoretical relation to some other observable with measurable average. The average is obtained, for example, by a "Larmor clock," using a weak homogeneous magnetic field in the region D and the amount of spin rotations of an incident spin$-\frac{1}{2}$ particle [2, 41, 6]. An optical analog is provided by the "Rabi clock" [4]. It can also be deduced from average passage times at the region boundaries [27], as well as by measuring the total absorption if a weak complex absorbing potential acts in the region [16, 17, 31]. This last setup could be implemented with cold atoms and lasers as described in [11, 40] and will be discussed in Sect. 5.4.2.

The chapter is organized as follows: the first sections are devoted to fundamental and formal aspects (Sect. 5.2), examples (Sects. 5.3 and 5.4), or extensions (Sect. 5.5) of the dwell-time concept and operator, whereas Sect. 5.6 tackles the relationship between the moments of the dwell-time operator and flux–flux correlation functions (ffcf) [35], generalizing an approach by Pollak and Miller [38]. They showed that the average stationary dwell time agrees with the first moment of a microcanonical ffcf. We shall see that this relation holds also for the second moment, but not for higher moments, and extend their analysis to the time-dependent (wavepacket) case. We shall also discuss a possible scheme to measure ffcf's, thus paving the way toward experimental access to quantum features of the dwell-time distribution.

5.2 The Dwell-Time Operator

Unlike other time quantities, there has been a broad consensus on the operator representation of the dwell time [13, 19]. For one particle evolving unitarily with Hamiltonian $\widehat{H}$ in region D, which we limit here to one dimension for simplicity, $D = \{x : x_1 \leq x \leq x_2\}$, it takes the form

$$\widehat{T}_D = \int_{-\infty}^{\infty} dt\, \widehat{\chi}_D(t) = \int_{-\infty}^{\infty} dt\, e^{i\widehat{H}t/\hbar} \chi_D(\widehat{x}) e^{-i\widehat{H}t/\hbar}, \qquad (5.1)$$

where $\widehat{\chi}_D(t)$ is the (Heisenberg) projector onto D and $\chi_D(\widehat{x}) = \widehat{\chi}_D(0) = \int_{x_1}^{x_2} dx |x\rangle\langle x|$. Without delving too far in the functional analysis definition, i.e., into the proper description of its domain and of its adjoint, we can see that $\widehat{T}_D$ will be self-adjoint (we shall come back to this issue after studying the specific example of the free particle Hamiltonian in Sect. 5.3). At any rate, it is clear that, at least formally, this operator commutes with the Hamiltonian, as can be seen from what follows[1]:

$$\widehat{T}_D e^{-i\widehat{H}t/\hbar} = \int_{-\infty}^{\infty} d\tau\, e^{i\widehat{H}\tau/\hbar} \chi_D(\widehat{x}) e^{-i\widehat{H}(\tau+t)/\hbar} \qquad (5.2)$$

$$= \int_{-\infty}^{\infty} d\tau\, e^{i\widehat{H}(\tau-t)/\hbar} \chi_D(\widehat{x}) e^{-i\widehat{H}\tau/\hbar} = e^{-i\widehat{H}t/\hbar} \widehat{T}_D.$$

The commutation of $\widehat{T}_D$ and the Hamiltonian leads us to search for the eigenfunctions of dwell time in the stationary eigenspaces of the latter. Let α be the degeneracy

---

[1] Time-limited versions of the dwell-time operator such as $\int_0^{\infty} dt\, e^{i\widehat{H}t/\hbar} \chi_D(\widehat{x}) e^{-i\widehat{H}t/\hbar}$ do *not* generally commute with $\widehat{H}$, see [13] or Sect. 5.5.2.

index for these eigenspaces, such that $\widehat{H}|E,\alpha\rangle = E|E,\alpha\rangle$. We easily obtain the matrix elements of $\widehat{T}_D$ in the corresponding eigenspace:

$$\widehat{T}_D|E,\alpha\rangle = 2\pi\hbar \sum_\beta \langle E,\beta|\chi_{D}(\widehat{x})|E,\alpha\rangle |E,\beta\rangle . \tag{5.3}$$

We may thus reduce the problem of eigenvalues and eigenfunctions of the dwell-time operator to a set of matrix diagonalization problems in each of the eigenspaces of the Hamiltonian.

Except in Sect. 5.5 we shall assume that the Hamiltonian holds a purely continuous spectrum with degenerate (delta-normalized) scattering eigenfunctions $|\phi_{\pm k}\rangle$ corresponding to incident plane waves $|\pm k\rangle$, with energy $E = k^2\hbar^2/(2m)$, normalized as $\langle k|k'\rangle = \langle \phi_k|\phi_{k'}\rangle = \delta(k - k')$.

Following the same manipulation done for the S-operator in 1D scattering theory [27], it is convenient to define an on-the-energy-shell 2×2 dwell-time matrix T by factoring out an energy delta,

$$\langle \phi_k|\widehat{T}_D|\phi'_k\rangle = \delta(E - E')\frac{|k|\hbar^2}{m}\mathsf{T}_{kk'} , \tag{5.4}$$

where $E' = k'^2\hbar^2/(2m)$ and

$$\mathsf{T}_{kk'} = \langle \phi_k|\chi_{D}(\widehat{x})|\phi_{k'}\rangle \frac{2\pi m}{|k|\hbar}, \quad E = E' . \tag{5.5}$$

In particular, T_{kk} is the average dwell time for a finite space region defined by Büttiker in the stationary regime [6]:

$$\mathsf{T}_{kk} = \frac{1}{|j(k)|} \int_{x_1}^{x_2} \mathrm{d}x \, |\phi_k(x)|^2 , \tag{5.6}$$

where $j(k)$ is the incoming flux associated with $|\phi_k\rangle$.

An intriguing peculiarity of the quantum dwell time is that the diagonalization of T at a given energy provides in general two distinct eigenvalues $t_\pm(k)$, $k > 0$, and corresponding eigenvectors $|t_\pm(k)\rangle$, even in cases in which only a single classical time exists, such as free motion or transmission above the barrier; some explicit examples are discussed below. A consequence is a broader variance of the quantum dwell-time distribution compared to the classical one.

The quantum dwell-time distribution for a state $|\psi\rangle = |\psi(t = 0)\rangle$ is formally given by

$$\Pi_\psi(\tau) = \langle \psi|\delta(\widehat{T}_D - \tau)|\psi\rangle , \tag{5.7}$$

since the self-adjointness of the dwell-time operator implies that the spectral theorem applies and that operator moments coincide with the moments of the distribution. We shall not consider in this work other sources of fluctuations such as mixed states or ensembles of Hamiltonians. In these two cases one could compute distributions of *average* dwell times, whereas here we shall be interested in the distribution of the dwell time itself, but only for pure states and a single Hamiltonian.

On computing the distribution of dwell times we run into the difficulty, mentioned above, that the dwell-time operator is multiply degenerate. It is useful in this case to profit from the spectral theorem to compute the generating function of the dwell-time distribution, defined as

$$f_\psi(\omega) = \int_0^\infty dt\, e^{i\omega t}\, \Pi_\psi(t) = \langle\psi|e^{i\omega \widehat{T}_D}|\psi\rangle\,. \tag{5.8}$$

Normalizing the dwell-time eigenvectors so as to have the resolution of the identity

$$\widehat{1} = \sum_\alpha \int_0^\infty dk\,|t_\alpha(k)\rangle\langle t_\alpha(k)|\,, \tag{5.9}$$

f_ψ can be written as

$$f_\psi(\omega) = \int_0^\infty dk\left[e^{i\omega t_+(k)}\langle\psi|t_+(k)\rangle\langle t_+(k)|\psi\rangle + e^{i\omega t_-(k)}\langle\psi|t_-(k)\rangle\langle t_-(k)|\psi\rangle\right]\,, \tag{5.10}$$

and hence,

$$\Pi_\psi(t) = \int_{-\infty}^\infty \frac{d\omega}{2\pi} e^{-i\omega t} f_\psi(\omega)$$
$$= \int_0^\infty dk\left[\delta(t - t_+)|\langle\psi|t_+\rangle|^2 + \delta(t - t_-)|\langle\psi|t_-\rangle|^2\right]\,. \tag{5.11}$$

The average of the dwell time for a wavepacket can be given in terms of the position probability density in correspondence (unlike its second moment, as shown below) to the expression for an ensemble of classical particles. It reads [18, 30]

$$\langle\psi|\widehat{T}_D|\psi\rangle = \int_{-\infty}^\infty dt \int_{x_1}^{x_2} dx\,|\psi(x, t)|^2 = \int_0^\infty dk\,|\langle k|\psi^{\text{in}}\rangle|^2 T_{kk}\,, \tag{5.12}$$

where $\psi(x, t) = \int_0^\infty dk\,\langle\phi_k|\psi\rangle\exp(-i\hbar k^2 t/2m)\phi_k(x)$ is the time-dependent wave packet and we assume, here and in the rest of the Chapter, incident wavepackets with positive momentum components. To write Eq. (5.12), use has been made of the standard scattering relation $\langle\phi_k|\psi\rangle = \langle k|\psi^{\text{in}}\rangle$, where $\langle x|k\rangle = (2\pi)^{-1/2}\exp(ikx)$ and ψ^{in} is the freely moving asymptotic incoming state of ψ. Space–time integrals of the form (5.12) had been used to define time delays by comparing the free motion to that with a scattering center and taking the limit of infinite volume [15].

The second moment takes the form, as before for wavepackets with incident positive momentum components,

$$\langle \widehat{T}_D^2 \rangle = \int_0^\infty dk (T^2)_{kk} |\langle k | \psi^{in} \rangle|^2 = \int_0^\infty dk (|T_{kk}|^2 + |T_{k-k}|^2) |\langle k | \psi^{in} \rangle|^2$$

$$= \int_0^\infty dk \frac{4\pi^2 m^2}{\hbar^2 k^2}$$

$$\times \left[\left(\int_{x_1}^{x_2} dx \, |\phi_k(x)|^2 \right)^2 + \left| \int_{x_1}^{x_2} dx \, \phi_k^*(x) \phi_{-k}(x) \right|^2 \right] |\langle k | \psi^{in} \rangle|^2 , \quad (5.13)$$

with a term without classical counterpart.

5.3 The Free Particle Case

In order to get a better grasp of the abstract results, it is adequate to illustrate them with the explicitly solvable free particle case for a region D that extends from $x_1 = 0$ to $x_2 = L$. Indeed the humble freely moving particle turns out to be rather interesting and surprisingly complex with regard to the dwell time. The free particle Hamiltonian is doubly degenerate; we can choose the degeneracy indices to coincide with the sign of the momentum, and, on computation, with due attention to the different normalization of the energy and the momentum eigenfunctions, we find the following generalized eigenfunctions of the free dwell-time operator:

$$|t_\pm(k), D\rangle = \frac{1}{\sqrt{2}} \left[|k\rangle \pm e^{ikL} | - k \rangle \right] , \quad (5.14)$$

where

$$t_\pm(k) = \frac{mL}{k\hbar} \left(1 \pm \frac{1}{kL} \sin kL \right) \quad (5.15)$$

are the corresponding eigenvalues, in clear contrast to the classical time $t_{class} = mL/(|k|\hbar)$, see Fig. 5.1. It is important to notice that the inverse function is multivalued. Due to this multivaluedness we have kept generalized eigenfunctions with the dimensionality of $|k\rangle$, instead of normalizing them to the delta function in dwell times. With this normalization it is easy to check that the resolution of the identity has the form (5.9).

Interestingly, $t_-(k)$ tends to 0 as $k \to 0$, one more very nonclassical effect that is better understood from the coordinate representation of the eigenvectors,

$$\langle x | t_+ \rangle = \frac{e^{ikL/2}}{\pi^{1/2}} \cos[k(x - L/2)] ,$$

$$\langle x | t_- \rangle = \frac{ie^{ikL/2}}{\pi^{1/2}} \sin[k(x - L/2)] , \quad (5.16)$$

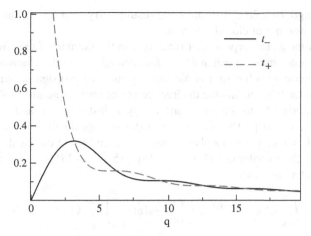

Fig. 5.1 (Color online) Eigenvalues of the dwell-time operator as a function of $q = kL$, in units $mL^2/\hbar$, for a freely moving particle in a region of length L

symmetric and antisymmetric with respect to the interval center. $\langle x|t_-\rangle$ vanishes at $L/2$, so the particle density tends to vanish in D with longer wavelengths as $k \to 0$.

The dwell-time operator for an interval $D' = [a, a + L]$ is obtained from the one above by translation:

$$\widehat{T}_{D'} = \mathrm{e}^{-ia\widehat{k}}\widehat{T}_D \mathrm{e}^{ia\widehat{k}} , \tag{5.17}$$

and, as a consequence, the eigenvalues are not modified, while the new eigenfunctions are easily computed as

$$|t_\pm(k), D'\rangle = \mathrm{e}^{-ia\widehat{k}}|t_\pm(k), D\rangle$$
$$= \frac{1}{\sqrt{2}}\left[\mathrm{e}^{-iak}|k\rangle \pm \mathrm{e}^{ik(a+L)}|-k\rangle\right] . \tag{5.18}$$

It is straightforward to compute directly the action of $\widehat{T}_D$ in the wavenumber representation,

$$\langle k|\widehat{T}_D|\psi\rangle = \frac{mL}{|k|\hbar}\left[\widetilde{\psi}(k) + \mathrm{e}^{-ikL}\frac{1}{kL}\sin(kL)\,\widetilde{\psi}(-k)\right] , \tag{5.19}$$

where $\widetilde{\psi}(k) = \langle k|\psi\rangle$. This allows us to study the functional aspects of the operator. In particular, it is easy to check, using the requirement of normalizability, that the domain of the operator is given by functions such that $\widetilde{\psi}(k)/k \to 0$ as $k \to 0$. The requirement of symmetry does not add further limitations to the domain. As to the deficiency indices, they are computed to be $(0, 0)$. These computations are carried out for the interval $D = [0, L]$, but, given the unitary equivalence of other intervals of the same length (see Eq. 5.17), the same results carry over for regions composed of an arbitrary number of closed intervals.

It should come as no surprise that functions in the domain of the (free particle) dwell-time operator must vanish at $k = 0$ fast enough, since the characteristic evolution of a generic wavefunction is dictated by the free propagator, with its $t^{-1/2}$ temporal behavior. This entails the divergence of the dwell time unless the state has no $k = 0$ component and the amplitude decay is faster than k as $k \to 0$ (for a simple analysis, see [10]). The divergence of the average dwell time for states with nonvanishing $k = 0$ components also occurs for ensembles of classical particles.

To calculate the distribution $\Pi_\psi(t)$, see Eqs. (5.10) and (5.11), we have in this case the characteristic function

$$
f_\psi(\omega) = \int_{-\infty}^{\infty} \mathrm{d}k \, \mathrm{e}^{i\omega m L/(|k|\hbar)} \cos\left[\frac{\omega m}{k^2 \hbar} \sin(kL)\right] |\widetilde{\psi}(k)|^2
$$
$$
+ i \int_{-\infty}^{\infty} \mathrm{d}k \, \mathrm{e}^{i\omega m L/(|k|\hbar)} \sin\left[\frac{\omega m}{|k|k\hbar} \sin(kL)\right] \mathrm{e}^{-ikL} \widetilde{\psi}(k)\overline{\widetilde{\psi}(-k)} .
$$

$$(5.20)$$

The distribution $\Pi_\psi(t)$ has support only on the positive semi-axis, as can be seen from the explicit expression for the eigenvalues. This, in turn, is a nontrivial check of the correctness of the definition.

A further initially surprising result of this analysis is that for highly monochromatic wavepackets (i.e., highly concentrated in one point in the momentum representation), the probability density for dwell times is generically bimodal because of the two eigenvalues $t_\pm(k)$ expounded in Eq. (5.15). An experimental verification of the eigenvalues and of the quantum nature of the dwell time, could be realized with the aid of a matter wave mirror located at a point $X > L$, reflecting an incident plane wave $|k\rangle$. The resulting standing wave would take the form, up to a global phase factor,

$$
|\psi_x\rangle = |k\rangle - \mathrm{e}^{-2ikX}| - k\rangle .
$$

$$(5.21)$$

We may now compute the average dwell time between 0 and L,

$$
\langle \psi_x|\mathsf{T}|\psi_x\rangle = \mathsf{T}_{kk} + \mathsf{T}_{-k-k} - 2\mathrm{Re}(\mathrm{e}^{-2ikX} \mathsf{T}_{k-k})
$$
$$
= 2\frac{Lm}{|k|\hbar}\left\{1 - \cos[k(L + 2X)]\frac{\sin(kL)}{kL}\right\} ,
$$

$$(5.22)$$

which oscillates between the maximum and minimum values $2t_\pm$, see (5.15); they occur at specific locations of the mirror, namely

$$
X_- = -\frac{L}{2} + \frac{\pi n}{k} ,
$$

$$(5.23)$$

$$
X_+ = -\frac{L}{2} + \frac{\pi n}{k} + \frac{\pi}{2k}
$$

$$(5.24)$$

(n integer such that $X > L$), for which the standing wave $|\psi_x\rangle$ becomes proportional to $|t_\pm\rangle$. The proportionality factor $2^{1/2}$ accounts for the fact that the extrema correspond to *twice* the eigenvalues. This is easy to interpret physically with reference to a classical particle which, under similar circumstances, would traverse the region twice, first rightward and then leftward. For any X between the privileged values $X_\pm$ given above, $|\psi_x\rangle$ is a linear superposition of the dwell-time eigenvectors and thus the resulting average dwell time lies in a continuum between twice the eigenvalues. In a proposed experiment a highly monochromatic continuous beam would be sent toward the mirror and the particle density could be measured between 0 and L by fluorescence or other means. The oscillations of the signal as a function of X would be in sharp contrast to the classical case, for which the beam density and dwell time would remain unaffected by a change of the mirror's position.

5.3.1 A Comparison with Classicality

We have already noticed some of the peculiarities of the quantum dwell time compared to the classical dwell time. Here we elaborate this comparison further. Consider a classical ensemble of free particles, described by the initial probability density on phase space, $F(x, p)$. The dwell-time distribution for this case is given by

$$\Pi_{\text{class}}(t) = \int dx\, dp\, \delta\left(t - \frac{mL}{|p|}\right) F(x, p)$$
$$= \frac{mL}{t^2} \int dx\, \left[F\left(x, \frac{mL}{t}\right) + F\left(x, \frac{mL}{-t}\right) \right], \tag{5.25}$$

that is to say, the marginal momentum distribution evaluated at mL/t and $-mL/t$ and multiplied by the normalization factor mL/t^2. On the other hand, distribution (5.11) obtained from Eq. (5.20) includes effects of interferences between positive and negative momentum components in two different ways. In the first place, there is the obvious interference of the last line of (5.20); but in addition to this, the argument of the cosine also reveals these effects. In order to see better this point, let us examine a different operator,

$$\widehat{t_D} := \widehat{\lambda}_+ \widehat{T_D} \widehat{\lambda}_+ + \widehat{\lambda}_- \widehat{T_D} \widehat{\lambda}_- = mL/|\widehat{p}|, \tag{5.26}$$

where $\widehat{\lambda}_\pm$ are the projectors onto the positive/negative momentum subspaces. Notice that $\widehat{\lambda}_\pm$ do not commute with $\widehat{T_D}$. The last form in Eq. (5.26), specific of free motion, is particularly transparent and reproduces the one of the classical dwell time. The eigenfunctions are $|\pm k\rangle$, $k > 0$, and the corresponding eigenvalues are twofold degenerate and equal to the classical time, $mL/\hbar|k|$. The distribution of dwell times for this operator, and for positive-momentum states, is given by

$$\pi_\psi(\tau) = \frac{mL}{\hbar\tau^2} \left| \widetilde{\psi}\left(\frac{mL}{\hbar\tau}\right) \right|^2 , \tag{5.27}$$

which coincides with the classical distribution for initial wave functions whose support is limited to positive momenta. In this manner we see that distribution (5.11) incorporates interferences even if only positive momenta are present. To be even more explicit, observe that, for a state ψ with only positive momentum components, the second moment of the dwell-time distribution for operator (5.1) is given by

$$\langle \widehat{T}_D^2 \rangle = \int_0^\infty dk (T^2)_{kk} |\widetilde{\psi}(k)|^2 = \int_0^\infty dk (|T_{kk}|^2 + |T_{k-k}|^2) |\widetilde{\psi}(k)|^2$$

$$= \int_0^\infty dk \frac{m^2 L^2}{k^2 \hbar^2} \left[1 + \frac{1}{k^2 L^2} \sin^2(kL) \right] |\widetilde{\psi}(k)|^2 , \tag{5.28}$$

which contrasts with

$$\langle \widehat{T}_D^2 \rangle = \int_0^\infty dk \frac{m^2 L^2}{k^2 \hbar^2} |\widetilde{\psi}(k)|^2 , \tag{5.29}$$

thus indicating that indeed the second term is due to quantum interference.

As said above, there is another rather striking way of seeing this point by considering a highly monochromatic wavepacket. The classical distribution would have a very sharply defined peak around mL/p_0, where p_0 is the central momentum of the wavefunction, $p_0 = k_0\hbar$; on the other hand, the quantum distribution would have two sharp peaks centered on $t_+(k_0)$ and $t_-(k_0)$. The distance between peaks goes to 0 as p_0 increases, as was to be expected in the classical limit.

The on-the-energy-shell version of $\widehat{t}_D$, t, is also worth examining. By factoring out an energy delta function as in Eq. (5.4) we get for a plane wave $|k\rangle$ the average $t_{kk} = mL/(\hbar k)$, which is equal to T_{kk}, but the second moment differs, $(t^2)_{kk} = (t_{kk})^2 = (T_{kk})^2 \le (T^2)_{kk}$; in other words, the variance on the energy shell is 0 since only one eigenvalue is possible for t. Contrast this with the extra term in Eq. (5.28), which again emphasizes the nonclassicality of the dwell-time operator $\widehat{T}_D$ and its quantum fluctuation.

5.4 The Scattering Case

Let us now study the dwell time in a potential barrier or well without bound states, that is, the dwell time in an interval which coincides with the finite support of the potential $V(x)$, which presents no bound states. For simplicity, we shall assume as before that this interval starts at point $x = 0$ and has length L. We shall use the complete basis of incoming scattering stationary states, $|k^+\rangle$, with the following position representation, for $k > 0$,

$$\langle x|k^+\rangle = \frac{1}{\sqrt{2\pi}} \begin{cases} e^{ikx} + R^l(k)e^{-ikx} & \text{for } x < 0 \\ T^l(k)e^{ikx} . & \text{for } x > L \end{cases}, \tag{5.30}$$

whereas, for $k < 0$, we have

$$\langle x|k^+\rangle = \frac{1}{\sqrt{2\pi}} \begin{cases} T^r(-k)e^{ikx} & \text{for } x < 0 \\ e^{ikx} + R^r(-k)e^{-ikx} & \text{for } x > L \end{cases} \tag{5.31}$$

(the superscripts l, r in T^l, R^l and T^r, R^r stand for left and right incidence, respectively). We have omitted any explicit expression in the interval $[0, L]$ because of the potential dependence. The eigenvalues and eigenstates of the dwell-time operator can be formally computed using $\widehat{H}|k^+\rangle = (k^2\hbar^2/2m)|k^+\rangle$. To that end define

$$\xi(k) = \frac{\langle k^+ |\chi_D(\widehat{x})| k^+\rangle}{2\sigma(k)}, \tag{5.32}$$

where

$$\sigma(k) = |\langle -k^+ |\chi_D(\widehat{x})| k^+\rangle| \tag{5.33}$$

and

$$e^{i\varphi(k)} = \frac{\langle -k^+ |\chi_D(\widehat{x})| k^+\rangle}{\sigma(k)} . \tag{5.34}$$

Additionally, for the sake of compactness in later formulae, let

$$\mu(k) = \frac{1}{2}[\xi(-k) - \xi(k)] . \tag{5.35}$$

We then have that the eigenvalues of the dwell-time operator are given by

$$t_\pm(k) = \frac{2\pi m\sigma(k)}{|k|\hbar} \left[\frac{\xi(k) + \xi(-k)}{2} \pm \sqrt{1 + \mu^2(k)}\right], \tag{5.36}$$

while the eigenfunctions are

$$|t_\pm(k)\rangle = N_\pm \left\{|k^+\rangle + e^{i\varphi(k)}\left[\mu(k) \pm \sqrt{1 + \mu^2(k)}\right]|-k^+\rangle\right\} \tag{5.37}$$

and the normalization is given by

$$N_\pm = \frac{1}{\sqrt{2}}\left[1 + \mu^2 \pm \mu\sqrt{1 + \mu^2(k)}\right]^{-1/2} . \tag{5.38}$$

In fact, we can relate the quantities $\xi(k)$, $\sigma(k)$, and $\varphi(k)$ to scattering data whenever the support of the scattering potential is completely included in the region D. Let $\bar{\chi}_D(x)$ be the complementary function to $\chi_D(x)$. We can compute $\langle k'^+ | \bar{\chi}(\hat{x}) | k^+ \rangle$ explicitly, and using unitarity and the conditions this imposes on the scattering amplitudes, we are led to the following expressions (where we omit the arguments of the scattering amplitudes, since all are evaluated at k, and denote derivative with respect to k as ∂_k):

$$\langle k^+ | \chi_D(\hat{x}) | k^+ \rangle = \frac{L}{2\pi} |T^l|^2 + \frac{i}{2\pi} \left[R^l \partial_k \bar{R}^l + T^l \partial_k \bar{T}^l \right] + \frac{i}{4\pi k} \left[\bar{R}^l - R^l \right] ,$$

$$\langle -k^+ | \chi_D(\hat{x}) | -k^+ \rangle = \frac{L}{2\pi} \left[1 + |R^r|^2 \right] + \frac{i}{2\pi} \left[R^r \partial_k \bar{R}^r + T^r \partial_k \bar{T}^r \right]$$
$$+ \frac{i}{4\pi k} \left[e^{-2ikL} \bar{R}^r - e^{2ikL} R^r \right] ,$$

$$\langle -k^+ | \chi_D(\hat{x}) | k^+ \rangle = \frac{L}{2\pi} T^l \bar{R}^r + \frac{i}{2\pi} \left[R^l \partial_k \bar{T}^r + T^l \partial_k \bar{R}^r \right]$$
$$+ \frac{i}{4\pi k} \left[\bar{T}^r - e^{2ikL} T^l \right] . \tag{5.39}$$

Notice that the second term of each of the previous expressions can be identified with the corresponding on-shell matrix elements of $-(im/k\hbar)S^\dagger \partial_k S$. That is, with Smith's delay time.

5.4.1 The Square Barrier

For the specific case of a square barrier in which $V(x) = V_0[\theta(x - L) - \theta(x)]$, with $\theta(x)$ being Heaviside's unit step function, we can put to good use the symmetries of the Hamiltonian, namely $x \to L - x$ and $k \to -k$. These are realized on the scattering states as

$$\langle L - x | k^+ \rangle = e^{ikL} \langle x | -k^+ \rangle , \tag{5.40}$$

whence $\xi(k) = \xi(-k)$ and $\mu(k) = 0$. Furthermore,

$$\langle k^+ | \chi_D(\hat{x}) | k^+ \rangle = \frac{L |T(k)|^2}{2\pi \kappa^2} \left\{ k^2 - \frac{mV_0}{\hbar^2} \left[1 + \frac{1}{2\kappa L} \sin(2\kappa L) \right] \right\} , \tag{5.41}$$

where $\kappa = \sqrt{k^2 - 2mV_0/\hbar^2}$, real and positive for positive above-the-barrier momenta and with positive imaginary part for tunneling momenta. After some cumbersome algebra, one readily obtains

$$t_\pm(k) = \frac{|T(k)|^2 mL}{2\hbar |k| \kappa^2} \left[k^2 + \kappa^2 \pm (\kappa^2 - k^2) \cos \kappa L \right] \left[1 \pm \frac{1}{kL} \sin kL \right] , \tag{5.42}$$

which is to be compared with (5.15) namely the free particle result gets modulated by the transmission amplitude and a potential-dependent oscillatory factor. For high energies the modulating term tends to 1, and we recover the free particle case, as one should. For small momenta, both eigenvalues tend linearly to 0 with the momentum, unlike the free case where one of them diverges. It is also relevant that the dwell-time eigenvalues are bounded above, and therefore the average value is also bounded! In order to have a better grasp of this result, it is convenient to express the eigenvalues in dimensionless terms ($k = q/L$, $V_0 = \hbar^2 Q^2/(2mL^2)$):

$$t_\pm(k) = \frac{2mL^2}{\hbar} \frac{q \pm \frac{q}{\sqrt{q^2 - Q^2}} \sin\sqrt{q^2 - Q^2}}{2q^2 - Q^2 \pm Q^2 \cos\sqrt{q^2 - Q^2}}, \tag{5.43}$$

see Fig. 5.2.

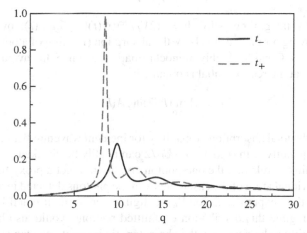

Fig. 5.2 (Color online) Dwell-time eigenvalues for the square barrier, in units of $mL^2/\hbar$, for $Q = 8$

5.4.2 Average Dwell Time from Fluorescence Measurements

In this section we shall model the measurement of the average dwell time $\tau_D = \langle \widehat{T}_D \rangle$ of an ultra-cold atom at a square barrier created by a laser shining perpendicular to its motion with homogeneous intensity between 0 and L.

Consider a two-level system coupled in a spatial region to an off-resonance laser with large detuning, $\Delta \gg \gamma, \Omega$, where Δ is defined as the laser frequency minus the frequency of the atomic transition, γ is the decay constant (inverse lifetime or Einstein's coefficient), and Ω is the Rabi frequency.

The amplitude for the atomic ground state up to the first photon detection is governed then by the following effective potential, see [25, 36] or Chap. 4:

$$V(x) = V_R - iV_I = \frac{\hbar\Omega^2}{4\Delta} - i\frac{\hbar\gamma\Omega^2}{8\Delta^2} , \tag{5.44}$$

so that the average detection delay (lifetime of the ground state if the atom at rest is put in the laser-illuminated region) is, see, e.g., [34], $4\Delta^2/\Omega^2\gamma$. Whereas γ is fixed for the atomic transition, Ω and Δ may be controlled experimentally, and the ratio Ω^2/Δ can always be chosen so that the real part of V remains constant. This still leaves some freedom to fix their exact values which we may use to set the imaginary part. If we do so making sure that at most one fluorescence photon is emitted per atom, i.e., $\tau_D \gg 4\Delta^2/\Omega^2\gamma$, the fluorescence signal produced by an atomic ensemble will be proportional to the absorption probability A. This signal provides, after calibration to take into account the detector solid angle and efficiency, an approximation for the derivative (5.46) and therefore the average dwell time for the potential (5.44) is

$$\tau_D \approx \hbar A/(2V_I) . \tag{5.45}$$

This follows by integrating $-dN/dt = (2V_I/\hbar)\langle\psi(t)|\chi_D(\widehat{x})|\psi(t)\rangle$ over time, N being the surviving norm, and $A = 1 - N$ the absorption (fraction of atoms detected). In the limit $V_I \to 0$ and for highly monochromatic incidence, the average (stationary) dwell time at the real potential is obtained:

$$\mathsf{T}_{kk} = \lim_{V_I \to 0} (\hbar/2)\partial_{V_I} A(k) , \tag{5.46}$$

where $A(k)$ is the total absorption probability for incident wavenumber k. The equivalence of this quantity with $(t_+(k) + t_-(k))/2$ can readily be checked.

One may think of relaxing the one-photon condition to get a proxy for the dwell time of an individual atom of the ensemble from the photons detected in an idealized one-atom-at-a-time experiment. For V_R negligible versus E, it could be expected that for some regime this distribution of emitted photons would also be bimodal. For the bimodality to be observed, the characteristic interval between modes ($\hbar/E$, where E is the particle's energy) should be greater than the characteristic interval between successive emission of fluorescence photons, but these conditions and $\Delta > \gamma$ are not compatible, and similar difficulties are found for on-resonance excitation. A pending task is the application of (deconvolution or operator normalization) techniques which have been successfully applied to the arrival time, at least in theory.

Figure 5.3 shows the exact dwell time and approximations for several values of V_I calculated for a transition of Cs atoms (the details are in the figure caption). A larger V_I implies larger errors but also a stronger signal. In practice the minimal signal requirements will determine the accuracy with which the dwell time can be measured. Figure 5.4 shows the relative error of the dwell-time maxima versus the corresponding absorption probability. In these figures the beam is monochromatic. We can check the reality of the quantum prediction at hand differing from the classical one, namely that for all ingoing waves the quantum mechanical dwell time is bounded, unlike the classical one.

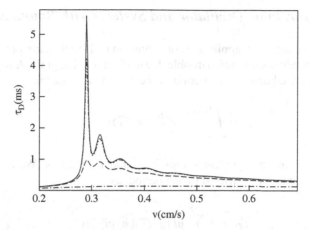

Fig. 5.3 Exact average dwell time (*solid line*) for Cs atoms crossing a square barrier of width 2 μm and height $8.2674 \times 10^3 \, s^{-1} \hbar$ (in velocity units, 0.28 cm/s) versus incident velocity. Approximations are calculated with Eq. (5.45) for $V_I = 3.307 \, s^{-1} \hbar = V_1$ (indistinguishable from the exact result, $\Delta = 2500\gamma$, $\Omega = 1.57\gamma$), 10 V_1 (*double dotted–dashed line*, $\Delta = 250\gamma$, $\Omega = 0.5\gamma$), $10^2 V_1$ (*dashed line*, $\Delta = 25\gamma$, $\Omega = 0.16\gamma$), and $10^3 V_1$ (*dotted–dashed line*, $\Delta = 2.5\gamma$, $\Omega = 0.05\gamma$). The transition is at 852 nm with $\gamma = 33.3 \times 10^6 \, s^{-1}$; Δ and Ω are obtained from Eq. (5.44)

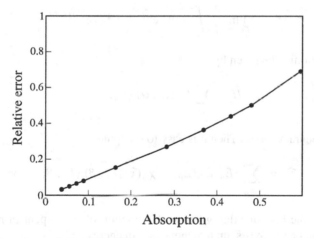

Fig. 5.4 Relative errors, calculated from the maxima of exact and approximate results, $|\tau_D(\text{exact}) - \tau_D(\text{approx})|/\tau_D(\text{exact})$, versus the absorption (i.e., detection) probability used to calculate $\tau_D(\text{approx})$ at the maximum. Same system as for the previous figure

5.5 Some Extensions

In this section we present miscellaneous extensions of the previous formalism and results: for bound states, for dwelling into states rather than in a spatial region, and for multiparticle systems.

5.5.1 The Harmonic Oscillator and Systems with Bound States

If we were to try the direct application of definition (5.1) to the case of the harmonic oscillator we would soon run into trouble due to divergent integrals. A natural option is to restrict the total time to one period of the system, as follows:

$$\widehat{T}_D = \int_{-T/2}^{T/2} d\tau \, e^{i\widehat{H}\tau/\hbar} \chi_D(\widehat{x}) e^{-i\widehat{H}\tau/\hbar} \,, \tag{5.47}$$

where the Hamiltonian is $\widehat{p}^2/2m + m\omega^2\widehat{x}^2/2$ and $T = 2\pi/\omega$. We can thus write the dwell-time operator, restricted to one period, as

$$\widehat{T}_D = T \sum_{n=0}^{\infty} \langle n \,|\chi_D(\widehat{x})|\, n \rangle \,|n\rangle \langle n| \,. \tag{5.48}$$

The eigenvalues, $T \langle n \,|\chi_D(\widehat{x})|\, n \rangle$, can be understood as the period times the proportion of that period spent by the particle in the stationary state in the region D. This suggests an alternative useful quantity for systems whose Hamiltonian has a purely point spectrum. Define

$$\widehat{\tau}_D = \lim_{T \to \infty} \frac{1}{T} \int_{-T/2}^{T/2} d\tau \, e^{i\widehat{H}\tau/\hbar} \chi_D(\widehat{x}) e^{-i\widehat{H}\tau/\hbar} \,. \tag{5.49}$$

Let the Hamiltonian be given by

$$\widehat{H} = \sum_{n,\alpha} E_n \,|E_n, \alpha\rangle \langle E_n, \alpha| \,, \tag{5.50}$$

with α a degeneracy index. Then it is easy to compute

$$\widehat{\tau}_D = \sum_{n,\alpha,\beta} |E_n, \alpha\rangle \langle E_n, \alpha| \chi_D(\widehat{x}) |E_n, \beta\rangle \langle E_n, \beta| \,, \tag{5.51}$$

that is to say, the operator that gives us the fraction of time spent in region D is diagonal in the energy basis, up to degeneracy in energy.

5.5.2 Times of Residence

The construction above suggests an extension to the time of residence, which is an analog of the dwell time valid for systems in which the concept of a region in space is not applicable. The question we now purport to answer is the following one: how much time has a system spent in a state $|\psi\rangle$? Or, alternatively, from time 0 to T, what is the proportion of time the system has spent in that state? Let then the projector associated with the state $|\psi\rangle$ be denoted by P and the unitary evolution operator by $U(t)$. We define the time-of-residence operator as

$$\widehat{\tau}_\psi(0; T) = \int_0^T dt\, U^\dagger(t) P U(t) . \tag{5.52}$$

For example [45], consider a two-level system with Hamiltonian

$$\widehat{H} = \frac{\hbar}{2} \begin{pmatrix} 0 & \Omega \\ \Omega & 0 \end{pmatrix} \tag{5.53}$$

and the state $|\psi\rangle = \begin{pmatrix} 1 \\ 0 \end{pmatrix}$. The time of residence gets written as

$$\widehat{\tau}_\psi(0; T) = \begin{pmatrix} \frac{T}{2} + \frac{1}{2\Omega} \sin(\Omega T) & \frac{-i}{2\Omega}(1 - \cos(\Omega T)) \\ \frac{i}{2\Omega}(1 - \cos(\Omega T)) & \frac{T}{2} - \frac{1}{2\Omega} \sin(\Omega T) \end{pmatrix} . \tag{5.54}$$

This entails that the measurement of this quantity would inevitably lead to one of the following two (eigen)values:

$$\tau_\pm = \frac{T}{2} \pm \frac{\sin(\Omega T/2)}{\Omega} . \tag{5.55}$$

The fact that only two values can be obtained in each realization of the experiment, and not a continuous distribution of time of presence in the state $|\psi\rangle = \begin{pmatrix} 1 \\ 0 \end{pmatrix}$, has been attributed to a predictive character of the measurement involved, as opposed to an observation that extends over time through the coupling with a weakly interacting clocking system [45].

As a matter of fact, the average time spent by a particle prepared at instant 0 in a generic state ranges from τ_- to τ_+ for the eigenstate of P. Represent a generic pure or mixed state ρ by a vector in or on the Bloch sphere, $\rho = (1 + \mathbf{r} \cdot \boldsymbol{\sigma})/2$; then the average time of residence in state $\begin{pmatrix} 1 \\ 0 \end{pmatrix}$ over the time interval $[0, T]$ is $T/2 + [(1 - \cos(\Omega T))r_y + \sin(\Omega T) r_x]/2\Omega$, which, as stated, ranges over $[\tau_-, \tau_+]$.

It should be noticed that the operator of time of residence from instant 0 to instant T is generically not stationary; for the example at hand we have

$$[\widehat{H}, \widehat{\tau}_\psi(0; T)] = \frac{i\hbar}{2} \begin{pmatrix} 1 - \cos(\Omega T) & i \sin(\Omega T) \\ -i \sin(\Omega T) & -1 + \cos(\Omega T) \end{pmatrix} . \tag{5.56}$$

Notice that whenever $T = 2n\pi/\Omega$, with n a natural number, the corresponding time of residence does indeed commute with the Hamiltonian; this is in keeping with the expressions of Sect. 5.5.1.

5.5.3 Multiparticle Systems

Definition (5.1) admits a straightforward extension to systems of many particles, using the formalism of second quantization, in which we perform the substitution

$$\chi_D(\widehat{x}) \rightarrow \widehat{n}_D = \int_D dx\, \widehat{a}_x^\dagger \widehat{a}_x\,, \qquad (5.57)$$

where $\widehat{a}_x$ is the annihilation operator at point x. For free particles, that is to say, for a Hamiltonian of the form

$$\widehat{H} = \int_{-\infty}^{\infty} dk\, \frac{\hbar^2 k^2}{2m} \widehat{n}_k = \int_{-\infty}^{\infty} dk\, \frac{\hbar^2 k^2}{2m} \widehat{a}_k^\dagger \widehat{a}_k\,, \qquad (5.58)$$

with $\widehat{a}_k$ the operator that annihilates a particle with wavenumber k, we can compute the dwell-time operator as

$$\widehat{T}_D = \int_{-\infty}^{\infty} dk\, \frac{mL}{|k|\hbar}\left[\widehat{a}_k^\dagger \widehat{a}_k + \frac{1}{kL} e^{ikL} \sin(kL)\widehat{a}_{-k}^\dagger \widehat{a}_k\right]. \qquad (5.59)$$

It is also feasible to write a dwell-time density operator, which for the free case reads

$$\widehat{\Pi}_a(t) = \int_{-\infty}^{\infty} dk\, \left\{\frac{1}{2}\left[\delta(t - t_+(k)) + \delta(t - t_-(k))\right]a_k^\dagger a_k \right.$$
$$\left. + \frac{e^{-ikL}}{2}\left[\delta(t - t_+(k)) - \delta(t - t_-(k))\right]a_k^\dagger a_{-k}\right\}. \qquad (5.60)$$

It is easy to see that $\langle\psi|\widehat{\Pi}_a(t)|\psi\rangle$ reduces to the dwell-time density for the one-particle marginal wavefunction. The real multiparticle aspect of this construction will only be accessible through dwell-time–dwell-time correlation functions, for which a suitable operational interpretation needs to be built.

In addition, we note that for multiparticle systems, the alternative question might be posed as what is the time during which a given number of particles n can be found in the region D. This naturally leads to the introduction of the dwell-time operator for n particles:

$$\widehat{T}_D(n) = \int_{-\infty}^{\infty} dt\, \widehat{U}^\dagger(t)\delta_{\widehat{n}_D,n}\widehat{U}(t)$$
$$= \frac{1}{2\pi}\int_{-\infty}^{\infty} dt \int_0^{2\pi} d\theta\, \widehat{U}^\dagger(t)e^{i\theta(\widehat{n}_D - n)}\widehat{U}(t)\,, \qquad (5.61)$$

where $\widehat{n}_D$ is the density operator restricted to the region D, defined in Eq. (5.57). Letting $\widehat{\rho}$ be the density matrix describing the state of the system, it is convenient to introduce the characteristic function

$$F(\theta; t) = \mathrm{Tr}[\widehat{\rho}\,\widehat{U}^{\dagger}(t)\mathrm{e}^{\mathrm{i}\theta\widehat{n}_D}\,\widehat{U}(t)] \, , \tag{5.62}$$

whose Fourier transform is the atom number distribution in the region D [21]:

$$P_D(n, t) = \frac{1}{2\pi} \int_0^{2\pi} \mathrm{e}^{-\mathrm{i}n\theta}\, F(\theta; t)\mathrm{d}\theta \, , \tag{5.63}$$

with $n \in \mathbb{N}$. The meaning of $P_D(n, t)$ is precisely the probability for n particles to be found in the spatial domain D at time t.

Knowledge of the atom number distribution can be used to compute the average dwell time for different number of particles, namely

$$\langle \widehat{T}_D(n) \rangle = \int_{-\infty}^{\infty} \mathrm{d}t\, \mathrm{Tr}[\widehat{\rho}(t)\delta_{\widehat{n}_D, n}] = \int_{-\infty}^{\infty} \mathrm{d}t\, P_D(n, t) \, . \tag{5.64}$$

5.6 Relation to Flux–Flux Correlation Functions

This section follows [35] and examines a link between the dwell time and flux–flux correlation functions (ffcf) that have been considered mostly in chemical physics to define reaction rates for microcanonical or canonical ensembles [24]. The motivation for this exercise is to relate the dwell-time distribution, and not just the average value, to other observables.

5.6.1 Stationary Flux–Flux Correlation Function

Pollak and Miller [38] have shown a connection between the average stationary dwell time and the first moment of an ffcf. They define a quantum microcanonical ffcf $C_{PM}(\tau, k) = \mathrm{Tr}\{\mathrm{Re}\,\widehat{C}_{PM}(\tau, k)\}$ by means of the operator

$$\begin{aligned}
\widehat{C}_{PM}(\tau, k) = 2\pi\hbar[&\widehat{J}(x_2, \tau)\widehat{J}(x_1, 0) + \widehat{J}(x_1, \tau)\widehat{J}(x_2, 0) \\
&- \widehat{J}(x_1, \tau)\widehat{J}(x_1, 0) - \widehat{J}(x_2, \tau)\widehat{J}(x_2, 0)]\delta(E - \widehat{H}) \, , \tag{5.65}
\end{aligned}$$

where $\widehat{J}(x, t) = \mathrm{e}^{\mathrm{i}\widehat{H}t/\hbar}\frac{1}{2m}[\widehat{p}\delta(\widehat{x} - x) + \delta(\widehat{x} - x)\widehat{p}]\mathrm{e}^{-\mathrm{i}\widehat{H}t/\hbar}$ is the quantum mechanical flux operator in the Heisenberg picture, and $\widehat{p}$ and $\widehat{x}$ are the momentum and position operators.

This definition is motivated from classical mechanics: Eq. (5.65) counts flux correlations of particles entering D through x_1 (x_2) and leaving it through x_2 (x_1) a time τ later. Moreover, particles may be reflected and may leave the region D through its entrance point. This is described by the last two terms, where the minus sign compensates for the change of sign of a back-moving flux. Note that these negative terms lead to a self-correlation contribution that diverges for $\tau \to 0$.

We shall first derive the average correlation time and show its equivalence with the average dwell time. We shall only consider positive incident momenta and define $\widehat{C}_{PM}^+$ by substituting $\delta(E - \widehat{H})$ by $\delta^+(E - \widehat{H}) := \delta(E - \widehat{H})\Lambda_+$, where Λ_+ is the projector onto the subspace of eigenstates of H with positive momentum incidence. By means of the continuity equation

$$-\frac{d}{dx}\widehat{J}(x, t) = \frac{d}{dt}\widehat{\rho}(x, t) , \tag{5.66}$$

where $\widehat{\rho}(x, t) = e^{i\widehat{H}t/\hbar}\delta(\widehat{x}-x)e^{-i\widehat{H}t/\hbar}$ is the (Heisenberg) density operator, $\widehat{C}_{PM}^+(\tau, k)$ can be written as

$$\widehat{C}_{PM}^+(\tau, k) = -2\pi\hbar\left(\frac{d}{d\tau}\widehat{\chi}_D(\tau)\right)\left(\frac{d}{dt}\widehat{\chi}_D(t)\right)_{t=0}\delta^+(E - \widehat{H}) . \tag{5.67}$$

By a partial integration and using the Heisenberg equation of motion, the first moment of the Pollak–Miller correlation function is given by

$$\text{Tr}\left\{\int_0^\infty d\tau \, \tau\widehat{C}_{PM}^+(\tau, k)\right\} = \text{Tr}\left\{2\pi\hbar\int_0^\infty d\tau\widehat{\chi}_D(\tau)\frac{1}{i\hbar}[\widehat{\chi}_D(0), \widehat{H}]\delta^+(E - \widehat{H})\right\} . \tag{5.68}$$

Boundary terms of the form $\lim_{\tau\to\infty} \tau^\gamma\widehat{\chi}_D(\tau)$, $\gamma = 0, 1, 2$, are omitted here and in the following. The contribution of these terms should vanish when an integration over stationary wavefunctions is performed to account for the wavepacket dynamics, as it is done explicitly in the next section. For potential scattering the probability density decays generically as τ^{-3}, which assures a finite dwell-time average, but for free motion it decays as τ^{-1} [32], making τ_D infinite, unless the momentum wave function vanishes at $k = 0$ sufficiently fast as k tends to 0 [11], as we have discussed before.

Writing the commutator explicitly and using the cyclic property of the trace give

$$\text{Tr}\left\{\int_0^\infty d\tau \, \tau \, \widehat{C}_{PM}^+(\tau, k)\right\}$$
$$= \text{Tr}\left\{2\pi\hbar\int_0^\infty d\tau \left(-\frac{d}{d\tau}\widehat{\chi}_D(\tau)\right)\widehat{\chi}_D(0)\delta^+(E - \widehat{H})\right\} , \tag{5.69}$$

and integration over τ yields the final result:

$$\text{Tr}\left\{\int_0^\infty d\tau \, \tau \, \widehat{C}_{PM}^+(\tau, k)\right\} = 2\pi\hbar\text{Tr}\left\{\widehat{\chi}_D(0)\delta^+(E - \widehat{H})\right\} = \mathsf{T}_{kk} . \tag{5.70}$$

Expressing the trace in the basis $|\phi_k\rangle$ gives back the stationary dwell time of Eq. (5.6), i.e., the diagonal element of the on-the-energy-shell dwell-time operator, T_{kk}.

The calculation of the average in [38] is different in some respects: (a) The coordinates x_1 and x_2 are taken to minus and plus infinity, but it can be carried out for finite values modifying Eq. (8) of [38] accordingly; (b) Formally there are no explicit boundary terms at infinity but a regularization is required in Eq. (16) of [38], which is justified for wave packets; (c) $\delta(E - \widehat{H})$ is used instead of $\delta^+(E - \widehat{H})$. This simply provides an additional contribution for negative momenta parallel to the one obtained here for positive momenta; (d) In our derivation the average correlation time is found to be real directly, even though $\widehat{C}^+_{PM}(\tau, k)$ is not self-adjoint, whereas in [38] the real part is taken. (The discussion of the imaginary time average in [38] is based on a modified version of Eq. (5.65).)

Next, we will show that the second moment of the Pollak–Miller ffcf equals the second moment of T. This was not observed in Ref. [38]. Proceeding in a similar way as above, we start with

$$\mathcal{I} = \text{Tr}\left\{ \int_0^\infty d\tau \, \tau^2 \widehat{C}^+_{PM}(\tau, k) \right\} . \tag{5.71}$$

Integrating by parts twice, neglecting the term at infinity, using Heisenberg's equation of motion and the fact that ϕ_k is an eigenstate of $\widehat{H}$, the real part is

$$\frac{\mathcal{I} + \mathcal{I}^*}{2} = \frac{2\pi m}{\hbar k} \int_0^\infty d\tau \, \langle \phi_k | [\widehat{\chi}_D(\tau) \widehat{\chi}_D(0) + \widehat{\chi}_D(0) \widehat{\chi}_D(\tau)] | \phi_k \rangle . \tag{5.72}$$

Introducing resolutions of the identity,

$$\text{Re}\mathcal{I} = \left\{ \frac{2\pi m}{\hbar k} \int_0^\infty d\tau \int_{-\infty}^\infty dk' \int_{x_1}^{x_2} dx \int_{x_1}^{x_2} dx' e^{i(E-E')\tau/\hbar} \phi_k^*(x) \phi_{k'}(x) \phi_{k'}^*(x') \phi_k(x') \right\}$$
$$+ c.c. , \tag{5.73}$$

where $c.c$ means complex conjugate. Making the changes $\tau \to -\tau$ and $x, x' \to x', x$ in the $c.c$-term, it takes the same form as the first one, but with the time integral from $-\infty$ to 0. Adding the two terms, the τ-integral provides an energy delta function that can be separated into two deltas which select $k' = \pm k$ to arrive at

$$\text{Tr}\left\{ \text{Re} \int_0^\infty d\tau \, \tau^2 \widehat{C}^+_{PM}(\tau, k) \right\} = \frac{4\pi^2 m^2}{\hbar^2 k^2} \left[\left(\int_{x_1}^{x_2} dx \, |\phi_k(x)|^2 \right)^2 \right.$$
$$\left. + \left| \int_{x_1}^{x_2} dx \, \phi_k^*(x) \phi_{-k}(x) \right|^2 \right] = (T^2)_{kk} . \tag{5.74}$$

In other words, the relation between dwell times and flux–flux correlation functions goes beyond average values and $C^+_{PM}(\tau, k)$ includes quantum features of the dwell time: note that the first summand in Eq. (5.74) is nothing but $(T_{kk})^2$, whereas the second summand is positive, which allows for a nonzero on-the-energy-shell dwell-time variance $(T^2)_{kk} - (T_{kk})^2$. We insist that the stationary state considered has positive momentum, $\phi_k(x)$, $k > 0$, but this second term implies the degenerate partner $\phi_{-k}(x)$ as well, and is generically nonzero.

We shall see in the next section with a more general approach that these connections do not hold for higher moments.

5.6.2 Time-Dependent Flux–Flux Correlation Function

A time-dependent version of the above flux–flux correlation function can be defined in terms of the operator

$$\widehat{C}(\tau) = \int_{-\infty}^{\infty} dt \left[\widehat{J}(x_2, t+\tau)\widehat{J}(x_1, t) + \widehat{J}(x_1, t+\tau)\widehat{J}(x_2, t) \right.$$
$$\left. - \widehat{J}(x_1, t+\tau)\widehat{J}(x_1, t) - \widehat{J}(x_2, t+\tau)\widehat{J}(x_2, t) \right], \tag{5.75}$$

which leads to the flux–flux correlation function

$$C(\tau) = \langle \text{Re}\,\widehat{C}(\tau) \rangle_\psi , \tag{5.76}$$

where the real part is taken to symmetrize the nonself-adjoint operator $\widehat{C}(\tau)$ as before.

As in the stationary case, Eq. (5.76) counts flux correlations of particles entering D through x_1 or x_2 at a time t and leaving it either through x_1 or x_2 a time τ later. Moreover we integrate over the entrance time t. It can be shown that the first moment of the classical version of Eq. (5.75), where $\widehat{J}$ is replaced by the classical dynamical variable of the flux, gives the average of the classical dwell time.

As in Eq. (5.67), we may rewrite $\widehat{C}(\tau)$ as

$$\widehat{C}(\tau) = -\int_{-\infty}^{\infty} dt \frac{d}{d\tau}\widehat{\chi}_D(\widehat{x}, t+\tau)\frac{d}{dt}\widehat{\chi}_D(\widehat{x}, t) . \tag{5.77}$$

From here we note that the ffcf $C(\tau)$ is not normalized:

$$\int_0^{\infty} d\tau \, \widehat{C}(\tau) = \int_{-\infty}^{\infty} dt \, \widehat{\chi}_D(t)\frac{d}{dt}\widehat{\chi}_D(t) = 0 , \tag{5.78}$$

as a result of the self-correlation.

Next we derive the average of the time-dependent correlation function. With a partial integration one finds

$$\int_0^{\infty} d\tau \, \tau\widehat{C}(\tau) = \int_{-\infty}^{\infty} dt \int_0^{\infty} d\tau \, \widehat{\chi}_D(\widehat{x}, t+\tau)\frac{d}{dt}\widehat{\chi}_D(\widehat{x}, t) .$$

A second partial integration with respect to t replacing d/dt by $d/d\tau$ and integrating over τ gives

$$\int_0^\infty d\tau\, \tau \widehat{C}(\tau) = \int_{-\infty}^\infty dt\, \widehat{\chi}_D^2(\widehat{x}, t) = \int_{-\infty}^\infty dt\, \widehat{\chi}_D(\widehat{x}, t) = \widehat{T}_D \,, \qquad (5.79)$$

where $\widehat{\chi}_D^2 = \widehat{\chi}_D$ has been used. Equation (5.79) generalizes the result of Pollak and Miller to time-dependent dwell times.

A similar calculation can be performed for the second moment of $C(\tau)$. After three partial integrations with vanishing boundary contributions to get rid of the factor τ^2 one obtains

$$\int_0^\infty d\tau\, \tau^2 \widehat{C}(\tau) = 2 \int_{-\infty}^\infty dt\, \int_0^\infty d\tau\, \widehat{\chi}_D(\widehat{x}, t + \tau)\widehat{\chi}_D(\widehat{x}, t) \,.$$

Making the substitutions $t + \tau \to t$ and $\tau \to -\tau$ in the complex conjugated term, we find

$$\mathrm{Re} \int_0^\infty d\tau\, \tau^2 \widehat{C}(\tau) = \widehat{T}_D^2 \,. \qquad (5.80)$$

5.6.3 Example: Free Motion

For a stationary flux of freely moving particles with energy E_k, $k > 0$, described by $\phi_k(x) = \langle x|k\rangle = (2\pi)^{-1/2}e^{ikx}$, the first three moments of the ideal dwell-time distribution on the energy shell are given by

$$\mathsf{T}_{kk} = \frac{mL}{\hbar k} \,, \qquad (5.81)$$

$$(\mathsf{T}^2)_{kk} = \frac{m^2 L^2}{\hbar^2 k^2} \left(1 + \frac{\sin^2(kL)}{k^2 L^2}\right) \,, \qquad (5.82)$$

$$(\mathsf{T}^3)_{kk} = \frac{m^3 L^3}{\hbar^3 k^3} \left(1 + 3\frac{\sin^2(kL)}{k^2 L^2}\right) \,. \qquad (5.83)$$

As proved above, the first two moments agree with the corresponding moments of the Pollak–Miller ffcf, but for the third moment we obtain instead

$$\mathrm{Tr}\left\{\mathrm{Re} \int_0^\infty d\tau\, \tau^3 \widehat{C}_{PM}^+(\tau, k)\right\} = \frac{m^3 L^3}{\hbar^3 k^3} \left[1 - \frac{3[1 + \cos^2(kL)]}{k^2 L^2} + \frac{3\sin(2kL)}{L^3 k^3}\right] \,.$$

$$(5.84)$$

In Fig. 5.5 the first three moments are compared. The agreement between $(\mathsf{T}^3)_{kk}$ and Eq. (5.84) is very good for large values of k, but they clearly differ for small k. Nevertheless, the agreement of the first two moments suggests a similar behavior of $\Pi(\tau)$ and $C(\tau)$.

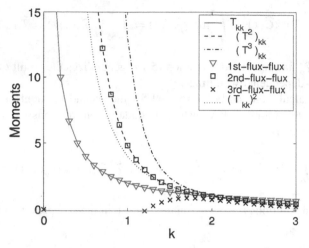

Fig. 5.5 (Color online) Comparison of the first three moments: T_{kk}, $(T^2)_{kk}$ and $(T^3)_{kk}$, with the corresponding moments of the flux–flux correlation function, for a free-motion stationary state with fixed k. $(T_{kk})^2$ is also shown. $\hbar = m = 1$ and $L = 2$

To calculate $\Pi_\psi(\tau)$ for a wavefunction $\widetilde{\psi}(k) := \langle k|\psi\rangle$ with only positive momentum components we use Eq. (5.11) and the explicit forms of the eigenvectors, Eq. (5.14) and eigenvalues $t_\pm$, Eq. (5.15):

$$\Pi_\psi(\tau) = \frac{1}{2} \sum_j \sum_{\gamma=\pm} \frac{|\widetilde{\psi}(k_j^\gamma(\tau))|^2}{|F_\gamma'(k_j^\gamma(\tau))|} , \tag{5.85}$$

where the j-sum is over the solutions $k_j^\gamma(\tau)$ of the equation $F_\gamma(k) \equiv t_\gamma(k) - \tau = 0$ and the derivative is with respect to k.

We use the following wavefunction [27]:

$$\widetilde{\psi}(k) = \mathcal{N}(1 - e^{-\alpha k^2}) e^{-(k-k_0)^2/[4(\Delta k)^2]} e^{-ikx_0} \Theta(k) , \tag{5.86}$$

where $\mathcal{N}$ is the normalization constant and $\Theta(k)$ the step function. For the free flux–flux correlation function we write

$$C(\tau) = \mathrm{Re} \int_0^\infty dk \int_0^\infty dk' \, \widetilde{\psi}^*(k)\widetilde{\psi}(k')\langle k|\widehat{C}(\tau)|k'\rangle , \tag{5.87}$$

and $C_{kk'}(\tau) = \langle k|\widehat{C}(\tau)|k'\rangle$ in the free case

$$C_{kk'}(\tau) = \frac{m}{2\pi\hbar k}\delta(k - k')\frac{d^2}{d\tau^2}\Big[2g(\hbar k\tau/m) - g(\hbar k\tau/m - L) - g(\hbar k\tau/m + L)\Big] , \tag{5.88}$$

where

$$g(x) = -2e^{imx^2/(2\hbar\tau)} \left(\frac{i\pi\,\hbar\tau}{2m}\right)^{1/2} + i\pi x \operatorname{erfi}\left(\sqrt{\frac{im}{2\hbar\tau}}\,x\right). \qquad (5.89)$$

The result is shown in Fig. 5.6. The ffcf shows a hump around the mean dwell time but it oscillates for small τ and diverges for $\tau \to 0$. As discussed above, this is due to the self-correlation contribution of wavepackets which are at x_1 or x_2 at the times t and $t + \tau$ *without* changing the direction of motion in between. A similar feature has been observed in a traversal-time distribution derived by means of a path integral approach [14].

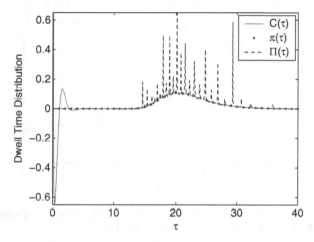

Fig. 5.6 (Color online) Comparison of dwell-time distribution $\Pi(\tau)$ (*dashed line*) and flux–flux correlation function $C(\tau)$ (*solid line*) for the freely moving wave packet (5.86). Furthermore, the alternative free-motion dwell-time distribution $\pi(\tau)$, Eq. (5.27), is plotted (*circles*). We set $\hbar = m = 1$ and $|x_0|$ large enough to avoid overlap of the initial state with the space region $D = [0, 45]$. $k_0 = 2$, $\Delta k = 0.4$, and $\alpha = 0.5$

In contrast, $\Pi_\psi(\tau)$ behaves regularly for $\tau \to 0$, but shows peaks in the region of the hump. This is because the denominator of Eq. (5.85) becomes zero if the slope of the eigenvalues $t_\pm(k)$ is zero, which occurs at every crossing of $t_+(k)$ and $t_-(k)$.

The distribution $\pi_\psi(\tau)$, see Eq. (5.27), is also computed: it agrees with $C(\tau)$ in the region near the average dwell time and it tends to 0 for $\tau \to 0$. However, it does not show the resonance peaks of $\Pi_\psi(\tau)$.

In the absence of a direct dwell-time measurement, the physical significance of $\widehat{T}_D$ and $\widehat{t}_D$ depends on their relation to other observables. The present results indicate that the second moment of the flux–flux correlation function is related to the former and not to the later, providing indirect support for the physical relevance of the dwell-time resonance peaks, but other observables could behave differently.

5.6.4 Approximations

While the previous results bring dwell-time information closer to experimental realization, the difficulty is translated onto the measurement of the ffcf, not necessarily
an easy task. A simple approximation is to substitute the expectation of the product of two flux operators by the product of their expectation values, which are the
current densities. Using the wave packet of Eq. (5.86), we have compared the times
obtained with the full expression (5.76) and with this approximation in Fig. 5.7. The
two results approach each other as $\Delta_k \to 0$, also by increasing L and/or k_0.

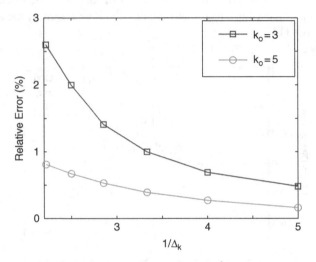

Fig. 5.7 (Color online) Comparison of the relative error of $\langle \widehat{T}_D^2 \rangle$ using the approximation $C_0(\tau)$
instead of $C(\tau)$ for free motion. $\alpha = 0.5$, $\hbar = m = 1$, and $L = 100$

The exact result can be approached systematically, still making use of ordinary
current densities, as follows: First decompose $\widehat{J}(x_i, t + \tau)\widehat{J}(x_j, t)$ by means of

$$\widehat{1} = \widehat{P} + \widehat{Q} , \tag{5.90}$$

$$\widehat{P} = |\psi\rangle\langle\psi| , \tag{5.91}$$

so that

$$\widehat{J}(x_i, t + \tau)\widehat{J}(x_j, t) = \widehat{J}(x_i, t + \tau)(|\psi\rangle\langle\psi| + \widehat{Q})\widehat{J}(x_j, t) . \tag{5.92}$$

It is useful to decompose $\widehat{Q}$ further in terms of a basis of states orthogonal to $|\psi\rangle$
and to each other, $\{|\psi_j^Q\rangle\}$,

$$\widehat{Q} = \sum_j |\psi_j^Q\rangle\langle\psi_j^Q| , \tag{5.93}$$

that could be generated by means of a Gram–Schmidt orthogonalization procedure. Now we can split Eq. (5.75):

$$\widehat{C}(\tau) = \widehat{C}_0(\tau) + \widehat{C}_1(\tau) , \qquad (5.94)$$

where $\widehat{C}_0(\tau)$ has the structure of $\widehat{C}$, but with P inserted between the two flux operators in each of the four terms. Similarly $\widehat{C}_1(\tau)$ has Q inserted and can be itself decomposed using Eq. (5.93).

We define $C(\tau) = C_0(\tau) + C_1(\tau)$ by taking the real part of $\langle \psi | \widehat{C}_0(\tau) + \widehat{C}_1(\tau) | \psi \rangle$. C_0 is the zeroth-order approximation discussed before and only involves ordinary, measurable current densities [9]. The nondiagonal terms from C_1, $\langle \psi | \widehat{J}(x_i, t) | \psi_j^Q \rangle$ $\langle \psi_j^Q | \widehat{J}(x_j, t + \tau) | \psi \rangle$, can also be related to diagonal elements of $\widehat{J}$ by means of the auxiliary states

$$\begin{aligned}
|\psi_1\rangle &= |\psi\rangle + |\psi_j^Q\rangle , \\
|\psi_2\rangle &= |\psi\rangle + i|\psi_j^Q\rangle , \\
|\psi_3\rangle &= |\psi\rangle - i|\psi_j^Q\rangle ,
\end{aligned} \qquad (5.95)$$

since

$$\begin{aligned}
\langle \psi | \widehat{J}(x, t) | \psi_j^Q \rangle &= \frac{1}{2} \langle \psi_1 | \widehat{J}(x, t) | \psi_1 \rangle - \frac{1}{4} \langle \psi_2 | \widehat{J}(x, t) | \psi_2 \rangle \\
&\quad - \frac{1}{4} \langle \psi_3 | \widehat{J}(x, t) | \psi_3 \rangle + \frac{i}{4} \langle \psi_3 | \widehat{J}(x, t) | \psi_3 \rangle \\
&\quad - \frac{i}{4} \langle \psi_2 | \widehat{J}(x, t) | \psi_2 \rangle .
\end{aligned} \qquad (5.96)$$

5.7 Final Comments

The quantum dwell-time distribution of a particle in a spatial region and its second moment present nonclassical features even for the simplest case of a freely moving particle, such as bimodality (due to two different eigenvalues for the same energy), with the strongest deviations from classical behavior occurring for de Broglie wavelengths of the order or larger than the region width. Progress in ultra-cold atom manipulation makes plausible the observation of these effects and motivates further effort to achieve an elusive direct measurement of the dwell times or to link the distribution and its moments to other observables. We have in this regard pointed out that the flux–flux correlation function provides access to the second moment. The potential impact on cold atom time–frequency metrology [42, 26] and other fields in which the dwell time plays a prominent role (such as conductivity [7] or chaos) remains an open question.

Acknowledgments We acknowledge discussions with M. Büttiker, J. A. Damborenea, and B. Navarro. This work has been supported by Ministerio de Educación y Ciencia (FIS2006-10268-C03-01), the Basque Country University (UPV-EHU, GIU07/40), EU Integrated Project QAP, and the EPSRC QIP-IRC (GR/S82176/0).

References

1. D. Alonso, R. Sala Mayato, J.G. Muga, Phys. Rev. A **67**, 032105 (2003). This paper states incorrectly that the distributions for bTD and btD coincide, see the last two equations in it and the related paragraph
2. A. Baz', Sov. J. Nucl. Phys. **4**, 182 (1967)
3. S. Boonchui, V. Sa-yakanit, Phys. Rev. **77**, 044101 (2008)
4. C. Bracher, J. Phys. B: At. Mol. Opt. Phys. **30**, 2717 (1997)
5. S. Brouard, R. Sala, J.G. Muga, Phys. Rev. A **49**, 4312 (1994)
6. M. Büttiker, Phys. Rev. B **27**, 6178 (1983)
7. M. Büttiker, H. Thomas, A. Prêtre, Phys. Lett. A **180**, 364 (1993)
8. C.A.A. de Carvalho, H.M. Nussenzveig, Phys. Rep. **364**, 83 (2002)
9. J.A. Damborenea, I.L. Egusquiza, G.C. Hegerfeldt, J.G. Muga, Phys. Rev. A **66**, 052104 (2002)
10. J.A. Damborenea, I.L. Egusquiza, J.G. Muga, J. Am. Phys. **70**, 738 (2002)
11. J.A. Damborenea, I.L. Egusquiza, J.G. Muga, B. Navarro, arXiv:quant-ph/0403081 (2004)
12. I.L. Egusquiza, J.G. Muga, A.D. Baute, in *Time in Quantum Mechanics*, vol. 1, J.G. Muga, R. Sala, I. Egusquiza (eds.), Lect. Notes Phys. **734** (Springer, Berlin, 2008), Chapter 10, pp. 305–332
13. H. Ekstein, A.J.F. Siegert, Ann. Phys. **68**, 509 (1971)
14. H.A. Fertig, Phys. Rev. Lett. **65**, 2321 (1990)
15. M.L. Goldberger, K.M. Watson, *Collision Theory* (Krieger, Huntington, 1975)
16. R. Golub, S. Felber, R. Gähler, E. Gutsmiedl, Phys. Lett. A **148**, 27 (1990)
17. Z. Huang, C.M. Wang, J. Phys. Condens. Matter **3**, 5915 (1991)
18. W. Jaworski, D.M. Wardlaw, Phys. Rev. A **37**, 2843 (1988)
19. W. Jaworski, D.M. Wardlaw, Phys. Rev. A **40**, 6210 (1989)
20. N.G. Kelkar, Phys. Rev. Lett. **99**, 210403 (2007)
21. V.V. Kocharovsky, Vl.V. Kocharovsky, M.O. Scully, Phys. Rev. Lett. **84**, 2306 (2000)
22. C.R. Leavens, W.R. McKinnon, Phys. Lett. A **194**, 12 (1994)
23. H. Lewenkopf, R.O. Vallejos, Phys. Rev. E **70**, 036214 (2004)
24. W.H. Miller, S.D. Schwartz, J.W. Tromp, J. Chem. Phys. **79**, 4889 (1983)
25. J.D. Miller, R.A. Cline, D.J. Heinzen, Phys. Rev. A **47**, R4567 (1993)
26. S.V. Mousavi, A. del Campo, I. Lizuain, J.G. Muga, Phys. Rev. A **76**, 033607 (2007)
27. J.G. Muga, in *Time in Quantum Mechanics*, vol. 1, J.G. Muga, R. Sala, I. Egusquiza (eds.), Lect. Notes Phys. **734** (Springer, Berlin, 2008), Chapter 2, pp. 305–332
28. J.G. Muga, G.R. Leavens, Phys. Rep. **338**, 353 (2000)
29. J.G. Muga, D. Wardlaw, Phys. Rev. E **51**, 5377 (1995)
30. J.G. Muga, S. Brouard, R. Sala, Phys. Lett. A **167**, 24 (1992)
31. J.G. Muga, S. Brouard, R. Sala, J. Phys. Condens. Matter **4**, L579 (1992)
32. J.G. Muga, V. Delgado, R.F. Snider, Phys. Rev. **52**, 16381 (1995)
33. J.G. Muga, R. Sala Mayato, I.L. Egusquiza (eds.), *Time in Quantum Mechanics*, vol. 1, Lect. Notes Phys. **734** (Springer, Berlin, 2008)
34. J.G. Muga, J. Echanobe, A. del Campo, I. Lizuain, J. Phys. B: At. Mol. Opt. Phys. **41**, 175501 (2008)
35. J. Muñoz, D. Seidel, J.G. Muga, Phys. Rev. A **79**, 012108 (2009)

36. B. Navarro, I.L. Egusquiza, J.G. Muga, G.C. Hegerfeldt, J. Phys. B: At. Mol. Opt. Phys. **36**, 3899 (2003)
37. J.P. Palao, J.G. Muga, S. Brouard, A. Jadczyk, Phys. Lett. A **233**, 227 (1997)
38. E. Pollak, W.H. Miller, Phys. Rev. Lett. **53**, 115 (1984)
39. A. Ruschhaupt, Phys. Lett. A **250**, 249 (1998)
40. A. Ruschhaupt, J.A. Damborenea, B. Navarro, J.G. Muga, G.C. Hegerfeldt, Europhys. Lett. **67**, 1 (2004)
41. V. Rybachenko, Sov. J. Nucl. Phys. **5**, 635 (1967)
42. D. Seidel, J.G. Muga, Eur. Phys. J. D **41**, 71 (2007)
43. F. Smith, Phys. Rev. **118**, 349 (1960)
44. D. Sokolovski, Phys. Rev. A **66**, 032107 (2002)
45. D. Sokolovski, Proc. R. Soc. Lond. A **460**, 1505 (2004)
46. D. Sokolovski, Phys. Rev. A **76**, 042125 (2007)
47. D. Sokolovski, in *Time in Quantum Mechanics*, vol. 1, J.G. Muga, R. Sala, I. Egusquiza (eds.), Lect. Notes Phys. **734** (Springer, Berlin, 2008), Chapter 7, pp. 305–332
48. D. Sokolovski, J.N.L. Connor, Phys. Rev. A **44**, 1500 (1991)
49. H.G. Winful, Phys. Rep. **436**, 1 (2006)
50. N. Yamada, Phys. Rev. Lett. **93**, 170401 (2004)

Chapter 6
The Quantum Jump Approach and Some of Its Applications

Gerhard C. Hegerfeldt

6.1 Introduction

Modern techniques allow experiments on a single-driven atom or a single system. The quantum jump approach was originally developed for the description of the temporal evolution of such a driven system and was later extended to more general situations like a moving particle coupled to a spatially confined laser beam or to spin-boson baths. In this chapter the underlying ideas are presented and illustrated by simple examples. Applications include the spectacular macroscopic light and dark periods in the fluorescence of a single atom, quantum counting processes, and arrival times.

The quantum jump approach[1] takes its name from the jump-like change of state when a measurement is performed. A simple example for such a jump-like change occurs in a two-level atom interacting with the quantized radiation field. If initially the atom is in the excited state and the radiation field in the no-photon state (vacuum), then the complete system evolves into a superposition of atomic and photon states. If at a specific time one detects a photon by absorption then, by standard theory, right thereafter the atom is in the ground state.

A more complicated example would be a single atom in a trap irradiated by a classical laser. Again, the complete system (atom plus radiation field) evolves under the joint time development into a complicated superposition of atomic and photon states, and again one could detect the fluorescence photons (outside the laser direction, laser light is not detected). Right after each photon detection the state of the atom is reset to the ground state for the two-level case and in general to

G.C. Hegerfeldt (✉)
Institut für Theoretische Physik, Universität Göttingen, Friedrich–Hund–Platz 1,
D-37077 Göttingen, Germany, hegerf@theorie.physik.uni-goettingen.de

[1] The quantum jump approach was developed in [35, 49, 23]. It is equivalent and simultaneous to the Monte Carlo wave function approach (MCWF) [17] and to the quantum trajectory approach [14].

Hegerfeldt, G.C.: *The Quantum Jump Approach and Some of Its Applications.* Lect. Notes Phys. **789**, 127–174 (2009)
DOI 10.1007/978-3-642-03174-8_6

more complicated states and even to density matrices. Moreover, the reset state may depend on the state prior to the detection and thus on the time a photon is detected. Of course, recoil can also change the atomic center-of-mass momentum.

The quantum jump approach (QJA) has been developed to study the temporal behavior of a single system coupled to a "bath," such as an atom coupled to the radiation field, driven by an external field like a laser. A spectacular example for a phenomenon that can be observed for a single system but not for an ensemble is the famous macroscopic quantum jumps of a three-level atom, predicted by Dehmelt [19]. He considered an atom with a ground state 1 and two excited states 2 and 3 where the former is strongly coupled to 1 and decays rapidly, while the latter is metastable (cf. Fig. 6.1). The 1–2 transition is strongly driven by a laser and the 1–3 transition is weakly driven.

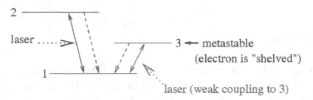

Fig. 6.1 (Color online) System with macroscopic light and dark periods. Three-level system in a V configuration. Level 2 decays rapidly while level 3 is metastable. The 1–2 transition is strongly driven by a laser, and the 1–3 transition is weakly driven

Semi-classically one would expect the electron to make rapid transitions between levels 1 and 2, accompanied by a stream of spontaneous photon emissions, in the order of 10^8 photons per second. These would form a visible light period. From time to time the weak driving of the 1–3 transition, however, would manage to put the electron into the metastable level 3 where it would stay for some time ("shelving"). During this time the stream of spontaneous photons would be interrupted, leading to a dark period. Then the electron would jump back to level 1, with a new light period beginning. This is depicted in Fig. 6.2, where each short line denotes the emission of a spontaneous photon.

Fig. 6.2 Each line represents an individual photon. Groups of photons represent a light period, with a dark period in between two such groups

Experimentally, the individual photons are not resolved, and in the fluorescence signal one just observes light and dark periods [10–13]. The light and dark periods can last from milli-seconds to minutes and their durations are random. In an ensemble of such atoms (e.g., gas with no cooperative effects) light and dark periods from different atoms would overlap, and consequently one would just see diminished fluorescence. Only light and dark periods from a single or a few atoms are directly observable.

The above semi-classical description is adequate when the atom is irradiated by incoherent light, with the level separations in the optical domain, since in this case one can use rate equations. However, when the atom is irradiated by coherent light as by a laser, quantum coherences can build up. Quantum mechanically the atom will then in general be in a superposition of the three states $|1\rangle$, $|2\rangle$, and $|3\rangle$ but never strictly in the state $|3\rangle$ for an extended period, i.e., there will always be a small admixture of $|2\rangle$. These coherences can change the situation considerably and can lead to quantitatively different predictions for the frequency and length of the light and dark periods. The small admixture of $|2\rangle$ could, in principle, even have the strange consequence that the first photon after a dark period could come from a 2–1 transition instead of a 3–1 transition as predicted by the semi-classical argument, although an experimental verification of this effect is presently beyond reach.

The QJA was specifically developed to treat problems like these. The QJA has turned out to be a practical tool for questions concerning a single system and often has technical and conceptual advantages. An extensive review [42] appeared sometime ago, and therefore this chapter will be more intent on trying to explain the basics and illustrate the approach by simple, but interesting, examples. In particular, the above light and dark periods will be treated in some detail and it will be shown that, with suitably chosen parameters, the first photon after a dark period can indeed come in about half of all cases from the 2–1 transition. More details as well as applications can be found in [35, 23, 29, 30, 5, 31, 8, 32, 33, 6, 34, 9, 7] and in the brief survey [24]. Also included are some more recent extensions of the approach to incorporate recoil (cf. e.g. [25]) and the possibility of replacing density matrices by pure states [26] as well as applications to arrival times and to spin-boson models which have been discussed in [27, 28].

6.2 Repeated Measurements on a Single System: Conditional Time Development, Reset Operation, and Quantum Trajectories

In this section the basic ideas underlying the QJA will be explained by means of an atomic system which is driven by a laser and radiates photons. For the example of a two-level system the resulting conditional Hamiltonian for the time development between photon detections will be calculated explicitly in Sect. 6.2.1. From the conditional Hamiltonian the probability density for the first photon is derived in Sect. 6.2.2. The reset operation which yields the atomic state right after a photon detection will also be explicitly determined for this example in Sect. 6.2.4. Although the result for the two-level system is obviously just the ground state, the explicit calculation for this simple example facilitates the understanding of the general case in Sect. 6.4. In Sect. 6.2.5 the notion of quantum trajectories will be explained by means of this example.

It is intuitively reasonable that it should make no difference physically whether or not the photons radiated by an atom are detected and absorbed once they are sufficiently far away. It therefore suggests itself to employ gedanken photon measurements, over all space and with ideal detectors, at instances a time Δt apart. For a single-driven atom this may look as in Fig. 6.3. Starting in some initial state with no photons (the laser field is considered as classical), at first one will detect no emitted photon in space and then, say at the n_1th measurement, a first photon will be detected (and absorbed), the next photon at the n_2th measurement, and so on.

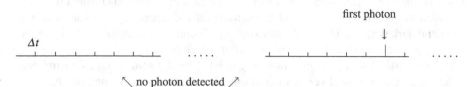

Fig. 6.3 Repeated photon measurements at times Δt apart. The first are null measurements where no photon is found

Ideally one would like to let $\Delta t \to 0$ to simulate continuous measurements. But this is impossible in the framework of quantum mechanics with ideal measurements due to the quantum Zeno effect [39, 36]. To get an idea how to choose Δt, intuitively it should be large enough to allow the photons to get away from the atom. On the other hand, Δt should also be short compared to level lifetimes so that these can be resolved in time. This leads to the requirement

$$\Delta t \cong 10^{-13} - 10^{-10}\text{s} . \tag{6.1}$$

For a similar requirement compare the book by Dirac [20]. For consistency one has of course to check that all results do not depend on the particular value of Δt when chosen in this interval. Also, for systems coupled to other baths, another range for Δt may be appropriate.

These measurements on a single atom can be translated into an ensemble description as follows. Let $\mathcal{E}$ be an ensemble of many atoms, each with its own quantized radiation field, of which the considered atom plus field is a member. At time $t_0 = 0$ the ensemble is described by the state $|0_{\text{ph}}\rangle|\psi_A\rangle$. Now one imagines that on each member of $\mathcal{E}$ photon measurements are performed at times $\Delta t, ..., n\Delta t, ...$. Now we denote, for $n = 1, 2, ...,$ by $\mathcal{E}_0^{(n\Delta t)}$ the subensemble which consists of all systems of $\mathcal{E}$ for which at the times $\Delta t, ..., n\Delta t$ no photon was detected. This is depicted in Fig. 6.4 where the individual system under consideration, atom plus radiation field, is denoted by a dot "$\cdot$" and it is a member of $\mathcal{E}_0^{(n\Delta t)}$.

Now one can proceed by ordinary quantum mechanics and the von Neumann–Lüders reduction rule [1]. Let $\mathcal{P}_0$ be the projector onto the no-photon subspace,

$$\mathcal{P}_0 \equiv |0_{\text{ph}}\rangle \mathbb{1}_A \langle 0_{\text{ph}}| , \tag{6.2}$$

and let $U(t, t_0)$ be the complete time-development operator, including the laser driving and the interaction of the atom with the quantized radiation field. Then the subensemble $\mathcal{E}_0^{(\Delta t)}$ is described by

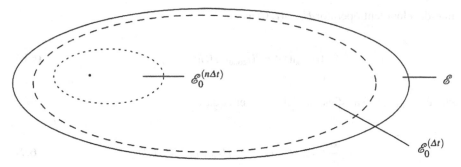

Fig. 6.4 Ensemble $\mathcal{E}$ and subensembles. $\mathcal{E}_0^{(n\Delta t)}$ denotes the subensemble with no photons found until, and including, the nth measurement. The *dot* denotes the single system actually under consideration for which it is assumed that no photon was found until the nth measurement so that it lies in $\mathcal{E}_0^{(n\Delta t)}$

$$\mathcal{P}_0 U(\Delta t, 0)|0_{\mathrm{ph}}\rangle|\psi_A(0)\rangle \tag{6.3}$$

and the subensemble $\mathcal{E}_0^{(n\Delta t)}$ by

$$\mathcal{P}_0 U(n\Delta t, (n-1)\Delta t)\mathcal{P}_0...\mathcal{P}_0 U(\Delta t, 0)|0_{\mathrm{ph}}\rangle|\psi_A(0)\rangle \equiv |0_{\mathrm{ph}}\rangle|\psi_{\mathrm{cond}}(t)\rangle , \tag{6.4}$$

where we have put $t \equiv \Delta t$. The relative size of the subensemble $\mathcal{E}_0^{(n\Delta t)}$ is the probability to find a member of $\mathcal{E}$ in $\mathcal{E}_0^{(n\Delta t)}$ and is given by the norm square of the above expression. Hence

$$P_0(t) \equiv \| \,|\psi_{\mathrm{cond}}(t)\rangle \|^2 \tag{6.5}$$

is the probability to find no photon *until* time $t = n\Delta t$. Since the decrease of $P_0(t)$ in the time interval Δt is the probability, $w_1(t)\Delta t$, of finding a photon in Δt, one has

$$w_1(t)\Delta t = -(P_0(t + \Delta t) - P_0(t)) . \tag{6.6}$$

This relation also results from the reset operation, as explained in Sect. 6.2.4.

To calculate $|\psi_{\mathrm{cond}}(t)\rangle$ we note that

$$\mathcal{P}_0 U(t' + \Delta t, t')\mathcal{P}_0 = |0_{\mathrm{ph}}\rangle\langle 0_{\mathrm{ph}}|U(t' + \Delta t, t')|0_{\mathrm{ph}}\rangle\langle 0_{\mathrm{ph}}| \tag{6.7}$$

and that the inner expression is a purely atomic operator which is easily obtained by second-order perturbation theory.

The time development of $|\psi_{\mathrm{cond}}(t)\rangle$ occurs under the condition that no photon is observed until time t, i.e., it is a *conditional time development*, and therefore the norm of $|\psi_{\mathrm{cond}}(t)\rangle$ decreases. It will be shown in the following that, on a coarse-grained timescale, $|\psi_{\mathrm{cond}}(t)\rangle$ can be written by means of a nonunitary conditional time-development operator $U_{\text{cond}}(t, 0)$ as

$$|\psi_{\text{cond}}(t)\rangle = U_{\text{cond}}(t, 0)|\psi_A(0)\rangle \,,\tag{6.8}$$

where $U_{\text{cond}}(t, 0)$ satisfies the Schrödinger equation

$$\frac{\mathrm{d}}{\mathrm{d}t} U_{\text{cond}}(t, 0) = -\frac{i}{\hbar} H_{\text{cond}}(t) U_{\text{cond}}(t, 0)\tag{6.9}$$

with a complex conditional Hamiltonian H_{cond}. The nonunitarity of $U_{\text{cond}}(t, 0)$ corresponds to the decrease in time of the probability $P_0(t)$ of finding no photon until time t. This will be illustrated now for the example of a two-level system. The general case will be treated in Sect. 6.4.

6.2.1 The Conditional Hamiltonian for a Two-Level System

In this section the atom will be assumed to be at rest and located at the origin. Its center-of-mass motion will be taken into account in Sect. 6.7. For optical wavelengths the atom can be treated as point-like. Let $\mathbf{E}_{\text{ext}}(0, t)$ denote an external field at the origin. Then the Hamiltonian for the complete system, atom plus quantized electromagnetic field $\hat{\mathbf{E}}$, is given in the Schrödinger picture by [37, 16]

$$H = \hbar\omega_0|2\rangle\langle 2| + \sum \hbar\omega_k a_{\mathbf{k}\lambda}^\dagger a_{\mathbf{k}\lambda} + e\hat{\mathbf{D}} \cdot \hat{\mathbf{E}}(0, 0) + e\hat{\mathbf{D}} \cdot \mathbf{E}_{\text{ext}}(0, t) \,,\tag{6.10}$$

where $\hat{\mathbf{D}}$ is the electronic dipole operator,

$$\hat{\mathbf{D}} = \sum_{ij} \langle i|\hat{\mathbf{X}}_e|j\rangle \, |i\rangle\langle j| \equiv \sum_{ij} \mathbf{D}_{21} \, |i\rangle\langle j| \,.\tag{6.11}$$

In the usual rotating-wave approximation the Hamiltonian becomes

$$H = \hbar\omega_0 \, |2\rangle\langle 2| + \sum \hbar\omega_k a_{\mathbf{k}\lambda}^\dagger a_{\mathbf{k}\lambda} + \left[e\hat{\mathbf{D}}^{(-)} \cdot \mathbf{E}_{\text{ext}}^{(+)}(t) + \text{h.c.} \right] + \left[e\hat{\mathbf{D}}^{(-)} \cdot \hat{\mathbf{E}}^{(+)} + \text{h.c.} \right]$$
$$\equiv H_A^0 + H_F^0 + H_{AL}(t) + H_{AF} \,,\tag{6.12}$$

where

$$H_A^0 = \hbar\omega_0 \,|2\rangle\langle 2| \,, \qquad H_F^0 = \sum_{\mathbf{k}\lambda} \hbar\omega_k a_{\mathbf{k}\lambda}^\dagger a_{\mathbf{k}\lambda} \,, \qquad (6.13)$$

$$\hat{\mathbf{D}}^{(-)} = \mathbf{D}_{21}\,|2\rangle\langle 1| \,, \qquad \mathbf{D}_{ij} = \langle i|\mathbf{X}|j\rangle \,, \qquad (6.14)$$

$$\hat{\mathbf{E}}^{(+)} = \sum_{\mathbf{k}\lambda} ie \left\{ \frac{\hbar\omega_k}{2\varepsilon_0 V} \right\}^{1/2} \boldsymbol{\varepsilon}_{\mathbf{k}\lambda} a_{\mathbf{k}\lambda} \,,$$

$$\mathbf{E}_{\mathrm{ext}}^{(+)}(t) = \frac{1}{\sqrt{2\pi}} \int_0^\infty d\omega \tilde{\mathbf{E}}_{\mathrm{ext}}(\omega) e^{-i\omega t} \,,$$

and a frequency cutoff is included. V is the quantization volume, later taken to infinity, $a_{\mathbf{k}\lambda}^\dagger$ and $a_{\mathbf{k}\lambda}$ are the photon creation and annihilation operators, and $\boldsymbol{\varepsilon}_{\mathbf{k}\lambda}$ a polarization vector. For an external laser field of frequency ω_L,

$$\mathbf{E}_L(t) = \mathrm{Re}\, \mathbf{E}_0 e^{-i\omega_L t} \,, \qquad (6.15)$$

and the Rabi frequency, Ω, is defined as

$$\Omega = e\mathbf{D}_{12} \cdot \mathbf{E}_0 \,. \qquad (6.16)$$

For an external laser field, $H_{AL}(t)$ then becomes, in the rotating-wave approximation,

$$H_{AL}(t) = \frac{\Omega}{2} \left\{ |2\rangle\langle 1|\, e^{-i\omega_L t} + \mathrm{h.c.} \right\} \,, \qquad (6.17)$$

where the Rabi frequency Ω plays the role of a coupling constant.

Going over to the interaction picture with respect to $H_A^0 + H_F^0$ one has

$$H_I(t) = H_{AL}^I(t) + H_{AF}^I(t) \,, \qquad (6.18)$$

which is obtained by replacing $|2\rangle\langle 1|$ and $a_{\mathbf{k}\lambda}$ in the original interaction Hamiltonian by $|2\rangle\langle 1|e^{i\omega_0 t}$ and $a_{\mathbf{k}\lambda}e^{-i\omega_k t}$, respectively. We now calculate, for $t_i \le t' < t_{i+1}$,

$$\langle 0_{\mathrm{ph}}| \frac{d}{dt'} U_I(t', t_i) |0_{\mathrm{ph}}\rangle \,. \qquad (6.19)$$

In the first-order contribution only $H_{AL}^I(t)$ remains since $H_{AF}^I(t)$ is linear in the creation and annihilation operators and hence

$$\langle 0_{\mathrm{ph}}| H_{AF}^I(t) |0_{\mathrm{ph}}\rangle = 0 \,. \qquad (6.20)$$

The second-order contribution is, by (6.20),

$$-\hbar^{-2} \int_{t_i}^{t'} dt'' \left\{ \langle 0_{\mathrm{ph}}| H_{AF}^I(t') H_{AF}^I(t'') |0_{\mathrm{ph}}\rangle + H_{AL}^I(t') H_{AL}^I(t'') \right\} \,. \qquad (6.21)$$

Now, if the external field $\mathbf{E}_{ext}(t)$ is smooth in time, e.g., given by laser, and *not* wildly fluctuating like a thermal or chaotic field, then the second part in (6.21) contributes a term of higher order in Δt and can therefore be omitted. For a chaotic external field, however, this part may give rise to a contribution of the order of Δt and then has to be retained. In this case one can no longer work with state vectors (wavefunctions) but has to use (conditional) density matrices. A particular example of this is treated in [30].

Thus, supposing a smooth external field, only terms of the form $aa^\dagger$ survive in the second-order contribution, which then becomes

$$-\hbar^{-2}\int_{t_i}^{t'} dt'' |2\rangle\langle 1|1\rangle\langle 2| \sum_{\mathbf{k}\lambda} \frac{e^2\hbar\omega_k}{2\varepsilon_0 V}(\mathbf{D}_{21}\cdot\boldsymbol{\varepsilon}_{\mathbf{k}\lambda})(\boldsymbol{\varepsilon}_{\mathbf{k}\lambda}\cdot\mathbf{D}_{12})e^{-i(\omega_\mathbf{k}-\omega_0)(t'-t'')}$$

$$= -\hbar^{-2}|2\rangle\langle 2| \int_0^{t'-t_i} d\tau \sum_{\mathbf{k}\lambda} \frac{e^2\hbar\omega_k}{2\varepsilon_0 V}(\mathbf{D}_{21}\cdot\boldsymbol{\varepsilon}_{\mathbf{k}\lambda})(\boldsymbol{\varepsilon}_{\mathbf{k}\lambda}\cdot\mathbf{D}_{12})e^{-i(\omega_\mathbf{k}-\omega_0)\tau} . \quad (6.22)$$

One can now use properties of the correlation function

$$\kappa(\tau) \equiv \sum_{\mathbf{k}\lambda} \frac{e^2\omega_\mathbf{k}}{2\varepsilon_0\hbar V}(\mathbf{D}_{21}\cdot\boldsymbol{\varepsilon}_{\mathbf{k}\lambda})(\boldsymbol{\varepsilon}_{\mathbf{k}\lambda}\cdot\mathbf{D}_{12})\, e^{-i(\omega_\mathbf{k}-\omega_0)\tau} . \quad (6.23)$$

With $V^{-1} = \Delta^3 k/(2\pi)^3$ one can perform the limit $V\to\infty$, and the sum over $\mathbf{k}$ becomes an integral over ω, with a suitable frequency cutoff, and an integral over the unit sphere. The correlation function has an effective width of the order of ω_0^{-1} around $\tau=0$, and for $t'-t_i \gg \omega_0^{-1}$ one can therefore extend the τ integration in (6.22) to infinity [2]. This amounts to the replacement

$$\int_0^{t'-t_i} d\tau e^{i(\omega_0-\omega_\mathbf{k})\tau} \cong \pi\delta(\omega_\mathbf{k}-\omega_0) + i\mathcal{P}\frac{1}{\omega_\mathbf{k}-\omega_0} \quad (6.24)$$

and corresponds to the usual Markov approximation in the derivation of the optical Bloch equations [38, 43]. The principal-value term is analogous to a level shift and will be omitted [16, 43]. For the second-order contribution one then obtains

$$-|2\rangle\langle 2| \int d^3k \frac{e^2\omega_\mathbf{k}}{(2\pi)^3\hbar 2\epsilon_0}\mathbf{D}_{21}\cdot\sum_{\lambda=1}^{2}\boldsymbol{\varepsilon}_{\mathbf{k}\lambda}\boldsymbol{\varepsilon}_{\mathbf{k}\lambda}\cdot\mathbf{D}_{12}\pi\delta(\omega_\mathbf{k}-\omega_0) . \quad (6.25)$$

The last integral is denoted by Γ and using

$$\sum_{\lambda=1}^{2}|\boldsymbol{\varepsilon}_{\mathbf{k}\lambda}\cdot\mathbf{D}_{12}|^2 = |\mathbf{D}_{12}|^2 - |\hat{\mathbf{k}}\cdot\mathbf{D}_{12}|^2 \quad (6.26)$$

one obtains

$$\Gamma \equiv \frac{e^2}{6\pi\varepsilon_0\hbar c^3}|\mathbf{D}_{21}|^2|\omega_0|^3 = A/2 , \tag{6.27}$$

where A is the usual Einstein coefficient.

Hence, integrating over t' from t_i to t_{i+1} and using $1 + \alpha \approx e^\alpha$ for small α, we obtain

$$\langle 0_{\text{ph}}|U_I(t_{i+1}, t_i)|0_{\text{ph}}\rangle = \exp\left\{-\frac{\mathrm{i}}{\hbar}\int_{t_i}^{t_{i+1}} \mathrm{d}t' \left\{H_{AL}^I(t') - \mathrm{i}\hbar\Gamma|2\rangle\langle 2|\right\}\right\} . \tag{6.28}$$

For small Δt this can be replaced by a time-ordered exponential,

$$\langle 0_{\text{ph}}|U_I(t_{i+1}, t_i)|0_{\text{ph}}\rangle = \mathcal{T}\exp\left\{-\frac{\mathrm{i}}{\hbar}\int_{t_i}^{t_{i+1}} \mathrm{d}t' \left\{H_{AL}^I(t') - \mathrm{i}\hbar\Gamma|2\rangle\langle 2|\right\}\right\}$$

$$\equiv U_{\text{cond}}^I(t_i, t_{i-1}) . \tag{6.29}$$

From this, one has, with $t = t_n$,

$$\prod_{i=1}^{n}\langle 0_{\text{ph}}|(U_I(t_i, t_{i-1})|0_{\text{ph}}\rangle = \mathcal{T}\exp\left\{-\frac{\mathrm{i}}{\hbar}\int_0^t \mathrm{d}t' \left\{H_{AL}^I(t') - \mathrm{i}\hbar\Gamma|2\rangle\langle 2|\right\}\right\}$$

$$= U_{\text{cond}}^I(t, 0) , \tag{6.30}$$

where the product sign on the l.h.s. includes an ordering in an obvious way.

Since

$$\langle 0_{\text{ph}}|U(t_i, t_{i-1})|0_{\text{ph}}\rangle = \mathrm{e}^{-\mathrm{i}H_A^0 t_i/\hbar}\langle 0_{\text{ph}}|U_I(t_i, t_{i-1})|0_{\text{ph}}\rangle \mathrm{e}^{\mathrm{i}H_A^0 t_{i-1}/\hbar} \tag{6.31}$$

and since, for $t = t_n = n\Delta t$,

$$U_{\text{cond}}(t, 0) = \prod_{i=1}^{n}\langle 0_{\text{ph}}|U(t_i, t_{i-1})|0_{\text{ph}}\rangle ,$$

we obtain, on a coarse-grained timescale, from (6.30) and (6.31)

$$U_{\text{cond}}(t, 0) = \mathcal{T}\exp\left\{-\frac{\mathrm{i}}{\hbar}\int_0^t \mathrm{d}t' \left\{H_A^0 + H_{AL}(t') - \mathrm{i}\hbar\Gamma|2\rangle\langle 2|\right\}\right\} , \tag{6.32}$$

which is the transformation of (6.30) back to the Schrödinger picture.

Thus the conditional Hamiltonian for a two-level atom with no photon emission until time t is given, on the coarse-grained timescale, by

$$H_{\text{cond}}(t) = H_A^0 + H_{AL}(t) - \mathrm{i}\hbar\frac{A}{2}|2\rangle\langle 2| . \tag{6.33}$$

It may be worthwhile to point out that (6.29) can be derived directly in a more elaborate way without recourse to perturbation theory by means of the Markov property alone. Therefore the possible errors involved in going from (6.19) to (6.29) do not add up.

6.2.2 First-Photon Probability Density for a Laser-Driven Two-Level System

For an external laser field one has

$$H_{\text{cond}}(t) = \hbar\omega_0 \, |2\rangle\langle 2| + \frac{\Omega}{2} \left\{ |2\rangle\langle 1| \, e^{-i\omega_L t} + \text{h.c.} \right\} - i\hbar\frac{A}{2} |2\rangle\langle 2| \, . \tag{6.34}$$

It is noteworthy that one can get rid of the time dependence by going over to a "laser-adapted interaction picture" by means of the operator

$$H_L^0 = \hbar\omega_L |2\rangle\langle 2| \tag{6.35}$$

instead of H_A^0. In this interaction picture the conditional Hamiltonian becomes

$$H_{\text{cond}}^{IL} = -\hbar\Delta \, |2\rangle\langle 2| + \frac{\Omega}{2} \left\{ |2\rangle\langle 1| + |1\rangle\langle 2| \right\} - i\hbar\frac{A}{2} |2\rangle\langle 2| \, , \tag{6.36}$$

where $\Delta = \omega_L - \omega_0$ is the detuning. In matrix form this reads

$$H_{\text{cond}}^{IL}/\hbar = \begin{pmatrix} 0 & \Omega/2 \\ \Omega/2 & -\Delta - iA/2 \end{pmatrix} \, . \tag{6.37}$$

In these expressions one clearly sees that the imaginary term leads to a decrease of the state vector norm – this corresponds to a decrease of the probability $P_0(t)$ of finding no photon until time t.

From (6.6) one then obtains for the probability density for the first photon

$$\begin{aligned} w(t) &= -\frac{\text{d}}{\text{d}t} P_0(t) \\ &= -\frac{\text{d}}{\text{d}t} || \exp\{-iH_{\text{cond}}t/\hbar\}|\psi_A(0)\rangle ||^2 \\ &= \frac{i}{\hbar} \langle\psi_{\text{cond}}(t)| H_{\text{cond}} - H_{\text{cond}}^\dagger |\psi_{\text{cond}}(t)\rangle \, . \end{aligned} \tag{6.38}$$

For the two-level system this becomes

$$w(t) = A \, |\langle 2| \exp\{-iH_{\text{cond}}^{IL}t/\hbar\}|\psi_A(0)\rangle|^2 \, . \tag{6.39}$$

If $\lambda_\pm$ denote the eigenvalues of $H_{\text{cond}}^{IL}/\hbar$ then the conditional time-development operator in this interaction picture is given by

$$\exp\{-iH_{\text{cond}}^{IL}t/\hbar\} = \frac{1}{\lambda_+ - \lambda_-}\left\{e^{-i\lambda_+t}\left[H_{\text{cond}}^{IL} - \lambda_-\right] - e^{-i\lambda_-t}\left[H_{\text{cond}}^{IL} - \lambda_+\right]\right\}. \tag{6.40}$$

This identity can be checked directly by applying both sides of the equation to the eigenvectors $|\lambda_\pm\rangle$. If one starts in the atomic ground state, $|\psi_A(0)\rangle = |1\rangle$, one obtains for $\omega_L = \omega_0$ (zero detuning, $\Delta = 0$)

$$\lambda_\pm = -\frac{iA}{4} \pm \frac{1}{4}\sqrt{4\Omega^2 - A^2}, \tag{6.41}$$

and for the first-photon probability density

$$w(t) = \frac{4A\Omega^2}{|4\Omega^2 - A^2|}e^{-At/2}\left|\sin\left(\frac{1}{4}\sqrt{4\Omega^2 - A^2}t\right)\right|^2. \tag{6.42}$$

This probability density is plotted in Fig. 6.5 for $\Omega = 2A$. It should be noted that $w(0) = 0$. This is connected to the so-called anti-bunching of photons as follows. One can imagine that at time $t = 0$ a photon was emitted, right after which the atom is in the ground state. Hence $w(0) = 0$ means that in light emitted from the

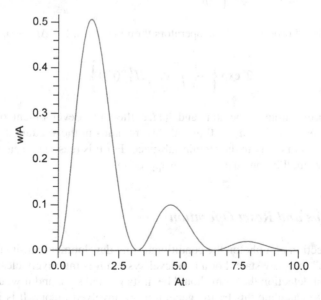

Fig. 6.5 (Color online) Probability density, $w(t)$, for the first photon as a function of time for a laser-driven two-level system, with $\Omega = 2A$. The vanishing of $w(t)$ for $t = 0$ implies anti-bunching of photons

atom there is only a small probability to find two photons in quick succession. This is in contrast to natural (chaotic) light where there is bunching, i.e., photons tend to come in pairs. Anti-bunching, which is usually derived by means of the optical Bloch equations, just reflects the fact that it takes some time for the light source to pump the atom back to the excited state before the next photon can be emitted.

From (6.42) the mean waiting time, τ, for the first photon is calculated as

$$\tau = \frac{2\Omega^2 + A^2}{A\Omega^2} . \tag{6.43}$$

This result will be used later in connection with the discussion of macroscopic light and dark periods.

6.2.3 Connection with the Quantum Zeno Effect

When one lets Δt become smaller and smaller the above derivation shows very nicely how and where the quantum Zeno effect [39] turns up in a very natural way. If Δt is chosen much smaller than the inverse optical frequencies, the last exponential in (6.22) can be replaced by 1, and the integral becomes proportional to $t' - t_i$. Equation (6.28) is then replaced by

$$\langle 0_{\text{ph}} | U_I(t_{i+1}, t_i) | 0_{\text{ph}} \rangle = \exp \left\{ -\frac{\mathrm{i}}{\hbar} \left\{ H_{AL}(t_i) \Delta t - \mathrm{i}\hbar \, \text{const} \, |2\rangle\langle 2| (\Delta t)^2 \right\} \right\} . \tag{6.44}$$

The time-ordered product of these operators then becomes, for $\Delta t \to 0$,

$$\mathcal{T} \exp \left\{ -\frac{\mathrm{i}}{\hbar} \int_0^t \mathrm{d}t' \left\{ H_{AL}^I(t') \right\} \right\} . \tag{6.45}$$

This is a purely atomic operator, and hence the time development of the field becomes frozen, i.e., for $\Delta t \to 0$ one always remains in the vacuum and the temporal change occurs only in the atomic subspace. For this reason one cannot choose Δt arbitrarily small in the quantum jump approach.

6.2.4 Jumps and Reset Operation

With the detection of a photon the conditional time development terminates and the atom "jumps" to a new state. For a two-level system it is intuitively clear that right after a photon detection the atom should be in its ground state and it would seem to use overkill to calculate this by the general more involved theory. It is instructive, however, to see how the machinery works for this simple system and it clarifies the procedure for the general case.

We consider an ensemble where no photons are present at some time t. At this instant the total ensemble is thus described by a state $|0_{ph}\rangle|\phi\rangle$ or more generally by a density matrix $\rho_{tot}(t) = |0_{ph}\rangle\rho\langle 0_{ph}|$ where ρ is the density matrix of the atoms. The (normalized) state of the subensemble for which photons are detected by a nonabsorptive measurement at time $t + \Delta t$ is given, in view of the von Neumann–Lüders projection postulate [1], by

$$\mathcal{P}_1\rho_{tot}(t + \Delta t)\mathcal{P}_1/\mathrm{tr}(\cdot) , \tag{6.46}$$

where

$$\mathcal{P}_1 \equiv \mathbb{1} - |0_{ph}\rangle\mathbb{1}_A\langle 0_{ph}| , \tag{6.47}$$

and where the trace of $\mathcal{P}_1\rho_{tot}(t + \Delta t)\mathcal{P}_1$ gives the probability of detecting a photon in the time interval Δt. Note that (6.46) still contains the photons.

After a photon measurement by absorption no photons are present any longer and it was argued in [23] that the resulting state is obtained from (6.46) by a partial trace over the photons, i.e., by

$$|0_{ph}\rangle\left(\mathrm{tr}_{ph}\mathcal{P}_1\rho_{tot}(t + \Delta t)\mathcal{P}_1\right)\langle 0_{ph}|/\mathrm{tr}(\cdot) . \tag{6.48}$$

The physical reason for this is that for the atomic description alone it should make no difference in infinite space whether or not the photons are absorbed, as long as they are sufficiently far away from the atom and no longer interacting with it.[2] We define the superoperator $\mathcal{R}$, which acts on atomic density matrices, by

$$\mathcal{R}\rho\, \Delta t \equiv \mathrm{tr}_{ph}\left(\mathcal{P}_1 U(t + \Delta t, t)|0_{ph}\rangle\,\rho\,\langle 0_{ph}|U^\dagger(t + \Delta t, t)\mathcal{P}_1\right) \tag{6.49}$$

and call $\mathcal{R}$ the reset operation and $\mathcal{R}\rho/\mathrm{tr}(\cdot)$ the (normalized) reset state. The atomic trace of $\mathcal{R}\rho\, \Delta t$ gives the probability of detecting a photon in the time interval Δt when initially there were no photons and the atom was in the state ρ.

Equation (6.49) can be calculated by perturbation theory for $U_I(t + \Delta t, t)$, as in Sect. 6.2.1. Now the first-order contribution suffices and in this order the external (laser) field drops out since $\mathcal{P}_1|0_{ph}\rangle = 0$. One obtains in a straightforward way

$$\mathcal{R}\rho\, \Delta t = e^{-iH_A^0 \Delta t/\hbar}|1\rangle\langle 2|\rho|2\rangle\langle 1|e^{iH_A^0 \Delta t/\hbar}$$
$$\times \int_0^{\Delta t} dt' \int_0^{\Delta t} dt'' \sum_{k\lambda} e^{i(\omega_k - \omega_0)(t' - t'')} \frac{e^2\omega_k}{2\varepsilon_0\hbar V}(\mathbf{D}_{12} \cdot \boldsymbol{\varepsilon}_{k\lambda})(\boldsymbol{\varepsilon}_{k\lambda} \cdot \mathbf{D}_{21}) . \tag{6.50}$$

To apply the Markov property, we decompose the rectangular integration domain over t' and t'' in (6.50) into two triangles, leading to

---

[2] For a cavity, however, where photons can return and revivals can occur, the results are different.

$$\int_0^{\Delta t} dt' \int_0^{t'} dt'' e^{i(\omega_k - \omega_0)(t' - t'')} + \int_0^{\Delta t} dt'' \int_0^{t''} dt' e^{i(\omega_k - \omega_0)(t' - t'')} . \qquad (6.51)$$

As in (6.22) and (6.24), by the properties of the correlation function, each of the inner integrals can be replaced by $\pi \delta(\omega_k - \omega_0)$ where the principal values cancel each other. In this way one obtains

$$\begin{aligned} \mathcal{R}\rho \, \Delta t &= A|1\rangle\langle 2|\rho|2\rangle\langle 1|\Delta t \\ &= A\rho_{22}|1\rangle\langle 1|\Delta t . \end{aligned} \qquad (6.52)$$

Thus, as expected, the probability density is given by the Einstein coefficient multiplied by the occupation of the excited level, and after detection of a photon the atom is in the ground state. The appearance of the Einstein coefficient and the ground state reflects the fact that one observes spontaneous photons only since the laser is treated classically. Anyway, if the laser photons were treated quantum mechanically it should be recalled that photons from stimulated emissions have the same direction as the incident, stimulating, laser photons and are therefore not observed.

Now, if instead of a normalized atomic density matrix one takes the conditionally time-developed state $|\psi_{\text{cond}}(t)\rangle$, then the atomic trace of

$$\mathcal{R}\left(|\psi_{\text{cond}}(t)\rangle\langle\psi_{\text{cond}}(t)|\right) \qquad (6.53)$$

gives the probability density for a photon at time t under the condition that no photon has been detected before, i.e., the probability density for the *first* photon after $t = 0$. This was denoted by $w(t)$ in (6.42) and therefore

$$w(t) = \text{tr}\big(\mathcal{R}\left(|\psi_{\text{cond}}(t)\rangle\langle\psi_{\text{cond}}(t)|\right)\big) . \qquad (6.54)$$

The general reset state for systems at rest has been determined in [23, 34] and is given in Sect. 6.4. It may depend on $|\psi_{\text{cond}}(t_1)\rangle$ where t_1 is the detection time of the photon.

6.2.5 Quantum Trajectories

One can now distinguish different steps in the temporal behavior of the single atom under the above gedanken measurements.

(i) Until the detection of the first photon, the atom belongs to the subensembles $\mathcal{E}_0^{(n\Delta t)}$ and hence is described by the (non-normalized) vector

$$|\psi_{\text{cond}}(t)\rangle = U_{\text{cond}}(t, 0)|\psi_A(0)\rangle . \qquad (6.55)$$

(ii) The first photon is detected at some (random) time t_1, according to the probability density

$$w(t) = -\frac{\mathrm{d}P_0(t)}{\mathrm{d}t} = -\frac{\mathrm{d}}{\mathrm{d}t} \| \, |\psi_{\mathrm{cond}}(t)\rangle \|^2 \, . \tag{6.56}$$

(iii) *Jump:* With the detection of a photon the atom has to be reset to the appropriate state. For example, a two-level atom will be in its ground state right after a photon detection.

(iv) From this reset state the time development then continues with $U_{\mathrm{cond}}(t, t_1)$, until the detection of the next photon at the (random) time t_2. Then one has to reset (jump), and so on.

In this way one obtains a *stochastic path in the Hilbert space of the atom*. The stochasticity of this path is governed by quantum theory, and the path is called a *quantum trajectory*. In general the reset state will not be a pure state but a density matrix, as explained in Sect. 6.4.2, and this may result in quantum trajectories with density matrices instead of pure states, even if one starts in a pure state. However, it is pointed out in Sect. 6.6 that one can replace such a trajectory by a trajectory consisting of pure states only. The stochastic process underlying the quantum trajectories is a jump process with values in a Hilbert space. If the reset state is always the same, e.g., the ground state, one has a renewal process. If the reset state depends on the conditional state before the jump, one has a Markov process only.

In which sense the parts of a trajectory between jumps can be regarded as an ensemble created by *repetition* from a single system at stochastic times will be discussed in the last section.

The steps (i)–(iv) above can be used for simulations of a trajectory. This will be discussed in more detail in Sect. 6.6 for the specific example of a three-level cascade system.

6.3 Application: Macroscopic Light and Dark Periods

The ideas of the preceding section will now be used to provide a direct quantum mechanical understanding of macroscopic light and dark periods without employing Bloch equations or a rate-equation approach. For the V system of Fig. 6.1 which employs two coherent light sources the QJA as described so far can be applied right away. For setups which in addition to a laser also have driving by a lamp the QJA has to be carried over to include incoherent driving. This has been done in [30], and those results can be used to discuss those experimentally realized systems in [10–13] which have driving by a lamp.

In this section the QJA will be applied to the V system of Fig. 6.1. A strong laser of frequency ω_{L1} drives the 1–2 transition, while the transition from 1 to the metastable state 3 is weakly driven by a laser of frequency ω_{L2}. It is assumed that the laser frequencies are close to the transition frequencies ω_2 and ω_3 and that the latter are far apart. Then instead of (6.12), (6.13), and (6.17) one has

$$H_A^0 = \hbar\omega_2 |2\rangle\langle 2| + \hbar\omega_3 |3\rangle\langle 3| \, ,$$

$$\mathbf{D}_{i1} = \langle i|\mathbf{X}|1\rangle \, , \qquad \hat{\mathbf{D}}^{(-)} = \mathbf{D}_{21}|2\rangle\langle 1| + \mathbf{D}_{31}|3\rangle\langle 1| \, , \tag{6.57}$$

$$H_{AL}(t) = \frac{\Omega_1}{2} \left\{ |2\rangle\langle 1| e^{-i\omega_{L1}t} + \text{h.c.} \right\} + \frac{\Omega_2}{2} \left\{ |3\rangle\langle 1| e^{-i\omega_{L2}t} + \text{h.c.} \right\} \, ,$$

where Ω_1 and Ω_2 are the respective Rabi frequencies characterizing the strength of the atom–laser interaction. One can now determine the conditional Hamiltonian for this system either as in the preceding section or by using the general expression in (6.83). Written in an obvious matrix form the result is of the form

$$H_{\text{cond}}/\hbar = \begin{pmatrix} 0 & e^{i\omega_{L1}t}\Omega_1/2 & e^{i\omega_{L2}t}\Omega_2/2 \\ e^{-i\omega_{L1}t}\Omega_1/2 & \omega_2 - iA_2/2 & -i\gamma_{12} \\ e^{-i\omega_{L2}t}\Omega_2/2 & -i\gamma_{21} & \omega_3 - iA_3/2 \end{pmatrix} \, , \tag{6.58}$$

where A_2 and A_3 are the respective Einstein coefficients and where the γ_{ij} terms will later be seen to be negligible by the rotating-wave approximation. Going over to a "laser-adapted interaction picture" by means of the operator

$$H_L^0 = \hbar\omega_{L1}|2\rangle\langle 2| + \hbar\omega_{L2}|3\rangle\langle 3|, \tag{6.59}$$

one obtains, with the detunings $\Delta_1 = \omega_{L1} - \omega_2$ and $\Delta_2 = \omega_{L2} - \omega_3$,

$$H_{\text{cond}}^{IL}/\hbar = \begin{pmatrix} 0 & \Omega_1/2 & \Omega_2/2 \\ \Omega_1/2 & \Delta_1 - iA_2/2 & 0 \\ \Omega_2/2 & 0 & \Delta_2 - iA_3/2 \end{pmatrix} \, , \tag{6.60}$$

where the rapidly oscillating terms $\gamma_{ij} \exp\{\pm i(\omega_{L2} - \omega_{L1})t\}$ can be, and have been, omitted. It will be assumed in the following that

$$\Omega_1^2 \gg \Omega_2^2, \; A_3A_2 \quad \text{and} \quad A_2 \gg A_3, \; \Omega_2 \tag{6.61}$$

as well as $\Delta_1 = 0$. From the inequalities it follows that the upper left 2×2 matrix dominates. (This becomes evident if one adds $\Delta_2 \mathbb{1}$ to $H_{\text{cond}}^{IL}/\hbar$ to get rid of $-\Delta_2$ in the 33 component of the matrix.) The upper left 2×2 matrix is the same as in (6.37) for a two-level system. Therefore, two of the eigenvalues, $\lambda_{1,2}$, of $H_{\text{cond}}^{IL}/\hbar$ are approximately given by (6.41) and the third by

$$\lambda_3 + \Delta_2 \sim -iA_3 \, . \tag{6.62}$$

Therefore, there are different orders of magnitudes present in the imaginary parts of the eigenvalues and one has

$$-\text{Im}\lambda_3 \ll -\text{Im}\lambda_{1,2} \, . \tag{6.63}$$

From this one obtains an understanding of the light and dark periods as follows. Let $|\lambda_i\rangle$ be (nonorthogonal) eigenvectors of $H^{IL}_{cond}/\hbar$ (in case of degeneracy one has to consider limits) and let $\langle \lambda^i|$ be the dual basis, defined by

$$\langle \lambda^i | \lambda_j \rangle = \delta_{ij} \,. \tag{6.64}$$

Then one can write, for initial state $|1\rangle$,

$$\exp\{-iH^{IL}_{cond}t/\hbar\}|1\rangle = \sum_{j=1}^{3} e^{-i\lambda_j t} |\lambda_j\rangle \langle \lambda^j | 1\rangle \,. \tag{6.65}$$

Since $|\lambda_{1,2}\rangle$ are mainly a superposition of $|1\rangle$ and $|2\rangle$ and $|\lambda_3\rangle$ is close to $|3\rangle$, the last term in the sum is negligible for times of the order A_2^{-1} and Ω_1^{-1}, while for times much larger than A_2^{-1} and Ω_1^{-1}, the first two terms become exponentially small, and the third term, though small, remains.

As in (6.38) the probability density for the first photon is

$$
\begin{aligned}
w(t) &= - P_0'(t) \\
&= \langle 1 | \exp\{iH^{IL\dagger}_{cond}t/\hbar\} \{A_2|2\rangle\langle 2| + A_3|3\rangle\langle 3|\} \exp\{-iH^{IL}_{cond}t/\hbar\}|1\rangle \qquad (6.66)\\
&= A_2 \left| \langle 2| \exp\{-iH^{IL}_{cond}t/\hbar\}|1\rangle \right|^2 + A_3 \left| \langle 3| \exp\{-iH^{IL}_{cond}t/\hbar\}|1\rangle \right|^2 \,.
\end{aligned}
$$

By the preceding argument, the first term is close to the two-level expression in (6.42). It dominates for small times and then becomes exponentially small. For large times the other, small, term remains which then slowly drops off with a very small exponential exponent.

For the two-level system the waiting time between two photons is of the order of A_2^{-1} and Ω_1^{-1} and waiting times much larger than this practically never happen. One can therefore now *define* what one can reasonably understand as a dark period. One can pick a time T_0 satisfying

$$A_2^{-1}, \ \Omega_2^{-1} \ll T_0 \ll A_3^{-1}, \ \Omega_3^{-1} \,. \tag{6.67}$$

Then, if the waiting time between two photons is larger than T_0, this is called a *dark period*.

For the temporal behavior of a single atom and its associated quantum trajectory one has now the following situation. One starts in $|1\rangle$, one quickly has a photon, is back in $|1\rangle$, again a quick photon, and so on. In a very rare event there is no photon within a time interval of length T_0. Once this has happened there is only a very small probability density for the next photon and the total waiting time is *much larger than T_0*.

The probability of no photon within time T_0 is

$$P_0(T_0) \cong \| e^{-i\lambda_3 T_0} |\lambda_3\rangle \langle \lambda^3 | 1\rangle \|^2 \,. \tag{6.68}$$

The exponential is close to 1 while the remainder is small. Thus, after each photon there is a small probability, i.e., $P_0(T_0)$, for a dark period. Therefore a dark period occurs after an average number of $1/P_0(T_0)$ photons; the latter form a "light period." Note that if $P_0(T_0)$ is too small a dark period may in practice never occur. This clearly depends on the parameters chosen.

The probability density, $w_D(t)$, for the length of a dark period is the same as the probability density for the first photon to occur after a time interval of length T_0. Thus one has from (6.66), after normalization,

$$
\begin{aligned}
w_D(t) &= w(t)\Theta(t - T_0)/ \int_{T_0}^{\infty} dt'\, w(t') \\
&\cong 2|\mathrm{Im}\lambda_3|e^{-2|\mathrm{Im}\lambda_3|t}\,\Theta(t - T_0)/e^{-2|\mathrm{Im}\lambda_3|T_0} ,
\end{aligned}
\tag{6.69}
$$

and this is essentially an exponential distribution. Here $\Theta(t)$ is the step function which vanishes for $t < 0$ and is 1 for $t > 0$. The mean duration, T_D, of a dark period is then

$$
T_D = T_0 + \frac{1}{2|\mathrm{Im}\lambda_3|} \cong \frac{1}{2|\mathrm{Im}\lambda_3|} .
\tag{6.70}
$$

Since after each photon from the 1–2 transition the probability for a dark period is $P_0(T_0)$, one needs on the average $P_0(T_0)^{-1}$ tries, i.e., photons, before a dark period occurs. Therefore the mean duration, T_L, of a light period is approximately the number of these needed photons multiplied by the mean waiting time between photons for the two-level system, i.e., by τ from (6.43). Hence one has

$$
T_L \cong \frac{2\Omega^2 + A^2}{A\Omega^2} \frac{1}{P_0(T_0)} .
\tag{6.71}
$$

For a quantitative treatment one has to calculate the eigenvalues λ_i which can, in principle, be done in a complicated closed form. Simpler approximate expressions are [49]

$$
\begin{aligned}
\lambda_{1,2} &\cong -\frac{iA_2}{4} \pm \frac{1}{4}\sqrt{4\Omega_1^2 - A_2^2} , \\
\lambda_3 + \Delta_2 &\cong -\frac{iA_3}{2} - i\Omega_2^2 \frac{A_2\Omega_1^2 + 2i\Delta_2(\Omega_1^2 - 4\Delta_2^2 - A_2^2)}{2(\Omega_1^2 - 4\Delta_2^2)^2 + 8A_2^2\Delta_2^2} .
\end{aligned}
\tag{6.72}
$$

With this one obtains from (6.70)

$$
T_D = \{A_3 + \frac{\Omega_2^2\Omega_1^2 A_2}{(\Omega_1^2 - 4\Delta_2^2)^2 + 4A_2^2\Delta_2^2}\}^{-1} .
\tag{6.73}
$$

For the parameters

$$\Omega_1 = A_2 = 10^8 \text{ s}^{-1}, \quad \Omega_2 = 10^4 \text{ s}^{-1}, \quad A_3 = 1 \text{ s}^{-1}, \text{ and } \Delta_2 = \Omega_1/2 \qquad (6.74)$$

this gives

$$T_D = 0.5 \text{ s} . \qquad (6.75)$$

In order to make sure that these long dark periods do not occur too seldom one has to make sure that $P_0(T_0)$ is not too small. To determine $P_0(T_0)$ one can avoid the calculation of the eigenvectors by using

$$|\lambda_3\rangle\langle\lambda^3| = \frac{(H_{\text{cond}}^{IL}/\hbar - \lambda_1)(H_{\text{cond}}^{IL}/\hbar - \lambda_2)}{(\lambda_3 - \lambda_1)(\lambda_3 - \lambda_2)} . \qquad (6.76)$$

This identity and its cyclic variants are easily checked by applying it to the eigen-vectors $|\lambda_i\rangle$. Neglecting the exponential in (6.68), which is close to 1, one finds

$$P_0(T_0) \cong \Omega_2^2 \frac{A_2^2 + 4\Delta_2^2}{(\Omega_1^2 - 4\Delta_2^2)^2 + 4A_2^2\Delta_2^2} . \qquad (6.77)$$

Intuitively one would expect a maximum for the frequent occurrence of dark periods if the driving is on resonance ($\Delta_2 = 0$), but contrary to this, when Ω_1 is large, the dark periods are most frequent for Δ_2 close to Ω_1. This was already noticed in [15]. For the above parameters the mean duration of a light period is $T_L = 1.5$ s so that light and dark periods have a similar mean duration.

From (6.65) and (6.76) another curious effect follows. In a dark period the atom is essentially in the state $|\lambda_3\rangle$ which can be calculated by applying (6.76) to $|1\rangle$ and then normalize the resulting vector to 1. The eigenvector $|\lambda_3\rangle$ has a very small second component and therefore the next photon after a dark period can also come from a 2–1 transition, not only from a 3–1 transition as suggested by a rate-equation approach. For the parameters in (6.74) the probability for this strange effect is an astonishing 1/2 [49]!

6.4 The General N-Level System and Optical Bloch Equations

6.4.1 The Conditional Hamiltonian

In the general case, the Hamiltonian for the complete system, N-level atom plus quantized electromagnetic field, is again given by an expression as in (6.10), only with $\hbar\omega_0|2\rangle\langle2|$ replaced by

$$H_A^0 = \sum_i \hbar\omega_i \, |i\rangle\langle i| . \qquad (6.78)$$

In the rotating-wave approximation the Hamiltonian then becomes

$$
\begin{aligned}
H &= \sum_i \hbar\omega_i \, |i\rangle \, \langle i| + H_F^0 + \left[e\hat{\mathbf{D}}^{(-)} \cdot \mathbf{E}_{\text{ext}}^{(+)}(t) + \text{h.c.} \right] + \left[e\hat{\mathbf{D}}^{(-)} \cdot \hat{\mathbf{E}}^{(+)} + \text{h.c.} \right] \\
&\equiv H_A^0 + H_F^0 + H_{AL}(t) + H_{AF} ,
\end{aligned}
\tag{6.79}
$$

where now

$$
\hat{\mathbf{D}}^{(-)} = \sum_{i>j} \mathbf{D}_{ij} |i\rangle \langle j| ,
$$

$$
\mathbf{D}_{ij} = \langle i|\mathbf{X}|j\rangle
\tag{6.80}
$$

and where $i > j$ means $\omega_i > \omega_j$. The expression for $\hat{\mathbf{E}}$ is the same as in (6.13). We introduce the notation $\omega_{ij} = \omega_i - \omega_j$.

One can now proceed as in Sect. 6.2. Instead of the single correlation functions $\kappa(\tau)$ in (6.23) one can introduce correlation functions of the form

$$
\kappa_{ji\ell m}(\tau) \equiv \sum_{\mathbf{k}\lambda} \frac{e^2 \omega_{\mathbf{k}}}{2\varepsilon_0 \hbar V} (\mathbf{D}_{ji} \cdot \boldsymbol{\varepsilon}_{\mathbf{k}\lambda})(\boldsymbol{\varepsilon}_{\mathbf{k}\lambda} \cdot \mathbf{D}_{\ell m}) \, e^{-\mathrm{i}(\omega_{\mathbf{k}}-\omega_{\ell m})\tau} .
\tag{6.81}
$$

The correlation function has an effective width of the order of $\omega_{\ell m}^{-1}$ around $\tau = 0$, and for $\Delta t \gg \omega_{\ell m}^{-1}$ one can, similarly as in the $N = 2$ case, extend a time integration in the second-order contribution to infinity [2]. This amounts to the replacement

$$
\int_0^{t'-t_i} \mathrm{d}\tau \, e^{\mathrm{i}(\omega_{\ell m}-\omega_{\mathbf{k}})\tau} \cong \pi \delta(\omega_{\mathbf{k}} - \omega_{\ell m}) + \mathrm{i}\mathcal{P}\frac{1}{\omega_{\mathbf{k}} - \omega_{\ell m}}
\tag{6.82}
$$

and corresponds to the usual Markov approximation in the derivation of the optical Bloch equations [38, 43]. Again, the principal-value term corresponding to a level shift will be omitted.

Then one finds that the conditional Hamiltonian for an N-level atom with no photon detection until time t is, on the coarse-grained timescale, given in the Schrödinger picture by

$$
H_{\text{cond}}(t) = H_A^0 + H_{AL}(t) - \mathrm{i}\hbar\hat{\boldsymbol{\Gamma}} ,
\tag{6.83}
$$

where the damping operator $\hat{\boldsymbol{\Gamma}}$ is defined by

$$
\hat{\boldsymbol{\Gamma}} \equiv \sum_{\substack{i\ell j \\ i,\ell>j}} \Gamma_{ijj\ell} |i\rangle\langle\ell|
\tag{6.84}
$$

and the generalized damping constants $\Gamma_{ijk\ell}$ by

$$
\Gamma_{ijk\ell} \equiv \frac{e^2}{6\pi\varepsilon_0 \hbar c^3} \mathbf{D}_{ij} \cdot \mathbf{D}_{k\ell} |\omega_{k\ell}|^3 .
\tag{6.85}
$$

Note that, similar to (6.27), for $i > j$, one has

$$\Gamma_{ijji} = A_{ij}/2 , \qquad (6.86)$$

where A_{ij} is the Einstein coefficient for the $i-j$ transition.

6.4.2 The Reset Operation

One can again proceed as in Sect. 6.2 to determine the state after a photon detection, but additional complications can arise unless all optical transition frequencies ω_{ij} and $\omega_{\ell m}$ are far apart or unless two optical transition frequencies are very close so that $|\omega_{ij} - \omega_{\ell m}|\Delta t \ll 1$. The latter case can lead to interesting coherence effects which were discussed for the Λ system in [31]. In general, oscillatory terms can appear, but it was shown in [34] that for photon counting these can be omitted. The result of [23, 34] is

$$\mathcal{R}\rho = \sum_{\substack{ij\ell m \\ i>j,\ell>m}} \left\{ \Gamma_{ji\ell m} + \Gamma_{\ell mji} \right\} |j\rangle\langle i|\rho|\ell\rangle\langle m| . \qquad (6.87)$$

In the sum, only transitions from higher to lower atomic levels appear. This is physically reasonable since, with a classical laser, one is detecting spontaneous photons only. Anyway, if the laser photons were also treated quantum mechanically then photons from stimulated emissions would have the same direction as the incident, stimulating, laser photons and would therefore not be observed.

As in Sect. 6.2.4, $\mathrm{tr}(\mathcal{R}\rho) \Delta t$ is the probability to find a photon in Δt, and the analog of (6.54) again gives the probability density for the first photon.

6.4.2.1 Examples

(i) *Two-level system.* Then $\mathbf{D}_{ij} = 0$ unless i or j equals 1, the ground state. Then $\mathcal{R}\rho = A \rho_{22}|1\rangle\langle 1|$, by (6.87). This means that after a photon detection the atom is in the ground state as expected and as already found in Sect. 6.2.4.

(ii) Λ-*system,* two ground states $| 1\rangle$ and $| 2\rangle$, excited state $| 3\rangle$ (cf. Fig. 6.6): only $\mathbf{D}_{13}, \mathbf{D}_{31}, \mathbf{D}_{23}$, and $\mathbf{D}_{32}$ are nonzero. Then one has, with the Einstein coefficients A_{31} and A_{32} and in matrix notation,

$$\mathcal{R}\rho = \rho_{33} \begin{pmatrix} A_{31} & \Gamma_{1332} + \Gamma_{3213} & 0 \\ \Gamma_{2331} + \Gamma_{3123} & A_{32} & 0 \\ 0 & 0 & 0 \end{pmatrix} . \qquad (6.88)$$

Therefore, the normalized reset matrix is always the same and given by the matrix divided by its trace. The off-diagonal terms vanish only if $\mathbf{D}_{31} \cdot \mathbf{D}_{23} = 0$. For levels 1 and 2 sufficiently far apart the off-diagonal terms can be neglected for most questions so that the atom can be taken to be either in state $| 1\rangle$ or $| 2\rangle$

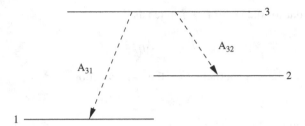

Fig. 6.6 The Λ configuration. The upper level 3 can decay to the lower levels 1 and 2. In addition there may be driving of the 1–3 and 2–3 transitions

after a photon detection. On the other hand, for levels 1 and 2 close together the off-diagonal terms lead to interesting coherence effects such as dark periods [29, 32] and quantum beats in the correlation function $g^{(2)}(\tau)$ even under illumination with *incoherent* light [30].

(iii) *Three-level cascade system.* This level system is shown in Fig. 6.7. In matrix notation one obtains from (6.87)

$$
\mathcal{R}\rho = \begin{pmatrix} A_{21}\rho_{22} & (\Gamma_{1232} + \Gamma_{3212})\rho_{23} & 0 \\ (\Gamma_{2321} + \Gamma_{2123})\rho_{32} & A_{32}\rho_{33} & 0 \\ 0 & 0 & 0 \end{pmatrix}. \tag{6.89}
$$

Here the normalized reset matrix depends on ρ. For a quantum trajectory the reset matrix will therefore in general depend on the time of the resetting since matrix elements of the conditional density matrix elements vary with time if both $\mathbf{D}_{21}$ and $\mathbf{D}_{32}$ are nonzero. The dependence on the state prior to detection, and thus on time, is easy to understand for large level separation. If, for example, the atom were in state $| 2 \rangle$ there would only be a transition to $| 1 \rangle$, while if it were in a superposition of $| 2 \rangle$ and $| 3 \rangle$ it could go to $| 1 \rangle$ or $| 2 \rangle$ or a mixture thereof.

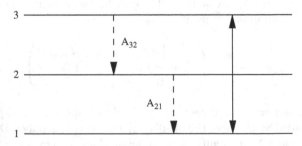

Fig. 6.7 The three-level cascade configuration

6.4.3 Quantum Trajectories and Recovering the Bloch Equations

It will now be shown that the collection of all quantum trajectories of an N-level system with the same initial state yields a solution of the usual Bloch equations for the system.

We consider a large ensemble of systems (atoms) interacting with the quantized radiation field and an external field. After taking a partial trace over the photons, the system alone is described by a density matrix, $\rho(t)$. To each individual atom there corresponds a quantum trajectory, each starting with the initial state $\rho(t_0)$ and with jumps at different times. Now consider a time t. At a fixed time t, each trajectory is in a particular mixed or pure state, which depends on its history. Let the incoherent weighted sum of these states be denoted by $\rho^{(\text{traj})}(t)$. Then in the time interval $(t, t + dt)$ a particular subensemble of systems will have no jumps (photon detections), and this subensemble is described, including its relative size, by

$$U_{\text{cond}}(t + dt, t)\, \rho^{(\text{traj})}(t)\, U_{\text{cond}}^{\dagger}(t + dt, t), \tag{6.90}$$

since the state of each individual trajectory satisfies such an equation and (6.90) is just the weighted incoherent sum of all these. Similarly, the subensemble of systems *with* a jump is described, including its relative size, by

$$\mathcal{R}(\rho^{(\text{traj})}(t))\, dt \,. \tag{6.91}$$

Hence, with

$$U_{\text{cond}}(t + dt, t) = \mathbb{1} - \frac{i}{\hbar} H_{\text{cond}} dt \tag{6.92}$$

one has

$$\rho^{(\text{traj})}(t + dt) = \rho^{(\text{traj})}(t) - \frac{i}{\hbar}\{H_{\text{cond}}\rho^{(\text{traj})} - \rho^{(\text{traj})} H_{\text{cond}}^{\dagger}\} dt + \mathcal{R}(\rho^{(\text{traj})}(t))\, dt \tag{6.93}$$

and this gives

$$\dot{\rho}^{(\text{traj})} = -\frac{i}{\hbar}\{H_{\text{cond}}\rho^{(\text{traj})} - \rho^{(\text{traj})} H_{\text{cond}}^{\dagger}\} + \mathcal{R}\rho^{(\text{traj})} \,. \tag{6.94}$$

This coincides with the quantum optical Bloch equations (which are usually written in a somewhat different way [38, 43]). Since the initial condition is the same, one has that $\rho^{(\text{traj})}(t)$ yields a solution of the optical Bloch equations.

Therefore, the QJA lends itself to a numerical solutions of the Bloch equations by simulations of quantum trajectories. If $\mathcal{R}$ always resets to pure states, one can work in dimension N instead of N^2 for the Bloch equations. For large N this is a tremendous numerical advantage. On the other hand, if $\mathcal{R}$ resets to a density matrix this advantage would be lost; however, as indicated in Sect. 6.6 one can always reset to (auxiliary) pure states.

There are quite efficient ways to derive master equations like the Bloch equations. Hence, for given Bloch equations, one can read off from (6.94) either H_{cond} or $\mathcal{R}\rho$ if the other is known, or one can make an educated guess for both from the form of the Bloch equations.

If one is only interested in simulating solutions of the Bloch equations there are alternative schemes to achieve this, e.g., trajectories of diffusion processes [22, 46]. Still another type of trajectories has been used in [3], see Chap. 10 to study photonic crystals and band gap materials where the Markov property does not hold.

6.5 Quantum Counting Processes

The QJA will now be used to determine the statistics of broadband photon counting for a general N-level atom and, more generally, for systems with jumps (with reset operation $\mathcal{R}$) and a conditional time development between jumps. To be completely general we use density matrices.

6.5.1 Counting Statistics

For a density matrix the conditional time development is given by a superoperator, $\mathcal{T}_{\text{cond}}$, defined by

$$\mathcal{T}_{\text{cond}}(t, t_0)\, \rho(t_0) \equiv U_{\text{cond}}(t, t_0)\, \rho(t_0)\, U_{\text{cond}}^{\dagger}(t, t_0)\,. \tag{6.95}$$

If $\rho(t_0)$ is normalized, its trace gives the probability $P_0(t) \equiv P_0(t; \rho(t_0))$ of finding no photon (jump) between t_0 and t.

The subensemble with (i) no photon found between t_0 and t and (ii) a photon at $t + \Delta t$ is described by $\mathcal{R}\, \mathcal{T}_{\text{cond}}(t, t_0)\, \rho(t_0)$ since $\mathcal{R}$ performs the resetting. The size of the subensemble relative to the original ensemble, and thus the probability $w_1(t, t_0; \rho(t_0))\Delta t$ to find the first photon after t_0 at $t + \Delta t$, is the trace of this times Δt, i.e.,

$$w_1(t, t_0; \rho(t_0)) = \text{tr}\big(\mathcal{R}\, \mathcal{T}_{\text{cond}}(t, t_0)\, \rho(t_0)\big)\,, \tag{6.96}$$

which is the analog of (6.54).

Analogously, the subensemble for which photons are found in the intervals $[t_1, t_1 + \Delta t], \ldots, [t_n, t_n + \Delta t]$ but none in between and none between $t_n + \Delta t$ and t is described by

$$\mathcal{T}_{\text{cond}}(t, t_n)\, \mathcal{R}\, \mathcal{T}_{\text{cond}}(t_n, t_{n-1})\, \cdots\, \mathcal{R}\, \mathcal{T}_{\text{cond}}(t_1, t_0)\, \rho(t_0)\,, \tag{6.97}$$

where $\mathcal{T}_{\text{cond}}(t_i, t_{i-1} + \Delta t)$ has been approximated by $\mathcal{T}_{\text{cond}}(t_i, t_{i-1})$. The size of this subensemble relative to the original one is the trace times $(\Delta t)^n$, and it gives the probability for this event. Going over to the coarse-grained timescale we have that the probability density $w(t_1, \ldots, t_n; [t_0, t])$ for finding exactly n photons at times $t_1 < t_2 \cdots < t_n$ in the interval $[t_0, t]$ is given by

$$w(t_1, \ldots t_n; [t_0, t]) = \text{tr}\big(\mathcal{T}_{\text{cond}}(t, t_n) \, \mathcal{R} \, \mathcal{T}_{\text{cond}}(t_n, t_{n-1}) \, \cdots \, \mathcal{R} \, \mathcal{T}_{\text{cond}}(t_1, t_0) \, \rho(t_0)\big) \, . \tag{6.98}$$

Putting $t = t_n$ the superoperator $\mathcal{T}_{\text{cond}}(t, t_n)$ drops out and one obtains the probability density, $w(t_1, \cdots, t_n)$, for finding n photons at $t_1 < \cdots < t_n$ and none in between,

$$w(t_1, \ldots t_n) = \text{tr}\big(\mathcal{R} \, \mathcal{T}_{\text{cond}}(t_n, t_{n-1}) \, \cdots \, \mathcal{R} \, \mathcal{T}_{\text{cond}}(t_1, t_0) \, \rho(t_0)\big) \, . \tag{6.99}$$

This can be rewritten as follows. We denote by $\hat{\rho}_i$ the normalized reset matrix at time t_i. Then one has, by (6.96),

$$\mathcal{R} \, \mathcal{T}_{\text{cond}}(t_1, t_0) \, \rho(t_0) = w_1(t_1, t_0; \rho(t_0)) \, \hat{\rho}_1 \, ,$$
$$\mathcal{R} \, \mathcal{T}_{\text{cond}}(t_2, t_1) \mathcal{R} \, \mathcal{T}_{\text{cond}}(t_1, t_0) \, \rho(t_0) = w_1(t_1, t_0; \rho(t_0)) \, \mathcal{R} \, \mathcal{T}_{\text{cond}}(t_2, t_1) \hat{\rho}_1 \tag{6.100}$$
$$= w_1(t_1, t_0; \rho(t_0)) \, w_1(t_2, t_1; \hat{\rho}_1) \, \hat{\rho}_2,$$

and so on. In this way one obtains

$$\mathcal{R} \, \mathcal{T}_{\text{cond}}(t_n, t_{n-1}) \cdots \mathcal{R} \, \mathcal{T}_{\text{cond}}(t_1, t_0) \, \rho(t_0) = w_1(t_1, t_0; \rho(t_0)) \cdots w_1(t_n, t_{n-1}; \hat{\rho}_{n-1}) \, \hat{\rho}_n \tag{6.101}$$

and therefore

$$w(t_1, \ldots t_n) = w_1(t_1, t_0; \rho(t_0)) \, w_1(t_2, t_1; \hat{\rho}_1) \, \cdots \, w_1(t_n, t_{n-1}; \hat{\rho}_{n-1}) \, . \tag{6.102}$$

This result can also be obtained directly by considering quantum trajectories and the probability densities after the jumps.

For the two-level system, the V system and the Λ system, in example (ii) of Sect. 6.4 the reset matrix is always the same. In general, if after each jump the system is always reset to the same normalized state or density matrix, $\hat{\rho}_r$ say, i.e., if for any ρ, one has

$$\mathcal{R}\rho = \lambda_\rho \, \hat{\rho}_r = \text{tr}(\mathcal{R}\rho) \, \hat{\rho}_r \, , \tag{6.103}$$

then the memory is lost after each detection. In this case (6.99) factorizes into single-photon probabilities:

$$w(t_1, \ldots, t_n) = \text{tr}\big(\mathcal{R} \, \mathcal{T}_{\text{cond}}(t_n, t_{n-1}) \hat{\rho}_r\big) \cdots \text{tr}\big(\mathcal{R} \, \mathcal{T}_{\text{cond}}(t_1, t_0) \rho(t_0)\big)$$
$$= w_1(t_n, t_{n-1}; \hat{\rho}_r) \, \cdots \, w_1(t_2, t_1; \hat{\rho}_r) \, w_1(t_1, t_0; \rho(t_0)) \, . \tag{6.104}$$

In this case the analysis becomes particularly easy and many quantities can be calculated by Laplace transform as in [49] for the case of a single ground state.

It is also of practical interest to consider the probability density, denoted by $p(t_1, \ldots, t_n)$, associated with trajectories for which jumps (i.e., photons from the corresponding atom) are found in the intervals $[t_1, t_1 + \Delta t], \ldots, [t_n, t_n + \Delta t]$ and arbitrarily many jumps in between. To describe this subensemble, denote the time-development superoperator for the optical Bloch equations by $\mathcal{T}_B(t, t_0)$, i.e.,

$$\rho(t) = \mathcal{T}_B(t, t_0) \rho(t_0) . \tag{6.105}$$

Then starting at time t_0 with $\rho(t_0)$, the time development goes with $\mathcal{T}_B(t, t_0)$ until $t = t_1$, then there is the reset operation $\mathcal{R}$ to those trajectories with a jump, after that the time development with $\mathcal{T}_B(t_2, t_1)$, reset, and so on. The subensemble of trajectories for which jumps are found in the intervals $[t_1, t_1 + \Delta t], \ldots, [t_n, t_n + \Delta t]$ and arbitrarily many jumps in between and later is therefore described at time $t > t_n$ by the density matrix

$$\mathcal{T}_B(t, t_{n-1}) \mathcal{R} \, \mathcal{T}_B(t_n, t_{n-1}) \cdots \mathcal{R} \, \mathcal{T}_B(t_1, t_0) \rho(t_0) , \tag{6.106}$$

and the relative size of the subensemble is given by the trace of this times $(\Delta t)^n$. Therefore the probability density, $p(t_1, \ldots, t_n)$, for finding jumps (photons from the atom) at times $t_1 < \cdots < t_n$, with arbitrarily many jumps in between and at later times, is given by

$$p(t_1, \ldots, t_n) = \text{tr}\big(\mathcal{R} \, \mathcal{T}_B(t_n, t_{n-1}) \cdots \mathcal{R} \, \mathcal{T}_B(t_1, t_0) \rho(t_0)\big) . \tag{6.107}$$

The superoperator $\mathcal{T}_B(t, t_{n-1})$ in (6.106) has dropped out since it conserves the trace.

The above probability densities determine a classical stochastic process whose sample paths are given by the photon detection times of a single radiating atom. Without external pumping these paths terminate. Ergodicity allows one to replace time averages over a single trajectory by ensemble averages. In many cases the latter can be computed analytically.

6.5.2 Converse: From Bloch Equations to the Conditional Time Development

In this section it will be shown that one can derive the conditional time development directly from the Bloch equations and the reset operation, without using the projection machinery from Sects. 6.2.1 and 6.4.1.

We start at time t_0 and consider an ensemble of trajectories of radiating atoms whose incoherent sum of states at time t is described by a density matrix $\rho(t)$. It is assumed that $\rho(t)$ obeys the Bloch equations. We denote by $\rho_0(t)$ the density matrix for the subensemble of trajectories for which no jump (no photon detection) has occurred until time t. The normalization is chosen in such a way that the trace gives the relative size of the subensemble. To begin with, in the time interval $[t_0, t_0 + \Delta]$, a photon will have been detected for some of the atoms, i.e., there will have been a jump on some trajectories, and this subensemble is described by $\mathcal{R}\rho(t_0)\Delta t$, including its relative size. The complementary subensemble is described by the difference of the original $\rho(t_0 + \Delta t)$ and $\mathcal{R}\rho(t_0)\Delta t$, so that

$$\rho_0(t_0 + \Delta t) = \rho(t_0 + \Delta t) - \mathcal{R}\rho(t_0)\Delta t . \tag{6.108}$$

This subensemble as a total would evolve until $t_0 + 2\Delta t$ with the Bloch equations, i.e., as $\mathcal{T}_B(t_0 + 2\Delta t, t_0 + \Delta t)\rho_0(t + \Delta t)$, and the sub-subensemble with no photon detected at time $t + 2\Delta t$ is again a difference,

$$\rho_0(t_0 + 2\Delta t) = \mathcal{T}_B(t_0 + 2\Delta t, t_0 + \Delta t)\rho_0(t_0 + \Delta t) - \mathcal{R}\rho_0(t_0 + \Delta t)\Delta t . \tag{6.109}$$

In the same vain one obtains for general t

$$\rho_0(t + \Delta t) = \mathcal{T}_B(t + \Delta t, t)\rho_0(t) - \mathcal{R}\rho_0(t) \Delta t . \tag{6.110}$$

If $\mathcal{L}_B$ denotes the superoperator which generates the time development of the Bloch equations, then $\mathcal{T}_B(t + \Delta t, t) = e^{\mathcal{L}_B \Delta t}$ and (6.110) can be written as

$$\rho_0(t + \Delta t) = (\mathbb{1} + \mathcal{L}_B \Delta t) \rho_0(t) - \mathcal{R} \rho_0(t) \Delta t . \tag{6.111}$$

This finally yields for the subensemble with no jump (photon detection) until time t the equation

$$\dot{\rho}_0(t) = (\mathcal{L}_B - \mathcal{R}) \rho_0(t) . \tag{6.112}$$

The difference in the brackets is just the time development given by the conditional Hamiltonian, as seen from (6.94).

It may seem from this derivation that one might be able to take the limit $\Delta t \to 0$ without running into the quantum Zeno effect. However, it is well known that the optical Bloch equations are not valid for arbitrarily small times because also in their derivation the Markov property is used, and this requires Δt to be larger than the correlation time. Hence for $\Delta t \to 0$ one would have difficulties with (6.111).

6.5.3 Connection with Continuous Measurements

In order to describe continuous measurements, Davies and Srinivas [47, 48] have extended the axiomatics of quantum mechanics by postulates for "homogeneous quantum counting processes." In particular, their postulates imply the existence of two superoperators, J and S_t, which map trace class operators to trace class operators and satisfy certain properties. For an individual system of an ensemble described by a density matrix ρ their meaning is as follows. $\mathrm{tr}(S_t\rho)$ is the probability of finding no counting event in $[0, t]$, and the probability density, $w_{\mathrm{DS}}(t_1, \ldots, t_n; [0, t])$, for finding a counting event exactly at the times $t_1, \ldots, t_n$ in $[0, t]$ is given by

$$w_{\mathrm{DS}}(t_1, \ldots, t_n; [0, t]) = \mathrm{tr}\left(S_{t-t_n} J S_{t_n-t_{n-1}} J \cdots J S_{t_2-t_1} J S_{t_1}\rho\right) . \quad (6.113)$$

For a particular system J and S_t have to be determined phenomenologically or by intuition.

A comparison with (6.98) shows that this has the same structure as that obtained by the QJA and that the unknown superoperators J and S_t have to coincide with the $\mathcal{R}$ and $\mathcal{T}_{\mathrm{cond}}$ constructed above. In contrast to the axiomatic theory of Davies and Srinivas, however, the superoperators are explicitly known in the QJA. In this way one has arrived, without the need of new axioms, at the results of the continuous measurement theory of [47, 48] for the general N-level atom within the usual framework of quantum mechanics and with the reduction rule of von Neumann and Lüders for demolition measurements, except that a temporal coarse graining has been used. Counting processes analogous to (6.99) have also been derived in [44].

6.6 How to Replace Density Matrices by Pure States in Simulations

In many cases the reset state is not a pure state even if one starts in a pure state. Moreover, the particular reset state in a quantum trajectory may depend on the conditional state preceding it and thus on the whole history before the resetting. If in a simulation one would work with density matrices instead of pure states, one would enormously complicate the numerics for large dimension N of the level space since density matrices lead to N^2 instead of N, as pointed out in Sect. 6.4.3. It will be shown elsewhere [26] that for simulations one can always go over to quantum trajectories with pure states which yield both the same jump statistics and the Bloch equations as the original trajectories. In simple situations this procedure has been used before [4]. For simulations this is of great numerical advantage if N is large. The general proof of this statement is somewhat involved and it may be more instructive to see how this works in a specific example which is simple enough but still exhibits all salient features.

To this end we consider a simple three-level cascade system as in Fig. 6.7, where level 2 may decay to level 1 and level 3 to level 2. Only the 1–3 transition is driven by a laser. To keep things simple we assume the laser to be on resonance, i.e., $\omega_L = \omega_3 - \omega_1$. In the Schrödinger picture the conditional Hamiltonian is then of the form

$$H_{\text{cond}} = \sum_1^3 \hbar\omega_i |i\rangle\langle i| + \frac{\hbar\Omega}{2}\left(|1\rangle\langle 3| e^{i\omega_L t} + |3\rangle\langle 1| e^{-i\omega_L t}\right) - \frac{i\hbar A_2}{2}|2\rangle\langle 2| - \frac{i\hbar A}{2}|3\rangle\langle 3| ,$$

$$(6.114)$$

where A_2 and A are the Einstein coefficients of level 2 and 3, respectively. By going over to the interaction picture we get rid of the time dependence and obtain

$$H_{\text{cond}}^I = \frac{\hbar\Omega}{2}\left(|1\rangle\langle 3| + |3\rangle\langle 1|\right) - \frac{i\hbar A_2}{2}|2\rangle\langle 2| - \frac{i\hbar A}{2}|3\rangle\langle 3| . \qquad (6.115)$$

The reset operation for this system is given in (6.89), and here we write it in the form

$$\begin{aligned}\mathcal{R}\rho = \quad &A_2\rho_{22}|1\rangle\langle 1| + \Gamma\rho_{23}|1\rangle\langle 2| \\ +&\Gamma^*\rho_{32}|2\rangle\langle 1| + A\rho_{33}|2\rangle\langle 2| ,\end{aligned} \qquad (6.116)$$

where we use the abbreviation $\Gamma \equiv \Gamma_{1332} + \Gamma_{3213}$. Note that $|\Gamma|^2 \leq A_2 A$ so that $\mathcal{R}\rho$ is a positive operator.

The operator H_{cond}^I consists of the part $-i\hbar(A_2/2)|2\rangle\langle 2|$, responsible for the decay of level 2, and a remainder operating in the 2D subspace spanned by $|1\rangle$ and $|3\rangle$. If $\hbar\lambda_\pm$ denote the eigenvalues of the 2D part, i.e.,

$$\lambda_+ = -\frac{iA}{4} \pm \frac{1}{4}\sqrt{4\Omega^2 - A^2} , \qquad (6.117)$$

then the conditional time development in the interaction picture is easily determined as in (6.40) to be given by

$$\begin{aligned}\exp\{-iH_{\text{cond}}^I t/\hbar\} =\,&e^{-A_2 t/2}|2\rangle\langle 2| \\ +&\frac{2}{\sqrt{4\Omega^2 - A^2}}\left\{e^{-i\lambda_+ t}\left[\frac{\Omega}{2}\left(|1\rangle\langle 3| + |3\rangle\langle 1|\right) - \lambda_-|1\rangle\langle 1| - \left(\lambda_- + \frac{iA}{2}\right)|3\rangle\langle 3|\right]\right. \\ &\left.-e^{-i\lambda_- t}\left[\frac{\Omega}{2}\left(|1\rangle\langle 3| + |3\rangle\langle 1|\right) - \lambda_+|1\rangle\langle 1| - \left(\lambda_+ + \frac{iA}{2}\right)|3\rangle\langle 3|\right]\right\} .\end{aligned}$$

$$(6.118)$$

In particular, one finds

$$\begin{aligned}\exp\{-iH_{\text{cond}}^I t/\hbar\}|1\rangle = \frac{2}{\sqrt{4\Omega^2 - A^2}}\Big\{&e^{-i\lambda_+ t}\left[-\lambda_-|1\rangle + \frac{\Omega}{2}|3\rangle\right] \\ &-e^{-i\lambda_- t}\left[-\lambda_+|1\rangle + \frac{\Omega}{2}|3\rangle\right]\Big\},\end{aligned} \qquad (6.119)$$

$$\exp\{-iH_{\text{cond}}^I t/\hbar\}|2\rangle = e^{-A_2 t/2}|2\rangle .$$

For an initial state $|\alpha\rangle$ the probability density, denoted by $w_1(t, t_0 = 0; |\alpha\rangle)$, for the first photon is, by (6.38),

$$w_1(t, t_0 = 0; |\alpha\rangle) = \frac{i}{\hbar}\langle\alpha|\exp\{iH^{\dagger}_{\text{cond}}t/\hbar\}\left(H_{\text{cond}} - H^{\dagger}_{\text{cond}}\right)\exp\{-iH_{\text{cond}}t/\hbar\}|\alpha\rangle .$$

(6.120)

For a superposition of $|1\rangle$ and $|2\rangle$,

$$|\alpha\rangle = \alpha_1|1\rangle + \alpha_2|2\rangle ,$$

(6.121)

one then obtains

$$w_1(t, t_0 = 0; |\alpha\rangle) = |\alpha_2|^2 A_2 e^{-A_2 t} + |\alpha_1|^2 \frac{4A\Omega^2}{|4\Omega^2 - A^2|}e^{-At/2}\left|\sin\left(\frac{1}{4}\sqrt{4\Omega^2 - A^2}t\right)\right|^2 .$$

(6.122)

Figure 6.8 shows the behavior of $w_1(t, t_0 = 0; |\alpha\rangle)$ for a particular set of parameters as detailed in the figure caption. For these parameters the mean time for the first photon is $\tau = 13/(8A)$. Figure 6.8 should be compared with the first-photon probability density for the two-level atom in Fig. 6.5. In Fig. 6.8 the probability density does not vanish for $t = 0$ since one starts in a superposition of ground state and first excited state so that the first photon can come more quickly.

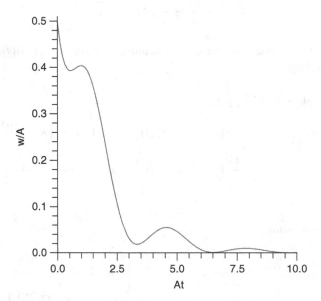

Fig. 6.8 (Color online) Probability density, $w_1(t, t_0 = 0; |\alpha\rangle)$, for the first photon as a function of time, for initial state $|\alpha\rangle = (|1\rangle + |2\rangle)/\sqrt{2}$, $A_2 = A$, $\Omega = 2A$, $|\Gamma|^2 = A_2 A/2$; the mean time for the first photon is $\tau = 13/(8A)$

In a simulation of a quantum trajectory with initial pure state $|\alpha\rangle$, one determines a time, t_1 say, for the first photon according to the probability density $w_1(t, t_0 = 0; |\alpha\rangle)$ by means of random number operators in a standard way. Then, after this first photon, the state has to be reset by the reset operation $\mathcal{R}$ in (6.116). For the present example this can be written in matrix form as

$$
\mathcal{R}\rho^{\,\text{cond}} = \begin{pmatrix} A_2 \rho_{22}^{\text{cond}} & \Gamma \rho_{23}^{\text{cond}} & 0 \\ \Gamma \rho_{32}^{\text{cond}} & A \rho_{33}^{\text{cond}} & 0 \\ 0 & 0 & 0 \end{pmatrix}, \tag{6.123}
$$

where $\rho^{\,\text{cond}}$ is given by the original state $|\alpha\rangle$ conditionally time-developed until time t_1 and written as a density matrix,

$$
\rho^{\text{cond}}(t_1) = \exp\{-iH_{\text{cond}}t_1/\hbar\}|\alpha\rangle\langle\alpha|\exp\{iH^{\dagger}_{\text{cond}}t_1/\hbar\}. \tag{6.124}
$$

Note that

$$
\text{tr}\rho^{\text{cond}} = w(t_1), \tag{6.125}
$$

the probability density for the first photon, as it should. The reset matrix normalized to trace 1 is denoted by $\hat{\rho}$,

$$
\hat{\rho} \equiv \mathcal{R}\rho^{\,\text{cond}}/\text{tr}(\mathcal{R}\rho^{\,\text{cond}}). \tag{6.126}
$$

If the initial state is either $|1\rangle$, $|2\rangle$, or $|3\rangle$, the reset matrix $\hat{\rho}$ after the first photon corresponds to a pure state, namely $|2\rangle$, $|1\rangle$, or $|2\rangle$, respectively. But for the initial state $|\alpha\rangle$ from (6.121) one has

$$
\begin{aligned}
\rho_{22}^{\text{cond}}(t_1) &= |\alpha_2|^2 e^{-A_2 t_1}, \\
\rho_{33}^{\text{cond}}(t_1) &= 4|\alpha_1|^2 \frac{\Omega^2}{\sqrt{4\Omega^2 - A^2}} e^{At_1/2} \sin^2 \sqrt{4\Omega^2 - A^2}\, t_1/4, \\
\rho_{23}^{\text{cond}}(t_1) &= 2i\bar{\alpha}_1\alpha_2 \frac{\Omega}{\sqrt{4\Omega^2 - A^2}} e^{-A_2 t_1/2 - At_1/4} \sin \sqrt{4\Omega^2 - A^2}\, t_1/4
\end{aligned} \tag{6.127}
$$

and then the reset matrix need not be pure state. Indeed, for it to be pure, the nonzero 2×2 submatrix has to have determinant 0, i.e.,

$$
A_2\rho_{22}^{\text{cond}} A\rho_{33}^{\text{cond}} - |\Gamma\rho_{23}^{\text{cond}}|^2 = 0. \tag{6.128}
$$

Since with the initial state $|\alpha\rangle$ from (6.121) one has $|\rho_{23}^{\text{cond}}|^2 = \rho_{22}^{\text{cond}}\rho_{33}^{\text{cond}}$ this requires $|\Gamma|^2 = A_2 A$ to have a pure reset state.

In case the latter condition is not fulfilled the reset matrix can correspond to a mixed state. To see how large the deviation from a pure state is if $|\Gamma|^2 < A_2 A$ we diagonalize $\hat{\rho}$. Its eigenvalues are 0 with eigenvector $|3\rangle$, p_+ and $p_- = 1 - p_+$, where

$$p_\pm = \frac{1}{2} \pm \frac{1}{2}\sqrt{1 + 4\frac{|\Gamma|^2|\rho_{23}^{\text{cond}}|^2 - A_2 A \rho_{22}^{\text{cond}} \rho_{33}^{\text{cond}}}{(A_2 \rho_{22}^{\text{cond}} + A \rho_{33}^{\text{cond}})^2}} \, , \qquad (6.129)$$

with corresponding normalized eigenvectors

$$|p_\pm\rangle = \begin{pmatrix} \Gamma \rho_{23}^{\text{cond}} \\ (A_2 \rho_{22}^{\text{cond}} + A \rho_{33}^{\text{cond}})p_\pm - A_2 \rho_{22}^{\text{cond}} \\ 0 \end{pmatrix} /\text{norm} \, . \qquad (6.130)$$

When $\rho_{23}^{\text{cond}}(t_1)$ vanishes $\hat{\rho}$ is diagonal and the eigenvectors are $|1\rangle$ and $|2\rangle$, respectively.

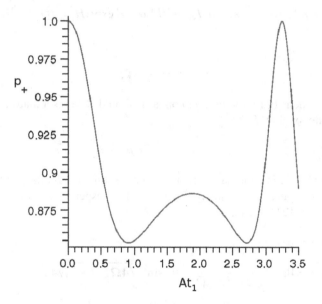

Fig. 6.9 (Color online) The eigenvalue p_+ of the reset matrix right after the first photon at time t_1, as a function of t_1, for the same initial state and parameters as in Fig. 6.8. In the center of the time interval, p_+ differs appreciatively from 1 and thus $\hat{\rho}$ considerably deviates from a pure state

In Fig. 6.9 we have plotted p_+ as a function of the time t_1 of the first photon, for the same initial state and parameters as in Fig. 6.8. At the initial time one has the initial pure state and hence $p_+ = 1$. Whenever $p_+ < 1$, $\hat{\rho}$ is a mixture, and the deviation from a pure state is the more pronounced the closer p_+ is to 1/2.

Figure 6.10 shows that the eigenvectors of the reset matrix $\hat{\rho}$ can deviate significantly from $|1\rangle$ and $|2\rangle$. In this figure we have plotted the absolute value, R, of the ratio of the first and second components of $|p_+\rangle$ as a function of the time t_1 of the first photon, for the same initial state and parameters as in Fig. 6.8. In the center of the depicted time interval the first and second components of $|p_+\rangle$ are of similar

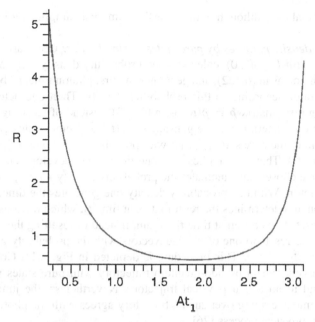

Fig. 6.10 (Color online) Absolute value, R, of the ratio of the first and second components of the eigenvector $|p_+\rangle$ of the reset matrix, as a function of the time t_1, for the same initial state and parameters as in Fig. 6.8. In the center of the time interval the first and second components of $|p_+\rangle$ are of similar magnitude, while at the boundary times it is predominantly in $|1\rangle$

magnitude, while at times close to the boundary of the figure $|p_+\rangle$ is close to $|1\rangle$ and $|p_-\rangle$ close to $|2\rangle$.

For a simulation one now has two quite different situations. When one starts in the atomic state $|1\rangle$, the atom is pumped to a superposition of $|3\rangle$ and $|1\rangle$. After the first photon it jumps to $|2\rangle$. After the second photon it jumps from $|2\rangle$ to $|1\rangle$, so after two photons one is back to the original starting state $|1\rangle$ and the simulation continues as before. When one starts in the atomic state $|2\rangle$, the atom decays to $|1\rangle$, and after that the simulation continues as before.

However, if one starts from a superposition of $|1\rangle$ and $|2\rangle$, which can be prepared, e.g., by a pulse from a second laser, the situation is completely different. In this case the atom will in general be reset to a nonpure state right after the first photon. In a simulation one would then have to continue the conditional time development until the next photon with a density matrix. For a higher dimensional atomic state space a simulation with density matrices is much more time consuming than with pure states. As noted at the beginning of this section, one can bypass this complication and go over to a sequence of (simulated) pure states, without changing the overall jump statistics in a trajectory and still generating the Bloch equations for an ensemble of such trajectories [26]. This tremendously reduces numerical effort in more complicated cases. We will demonstrate this procedure in the simple cascade model outlined above, although in this case the numerical simplifications are not so pronounced.

Replacing density matrices by pure states in simulations: One starts in the state $|\alpha\rangle$, develops with $U_{\text{cond}}(t, 0)$, calculates the probability density $w_1(t, t_0 = 0; |\alpha\rangle)$ for the first photon as in (6.122), and generates the first photon time, t_1, by a random number generator according to this probability density. Then one determines the corresponding reset matrix $\hat{\rho} \equiv \hat{\rho}(t_1)$ as in (6.123). Instead of resetting the atomic state to $\hat{\rho}(t_1)$ one determines its eigenvalues $p_{\pm}(t_1)$ and eigenstates $|p_{\pm}(t_1)\rangle$ and resets the atom to the pure state $|p_{+}(t_1)\rangle$ with probability $p_{+}(t_1)$ or to $|p_{-}(t_1)\rangle$ with probability $p_{-}(t_1)$. Then one applies the conditional time development $U_{\text{cond}}(t, t_1)$ to the pure state chosen and calculates the probability density $w_1(t, t_1; |p_{\pm}(t_1)\rangle)$ for the second photon. With this probability density one generates the time, t_2, for this second photon and determines the reset matrix at time t_2, which obviously depends on the pure reset state chosen at time t_1. Again, instead of resetting the atom to this reset matrix, one resets to one of its eigenvectors with the probability given by the corresponding eigenvalue. This procedure is depicted in Fig. 6.11. Continuing in this way, one generates a pseudo-quantum trajectory with pure states which does not correspond to an actual physical trajectory. Nevertheless, the jump statistics obtained by time averaging over such a trajectory agrees with the photon statistics of the original physical process [26].

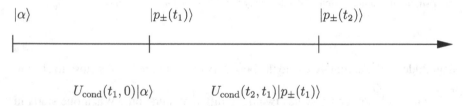

Fig. 6.11 Simulation of a trajectory consisting of pure states. One starts in $|\alpha\rangle$, develops with U_{cond}, generates the first jump time t_1, and determines the reset state for $U_{\text{cond}}(t_1, 0)|\alpha\rangle$ and its eigenvalues $p_{\pm}(t_1)$ and eigenvectors $|p_{\pm}(t_1)\rangle$. At t_1 one resets to $|p_{\pm}(t_1)\rangle$ with probability $p_{\pm}(t_1)$. Then one develops the chosen reset state with $U_{\text{cond}}(t, t_1)$, generates the next jump time t_2, and determines the reset state for $U_{\text{cond}}(t_2, t_1)|p_{\pm}(t_1)\rangle$ and its eigenvalues $p_{\pm}(t_2)$ and eigenvectors $|p_{\pm}(t_2)\rangle$. At t_2 one resets to $|p_{\pm}(t_2)\rangle$ with probability $p_{\pm}(t_2)$, and so on

Recovering the Bloch equations: Now one can repeatedly generate a large ensemble of such trajectories, always starting with the same initial state $|\alpha\rangle$. At a fixed time t, each trajectory is in a particular pure state, which depends on its history. Let the incoherent weighted sum of these states be denoted by $\rho_{|\alpha\rangle}^{(\text{sim})}(t)$. It can then be shown [26] that this density matrix satisfies the optical Bloch equations of the original problem with the same initial condition, i.e., $|\alpha\rangle\langle\alpha|$, and hence renders a solution of the Bloch equations of the original problem. In general, for a large number of levels, this can be numerically extremely advantageous.

6.7 Inclusion of Center-of-Mass Motion and Recoil

6.7.1 Conditional Hamiltonian and Reset Operation

Until now, the N-level system was assumed to be at rest. For atoms or ions in a trap this is often a good approximation. In general, however, the motion of the system should be taken into account, in particular when the laser intensity varies in space and when the center-of-mass (*cm*) motion should be treated quantum mechanically, as for lower velocities.

As an example we consider a two-level system with quantized *cm* motion, interacting with a classical laser and with the quantized radiation field. In the Schrödinger picture the corresponding Hamiltonian is of the form

$$H = \hat{\mathbf{P}}/2m + \hbar\omega_0 |2\rangle\langle 2| + \sum \hbar\omega_k a^\dagger_{\mathbf{k}\lambda} a_{\mathbf{k}\lambda} + e\hat{\mathbf{D}} \cdot \mathbf{E}(\hat{\mathbf{X}}, 0) + e\hat{\mathbf{D}} \cdot \mathbf{E}_L(\hat{\mathbf{X}}, t) , \quad (6.131)$$

where $\hat{\mathbf{X}}$ and $\hat{\mathbf{P}}$ denote *cm* position and momentum operator and $\hat{\mathbf{D}}$ the dipole operator as in (6.11). The classical laser field is of the form

$$\mathbf{E}_L(\mathbf{x}, t) = \operatorname{Re} \mathbf{E}_0(\mathbf{x}) e^{-i\omega_L t} , \quad (6.132)$$

where we allow for a position-dependent laser amplitude. We put

$$g_{k\lambda} \equiv ie\sqrt{\frac{\omega_k}{2\varepsilon_o \hbar V}} \boldsymbol{\varepsilon}_{\mathbf{k}\lambda} \cdot \mathbf{D}_{12}$$

and go over to a "laser-adapted *cm* field interaction picture" by means of the operator

$$\tilde{H}^0_L = \hat{\mathbf{P}}^2/2m + \hbar\omega_L |2\rangle\langle 2| + H^0_F \quad (6.133)$$

and denote the resulting Hamiltonian by H^{Lcm}_I. With the rotating wave approximation one then has in this interaction picture

$$
\begin{aligned}
H^{Lcm}_I = &- \hbar\Delta |2\rangle\langle 2| \\
&+ |1\rangle\langle 2| \sum g_{\mathbf{k}\lambda} a_{k,\lambda} \exp\{-i(\omega_k - \omega_L)t\} \exp\{i\mathbf{k} \cdot (\hat{\mathbf{X}} + \hat{\mathbf{P}}t/m)\} + \text{h.c.} \\
&+ \frac{\hbar}{2} |2\rangle\langle 1| \,\Omega(\hat{\mathbf{X}} + \hat{\mathbf{P}}t/m) + \text{h.c.} ,
\end{aligned} \quad (6.134)
$$

where $\Delta = \omega_L - \omega_0$ is the detuning and the position-dependent Rabi frequency is

$$\Omega(\mathbf{x}) = e\mathbf{D}_{12} \cdot \mathbf{E}_0(\mathbf{x}) .$$

As is (6.7) we consider the zero-photon matrix element of the time-development operator from t to $t + \Delta t$. In the Dyson series of the time-development operator we keep only terms proportional to Δt. Then only the laser contributes to the first order, which is

$$-\frac{i}{\hbar} \int_t^{t+\Delta t} dt' \left\{ \frac{\hbar}{2} |2\rangle\langle 1| \, \Omega(\hat{\mathbf{X}} + \hat{\mathbf{P}} t'/m) + \text{h.c.} \right\} \, .$$

The second-order term in the time-development operator becomes

$$\left(-\frac{i}{\hbar} \right)^2 \langle 0_{\text{ph}}| \int_t^{t+\Delta t} dt' \int_t^{t'} dt'' H_I^{Lcm}(t') H_I^{Lcm}(t'') |0_{\text{ph}}\rangle =$$

$$-\frac{1}{\hbar^2} |2\rangle\langle 2| \int_t^{t+\Delta t} dt' \int_t^{t'} dt'' \sum_{\mathbf{k}\lambda} |g_{\mathbf{k}\lambda}|^2 \exp\{i(\omega_k - \omega_L)(t' - t'')\}$$

$$\exp\{i\mathbf{k} \cdot (\hat{\mathbf{X}} + \hat{\mathbf{P}} t'/m)\} \exp\{-i\mathbf{k} \cdot (\hat{\mathbf{X}} + \hat{\mathbf{P}} t''/m)\} \, , \quad (6.135)$$

where again, as in Sect. 6.2.1, the laser terms have been omitted since they are proportional to $(\Delta t)^2$.

We note that

$$\hat{\mathbf{X}}(t') \equiv \hat{\mathbf{X}} + \hat{\mathbf{P}} t'/m = \hat{\mathbf{X}} + \hat{\mathbf{P}} t/m + \hat{\mathbf{P}}(t' - t)$$

is the time development of $\hat{\mathbf{X}}$ in the free Heisenberg picture of the *cm* motion of the system. With $\Delta t \approx 10^{-12}$ s as in (6.1) and the atomic velocity around 1 m/s one has $\Delta x \equiv v \Delta t = 10^{-12}$ m. With optical wavelengths one has $k \sim 2 \times 10^6$ m^{-1} so that $k \Delta x \sim 2 \times 10^{-6}$ and therefore the $\hat{\mathbf{P}}(t' - t)$ part can be neglected in the exponent. It can similarly be neglected in $\Omega(\hat{\mathbf{X}} + \hat{\mathbf{P}} t'/m)$ of the first-order term if $\Omega(x)$ does not vary significantly over a distance of Δx. Hence, under these conditions one has

$$\hat{\mathbf{X}}(t') \approx \hat{\mathbf{X}}(t'') \approx \hat{\mathbf{X}}(t) \, . \quad (6.136)$$

Then the last two exponentials in (6.135) cancel. As in Sect. 6.2.1 for the two-level case without *cm* motion this leads to the damping term $-\frac{1}{2} A |2\rangle\langle 2|$ and one obtains as conditional Hamiltonian in the laser-adapted *cm* interaction picture

$$H_{\text{cond}}^{Lcm} = -\hbar\Delta |2\rangle\langle 2| - \frac{i}{2}\hbar A |2\rangle\langle 2| + \frac{\hbar}{2}\Omega(\hat{\mathbf{X}}(t))|2\rangle\langle 1| + \text{h.c.} \, . \quad (6.137)$$

Reversing the interaction picture with respect to the free *cm* Hamiltonian, one obtains for the conditional Hamiltonian in the laser-adapted interaction picture, denoted as in (6.36) by H_{cond}^{IL},

$$H_{\text{cond}}^{IL} = \hat{\mathbf{P}}^2/2m - \hbar\Delta |2\rangle\langle 2| - \frac{i}{2}\hbar A |2\rangle\langle 2| + \frac{\hbar}{2}\Omega(\hat{\mathbf{X}})|2\rangle\langle 1| + \text{h.c.} \, . \quad (6.138)$$

In the Schrödinger picture this becomes

$$H_{\text{cond}} = \hat{\mathbf{P}}^2/2m + \hbar\omega_0|2\rangle\langle 2| - \frac{\mathrm{i}}{2}\hbar A|2\rangle\langle 2| + \frac{\hbar}{2}\Omega(\hat{\mathbf{X}})|2\rangle\langle 1|e^{-\mathrm{i}\omega_L t} + \text{h.c.} \,,$$

(6.139)

which is the analog of (6.34).

The reset operation looks somewhat different if *cm* motion is included. The atomic density operator ρ now describes both the *cm* motion and the internal two-level degrees of freedom. Therefore, if one takes matrix elements with momentum eigenvectors of the *cm* motion, then

$$\rho(\mathbf{p}, \mathbf{p}') \equiv \langle \mathbf{p}|\rho|\mathbf{p}'\rangle$$

(6.140)

becomes an operator for the internal degrees of freedom, which corresponds to a 2×2 matrix in case of a two-level system. The reset operation $\mathcal{R}$ is again given by the general expression in (6.49), but now with the Hamiltonian from (6.131). As before, in first-order perturbation theory the external field drops out and one obtains instead of (6.50)

$$\mathcal{R}\rho \,\Delta t = \int_0^{\Delta t} \mathrm{d}t' \int_0^{\Delta t} \mathrm{d}t''$$

$$\sum_{\mathbf{k}\lambda} e^{\mathrm{i}(\omega_k - \omega_0)(t' - t'')} \frac{e^2 \omega_k}{2\varepsilon_0 \hbar V} |\boldsymbol{\varepsilon}_{\mathbf{k}\lambda} \cdot \mathbf{D}_{21}|^2 e^{-\mathrm{i}\mathbf{k}\cdot\hat{\mathbf{X}}(t')}|1\rangle\langle 2|\rho|2\rangle\langle 1|e^{\mathrm{i}\mathbf{k}\cdot\hat{\mathbf{X}}(t'')} \,.$$

(6.141)

Similarly as before one can replace $\hat{\mathbf{X}}(t')$ and $\hat{\mathbf{X}}(t'')$ by $\hat{\mathbf{X}}$, by (6.136). Note that since $\langle 2|\rho|2\rangle$ is a *cm* operator, it does not commute with $\hat{\mathbf{X}}$. Now one can proceed as in (6.51) to split the double integral and use the property of the correlation function. Again this gives a $\pi\delta(\omega_k - \omega_0)$ which allows to perform the $k^2 dk$ integration. This gives

$$\mathcal{R}\rho = A|1\rangle\langle 1| \int \mathrm{d}\Omega_{\mathbf{k}} \frac{|\mathbf{D}_{21}|^2 - |\hat{\mathbf{k}} \cdot \mathbf{D}_{21}|^2}{|\mathbf{D}_{21}|^2} e^{-\mathrm{i}\mathbf{k}\cdot\hat{\mathbf{X}}}\langle 2|\rho|2\rangle e^{\mathrm{i}\mathbf{k}\cdot\hat{\mathbf{X}}} \,,$$

(6.142)

where the angular integration is over the unit vectors $\hat{\mathbf{k}}$. In momentum space one has

$$e^{\mathrm{i}\mathbf{k}\cdot\hat{\mathbf{X}}}|\mathbf{p}\rangle = |\mathbf{p} + \hbar\mathbf{k}\rangle$$

(6.143)

and this gives with (6.142)

$$\langle \mathbf{p}|\mathcal{R}\rho|\mathbf{p}'\rangle = A|1\rangle\langle 1| \int \mathrm{d}\Omega_{\mathbf{k}} \left(1 - \frac{|\hat{\mathbf{k}} \cdot \mathbf{D}_{21}|^2}{|\mathbf{D}_{21}|^2}\right) \langle 2|\langle \mathbf{p} + \hbar\omega_0\hat{\mathbf{k}}/c| \,\rho \,|\mathbf{p}' + \hbar\omega_0\hat{\mathbf{k}}/c\rangle|2\rangle \,.$$

(6.144)

The first factor in the integral is the usual dipole emission characteristics and the terms $\hbar\omega_0\hat{\mathbf{k}}/c$ yield momentum conservation after the photon emission. We note that the resulting reset matrix is a pure state only for internal degrees of freedom but not for the *cm* variables, even if the density matrix before the photon detection is a pure state.

Instead of asking for the detection of *any* photon one may ask for a photon detection in a given direction $\hat{\mathbf{k}}$. Then the reset operation, $\mathcal{R}_{\hat{\mathbf{k}}}$, is given by

$$\langle \mathbf{p} | \mathcal{R}_{\hat{\mathbf{k}}} \rho | \mathbf{p}' \rangle = A | 1 \rangle \langle 1 | (1 - \frac{|\hat{\mathbf{k}} \cdot \mathbf{D}|^2}{\mathbf{D}^2}) \langle 2 | \langle \mathbf{p} + \hbar \omega_0 \hat{\mathbf{k}} / c | \rho | \mathbf{p}' + \hbar \omega_0 \hat{\mathbf{k}} / c \rangle | 2 \rangle \,,$$

(6.145)

which reflects the dipole emission characteristics. If ρ is a pure state in momentum space this is again a pure state. The probability to detect a photon with direction in the solid angle $d\Omega_{\mathbf{k}}$ in the time interval dt is given by

$$\operatorname{tr} \mathcal{R}_{\hat{\mathbf{k}}} \rho \, d\Omega_{\mathbf{k}} \, dt \,.$$

(6.146)

Integrating (6.145) over all directions gives the reset matrix in (6.144).

Similarly, one may ask for a photon detection in a given direction $\hat{\mathbf{k}}$ and given polarization λ. The corresponding reset operation, $\mathcal{R}_{\hat{\mathbf{k}}\lambda}$, is then given by

$$\langle \mathbf{p} | \mathcal{R}_{\hat{\mathbf{k}}\lambda} \rho | \mathbf{p}' \rangle = A | 1 \rangle \langle 1 | \frac{|\boldsymbol{\varepsilon}_{\mathbf{k}\lambda} \cdot \mathbf{D}_{21}|^2}{\mathbf{D}^2} \langle 2 | \langle \mathbf{p} + \hbar \omega_0 \hat{\mathbf{k}} / c | \rho | \mathbf{p}' + \hbar \omega_0 \hat{\mathbf{k}} / c \rangle | 2 \rangle \,.$$

(6.147)

Again, if ρ is a pure state in momentum space this is again a pure state, and the analog of (6.146) holds.

In simulations of quantum trajectories of a two-level system with *cm* one can work with the reset operation $\mathcal{R}_{\hat{\mathbf{k}}}$ or with $\mathcal{R}_{\hat{\mathbf{k}}\lambda}$ and thus with pure states when one starts in a pure state [45]. In this case one does not need the more complicated procedure of Sect. 6.6.

6.7.2 Application to Quantum Arrival Times

An important open problem in quantum theory is the question of how to formulate the notion of "arrival time" of a particle, such as an atom, at a given location, i.e., the time instant of its first detection there. This is clearly a very physical question, but when the extension and spreading of the wave packet is taken into account, a satisfactory formulation is far from obvious. The problem of time in quantum mechanics, both for time instants and time durations such as dwell time, has received a great deal of theoretical attention recently [40]. When the translational motion of the particle can be treated classically, a full quantum analysis of arrival time is in fact not necessary. This is the case for fast particles, and therefore arrival times are presently measured mostly by means of time-of-flight techniques, whose analysis is carried out in terms of classical mechanics. Problems, though, arise for slow particles for which the finite extent of the wavefunction and its spreading become relevant, such as for cooled atoms dropping out of a trap. Then a quantum theoretical point of view is needed. It is therefore important to find out when the classical approximations fail and to propose measurement procedures for arrival times in the quantum case. Since the theoretical definition of a quantum arrival time is still subject to debate it is necessary to determine what exactly such measurement procedures are measuring and to compare such operational approaches with the existing, more abstract and axiomatic, theories.

An experimentally very natural approach to determine the arrival time of an atom is by quantum optical means. A region of space may be illuminated by a laser and upon entering the region an atom will start emitting photons. The first photon emission can be taken as a measure of the arrival time of the atom in that region.

It is easiest to first study the 1D case and use the corresponding equations of Sect. 6.7.1. Then (6.139) simply becomes 1D in p and x. Illuminating the half axis $x > 0$ perpendicularly by a laser, the Rabi frequency $\Omega(x)$ becomes a multiple of the step function $\Theta(x)$. In [18] the corresponding conditional time development has been solved explicitly and the distribution of the first-photon times have been calculated. If one deducts by a deconvolution technique the delays due to the finite pumping time, one obtains in the limit of weak pumping a surprising result for the arrival time distribution, namely the usual well-known quantum mechanical probability flux. More details can be found in Chap. 4.

Another model for arrival times will be discussed in Sect. 6.8. This model couples the center-of-mass motion to spins which in turn are coupled to bosons. The arrival of the particle induces a spin flip which in turn induces the emission of a boson.

6.8 Extension to Spin-Boson Models

In this section it will be shown in an example that the QJA can be extended to a system which is coupled to a bath I which in turn is coupled to another bath, bath II. Measurements are taken on bath II and from these one can infer properties of the small system. Bath I serves as an amplifier to enhance the signal and can be viewed as a part of the measuring apparatus, a part which is treated quantum mechanically.

The model to be considered here consists of a moving particle coupled to a spatial array of spins which in turn are coupled to bosons [21]. This model recently has been used to study arrival times as well as passage times [27, 28]. In places where the particle wavefunction overlaps with a spin there is a high spin-flip probability. Initially the spins are in the (metastable) up state. When the particle passes a spin there is a very high probability for a single spin flip, accompanied by a boson emission, as depicted in Fig. 6.12. The spin flip leads to a large energetic gain for neighboring spins and therefore to a large spin-flip probability of the neighboring spins. By a domino effect, this leads to a sudden flip of *all* spins, accompanied by a burst of bosons, which can be detected.

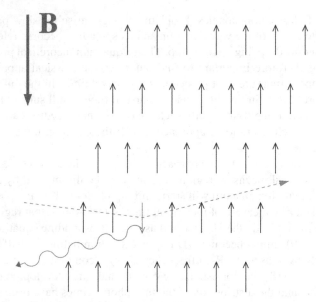

Fig. 6.12 (Color online) Array of spins in a metastable state in a weak magnetic field **B**. A passing particle produces a single spin flip, accompanied by a boson emission. By a domino effect, the spin flip causes all spins to flip, accompanied by an avalanche of bosons

The Hamiltonian is taken as

$$H = \frac{\hat{\mathbf{P}}^2}{2m} + \sum_j \frac{\varepsilon^{(j)}}{2} \hat{\sigma}_z^{(j)} - \sum_{j<j'} \frac{J^{(jj')}}{2} \hat{\sigma}_z^{(j)} \otimes \hat{\sigma}_z^{(jj')} + \hbar \sum_\ell \omega_l \hat{a}_\ell^\dagger \hat{a}_\ell$$

$$+\hbar \sum_j \sum_l \left(\gamma_l^{(j)} e^{i f_l^{(j)}} \hat{a}_\ell^\dagger \hat{\sigma}_-^{(j)} + \text{h.c.} \right)$$

$$+\hbar \sum_j \chi^{(j)}(\hat{\mathbf{x}}) \sum_\ell \left(g_l^{(j)} e^{i f_l^{(j)}} \hat{a}_\ell^\dagger \hat{\sigma}_-^{(j)} + \text{h.c.} \right) . \tag{6.148}$$

The individual terms are the free Hamiltonian for the particle, the spin Hamiltonian with ferromagnetic interaction, the free boson Hamiltonian, a small permanent spin-bath coupling which induces very rare spontaneous spin flips, and a further spin-bath coupling which is strongly enhanced if the particle is close to a spin, i.e., $|g_l^{(j)}| \gg |\gamma_l^{(j)}|$, where $\chi^{(j)}(x)$ is a sensitivity function, e.g., equal to 1 in a neighborhood of the jth spin, while $e^{i f_l^{(j)}}$ are possible phase factors which will later be put to 1.

It should be noted that the Hamiltonian conserves the excitation number, which is the sum of up spins and bosons. Hence a boson detection indicates a spin flip, which in turn indicates that the particle has passed close to a spin (unless one has an extremely rare exceptional spontaneous spin flip). Therefore, one can consider the time of a boson detection as a signal for the arrival of the particle at the spin array so that this model can be regarded as a model for arrival times [27, 28]. The motivation for this model comes from the desire to minimize the backreaction of the detection on the particle by employing the intermediary spin system.

To illustrate how the QJA works for this model we use a simple example in one space dimension, with a *single* spin and N discrete boson modes, where later N is taken to infinity. The modes ω_ℓ and coupling constants g_ℓ are taken as

$$\omega_\ell = \omega_M n/N, \quad n = 1, \cdots, N,$$
$$g_\ell = -iG\sqrt{\omega_\ell/N}, \tag{6.149}$$

where ω_M is the maximal boson frequency. The particle wave packet is assumed to come in from the left. The probability, $P_1^{\mathrm{disc}}(t)$ ("disc" stands for "discrete"), of finding the detector spin in state $|\downarrow\rangle$ at time t is given by

$$P_1^{\mathrm{disc}}(t) = \sum_\ell \int_{-\infty}^{\infty} dx \, |\langle x \downarrow 1_\ell | \Psi_t \rangle|^2 \tag{6.150}$$
$$\equiv 1 - P_0^{\mathrm{disc}}(t).$$

As long as no recurrences occur (i.e., no transitions $|\downarrow 1_\ell\rangle \mapsto |\uparrow 0\rangle$) one can regard

$$w_1^{\mathrm{disc}}(t) = \frac{d}{dt} P_1^{\mathrm{disc}}(t) = -\frac{d}{dt} P_0^{\mathrm{disc}}(t) \tag{6.151}$$

as the probability density for a spin flip (i.e., for a detection) at time t. In Fig. 6.13 the dots represent results of a numerical calculation of the spin-flip probability density of $N = 40$ spins, for a sensitivity function $\chi(x) = \Theta(x)$ and for

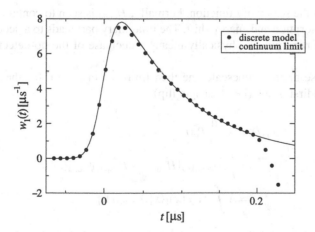

Fig. 6.13 *Dots*: spin-flip probability density for incoming Gaussian wavefunction, $N = 40$, $\chi(x) = \Theta(x)$, where Θ is the step function; *solid line*: QJA result for continuum limit ($N \to \infty$), the numerical evaluation is much faster

an incoming Gaussian wavefunction. Numerically the calculation is extremely time consuming.

Quantum jump approach: In the continuum limit, $N \to \infty$, the QJA can be employed and leads to significant numerical simplifications if the coupling constants are such that the Markov property holds. This means here that for the correlation function $\kappa(\tau)$, now defined as

$$\kappa(\tau) \equiv \sum_\ell |g_\ell|^2 e^{-i(\omega_\ell - \omega_0)\tau} , \qquad (6.152)$$

there is a small correlation time τ_c such that

$$\kappa(\tau) \approx 0 \text{ if } \tau > \tau_c .$$

Then one can define A and δ_{shift}, the analogs of Einstein coefficient and level shift, by

$$A \equiv 2 \operatorname{Re} \int_0^\infty d\tau \kappa(\tau) ,$$
$$\delta_{\text{shift}} \equiv 2 \operatorname{Im} \int_0^\infty d\tau \kappa(\tau) . \qquad (6.153)$$

Proceeding as in Sect. 6.2, but now with $\Delta t \gg \tau_c$, one obtains that the time development under the condition of no boson detection (i.e., no spin flip) is given by the conditional Hamiltonian

$$H_{\text{cond}} = \hat{P}^2/2m + \hbar/2 \, (\delta_{\text{shift}} - i \, A)\chi(\hat{x})^2 , \qquad (6.154)$$

where $\chi(x)$ is the sensitivity function. Formally, H_{cond} is seen to contain a position-dependent absorption and energy shift. The imaginary part leads to a decrease of the wavefunction norm which physically means a decrease of the no-detection probability.

On a coarse-grained timescale one then finds, as in (6.38), for the probability density of the first boson (i.e., first spin flip)

$$w_1(t) = -\frac{d}{dt} P_0(t)$$
$$= \frac{i}{\hbar} \langle \psi_{\text{cond}}(t) | H_{\text{cond}} - H_{\text{cond}}^\dagger | \psi_{\text{cond}}(t) \rangle \qquad (6.155)$$
$$= A \int dx \, \chi(x) |\langle x | \psi_{\text{cond}}(t) \rangle|^2 .$$

If $\chi(x)$ is the characteristic function of an interval, where the spin is located, this is just the decay rate A of the excited state of the detector multiplied by the probability that the particle is inside the detector but is not yet detected – a very physical result.

This result has been used to calculate $w_1(t)$, the solid curve in Fig. 6.13, for the continuous version of (6.149), with $\chi(x) = \Theta(x)$ and otherwise the same parameters as for the discrete case. Up to the time of revivals – due to the discrete nature of the bath – the agreement between the discrete version and the QJA result is excellent. The numerical evaluation of the QJA result, however, is much faster.

Incidentally, it should be noted that for the discrete example one did not calculate the probability for no boson detection *until* time t but rather the probability of no boson detection *at* time t. Figure 6.13 shows that until the occurrence of revivals for the discrete case and in the limit of infinite boson modes this makes no difference.

Reset operation: If ρ_{tot} denotes the density matrix for the total system when the particle is in the state ρ_{p}, the spins are up and no bosons present, then $\mathcal{R}$ is now given by

$$\mathcal{R}\rho_{\text{p}} \cdot \Delta t \equiv \text{tr}_{\text{spin}}\, \text{tr}_{\text{bath}}\, \mathbb{P}_1\, U(\Delta t, 0)\rho_{\text{tot}}\, U^\dagger(\Delta t, 0)\, \mathbb{P}_1 \,, \tag{6.156}$$

where $\mathbb{P}_1 \equiv \sum_\ell |1_\ell\rangle\langle 1_\ell|$ and where the partial trace is now over both spins and bosons.

For discrete case, with $N = 15$ boson modes and $\chi(x) = \Theta(x)$, the dots in Fig. 6.14 represent numerical results for $\langle x\, |\mathcal{R}\rho_{\text{p}}|\, x\rangle$ for a Gaussian state $\rho_{\text{p}} = |\psi\rangle\langle\psi|$. The dots are obtained by a very time-consuming calculation.

In the continuum limit, with the Markov property, and a pure state $\rho_{\text{p}} = |\psi\rangle\langle\psi|$ the reset state is again pure and one obtains

$$\mathcal{R}\rho_{\text{p}} = |\psi_{\text{reset}}\rangle\langle\psi_{\text{reset}}| \,, \tag{6.157}$$

with

$$|\psi_{\text{reset}}\rangle \equiv A^{1/2}\chi(\hat{x})|\psi\rangle \,. \tag{6.158}$$

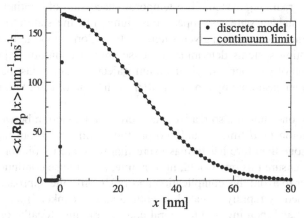

Fig. 6.14 *Dots*: $N = 15$ discrete boson modes, $\chi(x) = \Theta(x)$; reset state $\langle x\, |\mathcal{R}\rho_{\text{p}}|\, x\rangle$ for a Gaussian state $\rho_{\text{p}} = |\psi\rangle\langle\psi|$; *solid line*: Same with QJA for continuum limit, $N \to \infty$, the numerical evaluation is much faster

This means that after its detection, which is equivalent to the detection of a boson, the particle state is localized in the region of the detector (given by the sensitivity function $\chi(x)$). This is a physically very appealing result since a position measurement should localize the particle.

For the continuum limit of the discrete case in Fig. 6.14, the solid curve displays $\langle x | \mathcal{R}\rho_p | x \rangle = |\langle x | \psi_{\text{reset}} \rangle|^2$. The agreement between the discrete version and the QJA result is excellent. The numerical evaluation of the QJA result, however, is again not only much faster but almost trivial, by (6.158).

These results and their extensions to many spins have been applied to the study of arrival and passage times in [27, 28, 41] where more details can be found. In particular the backreaction of the detection of a boson on the particle has been studied by the QJA. Somewhat surprisingly it turns out that in spite of the presence of the spin system mediating between boson and particle this backreaction is not trivial and of similar nature as in the atom–photon model of Sect. 6.7.2.

6.9 Discussion

In general, a quantum mechanical description of the time development of a single system is not possible since quantum mechanics deals with ensembles. But for a single fluorescing system, like an atom in a trap, which is driven by a light source and on which photon detections are performed, this is at least partially feasible and it is just accomplished by the quantum jump approach (QJA). Technically it is achieved by reverting back to an ensemble description. Besides the statistical properties of the photon detections the QJA also allows one to calculate the temporal state of the single system depending on the, in principle unpredictable, stochastic detection times, with a smooth "conditional" time development by a conditional complex Hamiltonian between detections and a reset operation ("jump") of the state right thereafter. From this one can determine the statistical properties of a single radiating system. The temporal succession of continuously changing states and sudden state reset operations (jumps) of the state at random times form a quantum trajectory of a given single system. Correspondingly, an ensemble of many radiating systems determines an ensemble of quantum trajectories. The related question of whether a single quantum trajectory cannot just be viewed as a preparation of an ensemble by repetition after each jump or reset will be discussed below.

In principle one should also realize that there is a distinction between the statements "no photon *until* time t" and "no photon *at* time t." From the theoretical point of view one therefore has to make sure that there are no photons in between detections. To ensure this one would, in principle, have to use continuous measurements. In order to avoid this complication and to simplify the approach, the derivation is proceeded by rapidly repeated, hypothetical ("gedanken") measurements at times Δt apart and then invoked temporal coarse graining. Ideally, of course, one would like to let Δt tend to 0, but with the reduction rule of von Neumann and Lüders employed here this is not possible, due to the so-called quantum Zeno effect.

Therefore, one has to ensure that, within a certain range, the results are independent of the particular choice of Δt. This can indeed be verified if the Markov property holds for the system coupled to the photons (or to another bath). In the derivation presented in Sect. 6.2 this became particularly transparent because in second-order perturbation theory the interaction with the bath produced, by the Markov property, a term proportional to Δt and not to $(\Delta t)^2$. The latter form of dependence only sets in for $\Delta t \to 0$ and gives rise to the quantum Zeno effect. As mentioned in Sect. 6.2, however, the use of perturbation theory is not essential to the derivation and that therefore possible errors do not add up for the hypothetical measurements. Moreover, after the calculation is performed, the probability of finding no photon *until* and *at* time t turns out to be the same. Physically this can be understood as the fact that once the photon is away from the atom it is not reabsorbed. Again this can be shown mathematically to be a consequence of the Markov property, and it would not hold in a cavity where revivals can occur.

If the Markov property does not hold and if one uses a sequence of gedanken measurements which are some more or less arbitrary time Δt apart, one will usually be led to nonphysical results. As another example, in addition to cavities, the Markov property also does not hold for photonic crystals, due to a photonic band gap.

When one applies the QJA in simulations it is not necessary to simulate each of the rapidly repeated individual measurements at times Δt apart, but one can rather use the conditional Hamiltonian to calculate the waiting time distribution and generate the detection (or jump) times. Only if this is too complicated, e.g., for a large number of degrees of freedom coming from many levels or from inclusion of the atomic motion, one will simulate each time step Δt. The advantage of the QJA for simulations becomes particularly pronounced if one can work with pure states because this reduces the dimension to N, i.e., to the number of levels, compared to N^2 for density matrices as in the optical Bloch equations. As has been pointed out in Sect. 6.6 it is always possible to go over to pure states, even if the reset operation originally gave a density matrix. The ensemble of quantum trajectories corresponding to an ensemble of radiating systems provides a solution of the optical Bloch equation for the system, and hence in this respect simulations can be extremely useful for large N.

The hypothetical, gedanken, measurements employed in the derivation of the QJA are obviously highly idealized. A realistic photon detector misses many photons, be it by an efficiency less than 1, be it by an aperture less than 4π. The underlying assumption is that in such a case the experimental probability distribution is obtained from the ideal jump (detection) trajectory by assigning, as in [6], probabilities for the *recording* of the jumps and that one does not need to model the actual detector in detail.

A variant of the QJA arises in the investigation of the frequency spectrum in a light period of the Dehmelt system in Fig. 6.1. Here the theoretical problem is that in order to make sure that one is in a light period one has to detect the photons. This detection possibly changes the spectrum. In [33] this problem was solved by measuring the light period through photons emitted in a half space and using it to trigger a spectrometer in the other half space. This procedure bypasses the time–energy uncertainty relation.

There is an interesting application of the QJA to an experiment on the quantum Zeno effect. In the experiment [36] a large number of ions with a V configuration as in Fig. 6.1 were stored in a Penning trap. The time development was given by a so-called π pulse of length T_π connecting the ground state 1 and the metastable state 3. Intermittently the ground state population was measured by a very short pulse of a probe laser coupling level 1 with level 2, where emission of photons would indicate the ground as measurement outcome and level 3 otherwise. This was regarded in [36] as a measurement to which the projection postulate could be applied. With up to 64 probe pulses ("measurements") during a π pulse agreement was found, within the error bars, with the quantum Zeno predictions for the level populations. In an application of the QJA [5], on the other hand, the probe pulse was directly included into the dynamics and it was shown by means of the QJA that the probe pulse can indeed be regarded as a measurement which at least approximately satisfies the projection postulate. It was also possible to calculate the errors which arise when one replaces the probe pulse by an ideal measurement satisfying the projection postulate.

This is an example of a Heisenberg cut, where part of the measurement apparatus is included in the dynamics, the actual measurement then being on the photons. Of course, in the QJA one also employs the projection postulate, but only for the photons, and the photons are, so to speak, at the very end of the chain. The spin-boson detector model presented in Sect. 6.8 can be regarded as another example for a Heisenberg cut. Here the bath of spins can be either viewed as part of the apparatus or as part of the quantum mechanical system. The spin-boson detector model also illustrates the fact that the QJA can be applied to a variety of systems which are coupled to a chain of baths and where the gedanken measurements are performed on the last member of the chain.

When the resetting in a quantum trajectory is always done to the same state, e.g., to the ground state, and if the light source is constant in time, then the trajectory parts between jumps can be viewed as a preparation of an atomic ensemble by repetition, albeit at random times and of different lengths. There is also a non-Hermitian conditional time development in this part which is not given by the usual Hermitian Hamiltonian. Such a part of a quantum trajectory corresponds to a complete system consisting of the driven atom interacting with the quantized radiation field on which rapidly repeated photon measurements are performed with null outcome. The resetting of the system after the detection of a photon at some random time can be regarded as a new preparation. In view of the non-Hermitian time development between resettings this particular ensemble cannot be regarded as a usual ensemble of atoms plus field developing in time since the successive, rapidly repeated, null measurements between photon detections change the time development. When the reset operation is not always to the same state, but rather dependent on the conditional state immediately before a photon detection and thus dependent on the prior history of the trajectory, then the trajectory parts between jumps can also be regarded as an ensemble, but it is even stranger than the one before since the initial states are in general always different.

One may raise the question whether quantum trajectories can be interpreted realistically, as being "real," or whether they are just a convenient mathematical device. Now, if the hypothetical null and photon measurements in the derivation were actually performed then one probably would conclude that the atom follows this path. But what if the measurement are not, or only partially, performed? Can one assume, at least in principle and *consistently* and without running into contradictions, that there are individual photon emissions/detections at certain times, random though and not predictable? If this were so one might say that a quantum trajectory was realistic or objective, even without measurements actually being performed. However, with given sharp emission/detection times one would encounter, by the time–energy uncertainty relation, a much broader frequency spectrum than the usual one – a contradiction. So one should not assume, in spite of their physical interpretation, that the trajectories of the QJA are always realistic. Rather, they are adapted to particular questions such as photon statistics. The complete information is contained in the state vector or density matrix of the total system, i.e., system plus quantized radiation field or bath, and it is prudent to keep in mind this holistic view.

References

1. The projection postulate as commonly used nowadays is due to G. Lüders, Ann. Phys. **8**, 323 (1951). For observables with degenerate eigenvalues his formulation differs from that of J. von Neumann, *Mathematische Grundlagen der Quantenmechanik*, Springer (Berlin 1932) (English translation: *Mathematical Foundations of Quantum Mechanics*, Princeton University Press, 1955), Chapter V.1. The projection postulate intends to describe the effects of an ideal measurement on the state of a system, and it has been widely regarded as a useful tool.
2. For a simplified case one can see this directly as follows. The ω^2 inherent in d^3k and the ω_k in κ give a factor of ω^3. If this ω^3 is omitted in the definition of κ then the result can be seen by a straightforward calculation of the double integral in (6.22). The general case can be reduced to this by partial integration. We note that for ω_0 in the microwave range the condition $t\prime - t_i \gg \omega_0^{-1}$ does not hold. However, the radiative coupling of such levels is extremely small and is usually neglected in applications.
3. D. Alonso, I. de Vega, Phys. Rev. Lett. **94**, 200403 (2005)
4. A. Beige, Doctoral Dissertation, Universität Göttingen, Germany (1997)
5. A. Beige, G.C. Hegerfeldt, Phys. Rev. A **53**, 53 (1996)
6. A. Beige, G.C. Hegerfeldt, J. Mod. Opt. **44**, 345 (1997)
7. A. Beige, G.C. Hegerfeldt, J. Phys. A: Math. Gen. **30**, 1323 (1997)
8. A. Beige, G.C. Hegerfeldt, D.G. Sondermann, Quantum Semiclass. Opt. **8**, 999 (1996)
9. A. Beige, G.C. Hegerfeldt, D.G. Sondermann, Found. Phys. **27**, 1671 (1997)
10. J.C. Bergquist, R.G. Hulet, W.M. Itano, D.J. Wineland, Phys. Rev. Lett. **57**, 1699 (1986)
11. Th. Sauter, R. Blatt, W. Neuhauser, P.E. Toschek, Opt. Commun. **60**, 287 (1986)
12. W. Nagourney, J. Sandberg, H. Dehmelt, Phys. Rev. Lett. **56**, 2797 (1986)
13. J.C. Bergquist, W.M. Itano, R.G. Hulet, D.J. Wineland, Phys. Script. **T22**, 79 (1988)
14. H. Carmichael, *An Open Systems Approach to Quantum Optics*, Lect. Notes Phys. (Springer, Berlin, 1993)
15. C. Cohen-Tannoudji, J. Dalibard, Europhys. Lett. **1**, 441 (1986)

16. C. Cohen-Tannoudji, J. Dupont-Roc, G. Grynberg, *Atom-Photon Interactions* (John Wiley & Sons, New York, 1992)
17. J. Dalibard, Y. Castin, K. Mølmer, Phys. Rev. Lett. **68**, 580 (1992)
18. J.A. Damborenea, I.L. Egusquiza, G.C. Hegerfeldt, J.G. Muga, Phys. Rev. A **66**, 052104 (2002)
19. H.G. Dehmelt, Bull. Am. Phys. Soc. **20**, 60 (1975)
20. P.A.M. Dirac, *The Principles of Quantum Mechanics*, 4th edn (Clarendon Press, Oxford, 1959), p. 181
21. B. Gaveau, L.S. Schulman, J. Stat. Phys. **58**, 1209 (1990)
22. N. Gisin, I.C. Percival, J. Phys. A: Math. Gen. **25**, 5677 (1992)
23. G.C. Hegerfeldt, Phys. Rev. A **47**, 449 (1993)
24. G.C. Hegerfeldt, Fortschr. Phys. **46**, 596 (1998)
25. G.C. Hegerfeldt, in *Irreversible Quantum Dynamics*, F. Benatti, R. Floreanini (eds.), Springer Lect. Notes Phys. **622**, 233 (2003)
26. G.C. Hegerfeldt, in preparation
27. G.C. Hegerfeldt, J.T. Neumann, L.S. Schulman, J. Phys. A **39**, 14447 (2006)
28. G.C. Hegerfeldt, J.T. Neumann, L.S. Schulman, Phys. Rev. A **75**, 012108 (2007)
29. G.C. Hegerfeldt, M.B. Plenio, Phys. Rev. A **46**, 373 (1992)
30. G.C. Hegerfeldt, M.B. Plenio, Phys. Rev. A **47**, 2186 (1993)
31. G.C. Hegerfeldt, M.B. Plenio, Quantum Opt. **6**, 15 (1994)
32. G.C. Hegerfeldt, M.B. Plenio, Z. Phys. B **96**, 533 (1995)
33. G.C. Hegerfeldt, M.B. Plenio, Phys. Rev. A **53**, 1164 (1996)
34. G.C. Hegerfeldt, D.G. Sondermann, Quantum Semiclass. Opt. **8**, 121 (1996)
35. G.C. Hegerfeldt, T.S. Wilser, in *Classical and Quantum Systems*. Proceedings of the II International Wigner Symposium, July 1991, H.D. Doebner, W. Scherer, and F. Schroeck (eds.) (World Scientific, Singapore, 1992), p. 104
36. W.M. Itano, D.J. Heinzen, J.J. Bollinger, D.J. Wineland, Phys. Rev. A **41**, 2295 (1990)
37. R. Loudon, *The Quantum Theory of Light*, 3rd edn (Clarendon, London, 2000)
38. P.W. Milonni, Phys. Rep. **25 C**, 1 (1976)
39. B. Misra, E.C.G. Sudarshan, J. Math. Phys. **18**, 756 (1977)
40. J.G. Muga, R. Sala, I.L. Egusquiza (eds.), *Time in Quantum Mechanics* (Springer, Berlin, 2002), cf. also articles in the present book
41. J.T. Neumann, Doctoral Dissertation, Universität Göttingen, Germany (2007)
42. M.B. Plenio, P.L. Knight, Rev. Mod. Phys. **70**, 101 (1998)
43. M. Porrati, S. Putterman, Phys. Rev. A **39**, 3010 (1989)
44. R. Reibold, J. Phys. A: Math. Gen. **26**, 179 (1993)
45. C. Schön, A. Beige, Phys. Rev. A **64**, 023806 (2001)
46. D.G. Sondermann, in *Nonlinear, Deformed and Irreversible Quantum Systems*, H.-D. Doebner, V.K. Dobrev, P. Nattermann (eds.) (World Scientific, Singapore, 1995), p. 273
47. M.D. Srinivas, E.B. Davies, Opt. Acta **28**, 981 (1981)
48. M.D. Srinivas, E.B. Davies, Opt. Acta **29**, 235 (1982)
49. T.S. Wilser, Doctoral Dissertation, University of Göttingen, Germany (1991)

Chapter 7
Causality in Superluminal Pulse Propagation

Robert W. Boyd, Daniel J. Gauthier, and Paul Narum

7.1 Introduction

The theory of electromagnetism for wave propagation in vacuum, as embodied by Maxwell's equations, contains physical constants that can be combined to arrive at the speed of light in vacuum c. As shown by Einstein, consideration of the space–time transformation properties of Maxwell's equations leads to the special theory of relativity. One consequence of this theory is that no information can be transmitted between two parties in a time shorter than it would take light, propagating through vacuum, to travel between the parties. That is, the speed of information transfer is less than or equal to the speed of light in vacuum c and information related to an event stays within the so-called light cone associated with the event. Hypothetical faster-than-light (superluminal) communication is very intriguing because relativistic causality would be violated. Relativistic causality is a principle by which an event is linked to a previous cause as viewed from any inertial frame of reference; superluminal communication would allow us to change the outcome of an event after it has happened.

Soon after Einstein published the theory of relativity, scientists began the search for examples where objects or entities travel faster than c. There are many known examples of superluminal motion [1]. One example arises when observing radio emission in certain expanding galaxies known as superluminal stellar objects. This motion can be explained by considering motions of particles whose speed is just below c (i.e., highly relativistic) and moving nearly along the axis connecting the object and the observer [2]. Hence, these are not superluminal motions after all.

R.W. Boyd (✉)
The Institute of Optics and Department of Physics and Astronomy, University of Rochester, Rochester, NY 14627, USA, boyd@optics.rochester.edu

D.J. Gauthier
Department of Physics, Duke University, Durham, NC 27708, USA, gauthier@phy.duke.edu

P. Narum
Norwegian Defence Research Establishment, NO-2027 Kjeller, Norway, Paul.Narum@ffi.no

Boyd, R.W. et al.: *Causality in Superluminal Pulse Propagation*. Lect. Notes Phys. **789**, 175–204 (2009)
DOI 10.1007/978-3-642-03174-8_7

Explaining, in simple terms, why apparent superluminal motions do not violate the special theory of relativity or allow for superluminal communication can be exceedingly difficult. Also, approximations used to solve models of the physical world can lead to subtle errors, sometimes resulting in predictions of superluminal signaling. For these reasons, studying superluminal signaling can be an interesting exercise because it often reveals unexpected aspects of our universe or the theories we use to describe its behavior.

One example of apparent superluminal behavior occurs in the transfer of information encoded on optical pulses propagating through a dispersive material. Under conditions where the dispersion of the medium is anomalous over some spectral region, defined in greater detail below, it is possible to observe the peak of a pulse of light apparently leaving a dispersive material before a pulse peak enters! Does such a situation imply that information flows outside the light cone and thus that relativistic causality is violated?

The possibility of such "fast-light" behavior has been known for nearly a century and has been the source of continued controversy and confusion. Yet a rather simple mathematical proof shows that such behavior is completely consistent with Maxwell's equations describing pulse propagation through a dispersive material and hence does not violate Einstein's special theory of relativity. While the proof is straightforward, great care is needed in interpreting the special theory of relativity and in determining whether experimental observations are consistent with its predictions.

This chapter reviews the history of fast-light research, describes one approach for understanding how information encoded on optical beams flows through a dispersive material, and describes how these results can be interpreted within the framework of the special theory of relativity. In this chapter we do not discuss the issue of the optical analogue of the tunneling of particles through a potential barrier. It is known that this process can also lead to superluminal behavior, but for reasons quite distinct from the situation treated here, the propagation of light through dispersive media. Superluminal effects based on tunneling have been reviewed recently by Winful [3]; see also "Time in Quantum Mechanics – Vol. 1."

7.2 Descriptions of the Velocity of Light Pulses

Before we delve more deeply into the implications of superluminal propagation velocities, it is crucial to define exactly what we mean by the "velocity of light" in a dispersive material. Because a pulse disperses, its motion cannot be described rigorously using a single velocity. For this reason, there is in fact more than one way to define the velocity of light, depending on what aspect of light propagation is being considered. In this section, we review some of these definitions. Additional discussions of these points can be found in [4] or on page 58 of [5].

1. Oftentimes, the velocity of light is taken to mean the velocity of light in vacuum, the universal constant c. Since 1983, c has been defined to have the value 299,792,458 m/s. This is both the velocity at which points of constant phase move through the vacuum and the velocity at which disturbances in the electromagnetic field move.

2. In a material medium the points of constant phase move with the velocity $v_p = c/n$ where n is the refractive index and v_p is called the phase velocity.

3. When the refractive index varies with frequency, the velocity at which disturbances in the field move through the medium will, in general, be different both from the phase velocity and from c. One velocity that is often associated with the motion of a disturbance is called the group velocity and is given by $v_g = c/n_g$, where n_g is the group index which is related to the refractive index by $n_g = n + \omega \, dn/d\omega$. Clearly, the group index differs from the refractive index for a dispersive medium, that is, a medium for which the refractive index is frequency dependent. If the refractive index varies nearly linearly with frequency, at least over the frequency range of interest, the group velocity will itself be frequency independent. In this case, a pulse will propagate with negligible distortion, and the group velocity can be interpreted as the velocity with which the peak of the pulse moves. For the case of a medium with loss, there are some subtleties regarding the interpretation of the group velocity. We shall return to this point in Sect. 7.7.

4. There is also an energy velocity, defined by $v_E = S/u$, where S is the magnitude of the Poynting vector and u is the energy density. Loudon [6] shows that $v_E = c/[n' + n''(\omega/\Gamma)]$ for an absorbing medium comprised of a collection of Lorentz oscillators, where n' and n'' denote, respectively, the real and imaginary parts of the refractive index and Γ is the transition linewidth. The reason why one must specify the type of medium is that, near resonance, much of the energy density resides in the Lorentz oscillators. While it is not clear from naive inspection, the energy velocity reduces to the group velocity when damping is negligible (that is, when Γ goes to zero). Lysak [7] shows that the energy velocity is always less than c for a non-inverted atomic medium. Additional discussions of this topic have been presented by Sherman and Oughstun [8, 9].

5. There are three somewhat related velocities known as the front velocity, signal velocity, and information velocity. Let us suppose that initially the optical field vanishes in all space and at a certain moment of time t_0, it is suddenly turned on. The initial turn-on of the field will propagate at a velocity known as the front velocity. It can be shown theoretically that the front velocity is equal to c, because the abrupt turn-on must possess extremely high-frequency components that cannot induce a response in the optical medium.

   Let us now assume further that, following the front, the source emits a well-defined pulse, and that the field vanishes entirely both before the pulse begins and after the pulse ends. We refer to a pulse of this sort as a signal, and the velocity at which the peak of the pulse moves is known as the signal velocity. Under many practical circumstances, the signal velocity will equal the group velocity.

Any radiation that arrives before the main body of the pulse is known as a precursor. The precursors arrive after the arrival of the pulse front. A precursor consisting of high-frequency components is known as a Sommerfeld precursor, and it arrives at a time determined by the velocity $c/n(\infty) \approx c$. Any precursor associated with low-frequency components is called a Brillouin precursor, and it arrives at a time determined by the velocity $c/n(0)$. Here we symbolically represent the refractive index at high frequencies by $c/n(\infty)$ and the refractive index at low frequencies by $c/n(0)$.

Finally, the information velocity is the velocity at which the information content of the pulse is transmitted. From a practical point of view, useful information usually arrives at the peak of the pulse, which travels approximately at the group velocity. However, the information velocity is usually defined in terms of the earliest moment at which, even in principle, the information content of a pulse could be determined and hence is associated with the pulse front. This velocity is then associated with the velocity c. This thought can be made more precise by noting that information resides at points of discontinuity of an optical waveform [10, 11], which propagate at the vacuum velocity c because of their broad frequency content. This result follows from the argument that smooth parts of a pulse cannot carry information, since the future evolution of the pulse can, in principle, be predicted by performing a Taylor series expansion of the pulse amplitude. It is the difference between the practical and precise definitions of the information velocity that has given the greatest confusion regarding the tension between fast-light pulse propagation and the special theory of relativity.

6. There is still another velocity known as the centroid velocity. This approach goes back to Smith in 1970 [4], although the recent advocates have been Peatross et al. [12] and Cartwright and Oughstun [13]. The centroid velocity is the velocity with which the time center of mass moves through the material. The paper by Cartwright and Oughstun is especially interesting. They study the centroid velocity as a function of optical thickness of a material. They show that the centroid velocity is equal to the group velocity for thin media and is equal to the velocity of the Brillouin precursor (that is, $c/n(0)$) for thick media.

7.3 History of Research on Slow and Fast Light

Ideas about slow and fast light go back at least 170 years. The distinction between the group velocity and phase velocity was recognized by Hamilton [14] as early as 1839. A full theoretical treatment of the group velocity was presented by Lord Rayleigh [15, 16] in 1877. The classic book *Theory of Electrons* by Lorentz [17] provides formulas for the refractive index of an atomic vapor. Straightforward evaluation of these formulas shows that the group velocity can become very small (slow light) or very large (fast light). Early in the 20th century people were especially intrigued at the prediction of superluminal group velocities, as such velocities seemed at odds with the theory of relativity. Work by Brillouin and Sommerfeld helped to resolve this dispute by showing that, even though the group velocity could become superluminal, this did not mean that the front velocity (associated with the turn-on of a wave) could become superluminal. Interest in slow and fast light continued during the 1980s. Significant work during this period included the experiments [18, 19] aimed at verifying earlier predictions.

Many of the procedures that can be used to produce superluminal propagation are in fact variations of methods that were first used to slow down the velocity of light. We therefore review both slow- and fast-light methods in the remainder of this section.

A great impetus for more recent research was the experiment of Hau and coworkers [20] in 1999. This work captured the popular imagination by showing that light could be slowed down to the "human" speed of 17 m/s. The breakthrough in laboratory implementation behind this achievement was the use of electromagnetically induced transparency (EIT). EIT is a quantum coherence effect that induces transparency in a material while allowing it to retain large nonlinear properties. The use of EIT methods was crucial to this study. Without the use of EIT, the transmitted pulse of light, while significantly delayed, would have been attenuated so strongly as to be immeasurably weak. In order to implement EIT, Hau et al. applied both the signal field and a strong coupling field to their atomic medium. This coupling field induces a narrow transparency window that both allowed the signal to be transmitted and created a large spectral variation of the refractive index. The material medium used in the experiment of Hau et al. was an atomic ensemble in the form of a Bose–Einstein condensate. This experiment was soon followed by that of Kash [21], who showed that ultraslow-light speeds could also be obtained in a hot atomic vapor of rubidium. This observation dispelled the notion that the use of ultra-cold atoms was essential to ultraslow-light propagation.

More recently there has been enormous interest in developing additional tools for controlling the group velocity of light. In broad scope, there are two procedures that can be used to control the group velocity. One of these procedures is to make use of material resonances, such as the sharp absorption resonances of an atomic such as those used by Hau et al. Control can be achieved, for example, by applying a strong optical field to modify through nonlinear optical methods the optical response experienced by the signal field.

From a practical point of view, it is desirable to find means for producing slow light that avoid the need to use atomic ensembles held at exotic temperatures. One approach is to make use of slow light based on the concept of coherent population oscillations (CPO). This process is quite insensitive to the presence of dephasing processes, and thus can operate in room temperature solids, media that hold particular promise for use in practical applications. This CPO process leads to a very narrow spectral hole in the absorption profile of a saturable absorber and consequently to a very large value of the group index. Slow light based on the CPO effect was demonstrated first in ruby [22]. Later both slow and fast light based on the CPO effect was observed in an alexandrite crystal [23] and in an erbium-doped fiber amplifier [24].

In addition to EIT and CPO, a wide variety of other sorts of resonances have been used to produce slow-light effects. There has been particular success with the use of stimulated Brillouin scattering [25, 26] and stimulated Raman scattering [27]. In each of these processes, the strong gain feature induced by the presence of a strong pump field will also produce (as a consequence of Kramers–Kronig relations) a rapid spectral variation in the refractive index which in turn leads to strong slow-light effects.

The other procedure for producing slow and fast light is to use structural resonances to modify the optical response. For example, in a photonic crystal the group velocity of light can be slowed dramatically near the edge of the photonic Brillouin zone. This approach has also been quite exciting. One particular example of this approach is the work of Vlasov [28] in producing a group index of 300 by appropriate patterning of a silicon waveguide, with important implications for silicon photonics.

Historically, the greatest challenge in slow-light research has been to find situations in which a data packet could be delayed by many pulse lengths. Delay measured in units of pulse length is often referred to as the delay-bandwidth product. For several years, the largest delay-bandwidth product was limited to the value of approximately five observed in the initial experiment of Hau et al. More recently, a delay-bandwidth product of 80 was observed in one specific situation [29]. Nonetheless, the need to develop methods for producing large delay-bandwidth products at arbitrary wavelengths and for arbitrary pulse lengths remains an active area of research within the slow-light community.

Fast light possesses some similarities but some differences from slow light. First of all, fast light is conceptually very intriguing. People can accept the suggestion that many physical processes can be slowed down to an arbitrary degree. But it is more difficult to accept the suggestion that the same process can be speeded up to an arbitrary degree. For instance, there is no limit to how slowly one can walk across a lecture hall, but there are obvious limits to how quickly one can do so. But what is surprising is that, at a formal mathematical level, there seems to be almost complete symmetry between slow light and fast light. Since the group velocity is given by c/n_g where $n_g = n + \omega \, dn/d\omega$, we see that slow light occurs if $dn/d\omega$ is large and positive (known as normal chromatic dispersion) whereas fast light occurs if $dn/d\omega$ is large and negative (anomalous chromatic dispersion). Both types of behavior occur regularly in nature. The study of fast light is conceptually important as it allows us to examine the nature of the modification of the velocity of light. In addition, fast light can lead to applications of its own. One application of fast light is in the construction of regenerators for optical telecommunication. One form of regeneration requires that optical data pulses be actively centered in their time windows. The ability to advance as well as delay a data packet greatly facilitates this form of regeneration.

As described above, Sommerfeld and Brillouin investigated a step-modulated pulse propagating through a collection of Lorentz oscillators in a spectral region of anomalous dispersion. Based on this investigation, Brillouin suggested that the group velocity is not physically meaningful in this situation because the pulse becomes severely distorted [30]. For this reason, several textbooks on electromagnetism state that $v_g > c$ or $v_g < 0$ is unphysical.

An interesting twist to this story came about in 1970 when Garrett and McCumber [31] published a theoretical study on propagation of smooth-shaped pulses through a resonant absorber in a region of anomalous dispersion. They showed that the group velocity does have meaning even in the "fast light" case as long as the medium is thin enough so that pulse distortion is not too severe. In fact, they predicted that it is possible that the peak of a light pulse may exit the optical material before a pulse peak passes through the entrance face, which is the physical interpretation of a negative group velocity. The first indirect measurement of a group velocity exceeding c was made by Faxvog and collaborators [32, 33], who studied mode pulling in a self-mode-locked helium–neon laser containing a neon absorption cell.

Some years later, Chu and Wong [18] studied experimentally both slow and fast light for picosecond laser pulses propagating through a GaP:N crystal as the laser frequency was tuned through the absorption resonance arising from the bound A-exciton line. Both positive and negative group delays were observed. Because they were using short pulses, an autocorrelation method was used to measure pulse delay, which can obscure possible pulse distortions. Their experimental results were found to be in good agreement with theoretical predictions, which were obtained from a model that is a slight generalization of the model studied by Garrett and McCumber. Somewhat later, work by Ségard and Macke [34] on microwave pulse propagation through a resonant absorber made direct measurements of the field envelope, thus demonstrating directly that there was only minor pulse distortion. Negative group velocities have since been observed by others [35].

In these experiments, fast-light pulse advancement was accompanied by substantial pulse attenuation. Steinberg and Chiao [36] predicted that it is possible to use two adjacent gain lines to obtain fast light, where anomalous dispersion occurs when the carrier frequency of the pulse is set in the middle of the gain doublet. Chiao and collaborators published several other works that described why fast-light pulse propagation does not violate the special theory of relativity [37–39]. In particular, they focus on the idea that information is encoded on points of non-analyticity on optical waveforms, and it is these points that move at c. More recently, Parker and Walker [40] suggest that the very act of encoding information on a waveform necessarily creates points of non-analyticity.

The prediction of Steinberg and Chiao [36] was verified in an experiment by Wang et al. [41], where the gain doublet was produced in a laser-pumped cesium vapor. They observed measurable pulse advancement in combination with small pulse amplification. While they were careful to point out that their experimental observations were consistent with the special theory of relativity, they did not give a detailed explanation of why this was the case. Unfortunately, some of the popular press cast their experiment as violating Einstein's theory, giving rise to considerable confusion and controversy.

Soon thereafter, Stenner et al. [42] designed an experiment to measure directly the speed at which information propagates through a fast-light material. They used

an experimental setup similar to that used by Wang et al., but with large dispersion
that gave rise to larger pulse advancement. Figure 7.5 shows an example of their data
for the case of a smooth Gaussian-shaped pulse propagating through the fast-light
medium in comparison to the same pulse propagating through vacuum. The larger
advancement relative to the pulse width obtained in their experiment made it easier
to distinguish the different velocities describing pulse propagation. From this data,
they inferred that $n_g = -19.6 \pm 0.8$, indicating that they were operating in the highly
superluminal regime.

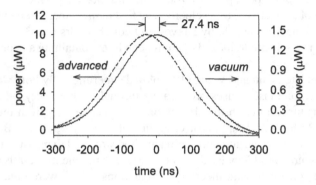

Fig. 7.1 Fast-light pulse propagation. Temporal evolution of a 263.4-ns-long (full width at half
maximum) pulse propagating through a laser-pumped potassium vapor (*dashed line*) and vacuum
(*solid line*) [42]

To determine the information velocity, Stenner et al. encoded new information on
the waveform at the top of the Gaussian-shaped pulse by rapidly turning the pulse
off or by switching it to a higher value. Such an approach enhanced their ability to
estimate the location of a non-analytic point in the presence of noise. The moment
when a decision was made to switch between the communication symbols (either
pulse high or pulse low) corresponded to the point of non-analyticity.

They detected the location of the point of non-analyticity by determining the
arrival of new information using a receiver that distinguished between symbols to
a desired level of certainty, characterized by the bit error rate (BER). Using this
method, they found that the information velocity is always less than but nearly equal
to c, even for a medium where n_g is highly superluminal. Thus, they demonstrated
that the peak of the advanced pulse at the exit face of the medium is not causally connected to the peak at the entrance face. Follow-up studies showed that information
also travels nearly at c even for a material where $v_g \ll c$ [43] and that information
propagation is connected to the front velocity and optical precursors [44].

Other analyses of the relation between superluminality and information transfer
have been reported as well in the literature. Diener [45] has concluded that superluminal group velocities do not imply superluminal information velocities because
the pulse shape can always be determined by analytic continuation of the pulse
shape within the light cone. Kurizki et al. [46] have shown that the injection of

spectrally narrow wavepackets into quantized amplifying media can give rise to transient tachyonic wavepackets. Kuzmich et al. [47] have studied limitations on information transfer in fast-light situations based on quantum effects. Wynne [48] has argued theoretically that information cannot be transmitted superluminally and that claims to the contrary are the result of incorrect reasoning. Tanaka et al. [49] have observed negative group velocities in a Rb vapor. Ruschhaupt and Muga [50] have shown theoretically that the peak of an electromagnetic pulse can arrive simultaneously at different positions in an absorbing waveguide. Clader et al. [51] have shown that instabilities often associated with superluminal propagation can be avoided through use of sufficiently short pulses.

The surprising behavior discussed above can be illustrated with some examples. Under conditions of sufficiently large anomalous dispersion, the group index can take on negative values (recall the definition $n_g = n + \omega \, dn/d\omega$). This possibility raises the question of what it means for a group velocity $v_g = c/n_g$ to become negative. Figure 7.2 shows a numerical simulation of the propagation of a pulse through a material possessing a negative value of the group velocity. The influence of gain, absorption, or group velocity dispersion is not included in this model, and thus the simulation is based simply on performing a numerical integration of the reduced wave equation

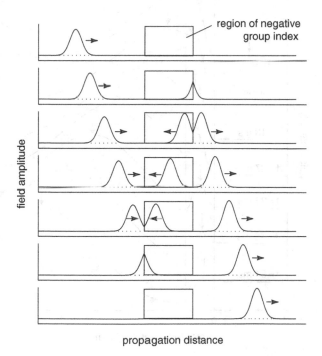

Fig. 7.2 Numerical simulation of pulse propagation through a material with a negative value of the group velocity

$$\frac{\partial A}{\partial z} - \frac{1}{v_g}\frac{\partial A}{\partial t} = 0 \qquad (7.1)$$

for the pulse amplitude $A(z, t)$ for the situation in which the group velocity is negative. One sees from Fig. 7.2 that the pulse appears to leave the material before it enters and that the pulse appears to move backward within the material. How behavior of this sort can possibly be physical and consistent with the concept of causality is a matter that we will deal with in later sections of this chapter. For the present, we simply point out that an input pulse in the form of a Gaussian waveform has wings that extend from plus infinity to minus infinity. In this sense, even in the top frame of Fig. 7.2, the input pulse already has a contribution at the output, and there is no possibility for a violation of causality to occur. Alternatively, we can consider superluminal pulse propagation to represent a special form of pulse reshaping, in which the pulse form is retained but shifted earlier in time.

Experimental verification of this sort of behavior has been reported by Gehring et al. [24] in an experiment that studied pulse propagation through an erbium-doped optical amplifier. A negative value of the group index was obtained by means of the CPO effect described above. Some of the experimental results are shown in Fig. 7.3.

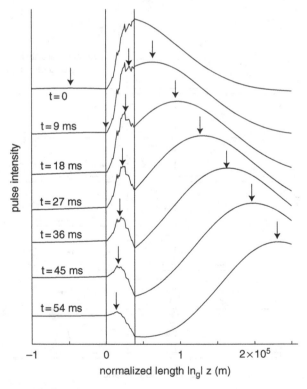

Fig. 7.3 Experimental results demonstrating the reality of negative group velocities [24]. The *arrows* point to the peak of each pulse

Note that the peak of the pulse clearly is moving in the backward direction (right to left) inside the material, even though outside the material the pulse is moving left to right. Because the experiment was performed through use of an amplifying medium, the output pulse is larger than the backward-going pulse within the material.

7.4 The Concept of Simultaneity

As we mentioned earlier, a primary concern of this chapter is to examine how the superluminal pulse propagation can be compatible with the concept of causality. But to understand what is meant by causality, one must first understand what it means for events to occur simultaneously. In this section, we present an examination of the concept of simultaneity. Much of the subtlety involved in considerations of causality involves the concept of what it means for two events to be simultaneous [52].

We begin by distinguishing local simultaneity from distant simultaneity. There are subtleties involved in each concept. We first consider the case of local simultaneity, as it is the more basic concept. One says that two events, A and B, located at the same point in space, are simultaneous if they occur at the same time. The subtlety of this definition is that it presupposes that one understands what one means by time. If the events are not simultaneous, either A occurs before B or it occurs after B. The distinction between "before" and "after" implies that there is a direction of the flow of time. We know from everyday experience that time flows in one direction (from the past to the future), but this point remains unsettling in part because it is not obvious what physical process breaks the symmetry between earlier and later times. This thought can be made more precise by noting that most of the laws of physics are symmetric upon the reversal of the sign of the time coordinate. The existence of certain laws which do not obey this property, such as the tendency of entropy to increase monotonically, may lead to some understanding of the origin of the direction of the flow of time [53].

In considerations of causality and simultaneity, one conventionally defines an "event" as a process that occurs at a given location with coordinates x, y, and z at a given time t. One can thus define an event in terms of its space–time coordinates (x, y, z, t). As noted above, the time coordinate of the event has a very different character from the spatial coordinates, in that there is a sense of directionality to the time coordinate not present in the spatial coordinates.

So how does one define time? One definition is that put forth by Kant, as reported by Jammer [52]. Kant says that if event A could cause event B, then A is said to occur before B. This thought is very much consistent with modern views of physical causality, but has the disadvantage that one cannot then examine the relation between causality and the flow of time if time has been defined in terms of causality. Perhaps a better procedure is to follow the lead of Einstein and define time simply to be what a clock measures. We assume that we place a "clock" at the point (x, y, z) and define the unit of time to be the interval between successive ticks of the clock. From this point of view, a good clock is one for which the laws of classical mechanics are well satisfied with time defined in this manner. To summarize this point, we conclude that two events both occurring at the same spatial point (x, y, z) are said to be simultaneous if both events occur at the same time t, with t defined as above.

Somewhat more subtle is the concept of distant simultaneity. The underlying question here is, what does it mean for two events to be called simultaneous if they do not occur at the same point in space? Philosophical discussion of this point goes back at least as far as the golden age of Greece. This issue certainly possesses a technological component (How could one hope to measure distant simultaneity without the help of reliable clocks?), but also addresses the conceptual issue of what it means for two separated events to be said to be simultaneous.

One impetus for these discussions occurred within the field of astrology, which holds that one's future depends on the configuration of the stars and planets at the time of one's birth. It thus became important to know the state of the heavens at the moment of a child's birth, even if the heavens were obscured by cloud cover or rendered unobservable by daylight. It is interesting to note that Saint Augustine argued against the validity of astrology by means of the following argument [54]. He considered the hypothetical situation in which two women located in different households were to give birth at approximately the same time. One woman had great wealth, whereas the other was a servant. The child of the wealthy woman would almost certainly be more successful in life than the child of the poor woman. If these children were born simultaneously, this occurrence would contradict the predictions of the laws of astrology. But how would one establish the simultaneity of the two births, occurring at separated points? In a manner that foreshadows that of Einstein some 1500 years later, Augustine proposes the following procedure. Two messengers are employed and they are selected so that they run at the same speed. One messenger is stationed near each expectant mother, and at the moment that the child is born the messenger is told to run to the other household to announce the birth. If the messengers meet en route, the exact location of their meeting is recorded, and if this spot is exactly equidistant between the two households the births are said to have occurred simultaneously.

Within modern physics, one defines distant simultaneity in terms of synchronized clocks. One assumes that two clocks of identical construction are located at spatial points A and B. Being of identical construction, these clocks are, therefore, assumed to run at the same rate. If the clocks can be synchronized, then the concept of distant simultaneity becomes meaningful, in the sense that two events are said to be simultaneous if the event at A occurs at a time measured by the clock at A, that is, the same time as the time of event at B as measured by the clock at B.

Eddington [55] describes two possible procedures for synchronizing distant clocks. One method is to transport clock A to point B, set the clocks to read the same time, and then transport clock A back to its original location. Of course, an auxiliary clock can alternatively be used for this purpose. Because of relativistic time dilation, the clock needs to be moved very slowly in order for this procedure to be valid. In principle, one can always perform this procedure, because time dilation effects are second order in the ratio v/c (here v is the velocity of the clock), whereas the time required for the transport scales as $1/v$. This method also presupposes that the clock maintains its accuracy during the time intervals of acceleration needed to change its velocity.

The second method, described by Einstein in his 1905 paper, is based on signaling using light beams. Several variations of this method exist. One is for A to send a light pulse to B, where it is reflected back to A. B sets its clock to some reference time (for instance $t = 0$) at the moment that the pulse arrives at B. A waits until the light pulse returns, and then sets its clock to the reference time at a moment exactly halfway between the time t_1 at which the pulse left A and the time t_2 at which the pulse returned. Many authorities have argued that while this procedure provides an acceptable procedure for synchronizing the two clocks, the synchronization thus achieved is simply one of arbitrary convention. They argue that A could set the reference time of its clock to any time in the interval (t_1, t_2) and thereby establish an entirely consistent form of clock synchronization. The reason for this arbitrariness is that there appears to be no definitive proof that the velocity of light c_+ along the positive x-direction (for instance) is the same as the velocity c_- along the negative x-direction. The argument is that measurements of the velocity of light actually yield only the average of c_+ and c_-, and it is this quantity that is conventionally known as c. This unexpected conclusion follows from the fact that it is possible to measure the one-way velocity of light only if one already has synchronized clocks at both ends of the beam path, which cannot be possible if one's intent is to develop a procedure for clock synchronization.

7.5 Causality and Superluminal Pulse Propagation

The key to understanding why fast-light pulse propagation is consistent with the special theory of relativity is to investigate what is meant by a signal, as described above in Sect. 7.2 and its connection to an event. In Einstein's public discussions of the theory [56], he focuses on the concept of an "event," such as a spark caused by a lightning bolt, and how the event (or multiple events) would be observed by people at various locations. He was especially interested in observers moving with respect to a coordinate system that is stationary with respect to the events. A detailed description of his findings is not needed for our present discussion, as it is necessary to consider only the properties of a single event in a single coordinate system.

A convenient way to discuss the flow of information from an event is to use a space–time diagram (Minkowski diagram), where the horizontal axis is a single spatial coordinate and time is plotted along the vertical axis (see Fig. 7.5). According to the special theory of relativity, the fastest way that knowledge of the event can reach an observer is if it travels at the speed of light in vacuum; the lines that connect points in a space–time diagram that follow vacuum speed-of-light propagation define the light cone – the shaded region in Fig. 7.5(a). The inverse of the slope of lines drawn in a space–time diagram is equal to the velocity. Observers at space–time points within the shaded cone (e.g., observer A in Fig. 7.5(b)) are able to see the event and those outside the cone cannot (e.g., observer B in Fig. 7.5(b)). Note that the cone extending for times preceding the event represents the space–time regions where light could reach the location of the event. That is, an observer (not shown) in this region can affect the event but cannot see the event.

On the other hand, hypothetical faster-than-light propagation of information is relativistically acausal. Acausal means that there is no direct time-ordered link between a cause and an effect. An example of a hypothetical faster-than-light communication scheme is shown in Fig. 7.5(c), where we assume that it is possible to transmit information with a speed that is less than zero (negative velocity). If such superluminal signal was possible, information could be transmitted from the positive-time light cone to a person at position D. This observer could change the outcome of the event (e.g., prevent it from happening) because she is located within the light cone leading to the event, but at a time before the event happens. Thus, she can change the outcome of the event.

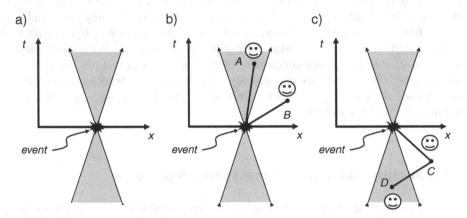

Fig. 7.4 The light cone associated with an event. (a) Space–time diagram of an event. (b) Observers at space–time points A and B. In a world that is relativistically causal, A observes the event, but B does not. (c) Communication in a hypothetical acausal world. If relativistic causality could be violated, a person at C could observe the event and transmit information to a person at D using a superluminal communication channel. The person at D could then change the outcome of the event

To address whether fast-light pulse propagation provides a mechanism for relativistically acausal communication, it is necessary to define a signal. As discussed briefly in Sect. 7.2, Sommerfeld [30] defined a "signal" as a wave that is initially zero and suddenly turns on to a finite value, which is known as a step-modulated pulse. In terms of the special theory of relativity, the moment that the wave turns on corresponds to the event.

In an analysis conducted by Sommerfeld and Brillouin [30], they used Maxwell's equations to predict the propagation of a step-modulated electromagnetic wave, which was coupled to a set of equations that described how it modifies the dispersive material. For the dispersive material, they assumed that it consisted of a collection of Lorentz oscillators. A Lorentz oscillator is a simple model for an atom that describes its resonant behavior when interacting with light. Each oscillator consists of a massive (immoveable) positive core and a light negative charge that experiences a restoring force obeying Hooke's Law. Furthermore, they assumed that the negative charge experiences a velocity-dependent damping so that any oscillations set in motion will eventually decay in the absence of an applied field.

Conceptually, an incident electromagnetic wave polarizes the material (causes a displacement of the negative charges away from their equilibrium position), and this polarization acts back on the electromagnetic field to change its properties (e.g., amplitude and phase). The coupled Maxwell–Lorentz-oscillator model is known to possess spectral regions of anomalous dispersion where the group velocity v_g takes on negative values or values greater than c thus should be able to address the controversy. The model is so good that it is still in use today for describing the linear optical response of dispersive materials.

Using asymptotic methods to solve the pertinent inverse Fourier integral, Sommerfeld was able to predict what happens to the propagated field for times immediately following the sudden turn-on of the wave, that is, what happens in the vicinity of the pulse front. He was able to show that the velocity of the front is always equal to c. In other words, the front of the pulse coincides with the boundary of the light cone shown in Fig. 7.5, even though the pulse is propagating through the dispersive dielectric material.

Sommerfeld gave an intuitive explanation for his prediction. When the electromagnetic field first starts to interact with the oscillators, they cannot immediately act back on the field via the induced polarization because they have a finite response time. Thus, for a brief moment after the front passes, the dispersive material behaves as if there is nothing there – as if it were vacuum. From the point of view of information propagation, one should be able to detect the field immediately following the front and hence observe information traveling precisely at c.

After the front passes, mathematical predictions are very difficult to make because of the complexity of the problem. Brillouin extended Sommerfeld's work to show that the initial step-modulated pulse, after propagating far into a medium with a broad resonance line, transforms into two wavepackets (now known as optical precursors) and is then followed by the bulk of the wave (what Brillouin called the "main signal"). They found that the precursors tend to be very small in amplitude and thus it would be difficult to measure information transmitted at c; rather, it would be easiest to detect at the arrival of the main signal, which they found travels slower than c. The term precursor is somewhat confusing because it implies that the wavepacket comes before something; in this usage, the precursors come before the main signal, but after the pulse front.

One aspect of Sommerfeld and Brillouin's result that can lead to confusion is the possible situation when one or more of these wavepackets travel faster than c. What is implied here is that a velocity can be assigned to the precursors and the main signal to the extent that they do not distort and that these velocities can all take on different values. In a situation where the velocity of a wavepacket exceeds c, it will eventually approach the pulse front (which travels at c), become much distorted (so

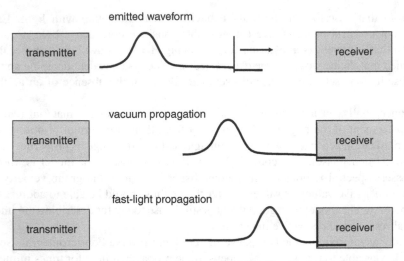

Fig. 7.5 In a typical experiment, the emitted waveform has the shape of a well-defined pulse. Nonetheless, the waveform has a front, the moment of time when the intensity first becomes non-zero. When such a waveform passes through a fast-light medium, the peak of the pulse can move forward with respect to the front, but can never precede the front. Thus, even though the group velocity is superluminal, no information can be transmitted faster than the front velocity, which is always equal to c

that assigning it a velocity no longer makes sense), and either disappear or pile up at the front.

Sommerfeld and Brillouin's research appeared to satisfy scientists in the early 1900s that "fast light" does not violate the special theory of relativity. Yet there continue to be researchers who question various aspects of their work. One point of contention is that some people believe that it is impossible to generate a waveform in the lab that has a truly discontinuous jump (e.g., there is no electromagnetic field before a particular time and then a field appears). Yet having something appear at a particular space–time point is precisely what is meant by an event described above. Thus, if one does not believe in discontinuous waveforms, then the very conceptual framework of the special theory of relativity and the associated light cone shown in Fig. 7.5 would need to be thrown out. Many scientists are unwilling to do so. Also, the existence of optical precursors has been questioned because Sommerfeld and Brillouin made some mathematical errors in their analysis concerning the prop-agated field for times well beyond the front, although recent research suggests that precursors can be readily observed in experimental setups similar to that used in recent fast-light research [44, 57].

So how can the data shown in Fig. 7.1 be consistent with the special theory of relativity? To answer this question, we need to make a connection between Som-merfeld's idea of a signal and the data shown in the figure. In the experiment, a pulse was generated by opening a variable-transmission shutter (an acousto-optic modulator); only a segment of the pulse is shown in the figure. At an earlier time not shown in the figure, the light was turned from the off state to the on state, but with very low amplitude. The moment the light first turns on coincides with the pulse front (the event). At a later time, the pulse amplitude grows smoothly to the peak of the pulse and then decays.

As far as information transmission is concerned, all the information encoded on the waveform is available to be detected at the pulse front (although it might be difficult to measure in practice). The peak of the pulse shown in Fig. 7.5 contains no new information. Thus, the fact that the peak of the pulse is advanced in time is not a violation of the special theory of relativity, so long as it never advances beyond the pulse front. Figure 7.6 shows a schematic of the light cone for such a fast-light experiment.

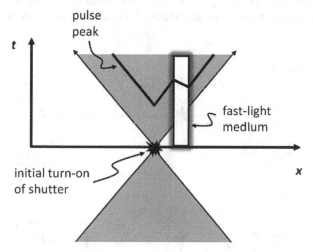

Fig. 7.6 Pulse propagation in a fast-light medium with a negative group velocity. The space–time diagram shows that the peak is advanced as it passes through the medium, but the pulse front is unaffected. The opening angle of the *light cone* is drawn differently from that in Fig. 7.5 for illustration purposes only

In our opinion, all experiments to date are consistent with the special theory of relativity, even though it may be difficult to show this. In some experiments, the pulse shape is such that it is exceedingly difficult to detect the pulse front and hence it may appear that the special theory has been violated. In other experiments especially designed to accentuate the pulse front, it has been shown that the information velocity is equal to c within the experimental uncertainties in both fast-light [42] and slow-light regimes [43].

7.6 Quantum Mechanical Aspects of Causality and Fast Light

Quantum mechanics provides a mechanism that at first glance seems to imply the possibility of superluminal communication, even for propagation through vacuum. This mechanism is the simultaneous collapse of the wave function at all points in space caused by a measurement performed on the system at one particular point in space. This is an effect that does not occur in classical physics and hence deserves further consideration with regard to the possibility of superluminal communication.

Let us consider a hypothetical superluminal communication system based on this effect [58–60], as illustrated in Fig. 7.7. We consider entangled particles generated by an Einstein–Podolsky–Rosen (EPR) source. For concreteness, we consider a system that generates two correlated photons that travel in opposite directions and carry zero total spin angular momentum. Furthermore, two observers, Alice and Bob, are located on opposite sides of and at large distances from the source. They are equipped with optical components that can analyze the state of polarization of the arriving photons. Bob is slightly further away from the source than Alice, and we want to establish a one-way superluminal communication link from Alice to Bob.

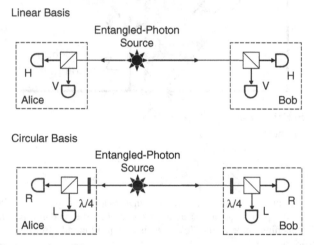

Fig. 7.7 Potential superluminal quantum communication scheme. A source produces pairs of entangled photons, where one photon is sent to Alice and the other to Bob. In the linear measurement basis, a polarizing beam splitter and two single-photon detectors determine whether the photons are horizontally (H) or vertically (V) polarized. In the circular measurement basis, a quarter-wave plate is inserted before the beam splitter, allowing for the measurement of *left* (L) or *right* (R) circularly polarized photons. Bits are communicated by inserting or not the quarter-wave plate in the setups, as described in the text

In one scenario, Alice places a polarizing beam splitter that spatially separates one state of linear polarization (say vertical, V) from the other state of polarization (horizontal, H). The output ports of the polarizing beam splitter are directed to single-photon detectors. Bob has an identical apparatus, which is at a great distance from Alice, and he aligns the axis of his polarizing beam splitter the same as Alice's. Because of the fact that their total angular momentum of the photons is zero, whenever Alice measures V, the bi-photon wave function collapses and Bob is assured of measuring H essentially instantaneously after Alice performs her measurement.

Similarly, Bob will measure V whenever Alice measures H. In this configuration, the polarizing beam splitters and single-photon detectors perform measurements in what we call the "linear" basis.

Alice and Bob can also perform measurements in the "circular" basis, where the analysis apparatus will determine whether the photons are left circular (LC) polarized or right circular (RC) polarized. This measurement can be performed by placing a quarter-wave plate in front of the polarizing beam splitters, where the optical axis of the plate is orientated at $45°$ to the axis of the linear polarizing beam splitter. With the wave plate in the system, Bob is assured to measure RC (LC) whenever Alice measures RC (LC).

The hypothetical superluminal communication scheme is based on a change of measurement basis. By inserting the wave plate into the setup or not, Alice can force Bob's photon to be either linearly or circularly polarized (more precisely, she can force Bob's photon to be an eigenstate of either linear or circular polarization). Thus, it appears as if Alice can transmit binary information to Bob by inserting or not inserting the wave plate into her apparatus. All Bob has to do is to determine with certainty whether Alice was using the linear or circular basis. The first problem with this scheme is a well-known classical result: it is only possible to measure whether an optical beam is linear or circular polarized by analyzing it both with linear and circular polarizers. In other words, Bob would have to send the photon through the linear-basis apparatus *and* the circular-basis apparatus. Unfortunately, one apparatus destroys the incident photon as a result of the measurement and hence it is unavailable to send on the other.

One way around this problem is for Bob to "clone" the incident photon so that there are two copies, where one copy will be sent to a linear-basis apparatus and the other is sent to a circular-basis apparatus [58, 59]. The process of stimulated emission of radiation can in a sense clone an incident photon, so one might think that an optical amplifier in the path of the photon would be useful for this communication scheme. Unfortunately, an optical amplifier adds additional photons to the beam path via the process of spontaneous emission. These additional photons have an arbitrary state of polarization [61]. They destroy the benefits of the amplifier and hence prevent Alice from communicating with Bob via the nonlocal characteristics of quantum mechanics.

The problem with the superluminal communication scheme is much deeper that it appears from this discussion. The very linearity of quantum mechanics prevents the cloning of an arbitrary quantum state, a result of the *no-cloning theorem* [62, 63]. Thus, any device – not just an optical amplifier – fails to clone the incident photon and hence the communication scheme fails.

Other researchers have wondered whether an imperfect copy of the incident photon would be sufficient for superluminal communication. The best or optimal quantum copying machine has been identified; even with the best possible copying apparatus, the quantum communication scheme just barely fails. This failure is nicely summarized by N. Gisin [64] in his 1998 paper: "Once again, quantum mechanics is right at the border line of contradicting relativity, but does not cross it. The peaceful coexistence between quantum mechanics and relativity is thus re-enforced."

7.7 Numerical Studies of Propagation Through Fast-Light Media

In order to explore further some of the features of fast-light propagation described above, we have performed numerical studies of the propagation of optical pulses through fast-light media. For these studies, we assume that the medium consists of a single Lorentzian absorption line set on a broad gain background. We assume the presence of the broad gain background to prevent the transmitted pulse from becoming so weak as to be immeasurable. The absorption coefficient of the material is thus taken to be of the form

$$\alpha(\delta) = \alpha_b + \frac{\alpha_l}{1 + (\delta^2/\gamma^2)} . \tag{7.2}$$

Here α_b is the value of the background absorption coefficient (assumed negative), α_l is the line-center absorption coefficient of the absorption line, γ is its linewidth, and δ is the frequency detuning from line center. According to the Kramers–Kronig relations, the refractive index associated with this absorption is given by

$$n(\delta) = n_b + \frac{\alpha_1 \lambda}{4\pi} \frac{\gamma/\delta}{1 + (\delta^2/\gamma^2)} , \tag{7.3}$$

where λ is the optical wavelength and n_b is the background refractive index, which we shall subsequently set equal to unity. From this result, we then find that the group index $n_g = n + \omega \, dn/d\omega$ is given by

$$n_g = 1 + \frac{\alpha_1 c}{2\gamma} \frac{1 + (\delta^2/\gamma^2)}{1 + [(\delta^2/\gamma^2)]^2} . \tag{7.4}$$

The expression for the group velocity $v_g = c/n_g$ then follows directly.

The input pulse is taken to be a transform-limited Gaussian pulse of the form

$$A(z, t) = A_0 e^{-t^2/T^2} . \tag{7.5}$$

Here T is the pulse width defined as the amplitude half-width to $1/e$ or the intensity half-width to $1/e^2$. Equivalently, we can describe the pulse in the frequency domain as

$$\tilde{A}(z, \omega + \delta) = \tilde{A}_0 e^{-\delta^2/\xi^2} , \tag{7.6}$$

where $\xi = 2/T$ is the frequency-domain pulse width. This pulse will be advanced compared to vacuum propagation as it passes through the medium. Neglecting dispersion in the group velocity and the fact that the frequency-varying absorption will cause spectral reshaping of the pulse, the amount of pulse advancement $\Delta T = L/v_g - L/c$ resulting from propagation through a length L of the medium is found to be

$$\Delta T = -\frac{\alpha_1 L}{2\gamma} \frac{1+(\delta^2/\gamma^2)}{1+[(\delta^2/\gamma^2)]^2} .$$ (7.7)

As can be seen, the amount of pulse advancement increases with increasing absorption $\alpha_l L$ and also with decreasing linewidth γ. However, a line that is too narrow compared to the spectral width of the input pulse will lead to pulse distortion either by group velocity dispersion or by spectral reshaping of the pulse [65–67]. In the cases studied here, pulse narrowing by spectral reshaping is the dominant effect. If the limit on pulse narrowing is set so that to the first order the pulse duration becomes infinitesimally small, we find by means of the procedure described in [66] that the following inequality must be satisfied:

$$2\alpha_l L \leq (\gamma T)^2 .$$ (7.8)

The allowable total integrated absorption $\alpha_l L$ is limited by two factors. First, in order to be able to detect the output pulse, the transmission at the center frequency of the pulse must not be too small. Second, in order to avoid instabilities, the gain of the background must not be too large. Taking (somewhat arbitrarily) the minimum allowable transmission at the center frequency of the pulse to be e^{-32} and the maximum allowable gain to be e^{32}, we find that the maximum relative pulse advancement $\Delta T/T$ that can be obtained in the system studied here is given according to Eqs. (7.7) and (7.8) by

$$\Delta T/T = 2\sqrt{2} .$$ (7.9)

This maximum pulse advancement is obtained when the line-center absorption of the absorption line is set equal to e^{-64} and the gain of the broadband background is set equal to e^{32}. Other authors have deduced similar limits on the maximum possible pulse advancement [68].

A more detailed description of the propagation of the optical pulses through the material medium can be obtained by performing a numerical simulation of the propagation process. Because we are considering the situation in which the material responds linearly to the optical pulse, we model the propagation by means of the following procedure. The input pulse is decomposed into a Fourier integral, and each frequency component is allowed to propagate through the medium, acquiring phase and amplitude modifications in accordance with its frequency-dependent refractive index (Eq. 7.3) and absorption coefficient (Eq. 7.2). The time evolution of the output pulse is then obtained by performing the inverse Fourier transform on the output spectrum.

Some of the results of this procedure are shown in Fig. 7.8. The parameters for the advanced pulse were chosen to give the maximum possible advancement in that sense that according to the first-order analytic model the pulse duration would shrink to zero. We see that the pulse has narrowed but not as much as the first-order theory would predict. In addition, the amount of pulse advancement is 12% smaller that predicted by Eq. (7.9). For comparison, in this figure we also

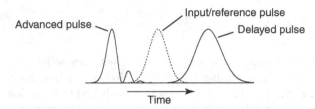

Fig. 7.8 Pulse advancement in a medium consisting of an absorption line with a total attenuation of e^{-64} set on a broad gain background with a gain of e^{64}. The ratio between the width γ (half-width at half maximum) of the Lorentzian line to the spectral width ξ (half-width to $1/e$ in amplitude) of the input pulse is 5.7. The input/reference pulse shows what the output pulse would be if the medium were replaced by an equal length of vacuum. Also shown for comparison is the pulse delay that experienced upon propagation through a slow-light medium consisting of an e^{64} gain line set on a broad e^{-32} absorptive background. Note that pulses tend to compress in time for fast-light propagation and broaden in time for slow-light propagation

give an example of pulse propagation through a slow-light medium. In the configuration studied here, there is no fundamental limit on how strong the broad absorptive background could be, and therefore there is no limit on how much the pulse can be delayed [66]. In choosing the parameters for this example, we required that the pulse broaden by no more than a factor of $\sqrt{2}$. Other configurations using multiple lines and operating far from line center can also give rise to large delays [65, 67, 69].

No other choice of line strengths and linewidths has been found that gives significantly more pulse advancement than that shown in Fig. 7.8. The reason for this behavior is that fast light occurs when the center frequency of the light pulse is at a local minimum of the transmission. Therefore, if one tries to obtain a larger pulse advancement, either the transmission at the center frequency will be too low for detection of the pulse at the output or the gain away from the center frequency will be so high that it leads to instabilities [19, 68].

The fact that the peak arrives at the output earlier than it would have arrived had the medium been replaced by an equal length of vacuum suggests that the peak of the pulse travels faster in the medium than the speed of light in vacuum. A situation of this sort is shown in Fig. 7.9. Here the material parameters are chosen such that the group velocity from first-order theory is twice the speed of light in vacuum. As can be seen, at $t = 0.4$ (in all of our numerical work we normalize the time in units of the vacuum transit time through the medium) the pulse peak has traveled approximately twice as far as the reference pulse. Due to spectral reshaping of the pulse by the frequency-varying absorption, the apparent pulse velocity then begins to slow down and pulse distortion starts to occur. This simulation is performed with a ratio of the width γ of the absorption line to the spectral width ξ of the input pulse of 4. For this ratio, we have found that the maximum allowable integrated line strength (that is, $\exp(-\alpha_l L)$) before the occurrence of significant pulse distortion by spectral reshaping is e^{32}. The integrated line strength of the example shown in Fig. 7.9 is e^{48} and consequently after propagating approximately two-thirds of the

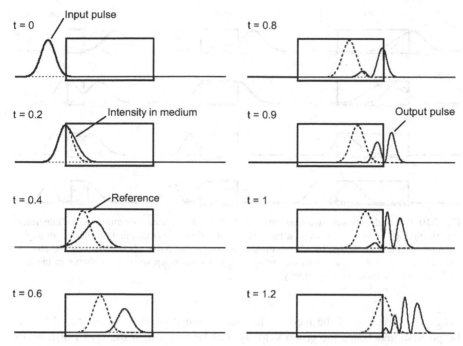

Fig. 7.9 A sequence of frames showing a Gaussian pulse propagating through a fast-light medium. The medium is comprised of a Lorentzian absorption line of integrated line strength e^{-48} set on a broad gain background with a gain of $e^{45.3}$ chosen to give an output pulse equal in amplitude to the input pulse. The ratio γ/ξ of the width of the absorption line to the spectral width of the input pulse is 4. For comparison we also show as the *dashed curve* how the pulse would propagate through an equal length of vacuum

way through the medium (at $t = 0.8$) severe pulse distortion sets in and the pulse breaks up into several parts.

The advancement ΔT in units of input pulse width T depends only on the integrated line strength $\exp(\alpha_l L)$ and on the linewidth γ but not on the physical length of the medium. Through use of the concept of group velocity, the pulse advancement shown in Fig. 7.8 can be considered to be the difference in the transit times of a pulse moving at c and of a pulse moving through the medium at the effective group velocity v'_g such that

$$\Delta T = \frac{L}{c} - \frac{L}{v'_g}. \tag{7.10}$$

This effective group velocity is, in general, different from the group velocity v_g given by first-order theory. Solving this equation for the effective group velocity, we find that

$$\frac{c}{v'_g} = 1 - \frac{\Delta T}{L/c}. \tag{7.11}$$

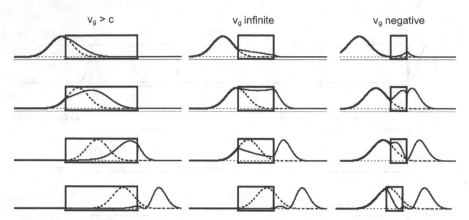

Fig. 7.10 Pulse advancement in a medium consisting of a Lorentzian absorption line of integrated line strength e^{-32} set on a broad gain background of gain $e^{30.7}$. Depending on the length of the medium compared to the spatial advancement of the pulse, the pulse advancement can be considered to be the result of a larger-than-c, infinite, or negative group velocity as shown in the *left*, *middle*, and *right columns*, respectively

When the length L of the medium lies in the range $c\Delta T < L < \infty$, ΔT will be positive and hence the group velocity will be larger than the speed of light in vacuum, for $c\Delta T = L$, the group velocity becomes infinite, and when L is smaller than $c\Delta T$ the group velocity is negative. Examples of these three situations are shown in Fig. 7.10.

When the group velocity is larger than c but finite, the pulse propagation takes the form of a distinct pulse that moves through the medium at a superluminal velocity. When the group velocity is infinite, the propagation through the medium occurs instantaneously in the sense that the peak at the output occurs at the same time that the peak of the input pulse reaches the input face of the medium. For negative group velocities, a well-defined pulse moves backward through the medium, as has been observed experimentally [24]. In this case the peak of the output pulse actually leaves the medium before the input pulse enters. For the three cases shown in Fig. 7.10, the group velocities obtained from the numerical simulation and from Eq. (7.11) are very close to the group velocity predicted by the first-order theory.

In the simulations described above, the strength of the broad gain background has been chosen such that the peak intensities of the input and output pulses are equal. When this is the case, we observe a distinct pulse propagating through the medium. We can then think of the pulse advancement as occurring as a result of superluminal pulse propagation. However, in many experimental situations, a large net absorption is experienced in passing through the medium. An example of such behavior is shown in Fig. 7.11. Here we assume the presence of an absorption line of integrated line strength e^{-32} and no gain background.

In this case the pulse decays approximately exponentially as it propagates through the medium, and the output peak is approximately e^{32} times smaller than that of the input. In order to make the pulse visible in the plots, the pulse intensities are

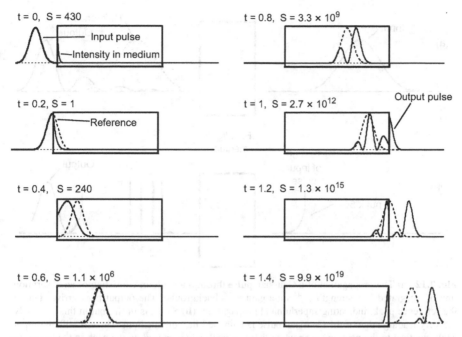

Fig. 7.11 Pulse propagation through a fast-light medium with large net absorption consisting of a absorption line of integrated line strength e^{-32}. In each panel the intensity is normalized to fill the vertical scale. The group velocity is 1.2 times c, but as a result of large absorption the peak in the medium lags behind the peak of the reference pulse until $t = 0.6$. For each panel the scale factor S gives the value by which the pulse intensity within the medium has been multiplied to ensure that it fills the entire vertical scale. The output pulse intensity has been multiplied by a factor of 2.1×10^{13} in all cases

normalized in each panel to fill the vertical scale. The intensity inside the medium initially decays exponentially, as would be expected because of the large absorption. Later, a clear pulse is formed. However, until $t = 0.6$ the pulse lags behind the reference pulse and in a sense propagates with a group velocity smaller than c. Later there are multiple peaks within the medium, and only at the output does a clear, relatively undistorted, advanced pulse appear. In this case the pulse advancement cannot be regarded as occurring as the result of superluminal propagation because there is no well-formed pulse propagating faster than c within the medium. Rather, the apparent superluminal behavior occurs as the result of spectral reshaping.

The appearance of an undistorted output pulse that propagates faster than the vacuum speed of light c might initially seem to imply a violation of the laws of the special theory of relativity. One can even find situations in which the peak of the output pulse leaves the medium before the peak of the input pulse reaches the medium, apparently violating causality.

The next two figures illustrate this sort of behavior. In Fig. 7.12 we show that the moderately distorted output pulse is not the result of the peak of the input pulse propagating faster than the speed of light in the medium, but rather is the result of

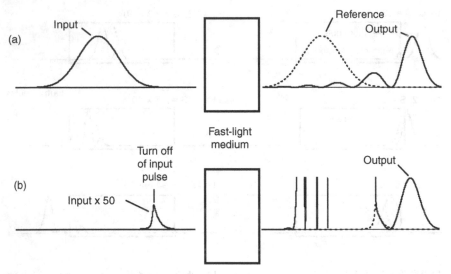

Fig. 7.12 (a) The propagation of a Gaussian pulse through a medium consisting of an absorption line of integrated line strength e^{-44} on a gain $e^{41.6}$ background. The output peak arrives before the reference peak, indicating superluminal propagation. (b) Same as in (a), but in this case only the very leading edge of the Gaussian pulse is launched into the medium. There is still an output peak identical to the peak in (a) showing that this peak is not caused by the peak in the Gaussian input pulse propagating faster than the vacuum speed of light c, but is the result of distortion and amplification of the leading edge of the pulse, a fully causal process

distortion and amplification of the early part of the input pulse. We first note that the Gaussian input pulse in Fig. 7.12(a) has, in principle, been in existence for all times, even if it falls off very rapidly as we move away from the peak. The second point to note is that we have been very careful in placing the center frequency of the input pulse at a frequency where the transmission through the medium is at a minimum and has the value 9%. Close to this frequency, the gain is very large, on the order of 10^{18}. Third, we note that, like previous workers, we use Gaussian input pulses. Gaussian pulses have the property of having a clear peak in the time domain while at the same time being well contained in the frequency domain. In Fig. 7.12(b) we show the result of launching only the very early parts of the input pulse. Up until the time labeled "turn off of input pulse" the input in Fig. 7.12(b) is the same as the Gaussian input in Fig. 7.12(a). After this point the Gaussian input is rapidly ramped down to zero as shown. On the output side of the medium, we see that the peak of Fig. 7.12(a) is still present, unchanged. This result indicates that this peak occurs as the result of the unchanged part of the input pulse, the leading edge, being amplified and distorted [42, 70]. Thus there is no causal connection between the peak of the output pulse and the peak of the input pulse. The relative gain required to turn the weak leading edge into the output peak shown is only 10^7, much below the 10^{18} gain that is available. The turning-off of the input pulse in Fig. 7.12(b) introduces frequency components on the input that are far from line center and see high gain on propagating through the medium, resulting in the later parts of the output going off scale in the plot shown.

From the type of simulation shown in Fig. 7.9, it is tempting to regard the pulse advancement as occurring as the result of the peak of the pulse propagating faster than c, moving gradually ahead of the reference pulse while suffering very little pulse distortion. This picture leads to concern about how this behavior can be consistent with the special theory of relativity. The resolution of this concern is that this is a very misleading picture. It is more correct to regard what happens on propagation as a continuous distortion of the pulse that moves the peak of the distorted pulse ahead of the peak of the undistorted reference pulse.

The information velocity can be considered to be the velocity at which the outcome of some "choice travels." In many circumstances the information velocity is, for practical purposes, equal to the group velocity. The simulation shown in Fig. 7.13 demonstrates that this is certainly not the case in a fast-light system. Here, the input vanishes up to the point marked "pulse front." Then a decision is made to launch a pulse into the medium with the shape and peak position as shown. The information velocity is the velocity within the medium at which this information travels. The decision to launch the pulse forces the input to jump from zero up to a value that is consistent with the rest of the pulse being Gaussian. This jump forms the pulse front [70, 42].

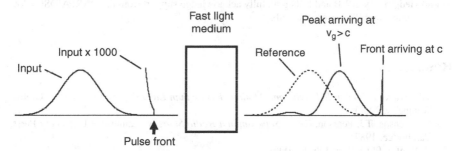

Fig. 7.13 Up to the point marked "pulse front," the input vanishes. At that point a decision is made to launch the pulse. The jump in pulse amplitude at this point forms the pulse front. This pulse front arrives at the output after propagating through the medium with velocity c and heralds the fact that the pulse has been launched. The subsequent arrival of the peak carries no new information, even though it can be thought of as arriving after superluminal propagation. These numerical results are consistent with the conceptual picture presented in Fig. 7.5

Under these conditions, we see that there are two peaks in the output, the "normal" advanced pulse and a second feature labeled "front." In this numerical simulation, the various parameters have been chosen such that these two peaks have equal amplitude. The first peak marks the arrival of the pulse front. This peak arrives with the same speed as the reference pulse, the speed of light in vacuum. Later the advanced pulse arrives. This peak arrives earlier than the peak of the pulse would have arrived had the medium been replaced by an equal length of vacuum. It can, therefore, be regarded as consequence of the group velocity being greater than c.

But this peak carries no new information, as we already know from the arrival of the front that the rest of the pulse must arrive. The arrival of the superluminal pulse does, therefore, not violate causality.

7.8 Summary

We have reviewed recent theoretical and experimental research that establishes that pulses can propagate through material systems with superluminal or even negative group velocities. Nonetheless, these exotic propagation effects are fully compatible with established notion of causality.

At a fundamental level, the nature of slow and fast light seems fairly well understood. But there still may be some surprises on the horizon. We noted in the body of this chapter that there seems to be no fundamental limit on how much one can delay a light pulse using slow-light methods, and in fact pulse delays as great as 80 pulse lengths have been observed [29]. Conversely, there seem to be severe limitations that limit the amount of advancement for a fast-light system to at most several pulse widths [71].[1] But can these limitations be overcome? This is an intriguing question that merits further examination.

Acknowledgments RWB and DJG gratefully acknowledge support from the DARPA/DSO Slow Light Program, and RWB from the NSF.

References

1. M. Fayngold, *Special Relativity and Motions Faster than Light* (Wiley-VCH Verlag GmbH, Weinheim, 2002)
2. J.A. Zensus, T.J. Pearson, (eds.), *Superluminal Radio Sources* (Cambridge University Press, Cambridge, 1987)
3. H. Winful, Phys. Rep. **436**, 1 (2006)
4. R. Smith, Am. J. Phys. **38**, 978 (1970)
5. P.W. Milonni, *Fast Light, Slow Light, and Left-Handed Light* (Institute of Physics Publishing, Bristol, 2005)
6. R. Loudon, J. Phys. A **3**, 233 (1970)
7. M. Lisak, J. Phys. A **9**, 1145 (1976)
8. G.C. Sherman, K.E. Oughstun, Phys. Rev. Lett. **47**, 1451 (1981)
9. G.C. Sherman, K.E. Oughstun, J. Opt. Soc. Am. B **12**, 229 (1995)
10. J. Aaviksoo, J. Lipmaa, J. Kuhl, J. Opt. Soc. Am. B **8**, 1631 (1988)
11. R.Y. Chiao, A.M. Steinberg, in *Progress in Optics 37*, E. Wolf (ed.) (Elsevier, Amsterdam, 1997).
12. J. Peatross, S.A. Glasgow, M. Ware, Phys. Rev. Lett. **84**, 2370 (2000)
13. N.A. Cartwright, K.E. Oughstun, J. Opt. Soc. Am. A **21**, 439 (2004)
14. W.R. Hamilton, Proc. R. Irish Acad. **1**, 341 (1839)

---

[1] N. of E.: Sect. 2.4.4 of Volume 1 discusses causal bounds for the negative time delays in the context of potential scattering, see also references therein.

15. L. Rayleigh (J.W. Strutt), Nature **24**, 382 (1881)
16. L. Rayleigh (J.W. Strutt), Nature **25**, 52 (1881)
17. H.A. Lorentz, *Theory of Electrons* (1909), reprinted by Dover Publications, New York (1952)
18. S. Chu, S. Wong, Phys. Rev. Lett. **49**, 1293 (1982)
19. B. Ségard, B. Macke, Phys. Lett. A **109**, 213 (1985)
20. L.V. Hau, S.E. Harris, Z. Dutton, C.H. Behroozi, Nature **397**, 594 (1999)
21. M.M. Kash, V.A. Sautenkov, A.S. Zibrov, L. Hollberg, G.R. Welch, M.D. Lukin, Y. Rostovtsev, E.S. Fry, M.O. Scully, Phys. Rev. Lett. **82**, 5229 (1999)
22. M.S. Bigelow, N.N. Lepeshkin, R.W. Boyd, Phys. Rev. Lett. **90**, 113903 (2003)
23. M.S. Bigelow, N.N. Lepeshkin, R.W. Boyd, Science **301**, 200 (2003)
24. G.M. Gehring, A. Schweinsberg, C. Barsi, N. Kostinski, R.W. Boyd, Science **312**, 985 (2006)
25. Y. Okawachi, M.S. Bigelow, J.E. Sharping, Z.M. Zhu, A. Schweinsberg, D.J. Gauthier, R.W. Boyd, A.L. Gaeta, Phys. Rev. Lett. **94**, 153902 (2005)
26. K.Y. Song, M.G. Herraez, L. Thevenaz, Opt. Express **13**, 82 (2005)
27. J.E. Sharping, Y. Okawachi, A.L. Gaeta, Opt. Express **13**, 6092 (2005)
28. Y.A. Vlasov, M.O. Boyle, H.F. Hamann, S.J. McNab, Nature **438**, 65 (2005)
29. R.M. Camacho, M.V. Pack, J.C. Howell, A. Schweinsberg, R.W. Boyd, Phys. Rev. Lett. **98**, 153601 (2007)
30. The work of A. Sommerfeld and L. Brillouin from 1914 is translated into English and collected in the book: L. Brillouin, Wave Propagation and Group Velocity (Academic Press, New York, 1960)
31. C.G.B. Garrett, D.E. McCumber, Phys. Rev. A **1**, 305 (1970)
32. F.R. Faxvog, J.A. Carruthers, J. Appl. Phys. **41**, 2457 (1970)
33. F.R. Faxvog, C.N.Y. Chow, T. Bieber, J.A. Carruthers, Appl. Phys. Lett. **17**, 192 (1970)
34. B. Ségard, B. Macke, Phys. Lett. **109**, 213 (1985)
35. A.M. Akulshin, A. Cimmino, G.I. Opat, Quantum Electron. **32**, 567 (2002)
36. A.M. Steinberg, R.Y. Chiao, Phys. Rev. A 49, 2071 (1994)
37. R.Y. Chiao, in *Amazing Light: A Volume Dedicated to Charles Hard Townes on His 80th Birthday*, R.Y. Chiao (ed.) (Springer, New York, 1996), p. 91
38. R.Y. Chiao, A.M. Steinberg, Tunneling times and superluminality, in *Progress in Optics XXXVII*, E. Wolf, (ed.) (Elsevier Science, Amsterdam, 1997), Ch. VI., pp. 345–405
39. J.C. Garrison, M.W. Mitchell, R.Y. Chiao, E.L. Bolda, Phys. Lett. A **245**, 19 (1998)
40. M.C. Parker, S.D. Walker, Opt. Commun. **229**, 23 (2004)
41. L.J. Wang, A. Kuzmich, A. Dogariu, Nature **406**, 277 (2000)
42. M.D. Stenner, D.J. Gauthier, M.A. Neifeld, Nature **425**, 695 (2003)
43. M.D. Stenner, D.J. Gauthier, M.A. Neifeld, Phys. Rev. Lett. **94**, 053901 (2005)
44. H. Jeong, A.M.C. Dawes, D.J. Gauthier, Phys. Rev. Lett. **96**, 143901 (2006)
45. G. Diener, Phys. Lett. A **223**, 327 (1996)
46. G. Kurizki, A. Kozhekin, A.G. Kofman, Europhys. Lett. **42**, 499 (1998)
47. A. Kuzmich, A. Dogariu, L.J. Wang, P.W. Milonni, R.Y. Chiao, Phys. Rev. Lett. **86**, 3925 (2001)
48. K. Wynne, Opt. Commun. **209**, 85 (2002)
49. H. Tanaka, H. Niwa, K. Hayami, S. Furue, K. Nakayama, T. Kohmoto, M. Kunitomo, Y. Fukuda, Phys. Rev. A **68**, 053801 (2003)
50. A. Ruschhaupt, J.G. Muga, Phys. Rev. Lett. **93**, 020403 (2004)
51. B.D. Clader, Q-Han Park, J.H. Eberly, Opt. Lett. **31**, 2921 (2006)
52. M. Jammer, *Concepts of Simultaneity* (Johns Hopkins University Press, Baltimore, 2006)
53. R.P. Feynman, R.B. Leighton, M. Sands, *The Feynman Lectures on Physics*, vol. 1 (Addison-Wesley, Reading, 1963). See Section 46–4.
54. St. Augustine, *Confessions*, book 7, chapter 6 (ca. 397 AD)
55. A.S. Eddington, *The Mathematical Theory of Relativity* (Cambridge University Press, Cambridge, 1923)
56. A. Einstein, *Relativity: The Special and the General Theory – A Clear Explanation that Anyone Can Understand* (Gramercy, New York, 1988)

57. W.R. LeFew, S. Venakides, D.J. Gauthier, preprint (2008) Available at: http://arxiv.org/PScache/arxiv/pdf/0705/0705.4238v2.pdf
58. N. Herbert, Found. Phys. **12**, 1171 (1982)
59. N. Herbert, *Faster Than Light* (New American Library Books, Penguin Inc., New York, 1998)
60. D. Dieks, Phys. Lett. **92A**, 271 (1982)
61. R.J. Glauber, in *New Techniques and Ideas in Quantum Measurement Theory*, D.M. Greenberg (ed.), Ann. N.Y. Acad. Sci. **480**, 336 (1986)
62. W.K. Wootters, W.H. Zurek, Nature **299**, 802 (1982)
63. V. Buzek, M. Hillery, Phys. Rev. A **54**, 1844 (1996)
64. N. Gisin, Phys. Lett. A **242**, 1 (1998)
65. H. Cao, A. Dogariu, L.J. Wang, IEEE J. Sel. Top. Quantum Electron. **9**, 52 (2003)
66. R.W. Boyd, D.J. Gauthier, A.L. Gaeta, A.E. Willner, Phys. Rev. A **71**, 023801 (2005)
67. M.D. Stenner, M.A. Neifeld, Z. Zhu, A.M.C. Dawes, D.J. Gauthier, Opt. Express **13**, 9995 (2005)
68. B. Macke, B. Ségard, F. Wielonsky, Phys. Rev. E **72**, 035601(R) (2005)
69. Z. Shi, R. Pant, Z. Zhu, M.D. Stenner, M.A. Neifeld, D. J. Gauthier, R.W. Boyd, Opt. Lett **32**, 1986 (2007)
70. M.W. Mitchell, R.Y. Chiao, Phys. Lett. A **230**, 133 (1999)
71. R.W. Boyd, P. Narum, J. Mod. Opt. **54**, 2403 (2007)

Chapter 8
Experiments on Quantum Transport of Ultra-Cold Atoms in Optical Potentials

Martin C. Fischer and Mark G. Raizen

8.1 Introduction

In this chapter, we describe our experiments with ultra-cold atoms in optical potentials and show how we can address fundamental issues of time in quantum mechanics. The high degree of experimental control and the conceptual simplicity are the main advantages of our system. We start with an overview of the basic interaction of atoms and light and make the connection between atoms in optical lattices and solid state physics. While this latter connection has evolved into a major theme in physics over the past decade, at the time of this work it was still new and unexplored. After introduction of the theoretical model and the basic equations, we introduce the experimental apparatus. We then review our experiments to observe the Wannier–Stark ladder in an accelerating lattice. This system was used to study quantum tunneling where short-time non-exponential decay was first observed for an unstable quantum system. We then describe our experiments to observe the quantum Zeno and anti-Zeno effects for an unstable system that is repeatedly interrogated. We conclude this chapter with a brief outlook into the future.

8.1.1 The Interaction of Atoms and Light

The manipulation of the motional state of individual atoms with light fields was observed as early as 1930, when Frisch measured the deflection of an atomic beam with resonant light from a sodium lamp [15]. The measured deflection was caused by the recoil momentum that an atom acquires when absorbing or emitting a single

M.C. Fischer (✉)
Department of Chemistry, Duke University, Durham, NC 27708, USA,
Martin.Fischer@duke.edu

M.G. Raizen
Center for Nonlinear Dynamics and Department of Physics, The University of Texas at Austin,
Austin, TX 78712, USA, raizen@physics.utexas.edu

Fischer, M.C., Raizen, M.G.: *Experiments on Quantum Transport of Ultra-Cold Atoms in Optical Potentials*. Lect. Notes Phys. **789**, 205–237 (2009)
DOI 10.1007/978-3-642-03174-8_8   © Springer-Verlag Berlin Heidelberg 2009

photon of light. When an atom absorbs a photon from a beam of light, it acquires momentum in the direction of the light beam. Since scattered photons are emitted without preferred direction, the momentum acquired during the emission averages to zero over many cycles. This leads to a net force on the atom which is called the *spontaneous force*, or *radiation pressure*. The spontaneous force scales with the scattering rate and for large detunings falls off quadratically with the detuning Δ_L of the light from the atomic resonance [9]:

$$F_{\text{spont}} \propto \frac{I}{\Delta_L^2} \,, \tag{8.1}$$

where I is the laser intensity.

Another type of force is based on the coherent scattering of photons. The oscillating electric field of light can induce a dipole moment in the atom. If the induced dipole moment is in phase with the electric field, the interaction potential is lower in regions of high field and the atom will experience a force toward those regions. If it is out of phase, a force pointing away from regions of high field results. This force is called the *dipole force*. As opposed to the spontaneous force, the dipole force only falls off linearly with the detuning from the atomic resonance in the limit of large detuning [9]

$$F_{\text{dipole}} \propto \frac{\nabla I}{\Delta_L}. \tag{8.2}$$

From the scaling laws for the two types of forces it is clear that with sufficient laser intensity, the spontaneous force can be made negligibly small while still generating an appreciable dipole force. As early as 1970, Ashkin succeeded in trapping small particles with a pair of opposing, focused laser beams, making use of both types of forces. However, only the relatively recent development of laser cooling and trapping techniques have created the conditions for controlled manipulation of atoms with the dipole force alone [7]. While the laser cooling and trapping required to prepare our atomic sample utilized near-resonant light and thus both types of light forces, the optical lattices were composed of far-detuned light, so that only the dipole interaction was important.

8.1.2 Optical Lattices and the Connection to Solid State Physics

In our experiments we created a periodic optical potential by spatially overlapping two laser light beams. The periodicity of the resulting standing wave was determined by the interference pattern in the region of overlap. In the nodes of a standing wave, the electric field of the light interferes destructively and atoms at those positions are unaffected by the light. Away from the nodes, the dipole interaction causes a light-induced shift of the atomic energy levels, which is maximal at the anti-nodes. This shift of the energy levels – which is another way of describing the aforementioned dipole force – is periodic in space. The system of a particle in a periodic potential is the textbook model of an electron in a crystal lattice and has been studied in great detail. In the 1920s Bloch arrived at the conclusion that due to the periodicity of the lattice, the eigenstates are plane waves modulated by periodic functions of position [4]. The implications of these findings on the description of transport in periodic potentials under the influence of externally applied fields are profound. Some of the resulting effects, such as Bloch oscillations and Wannier–Stark states, are treated in more detail in Sect. 8.4. Experimental verification of those predicted effects in crystal lattices, however, has been hindered by extremely short relaxation times. Electrons in a crystal lattice can scatter on impurities, dislocations, phonons, and even on other electrons. If the scattering occurs on a timescale faster than the timescale for coherent evolution of the system, coherent transport effects are destroyed. Advances in the production of very high purity superlattice structures in the 1970s allowed the experimental investigation of some of those coherent effects for the first time [31]. However, the ratio of the relaxation time to the characteristic timescale for coherent evolution in those systems was still only on the order of unity. In our system we can achieve a ratio on the order of 10^3. The relaxation time is mainly limited by spontaneous emission during the interaction, which can be made very small by detuning far from resonance. This high ratio and the ability to dynamically control the interaction potential in real time during the experiment allowed us to observe many of the coherent effects which are inaccessible in solid state systems. A more detailed comparison of the solid state and atom optics system is given in a recent overview article [38].

8.1.3 Interaction Hamiltonian

In this section, we derive the effective Hamiltonian for a two-level atom in a standing wave of far-detuned light, closely following Graham et al. [16]. The atom is assumed to have a ground state $|g\rangle$ and an excited state $|e\rangle$, separated in energy by $\hbar\omega_0$. For a single atom of such type in a classical light field $\mathbf{E}(\mathbf{r}, t)$, the Hamiltonian is the sum of three contributions: the kinetic energy of the center of mass, the internal energy, and the interaction energy [25]

$$H = H_{\text{CM}} + H_{\text{internal}} + H_{\text{interaction}} , \tag{8.3}$$

where

$$H_{\text{CM}} = \frac{p^2}{2M} , \tag{8.4}$$

$$H_{\text{internal}} = \frac{1}{2}\hbar\omega_0\,\sigma_z, \quad \text{and} \tag{8.5}$$

$$H_{\text{interaction}} = -\mathbf{d} \cdot \mathbf{E}(\mathbf{r}, t) = -\left(\langle e|\mathbf{d} \cdot \mathbf{E}|g\rangle\,\sigma^+ + \langle g|\mathbf{d} \cdot \mathbf{E}|e\rangle\,\sigma^-\right) . \tag{8.6}$$

The symbols $\sigma^{\pm}$ and σ_z denote the Pauli spin matrices. For a linear polarization vector $\hat{\varepsilon}$ of the light, we can define the resonant Rabi frequency as

$$\Omega = -\frac{\langle e|\mathbf{d}\cdot\mathbf{E}|g\rangle}{\hbar} = -\frac{\langle g|\mathbf{d}\cdot\mathbf{E}|e\rangle}{\hbar} = -\frac{\langle g|\mathbf{d}\cdot\hat{\varepsilon}|e\rangle}{\hbar}E , \qquad (8.7)$$

where we have assumed a slow variation of the field amplitude $E(\mathbf{r}, t)$.

We create an optical lattice by overlapping two laser beams with identical linear polarization vectors $\hat{\varepsilon}$. The electric field is then of the form

$$\mathbf{E}(\mathbf{r}, t) = \frac{1}{2}\hat{\varepsilon}\left(E_1\, e^{i(\mathbf{k}_1\cdot\mathbf{r}-\omega_1 t)} + E_2\, e^{i(\mathbf{k}_2\cdot\mathbf{r}-\omega_2 t)}\right) + c.c. \qquad (8.8)$$

Using this light field, we find that the interaction term is

$$H_{\text{interaction}} = \sum_{n=1,2} \hbar\frac{\Omega_n}{2}\left(\sigma^- e^{-i(\mathbf{k}_n\cdot\mathbf{r}-\omega_n t)} + \sigma^+ e^{i(\mathbf{k}_n\cdot\mathbf{r}-\omega_n t)}\right) , \qquad (8.9)$$

where we have used the rotating wave approximation to drop the counter-rotating terms $\sigma^+ e^{+i\omega t}$ and $\sigma^- e^{-i\omega t}$ [25]. To separate the center-of-mass motion of the atoms from their internal state, we write the atomic state as

$$|\Psi(\mathbf{r}, t)\rangle = c_g(\mathbf{r}, t)|g\rangle + c_e(\mathbf{r}, t)|e\rangle . \qquad (8.10)$$

We insert $|\Psi\rangle$ into the time-dependent Schrödinger equation with the Hamiltonian in Eq. (8.3) to obtain propagation equations for c_g and c_e. Following Graham et al. [16] for a sufficiently large detuning from resonance, we can adiabatically eliminate the excited state amplitude and remain with an equation for the (phase transformed) ground state amplitude $\tilde{c}_g$

$$i\hbar\partial_t\tilde{c}_g = \left[\frac{p^2}{2M} + \hbar\frac{\Omega_1\Omega_2}{4\Delta_L}\left(e^{i(\mathbf{q}\cdot\mathbf{r}-\delta t)} + e^{-i(\mathbf{q}\cdot\mathbf{r}-\delta t)}\right)\right]\tilde{c}_g , \qquad (8.11)$$

where $\mathbf{q} = \mathbf{k}_2 - \mathbf{k}_1$, $\delta = \omega_2 - \omega_1$ (the frequency difference between the two beams), and $\Delta_L = \frac{\omega_1+\omega_2}{2} - \omega_0$ (their average detuning from resonance). This leads to an effective Hamiltonian for an atom in the ground state

$$H = \frac{p^2}{2M} + V_0\cos(\mathbf{q}\cdot\mathbf{r} - \phi(t)) , \qquad (8.12)$$

where for generality we have introduced an arbitrary time-dependent phase $\phi(t)$ (the instantaneous frequency difference is $\delta(t) = \frac{d\phi}{dt}$). The amplitude of the potential term is

$$V_0 = \hbar\frac{\Omega_1\Omega_2}{2\Delta_L} . \qquad (8.13)$$

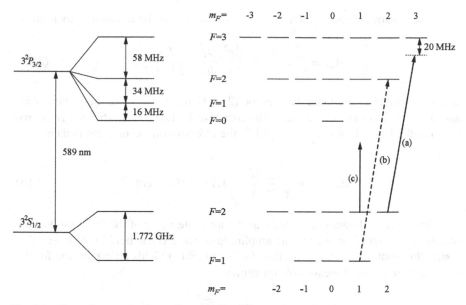

Fig. 8.1 Term diagram for the sodium D$_2$ line. The nuclear spin of sodium is $I = 3/2$, and so the ground state of sodium $3S_{1/2}$ has two hyperfine levels $F = 1, 2$. For the $3P_{3/2}$ excited state, we have $J = 3/2$ so that $F = 0, 1, 2, 3$. The $2F + 1$ magnetic sublevels are also shown. Representative examples of (**a**) the cooling and trapping light, (**b**) the optical pumping sideband, and (**c**) the far-detuned optical lattice light are shown as *arrows*.

The expression for the well depth V_0 contains the resonant Rabi frequencies Ω_1 and Ω_2. The calculation of these frequencies is complicated by the fact that sodium is by no means a system with a two-level structure, as can be seen in the term diagram for the levels contributing to the sodium D$_2$ line in Fig. 8.1. However, several factors make a determination of the well depth possible. Our initial condition (before the atoms interact with the light) is such that almost all atoms populate the hyperfine $F = 2$ level in the lower manifold. For linearly polarized light, all of the (nearly) degenerate m_F levels experience the same level shift in the far-detuned regime. Therefore the entire sample experiences the same effective potential [9, 32]. The actual dipole coupling for a particular ground state sublevel $|F\ m_F\rangle$ is obtained by summing over its couplings to all of the available excited states. When the detuning is large compared to the excited state frequency splittings, all of the excited states participate, and the detuning for each excited state is approximately the same. In addition, the dipole coupling summed over all excited states and all polarizations is independent of the m_F sublevel considered [32, 40]. Because of the spherical symmetry of the dipole operator, the three Cartesian components in this sum are equal and therefore the effective dipole coupling for the case of linearly polarized light and large detuning, regardless of the ground state population, is one-third the square of the dipole matrix element for the full D$_2$ ($J = 1/2 \leftrightarrow J' = 3/2$) transition

$$|d_{\text{effective}}|^2 = \frac{e^2|D_{12}|^2}{3} \, . \tag{8.14}$$

The dipole matrix element $e^2|D_{12}|^2$ can be obtained from the Einstein A coefficient,

$$A_{21} = \Gamma = \frac{1}{\tau} = \frac{\omega_0^3 e^2 |D_{12}|^2}{3\pi\varepsilon_0\hbar c^3} \frac{2J+1}{2J'+1} , \qquad (8.15)$$

which is related to the radiative lifetime [25]. Here, $J = 1/2$ is the ground state and $J' = 3/2$ is the excited state. The radiative lifetime, $\tau = 16.2$ ns, is known empirically. Using Eqs. (8.14) and (8.15), the effective dipole moment is then

$$d_{\text{effective}} = \sqrt{\frac{\varepsilon_0\hbar\lambda_L^3}{4\pi^2\tau}} = 1.71 \times 10^{-29} \text{ cm} . \qquad (8.16)$$

The time-averaged intensity (defined as the absolute value of the Poynting vector) of a beam of light is related to the amplitude of the electric field by $I = \frac{1}{2}c\varepsilon_0 E^2$. Using this relation together with Eqs. (8.7) and (8.13) yields an expression for the well depth in terms of measurable quantities

$$V_0 = \frac{2\pi c^2}{\tau\omega_0^3} \frac{\sqrt{I_1 I_2}}{\Delta_L} , \qquad (8.17)$$

where I_1 and I_2 are the intensities of the traveling wave components.

8.1.4 Spontaneous Emission Rate

In deriving the Hamiltonian for our system, we made the assumption that spontaneous emission can be neglected. Since spontaneous emission is the largest source of decoherence, this statement needs to be quantitatively verified. The total spontaneous photon scattering rate is given by the product of the lifetime and the (steady state) excited state population. Ignoring collisional relaxation we have for the scattering rate [25, 40]

$$R_{\text{sc}} = \left(\frac{\Gamma}{2}\right) \frac{S}{1 + S + 4(\Delta_L/\Gamma)^2} , \qquad (8.18)$$

where the *saturation parameter* is given by

$$S = \frac{I}{I_{\text{sat}}} = 2\left(\frac{\Omega}{\Gamma}\right)^2 . \qquad (8.19)$$

Using Eq. (8.7) the saturation intensity I_{sat} can be expressed as

$$I_{\text{sat}} = \frac{c\varepsilon_0\Gamma^2\hbar^2}{4d_{\text{effective}}^2} . \qquad (8.20)$$

For a linearly polarized far-detuned light beam, we can use the effective dipole matrix element defined in Eq. (8.16) and obtain $I_{sat} = 9.39$ mW/cm^2. For a large detuning we can also approximate the scattering rate as

$$R_{sc} \approx \frac{\pi \Gamma}{\Delta_L} \frac{V_0}{h},$$ (8.21)

where we have used the definition of the well depth for equal beam intensities as in Eq. (8.13). For typical experimental parameters of $V_0/h = 80$ kHz and $\Delta_L = 2\pi \times 40$ GHz, we get $R_{sc} = 60$ s^{-1} or roughly one event every 20 ms. For the tunneling experiments of Sect. 8.5, where the requirements on the spontaneous emission were the most stringent, the relevant interaction duration (the time of large acceleration) was at most 100 μs. In this time, less than 1% of the atoms scattered a spontaneous photon.

8.2 Experimental Apparatus

Three important steps were necessary to perform our experiments: the preparation of the initial condition, the generation and application of the interaction potential, and the measurement of the final state of the atoms. To outline the experimental sequence, a simplified schematic is shown in Fig. 8.2. We will give only a brief

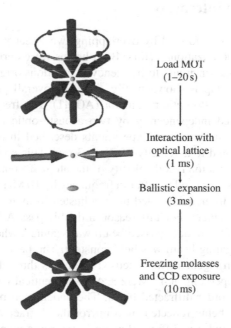

Load MOT
(1–20 s)

Interaction with
optical lattice
(1 ms)

Ballistic expansion
(3 ms)

Freezing molasses
and CCD exposure
(10 ms)

Fig. 8.2 Schematic of the experimental sequence. First the atoms are collected and cooled in a magneto-optic trap. The trapping fields are extinguished and the optical interaction potential is introduced. After interacting with the optical lattice, the atoms are allowed to expand freely in the dark. Finally, the cooling beams are turned on, freezing the atoms in place, and the fluorescence is imaged onto a charge-coupled device (CCD) camera

description of these steps; more details can be found in [11]. The starting point for the interaction was an atomic cloud that was trapped and cooled in a magneto-optic trap (MOT) in the standard $\sigma^+ - \sigma^-$ configuration [6, 37]. Loading sodium atoms from the thermal background into the MOT typically resulted in a cloud of 3×10^5 atoms with a final Gaussian distribution with a width of $\sigma_x = 0.3$ mm in position and $\sigma_p = 6\,\hbar k_L$ in momentum, where $\hbar k_L$ is the momentum of a single photon of resonant light. The trapping and cooling fields were then switched off and the interaction beams were turned on. The details of the generation of the interaction potential is given below. After a typical interaction duration of not more than a few milliseconds, the light beams were turned off and the atoms were allowed to expand freely. During this period of ballistic expansion, each atom moved a distance proportional to its velocity. This allowed us to determine the velocity distribution by recording the spatial distribution of the atomic cloud. For this purpose the resonant light was turned on after the free drift period to produce a viscous optical molasses that halted the ballistic motion of the atoms and provided spontaneously scattered resonant light for detection. This light was imaged onto a charge-coupled device camera (CCD) to obtain the desired spatial information. Nonuniform detection efficiencies within the optical molasses were measured and compensated for during data analysis.

8.3 Details of the Interaction

The optical potential was formed by overlapping two linearly polarized traveling waves with parallel polarization vectors. Both beams were derived from the same laser in order to reduce sensitivity to frequency fluctuations originating in the laser. A schematic of the setup is shown in Fig. 8.3. The overall power of the beams was adjusted by an acousto-optic modulator (AOM1). The frequencies of the two beams were controlled independently by two acousto-optic modulators (AOM2 and AOM3). During the tunneling experiments described in Sect. 8.5 the atoms needed to be accelerated to a velocity of up to 3 m/s. This corresponds to $100\,v_r$, where v_r is the single photon recoil velocity of the atom. To reach this velocity, the counterpropagating beams need to differ in frequency by 10 MHz. During the experiment the frequency difference needed to be adjusted from zero to this maximum value without misalignment. For this reason a double-pass AOM setup was chosen. The frequency of the double-passed beam was scanned, whereas the frequency of the counter-propagating beam was held constant. The beam in the variable frequency arm of the arrangement was focused by a lens through the acousto-optic modulator (AOM3) operating at 40 MHz $\pm\Delta\nu$. An identical lens was placed after the AOM in the first-order diffracted beam. The undeflected portion of the beam was discarded. After being reflected by a mirror the diffracted beam retraced its path through AOM3 and was diffracted again in the same manner. The beam was deflected twice on its path through the AOM, and the frequency was therefore downshifted by twice the drive frequency. Any change in the drive frequency of AOM3 led to an angle change of the first-order diffracted beam, but the beam completing

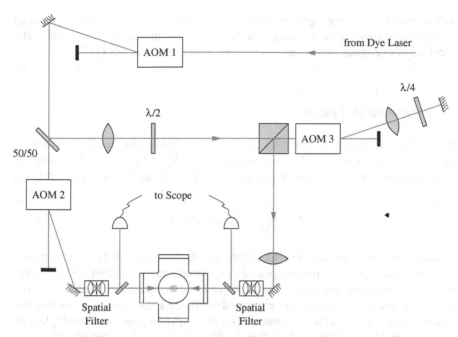

Fig. 8.3 Schematic of the interaction beam setup. AOM1 (40 MHz) provides the global control of the intensity. AOM3 is a double-passed, 40 MHz AOM shifting the beam frequency down by twice its drive frequency without leading to an appreciable angular deflection. AOM2 is in the single-passed configuration shifting down the beam frequency by 80 MHz

both passes through the AOM was still overlapped with the incoming beam regardless of the deflection angle. To separate the backreflected from the incoming beam the polarization was rotated along the path with a quarter-wave plate ($\lambda/4$) so that a polarization beam splitter cube could be used for separation. To compensate for the frequency offset of 80 MHz introduced by AOM3, the frequency in the second arm was down-shifted by AOM2, also by 80 MHz. The frequency difference between both beams was therefore $2\Delta\nu$. After passing through the acousto-optic modulators each beam was spatially filtered. The resulting transverse beam intensity profiles were approximately Gaussian with a beam radius of about 2 mm. The size and divergence of the beams were matched to avoid transverse spatial interference fringes, which could have created local variations of the well depth. A small part of each beam was diverted onto a photodiode to measure the optical power (the calibration accuracy was about 10%).

8.4 Quantum Transport

The system of ultra-cold atoms in a periodic optical potential offers a unique means of studying solid state effects with quantum optics tools. In order to gain insight into the possibilities for experiments, some of the basic properties of this system

will be reviewed. A thorough treatment of the fundamental properties can be found
in many solid state textbooks, such as Ashcroft and Mermin [2] or Marder [30]. The
specifics of our system are described more thoroughly by Fischer [11].

8.4.1 Stationary Lattice

We created the optical potential by spatially overlapping two counterpropagating
light beams ($\mathbf{k_L} \equiv \mathbf{k_2} = -\mathbf{k_1}$), which yields $\mathbf{q} = \mathbf{k_2} - \mathbf{k_1} = 2\mathbf{k_L}$. Choosing the
same frequency for both beams simplifies the effective Hamiltonian in Eq. (8.12) to

$$H = \frac{p^2}{2M} + V_0 \cos\left(2k_L x\right) , \tag{8.22}$$

assuming that beam propagation is along the x-axis. This form of the Hamiltonian is
a textbook example for a particle placed in a spatially periodic potential, and many
general properties of this system can be derived by symmetry arguments alone. The
most fundamental properties are expressed in *Bloch's theorem*. It states that the
eigenstates $\psi(x)$ of this Hamiltonian take on the form of a plane wave multiplied by
a function $u(x)$ of periodicity $d = \frac{\pi}{k_L} = \frac{\lambda}{2}$ (the periodicity of the potential):

$$\psi_{n,k}(x) = e^{ikx} u_{n,k}(x) , \tag{8.23}$$

where k is the quasi-momentum of the particle. The index n is called the *band
index* and appears in Bloch's theorem because for a given k there are many solutions to the Schrödinger equation. An important consequence of Bloch's theorem
is that the wave functions and the energy dispersion of the particle are periodic in
quasi-momentum (reciprocal) space with a periodicity of $K = \frac{2\pi}{d} = 2k_L$ in reciprocal space. Another property of paramount importance concerns the mean velocity
of a particle in a particular Bloch state $\psi_{n,k}$. It can be shown that the velocity is
determined by the energy dispersion relation as

$$v_n(k) = \frac{1}{\hbar} \frac{\partial E_n(k)}{\partial k} , \tag{8.24}$$

in analogy to the free particle case [2].

The problem of finding the energy eigenstates of H, that is solving $H|\psi\rangle =
E|\psi\rangle$, is equivalent to solving Mathieu's equation [1]. Sample dispersion curves
(energy versus the quasi-momentum k) were calculated and plotted in Fig. 8.4. For
a vanishing well depth V_0 the dispersion curve is the free particle energy parabola
$E(k) = \frac{\hbar^2 k^2}{2M}$. For a finite well depth V_0, the lowest crossing points of the free energy
parabolas at $k = \pm k_L$ develop a level repulsion due to the coupling of the levels by
the potential term. The amount of repulsion in this *avoided level crossing* can easily

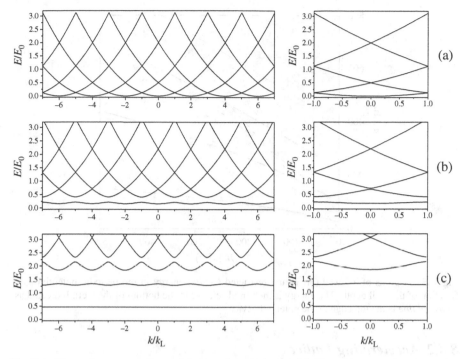

Fig. 8.4 Dispersion curves of a particle in a sinusoidal potential. Plotted here are the energy (in units of $E_0 = 8\,\hbar\omega_r$, $\hbar\omega_r$ being the recoil energy) versus the quasi-momentum k (in units of k_L) in the repeated-zone scheme (*left panels*) and in the reduced-zone scheme (*right panels*). The well depth V_0/h of the potential is (**a**) 0, (**b**) 40 kHz, and (**c**) 200 kHz

be estimated by first-order degenerate perturbation theory. The eigenenergies of the coupled system are

$$E_{1,2} = \frac{\hbar^2 k_L^2}{2M} \pm \frac{1}{2}V_0 \; . \tag{8.25}$$

The energy splitting for the first crossing and therefore the width of the first band gap is, to first order in V_0, equal to V_0 itself. The coupling term $\cos(2k_L x)$ connects only states with a difference in momentum of $2\,\hbar k_L$. For the calculation of the splitting at higher crossing points, we therefore need to resort to perturbation expansions of higher order.

The energy values evaluated at the band edges as a function of the well depth V_0 have been determined numerically and are displayed in Fig. 8.5. From this figure one can see that the energy bands evolve from a continuum of allowed energies, for a vanishing well depth, into the linearly spaced discrete energy levels of a harmonic oscillator, in the limit of large well depth.

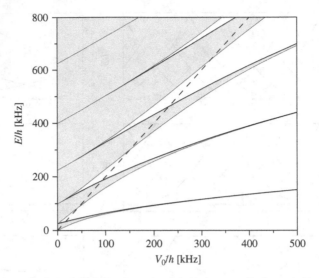

Fig. 8.5 Regions of allowed (*gray*) and restricted (*white*) energy in the periodic potential as a function of the well depth. The energy is measured relative to the bottom of the well. Indicated as a *dashed line* is the top edge of the potential ($2V_0$)

8.4.2 Accelerating Lattice

For the system of electrons in a crystal lattice, the most commonly encountered perturbation is an applied static electric field. This seemingly simple perturbation leads to a very rich system, whose properties were controversial for quite some time. Experimental tests in the field of solid state physics were hindered by decohering processes such as scattering of the electrons on impurities in the crystal lattice or scattering among themselves. These effects are negligible in our atom optics system and we were able to contribute to this field by studying some of the effects previously inaccessible to experiment.

8.4.2.1 Semiclassical Equations of Motion

A static electric field, which exerts a strong force on the electrons in a crystal, does not have the desired effect on a neutral atom in an optical potential. However, we can simulate the corresponding force by introducing an appropriate time dependence of the optical lattice. Let us consider an optical lattice composed of two counterpropagating light beams of unequal frequencies. A constant acceleration of the "standing" wave pattern is generated by linearly chirping the frequency difference of these counterpropagating beams. This is described by $\phi(t) = k_{\mathrm{L}}at^2$ in Eq. (8.12), where a is the acceleration. Inserting this into the Hamiltonian yields

$$H = \frac{p^2}{2M} + V_0 \cos\left[2k_{\mathrm{L}}\left(x - \frac{1}{2}at^2\right)\right]. \tag{8.26}$$

To make the connection to the solid state system, one can transform Eq. (8.26) to the frame of reference accelerated with the potential by applying a unitary transformation, following Peik et al. [36], resulting in

$$\tilde{H} = \frac{p^2}{2M} + V_0 \cos(2k_L x) + Max .$$
(8.27)

The last term containing the mass M of the atom is an inertial term, resulting from the transformation to an accelerating frame of reference. It mimics the role of the interaction potential $U_{el} = \mathcal{E}ex$ between an electric field $\mathcal{E}$ and the electron of charge e. Having established this connection, we can directly apply the results for the solid state system to an atom in the accelerated optical potential. A derivation of the semiclassical equations of motion for small electric fields can be found in standard textbooks [2, 30] and are simply stated here without proof. They express the relationship of the state's quasi-momentum k, band index n, energy $E_n(k)$, and mean velocity $v_n(k)$. By replacing the force $F = e\mathcal{E}$ with $F = Ma$, we obtain the following statements:

1. The band index n is a constant of motion.
2. The expression for the velocity in Eq. (8.24) remains unchanged and the evolution of the quasi-momentum is described by

$$\dot{k}_n(t) = -\frac{1}{\hbar} Ma .$$
(8.28)

3. The form of the band structure $E_n(k)$ is unchanged.

The restriction of *small fields* deserves special attention. The statement that the band index is a constant of motion indicates that inter-band transitions are being neglected. However, for larger fields electrons can tunnel across the band gap. An estimate for a "small" field strength is given by Ashcroft and Mermin as $E \ll \frac{E_g^2}{e\hbar v_F}$, with v_F being the typical electron velocity in the originating band and E_g being the minimum energy separation of the perturbed levels [2]. In our system this transforms to a condition for the acceleration

$$a \ll \frac{E_g^2}{\hbar^2 k_L} ,$$
(8.29)

where $v_r = \hbar k_L/M$ serves as the typical velocity at the edge of the Brillouin zone. Since for higher band indices the gaps get smaller and the velocity gets higher, a dramatic increase in the tunneling probability is to be expected. A more detailed study of tunneling across band gaps will be provided in Sect. 8.5.

8.4.2.2 Bloch Oscillations and Wannier–Stark States

One remarkable consequence of the equations of motion stated above is that particles exposed to a static field are predicted to oscillate in space rather than increase their velocity steadily. As can be seen by integrating Eq. (8.28), the quasi-momentum increases linearly with time as

$$k(t) = k_0 - \frac{Mat}{\hbar} . \tag{8.30}$$

The velocity of the particle with a given quasi-momentum k is given by Eq. (8.24) as the derivative of the dispersion curve at the point k. Since $E_n(k)$ is oscillatory in reciprocal space and k varies linearly with time, the velocity $v_n(t)$ is oscillatory in time. The period of oscillation τ_B is the time it takes for a particle to traverse the Brillouin zone of width $K = 2k_L$ and calculates to

$$\tau_B = \frac{2\hbar k_L}{Ma} = \frac{2v_r}{a} . \tag{8.31}$$

A sketch of these *Bloch oscillations* is graphically depicted in Fig. 8.6(a). An atom starting in the lowest band of the potential will increase its quasi-momentum k due to the applied force, as given by Eq. (8.30). As it approaches the edge of the Brillouin zone at a constant rate $\partial_t k$, the velocity decreases as the slope of the dispersion curve decreases. At $k = k_L$ the derivative $\partial_k E_0(k)$ is zero and according to Eq. (8.24) the particle is at rest. It will then reverse its velocity and continue its motion, until the velocity is reversed again at the next minimum of the dispersion curve. The reversal of its velocity at $k = k_L$ can be viewed as a first-order Bragg reflection of the particle wave by the periodic potential. The arguments above also hold for atoms in higher bands. They oscillate at the same Bloch frequency. However, the velocity reversal in higher bands corresponds to a higher order Bragg scattering process. It is important to note that this reversal of the atomic velocity occurs relative to the accelerated frame. In the laboratory frame the constant acceleration of the potential is superimposed on the oscillation of the atom.

For a higher field strength (or acceleration, in the atom optics system) the particle might not be able to follow the dispersion curve adiabatically as it approaches the edge of the Brillouin zone. It can cross the band gap and continue its motion in a higher band, as indicated in Fig. 8.6(b). This corresponds to a tunneling process through the band gap, in which case the semiclassical equations stated above no longer hold. For a particle undergoing tunneling, the transformation back to the laboratory frame reveals no change of velocity at all. The particle is simply lost out of the potential and can no longer track the acceleration.

The Bloch bands of an atom in a stationary potential are, by definition, continuous regions in the energy spectrum. Bloch oscillations in an accelerated lattice reveal themselves in the energy spectrum as discrete peaks with an energy separation of $h\nu_B$, where $\nu_B = 1/\tau_B$. This is a consequence of the Bloch bands splitting up into discrete *Wannier–Stark states*. A physical interpretation of these states can be

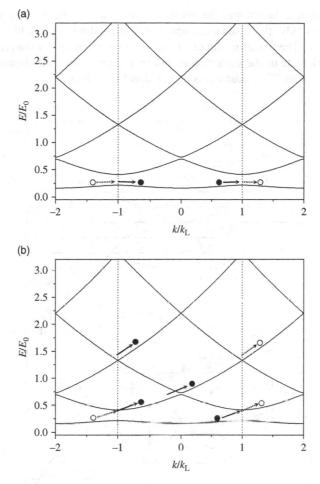

Fig. 8.6 Sketch of a particle trajectory in reciprocal space. In the *upper panel* (**a**) the rate of change of the quasi-momentum is slow enough for the particle to follow the dispersion curve adiabatically across the Brillouin zone boundary. This is equivalent to discontinuing the motion at one edge of the Brillouin zone and emerging from the other side in the same band. The *lower panel* (**b**) illustrates a case for a larger force, where the particle cannot follow the curve and tunnels through the band gap

obtained by regarding the transition between bands as a temporal interference effect. Quantum mechanically, atoms can tunnel between bands at all positions within the Brillouin zone. Since Bloch oscillations lead to multiple passes through the Brillouin zone, transition amplitudes can interfere constructively or destructively, depending on the rate at which the particle traverses the Brillouin zone. This is in analogy to the optical interference pattern generated by a plane wave of light illuminating an array of slits or a grating. The temporal interference produces sharp resonances spaced at the (temporal) grating period τ_B. The more traversals of the Brillouin zone the particle completes, the sharper the resonance becomes. If the particle tunnels out of the band quickly, the resonances are broad, indicating a short lifetime of the associated state. The tunneling out of a bound state is enhanced by the presence of a Wannier–Stark state of the same energy, but in a higher band and displaced by one or more lattice sites. This situation is depicted in Fig. 8.7(a).

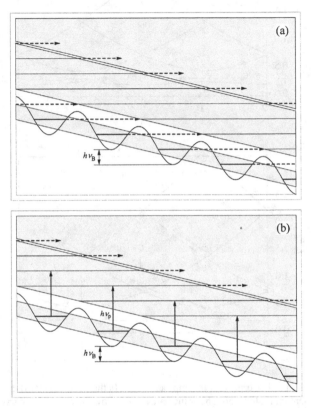

Fig. 8.7 Schematic of the Wannier–Stark ladder within the bands. In (**a**) the tunneling process is indicated. The presence of a Wannier–Stark state in the continuum of the higher band enhances the tunneling probability across the gap. In (**b**) a weak spectroscopic drive couples the states and introduces transitions. In either case, once an atom is in the second band, it can easily tunnel across successive band gaps into higher bands

Krieger and Iafrate [23] also consider the possibility of driving transitions between bands with an external alternating probe field. Assuming that the transition due to the probe drive is the dominant loss process from the first band (neglecting tunneling), they obtain a resonance condition for the drive frequency ν_p

$$\nu_p = \frac{\bar{E}_g}{h} + n\nu_B ,$$

(8.32)

where $\bar{E}_g$ is the average band separation. Here, the driving field provides a direct spectroscopic tool to probe the lattice structure of the Wannier–Stark states by allowing transitions between the states, as indicated in Fig. 8.7(b).

8.4.3 Band Spectroscopy and Wannier–Stark Ladders

In our experiments the initial atomic distribution was approximately Gaussian with a width of $\sigma_x = 0.3$ mm in position and $\sigma_p = 6\,\hbar k_L$ in momentum. However, to be able to study tunneling and transitions between single bands, an initial condition with only one populated band, preferably the lowest, was desired. If we suddenly turn on the optical potential within the atomic distribution, only a fraction of the atoms are transferred into the lowest band [35]. Most atoms will be projected into higher index bands. The location of the bands relative to the potential is indicated in Fig. 8.5. For a typical well depth of $V_0/h = 70$ kHz, we can see that atoms in the lowest band are trapped within the potential wells, whereas atoms in the second band are only partially trapped. Atoms in even higher bands have energies well above the potential and hence are effectively free. The location of the bands with respect to the potential well can be regarded as an indicator for the tunneling rates between bands when an acceleration is applied. Bands that lie entirely within the wells have a much smaller tunneling rate than bands outside the range of the potential. To empty all but the lowest band, we took advantage of this difference in tunneling rates across successive band gaps. After turning on the standing wave, it was accelerated to a velocity of $v_0 = 40\,v_r$, as indicated in Fig. 8.8. During this acceleration the atoms in the first band performed a sequence of Bloch oscillations within the potential and were accelerated in the laboratory frame. Atoms in higher bands could tunnel through the successively smaller band gaps and were lost out of the potential. The transport acceleration a_{trans} was chosen to maximize tunneling out of the second band while minimizing losses from the first trapped band. For typical experimental parameters of $V_0/h = 70$ kHz and $a_{\text{trans}} = 2000$ m/s^2, the Landau–Zener expression derived in Sect. 8.5 for the lifetime of the first and second band yields 24 ms and 40 μs, respectively. This ensured that after 600 μs of acceleration only the first band still contained a significant number of atoms.

For band spectroscopy experiments the frequency chirp was stopped after reaching the velocity v_0 and the frequency difference was held constant. At that point, a phase modulation at the frequency of v_p was added to one of the two counter-propagating beams forming the standing wave, as indicated in Fig. 8.8. This phase modulation could drive transitions between bands, if the band separation for some value of k was close to $E = h v_p$. The modulation typically lasted for 500 μs and was switched on and off smoothly over 16 μs to avoid any discontinuous phase changes in the potential that could induce transition to higher bands. The amplitude of the modulation was chosen to be small enough to not perturb the band structure.

In order to study Wannier–Stark states experimentally a constant acceleration of the optical potential was necessary. Therefore the frequency chirp was not stopped

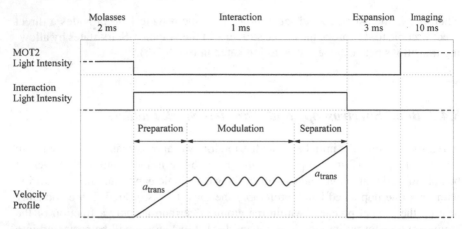

Fig. 8.8 Interaction beam timing diagram for the band spectroscopy experiments. After the molasses stage the resonant light is turned off and the optical lattice was turned on. A subset of atoms is projected into the fundamental band and separated in velocity by an acceleration a_{trans}. After this preparation stage, the optical lattice position and amplitude are varied to realize the potential under study. This step is followed by separating the atoms in the lowest band from those in higher bands by the same acceleration a_{trans}. The atoms are then allowed to expand freely in the dark and the spatial distribution is illuminated with the resonant molasses light

during the modulation time but was adjusted to yield the desired value of the acceleration a. To spectroscopically investigate the states, we superimposed the phase modulation at frequency ν_p onto this frequency chirp.

After a fixed time interval the modulation was turned off and the frequency chirping resumed at a rate corresponding to a_{trans}. This separated in momentum space the remaining trapped atoms in the lowest band from those having made the transition into higher bands. After reaching a final velocity $v_{final} = 80\,v_r$, the interaction beams were switched off suddenly.

In the detection phase we needed to distinguish three classes of atoms: (1) atoms that were not initially trapped in the lowest band and immediately tunneled out of the well during the initial acceleration, (2) atoms which were trapped in the first band at the beginning of the interaction but were driven out by the modulation, and (3) atoms that remained in the first band during the entire sequence. Since the atoms in different classes had left the trapping potential at different stages of the experimental sequence, they were accelerated to different velocities. Therefore, after drifting in the dark for 3 ms, these classes separated in space and could be distinguished by recording their position. For this purpose the atoms were imaged in the "freezing molasses" as described in Sect. 8.2. A typical fluorescence image of the atoms is shown in Fig. 8.9(a). The two-dimensional image was then integrated in the direction perpendicular to the axis of the interaction beams to obtain a one-dimensional distribution along the beam direction, containing all three classes of atoms. The corresponding integrated distribution is shown in Fig. 8.9(b). In order to reduce sensitivity to fluctuations of the number of atoms in the MOT, the number of survivors (atoms in class (3)) was normalized by the total number of atoms initially

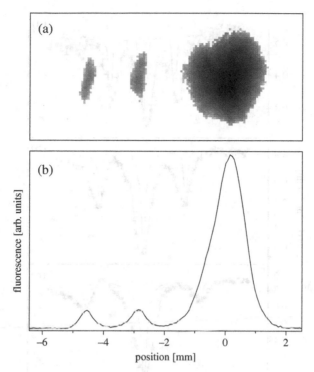

Fig. 8.9 Part (**a**) shows a fluorescence image from an atomic distribution acquired after a time of ballistic expansion. Part (**b**) shows the distribution integrated in the vertical direction. The large peak on the *right* is the part of the atomic cloud that was not trapped during the initial acceleration. The *center* peak indicates the atoms that were initially trapped in the first band but were driven out by the modulation. The *left* peak corresponds to atoms that remained trapped during the entire sequence. The survival probability is the area under the *left* peak normalized by the sum of the areas under the *left* and *center* peak

trapped in the first band, which was obtained by summing the contributions of class (2) and class (3).

To observe the temporal evolution of the fundamental band population, we repeated the sequence in Fig. 8.8 for various modulation durations, holding the probe frequency ν_p and amplitude m fixed. These studies resulted in the observation of Rabi oscillations between Bloch bands [13]. For large amplitudes of the modulation, we observed a *dynamical suppression* of the band structure, effectively turning off Bloch tunneling [28].

To obtain a spectrum of the Wannier–Stark states, we applied the modulation during a period of constant acceleration and repeated the sequence for various probe modulation frequencies, holding the modulation amplitude m and the duration fixed. Figure 8.10 shows three measured spectra for the accelerations of 947, 1260, and 1680 m/s^2, which correspond to the Bloch frequencies $\omega_B/2\pi = 16.0$, 21.4, and 28.5 kHz, respectively. The spectra were obtained at a fixed well depth of $V_0/h = 91.6$ kHz and a fixed probe modulation amplitude of $m = 0.05$. For a well depth

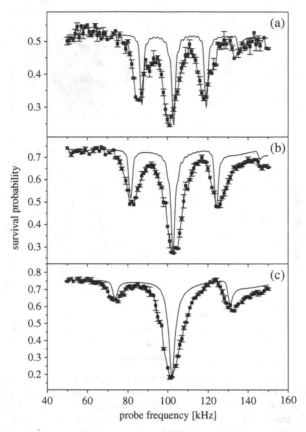

Fig. 8.10 Wannier–Stark ladder resonances for a well depth of $V_0/h = 91.6$ kHz and accelerations of (**a**) 947 m/s^2, (**b**) 1260 m/s^2, and (**c**) 1680 m/s^2, which correspond to the Bloch frequencies $\omega_B/2\pi = 16.0$, 21.4, and 28.5 kHz, respectively. For the chosen well depth, the average band spacing is $\bar{E}_g/h = 104$ kHz which is in good agreement with the location of the central resonance. The points are connected by *thin solid lines* for clarity. The *thick solid lines* show the results of numerical simulations using the experimental parameters. Figure from [29]; Copyright 1999 by the American Physical Society

of $V_0/h = 91.6$ kHz, the average band spacing is $\bar{E}_g/h = 104$ kHz, which is in good agreement with the location of the central resonance in the three spectra of Fig. 8.10.

Also shown in Fig. 8.10 is the result of a numerical integration of the time-dependent Schrödinger equation using the experimental parameters. We believe that phase noise in the interaction beams prevented the survival probability from reaching unity, when the probe was far from resonance, and reduced the depth of the spectral features by a constant factor. For this reason, the y-values of the theory curves were shifted and scaled to match the baseline and amplitude of the central resonance. In addition, the value for the probe modulation amplitude m was adjusted in the numerical simulations from 0.05 to 0.035 to reproduce the relative peak heights.

The spectral width of the resonances is fundamentally determined by the finite lifetime of the Wannier–Stark states due to tunneling. However, a number of experimental mechanisms (e.g., phase noise in the standing wave beams, variations in the well depth, or the finite transverse extent of the optical potential) contributed to the measured width being substantially broader than that predicted by the simulations.

8.5 Quantum Tunneling

In the previous section we studied the spectral features of Bloch states and Wannier–Stark states by driving transitions between those states. These inter-band transitions were imposed externally by a modulation of the potential. Without modulation the band index was conserved. The accelerations that transported the atoms through reciprocal space were small enough to preserve the validity of the semiclassical equations of motion. In this section we investigate the effect of a large acceleration of the optical potential. In this case the semiclassical equations no longer hold and inter-band tunneling can occur. The atoms can leave the trapping potential via tunneling into the continuum of free states. The system is therefore unstable and the number of trapped atoms decays with time. By adjusting the acceleration the stability of the system can be altered dynamically and the decay rates vary over a wide range. In this system, short-time deviations from the universal exponential decay law are observed [42]. In addition, we study the fundamental effects of measurements on the decay rate and report on the first observation of the quantum Zeno and anti-Zeno effects in an unstable system [12].

8.5.1 Classical Limit

As derived in Sect. 8.4, atoms in an accelerated standing wave are subject to a potential

$$V(x) = V_0 \cos(2k_{\mathrm{L}} x) + Max .$$ (8.33)

This potential is stated in the reference frame accelerated with the potential as given in Eq. (8.27). For a small enough acceleration a particle can be classically trapped within the wells of this "washboard" potential. In this case the particle will accelerate along with the potential. For a larger acceleration the potential wells become increasingly asymmetric up to a point where the particle is no longer confined by the potential. The critical acceleration $a_{\mathrm{c,class}}$, for which the potential loses its ability to confine the particle, can be found by solving for extrema of the potential in Eq. (8.33), which only exists for

$$|a| < a_{\mathrm{c,class}} = \frac{2k_{\mathrm{L}} V_0}{M} .$$ (8.34)

For accelerations smaller than $a_{c,class}$ the particle gets accelerated along with the potential whereas for larger accelerations there are no local potential minima.

8.5.2 Landau–Zener Tunneling

8.5.2.1 Tunneling Rates

In this section we provide a short description of the Landau–Zener tunneling process based on diabatic transitions in momentum space [35, 44]. An alternative description can be derived in the position representation [26, 45]. As a starting point we consider the semiclassical equations of motion describing the time evolution of the quasi-momentum in reciprocal space. In order to allow for inter-band transitions, we must now abandon the condition that the band index be a constant of motion. The shape of the Bloch bands and the time evolution equation for the quasi-momentum are still assumed to be valid. The stationary periodic potential causes the free particle energy levels to undergo a level repulsion. This shift is most pronounced at the edges of the Brillouin zone. A particle approaching the avoided level crossing might not be able to follow the dispersion curve adiabatically, in which case it continues its motion and diabatically changes levels across the energy gap. In 1932 Zener derived an expression for the probability P of diabatic transfer between two repelled levels [44]

$$P = \exp\left(-\frac{\pi}{2\hbar}\frac{E_g^2}{\frac{d}{dt}(\varepsilon_1 - \varepsilon_2)}\right), \tag{8.35}$$

where E_g is the minimum energy separation of the perturbed levels and $\varepsilon_{1,2}$ are the unperturbed energy eigenvalues of level 1 and 2, respectively. In our case the unperturbed energy curve is simply the free particle kinetic energy dispersion $E_p = p^2/2M$. Using the semiclassical equation of motion for the quasi-momentum, we obtain for the probability of transfer

$$P = e^{-a_c/a}, \tag{8.36}$$

where the critical acceleration a_c is given by

$$a_c = \frac{\pi}{4}\frac{E_g^2}{n\,\hbar^2 k_L}. \tag{8.37}$$

We let N denote the number of particles populating the lowest band within the first Brillouin zone. The rate of atoms crossing the band gap is equal to the rate of atoms approaching the transition region times the probability of tunneling if we assume the band to be uniformly populated. We obtain an exponential decay of the population N in the band under consideration as

$$N = N_0\, e^{-\Gamma_{LZ} t}, \tag{8.38}$$

with the Landau–Zener (LZ) decay rate Γ_{LZ} given by

$$\Gamma_{LZ} = \frac{a}{2v_r}\, e^{-a_c/a} . \tag{8.39}$$

Experimental studies of the tunneling rates out of the lowest band were performed in our group and the decay rates were compared to the Landau–Zener prediction [3, 27].

8.5.2.2 Deviations from Landau–Zener Tunneling

The expression for the LZ tunneling rate derived above is based on a single transit of the atom through the region of an avoided crossing. However, for small tunneling probability the atom can undergo Bloch oscillations within a given band, leading to multiple passes through the Brillouin zone. The tunneling amplitudes can interfere constructively or destructively depending on the rate at which the atom traverses the Brillouin zone. This mechanism is responsible for the formation of tunneling resonances. For small accelerations the tunneling rate is small and the atoms can perform many Bloch oscillations before leaving the band. Therefore large deviations from the Landau–Zener prediction for the tunneling rate are to be expected. For a larger acceleration the atom leaves the band quickly and the interference effects are less pronounced. For those cases the LZ prediction is a good approximation for the actual tunneling rate. These statements are in agreement with the observed tunneling rates [3, 27].

8.5.3 Non-exponential Decay

8.5.3.1 Theoretical Description

An exponential decay law is the universal hallmark of unstable systems and is observed in all fields of science. This law is not, however, fully consistent with quantum mechanics and deviations from exponential decay have been predicted for short as well as long times [20, 43, 14]. In 1957 Khalfin showed that if H has a spectrum bounded from below, the survival probability is not a pure exponential but rather of the form

$$\lim_{t \to \infty} P(t) \approx \exp(-ct^q) \qquad q < 1, c > 0 . \tag{8.40}$$

Later Winter examined the time evolution in a simple barrier-penetration problem [43]. He showed that the survival probability begins with a non-exponential, oscillatory behavior. Only after this initial time does the system start to evolve according to the usual exponential decay of an unstable system. Finally, at very long times, it decays like an inverse power of the time. The initial non-exponential decay behavior is related to the fact that the coupling between the decaying system and the reservoir is reversible for short enough times. Moreover, for these short times, the decayed and undecayed states are not yet resolvable, even in principle.

A simple argument will illustrate this point. We assume that the system is initially in the undecayed state $|\Psi_0\rangle$ at $t = 0$, and that the state evolves under the action of the Hamiltonian H,

$$|\Psi(t)\rangle = e^{-iHt/\hbar}|\Psi_0\rangle = A(t)|\Psi_0\rangle + |\Phi(t)\rangle , \qquad (8.41)$$

where $A(t)$ is the probability amplitude for remaining in the undecayed state and the state $|\Phi(t)\rangle$ denotes the decayed state with $\langle\Psi_0|\Phi(t)\rangle = 0$. The probability of survival P in the undecayed state is therefore $P(t) = |A(t)|^2$. Acting with the time evolution operator $e^{-iH(t+t')/\hbar}$ on the state $|\Psi_0\rangle$ yields

$$A(t + t') = A(t)A(t') + \langle\Psi_0|e^{-iHt'/\hbar}|\Phi(t)\rangle . \qquad (8.42)$$

If it were not for the last term, the equation above would generate the characteristic exponential decay law of an unstable system. However, the term under consideration describes the possibility for the decayed state $|\Phi(t)\rangle$ to re-form the initial state $|\Psi_0\rangle$ under the time evolution operator for time t'.

For very short times we can make a general prediction about the time evolution of the survival probability P. Given that the mean energy of the decaying state is finite and that H has a spectrum that is bounded from below, one can show following the arguments of Fonda et al. [14] that

$$\left.\frac{dP(t)}{dt}\right|_{t\to 0} = 0 . \qquad (8.43)$$

As outlined by Grotz and Klapdor [18] we can expand $A(t)$ in a power series

$$A(t) = 1 - i\frac{t}{\hbar}\langle\Psi_0|H|\Psi_0\rangle - \frac{t^2}{2\hbar^2}\langle\Psi_0|H^2|\Psi_0\rangle + O(t^3) . \qquad (8.44)$$

Using this expansion results in an expression for the survival probability

$$P(t) = |A(t)|^2 = 1 - \frac{t^2}{\hbar^2}\langle\Psi_0|(H - \bar{E})^2|\Psi_0\rangle + O(t^4) , \qquad (8.45)$$

where $\bar{E} = \langle\Psi_0|H|\Psi_0\rangle$. This form indicates a population transfer beginning with a flat slope and suggests an initial quadratic time dependence.

The results stated here are general properties independent of the details of the interaction. However, the timescale over which the deviation from exponential behavior is apparent depends on the particular timescales of the decaying system. Greenland and Lane point out a number of timescales which are relevant [17]. The first timescale τ_e is given by the time that it takes the decay products to leave the bound state region. This time can be estimated as

$$\tau_e = \frac{\hbar}{E_0} , \tag{8.46}$$

where E_0 is the energy released during the decay. It determines the amount of time required to pass before the decayed and undecayed states can be resolved. The second timescale τ_w is related to the bandwidth ΔE of the continuum to which the state is coupled

$$\tau_w = \frac{\hbar}{\Delta E} . \tag{8.47}$$

The phases of all states in the continuum evolve at a rate corresponding to their energy. Thus after the time τ_w the phases of these states have spread over such a wide range as to prevent the reformation of the initial undecayed state. After this dephasing time, the coupling is essentially irreversible.

Although these predictions are of general nature and applicable in every unstable system, deviations from exponential decay have not been observed experimentally in any other system than the one described here [42]. The primary reason is that these characteristic timescales in most naturally occurring systems are extremely short. For the decay of a spontaneous photon, the time τ_e it takes a photon to traverse the bound state size is approximately an optical period, 10^{-15} s. For a nuclear decay this timescale is orders of magnitude shorter, about 10^{-21} s. By contrast, the dynamical timescale for an atom bound in an optical lattice is just the inverse band gap energy, which in our experiments is on the order of several microseconds.

Niu and Raizen [34] performed a more detailed investigation of a two-band model of our system. They find an initial non-exponential regime that starts with a quadratic time dependence, then becomes a damped oscillation, and finally settles into an exponential decay. The timescale for which the coherent oscillations damp out and the exponential decay behavior sets in is identified as the crossover time t_c equal to

$$t_c = \frac{E_g}{a} \frac{1}{2\hbar k_L} . \tag{8.48}$$

For a typical value for the acceleration of $a = 10,000$ m/s^2 and a band gap of $E_g/h = 80$ kHz, the crossover time calculates to $t_c = 2$ μs.

8.5.3.2 Experimental Realization

The preparation of the initial state was done as described previously. After turning on the interaction beams, a small acceleration of $a_{trans} = 2000$ m/s^2 was imposed to separate those atoms projected into the lowest band from the rest of the distribution. After reaching the velocity $v_0 = 35\, v_r$, the acceleration was suddenly increased to a value a_{tunnel} where appreciable tunneling out of the first band occurred. Unlike in the band spectroscopy experiments no phase modulation was added to induce transitions between the bands. The large acceleration a_{tunnel} was maintained for a period of time t_{tunnel}, after which time the frequency chirping continued again at the decreased rate corresponding to a_{trans}. This separated in momentum space the atoms that were still trapped in the lowest band from those in higher bands. After reaching a final velocity of $v_{\text{final}} = 80\,v_r$, the interaction beams were switched off suddenly. A diagram of the velocity profile versus time is shown in Fig. 8.11(a).

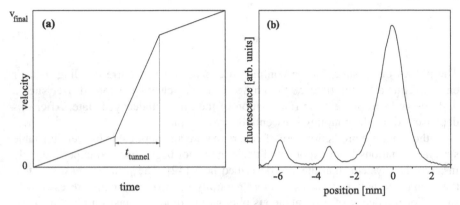

Fig. 8.11 Part (**a**) shows a diagram of the acceleration sequence to study tunneling out of the lowest band. Part (**b**) shows a typical integrated spatial distribution of atoms after ballistic expansion. The large peak on the *right* is the part of the atomic cloud that was not trapped during the initial acceleration. The *center* peak indicates the atoms that tunneled out of the potential during the fast acceleration period. The *leftmost* peak corresponds to atoms that remained trapped during the entire sequence. Figure from [12]; Copyright 2001 by the American Physical Society

In the detection phase we determined the number of atoms that were initially trapped and what fraction remained in the first band after the tunneling sequence. After an atom tunneled out of the potential during the sequence, it would maintain the velocity that it had at the moment of tunneling. During the period of free ballistic expansion the difference in final velocity between trapped and tunneled atoms led to their spatial separation (Fig. 8.11(b)). To observe the temporal evolution of the fundamental band population, we repeated a sequence such as in Fig. 8.11(a) for various tunneling durations t_{tunnel}, holding the other parameters of the sequence fixed.

Figure 8.12 shows the probability of survival in the accelerated potential as a function of the duration of tunneling for various values of the tunneling acceleration a_{tunnel} between 6000 and 20,000 m/s^2. The value for the well depth for all curves was $V_0/h = 92$ kHz. Initially, the survival probability shows a flat region, owing to the reversibility of the decay process for short times. At intermediate times the decay shows a damped oscillation that for long times evolves into the characteristic exponential decay law. By this time the coupling is essentially irreversible and reformation of the undecayed state is prohibited. As a comparison we also show the results of quantum mechanical simulations of the entire experimental sequence as solid lines in the same graph. The tunneling rates depend strongly on the well depth

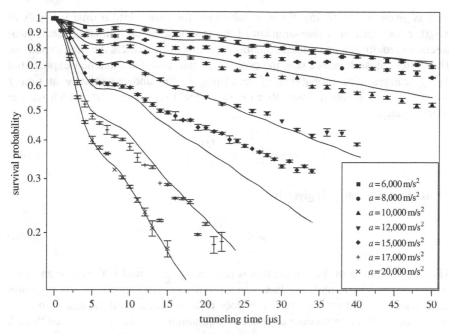

Fig. 8.12 Probability of survival in the accelerated potential as a function of duration of the tunneling acceleration. Data points for different values of the large acceleration a_{tunnel} are shown. Each point represents the average of five experimental runs, and the error bar denotes the error of the mean. These data were recorded for a well depth of $V_0/h = 92$ kHz and have been normalized to unity at $t_{\text{tunnel}} = 0$ to compare to the quantum mechanical simulations (shown in *solid lines* with no adjustable parameters)

of the potential. Considering the uncertainty of 10% in the calibration of the power in the interaction beams, the simulations match the observed data quite well.

8.5.4 Quantum Zeno and Anti-Zeno Effects

The universal phenomenon of non-exponential decay of unstable systems led Misra and Sudarshan in 1977 to the prediction that frequent measurements during this non-exponential period could inhibit decay entirely [33, 5, 41]. They named this effect the *quantum Zeno effect* after the Greek philosopher, famed for his paradoxes and puzzles. In his most famous paradox, Zeno considers an arrow flying through the air. The time of flight can be subdivided into infinitesimally small intervals during which the arrow moves only by infinitesimal amounts. Assuming the summation of infinitesimal terms amounts to nothing led Zeno to believe that motion is impossible and is merely an illusion. The version put forth by Misra and Sudarshan is the quantum mechanical version of the paradox.

To illustrate their main point, we consider the time evolution of a system in the non-exponential regime, where the probability of remaining in the undecayed state is given by Eq. (8.45). We now subdivide the time t into n time intervals of length τ and perform a measurement of the system after each interval. Each measurement redefines a new initial condition and effectively resets the time evolution. The system must therefore start the evolution again with the same non-exponential decay features. The probability of remaining in the undecayed state at time t (after n measurements at intervals τ) is therefore $P(t) = [P(\tau)]^n$, which we can approximate as

$$P(t) = \exp\left(-n\,\tau^2\frac{\langle H^2\rangle}{\hbar^2}\right) = e^{-\gamma t}\,, \tag{8.49}$$

where the decay rate γ is given by

$$\gamma = \tau\,\frac{\langle H^2\rangle}{\hbar^2}\,. \tag{8.50}$$

The time evolution of the system that is repeatedly measured is therefore an exponential decay. The remarkable fact is that the decay rate depends on the measurement interval τ and tends to zero as τ goes to zero. Reviews of the quantum Zeno effect can be found in modern textbooks of quantum mechanics [39]. Even though measurement-induced suppression of the dynamics of a two-state driven system has been observed [19, 24], no such effect was ever measured on an unstable system.

Whereas in the previous section we established the non-exponential time dependence, the focus of this section is the effect of measurements on the system decay rate. The quantity to be measured was the number of atoms remaining trapped in the potential during the tunneling segment. This measurement could be realized by suddenly interrupting the tunneling duration by a period of reduced acceleration a_{interr}, as indicated in Fig. 8.13(a). During this interruption tunneling was negligible and the atoms were therefore transported to a higher velocity without being lost out of the well. This separation in velocity space enabled us to distinguish the remaining atoms from the ones having tunneled out up to the point of interruption, as can be seen in Fig. 8.13(b). By switching the acceleration back to a_{tunnel}, the system was then returned to its unstable state. The measurement of the number of atoms that remained trapped defined a new initial state with the remaining number of atoms as the initial condition. The requirements for this interruption section were very similar to those during the transport section, namely the largest possible acceleration while maintaining negligible losses for atoms in the first band. Hence a_{interr} was chosen to be the same as a_{trans}.

Figure 8.14 shows the dramatic effect of frequent measurements on the decay behavior. The hollow squares indicate the decay curve without interruption. The solid circles in Fig. 8.14 depict the measurement of the survival probability in which after each tunneling segment of 1 μs an interruption of 50 μs duration was inserted. Only the short tunneling segments contribute to the total tunneling time. The survival probability clearly shows a much slower decay than the corresponding system measured without interruption. Care was taken to include the limited time response

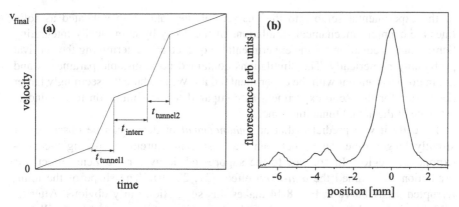

Fig. 8.13 Part (**a**) shows a diagram of the interrupted acceleration sequence. The total tunneling time is the sum of all the tunneling segments. Part (**b**) shows a typical integrated spatial distribution of atoms after ballistic expansion. The peaks can be identified as in Fig. 8.11. However, the area containing the tunneled fraction of the atoms is now composed of two peaks. Atoms that left the well during the first tunneling segment are offset in velocity from the ones having left during the second period of tunneling. The amount of separation is equal to the velocity increase of the well during the interruption. Figure from [12]; Copyright 2001 by the American Physical Society

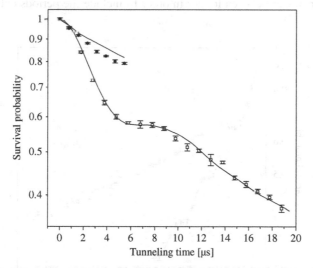

Fig. 8.14 Probability of survival in the accelerated potential as a function of duration of the tunneling acceleration. The *hollow squares* show the non-interrupted sequence, the *solid circles* show the sequence with interruptions of 50 μs duration every 1 μs. The error bars denote the error of the mean. The data have been normalized to unity at $t_{\text{tunnel}} = 0$ in order to compare to the quantum mechanical simulations (*solid lines*; no adjustable parameters). For these data the parameters were $a_{\text{tunnel}} = 15,000$ m/s^2, $a_{\text{interr}} = 2000$ m/s^2, $t_{\text{interr}} = 50$ μs, and $V_0/h = 91$ kHz. Figure from [12]; Copyright 2001 by the American Physical Society

of the experimental setup into the analysis of the data. Also indicated as solid lines are quantum mechanical simulations of the decay by numerically integrating Schrödinger's equation for the experimental sequence and determining the survival probability numerically. The simulations contained no adjustable parameters and are in good agreement with the experimental data. We attribute the seemingly larger decay rate for the Zeno experiment as compared to the simulation to the underestimate of the actual tunneling time.

Recently it was predicted that an *enhancement* of decay can be observed for slightly longer time delays between successive measurements during the non-exponential region. In contrast to the suppressed decay for the Zeno effect this prediction was named the *anti-Zeno* effect [21, 22, 10]. The shape of the uninterrupted decay curve in Fig. 8.14 makes this suggestion fairly obvious. After an initial period of slow decay the curve shows a steep drop as part of an oscillatory feature, which for longer time damps away to show the well-known exponential decay. If the system was interrupted right after the steep drop, one would expect an overall decay that is faster than the uninterrupted decay [22]. The solid circles in Fig. 8.15 show such a decay sequence, where after every 5 μs of tunneling the decay was interrupted by a slow acceleration segment. As in the Zeno case, these interruption segments force the system to repeat the initial non-exponential decay behavior after every measurement. Here, however, the tunneling segments between the measurements are chosen longer in order to include the periods exhibiting fast

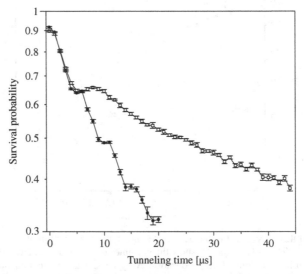

Fig. 8.15 Survival probability as a function of duration of the tunneling acceleration. The *hollow squares* show the non-interrupted sequence, the *solid circles* show the sequence with interruptions of 40 μs duration every 5 μs. The error bars denote the error of the mean. The experimental data points have been connected by *solid lines* for clarity. For these data the parameters were $a_{\text{tunnel}} = 15,000 \, \text{m/s}^2$, $a_{\text{interr}} = 2800 \, \text{m/s}^2$, $t_{\text{interr}} = 40 \, \mu\text{s}$, and $V_0/h = 116 \, \text{kHz}$. Figure from [12]; Copyright 2001 by the American Physical Society

decay. The overall decay is much faster than for the uninterrupted case, indicated by the hollow squares in the same figure.

The key to observing the Zeno and anti-Zeno effects is the ability to measure the state of the system in order to repeatedly redefine a new initial state. In our case the measurement is done by separating in momentum space the atoms still left in the unstable state from the ones that decayed into the reservoir. In order to distinguish the two classes of atoms, they must have a separation of at least the size of the momentum distribution of the unstable state, which in our case is the width of the first Brillouin zone of $\Delta p = 2\hbar k_L$. The time it takes for an atom to be accelerated in velocity by this amount is the Bloch period $\tau_B = 2v_r/a_{\text{interr}}$, assuming an acceleration of a_{interr}. An interruption shorter than this time will not resolve the tunneled atoms from those still trapped in the potential and therefore results in an incomplete measurement of the atom number. To investigate the effect of the interruption duration, we repeated a sequence to measure the anti-Zeno effect for varying interruption durations while holding all other parameters constant. Figure 8.16 displays the results of this measurement, interrupting the decay every 5 μs with an acceleration of a_{interr} of 2000 m/s². The hollow squares show the uninterrupted decay sequence as a reference. For an interruption duration smaller than the Bloch period of 30 μs, the measurement of the atom number is incomplete and has little or no effect. For a duration longer than the Bloch period, the effect saturates

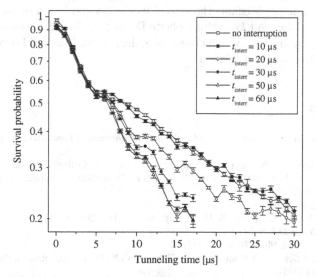

Fig. 8.16 Survival probability as a function of duration of the tunneling acceleration. The *hollow squares* show the non-interrupted sequence, other *symbols* indicate the sequence with a finite interruption duration after every 5 μs of tunneling. The error bars denote the error of the mean. A further increase of the interruption duration than as indicated does not result in a further change of the decay behavior. The experimental data points have been connected by *solid lines* for clarity. For these data the parameters were $a_{\text{tunnel}} = 15,000$ m/s², $a_{\text{interr}} = 2000$ m/s², and $V_0/h = 91$ kHz, leading to a Bloch period of $\tau_B = 30$ μs. Figure from [12]; Copyright 2001 by the American Physical Society

and results in a complete restart of the decay behavior after every interruption. Even though this method of interruption is not an instantaneous measurement of the state of the unstable system, we can still accomplish the task of redefining the initial state by first switching the system from an unstable to a stable one, then in a finite time perform the measurement, and finally switching the system back to being unstable again.

8.6 Conclusions

In conclusion, we have completed a detailed study of the onset of irreversibility in an unstable quantum system and its control by repeated interrogation. We end this chapter by making some comments about possible future directions. The development of new tools to control many-body systems is a promising direction to follow. In particular, our group has been working toward the experimental realization of few-body number states [8]. These states of a definite number of atoms in the ground state of a well are an ideal starting point for the study of few-body tunneling and the onset of irreversibility. Future work in our group will focus on this problem. Finally, it is a great pleasure to acknowledge and thank the many people who have collaborated with us on this work over the years. The experiments were carried out together with Kirk Madison, Steven Wilkinson, Cyrus Bharucha, Patrick Morrow, and Braulio Gutiérrez-Medina. Theoretical work was conducted in parallel to our experiments together with Qian Niu, Roberto Diener, and Bala Sundaram. This work was supported by the R. A. Welch Foundation, the National Science Foundation, and the Sid. W. Richardson Foundation.

References

1. M. Abramowitz, I.A. Stegun (eds.), *Handbook of Mathematical Functions* (Dover, New York, 1965)
2. N.W. Ashcroft, N.D. Mermin, *Solid State Physics* (Saunders College, Philadelphia, 1976)
3. C.F. Bharucha, K.W. Madison, P.R. Morrow, S.R. Wilkinson, B. Sundaram, M.G. Raizen, Phys. Rev. A **55**, R857 (1997)
4. F. Bloch, Z. Phys. **52**, 555 (1928)
5. C.B. Chiu, E.C.G. Sudarshan, B. Misra, Phys. Rev. D **16**, 520 (1977)
6. S. Chu, Science **253**, 861 (1991)
7. S. Chu, Rev. Mod. Phys. **70**, 685 (1998)
8. C.S. Chuu, F. Schreck, T.P. Meyrath, J.L. Hanssen, G.N. Price, M.G. Raizen, Phys. Rev. Lett. **95**, 260403 (2005)
9. C. Cohen-Tannoudji, J. Dupont-Roc, G. Grynberg, *Atom-Photon Interactions* (Wiley and Sons, New York, 1992)
10. P. Facchi, H. Nakazato, S. Pascazio, Phys. Rev. Lett. **86**, 2699 (2001)
11. M.C. Fischer, *Atomic Motion in Optical Potentials*, Ph.D. thesis, The University of Texas at Austin (2001)
12. M.C. Fischer, B. Gutiérrez-Medina, M.G. Raizen, Phys. Rev. Lett. **87**, 040402 (2001)
13. M.C. Fischer, K.W. Madison, Q. Niu, M.G. Raizen, Phys. Rev. A **58**, R2648 (1998)
14. L. Fonda, G.C. Ghirardi, G.C. Rimini, Rep. Prog. Phys. **41**, 587 (1978)

15. O.R. Frisch, Z. Phys. **86**, 42 (1933)
16. R. Graham, M. Schlautmann, P. Zoller, Phys. Rev. A **45**, R19 (1992)
17. P.T. Greenland, A.M. Lane, Phys. Lett. A **117**, 181 (1986)
18. K. Grotz, H.V. Klapdor, Phys. Rev. C **30**, 2098 (1984)
19. W.M. Itano, D.J. Heinzen, J.J. Bollinger, D.J. Wineland, Phys. Rev. A **41**, 2295 (1990)
20. L.A. Khalfin, JETP **6**, 1053 (1958)
21. A.G. Kofman, G. Kurizki, Phys. Rev. A **54**, R3750 (1996)
22. A.G. Kofman, G. Kurizki, Nature **405**, 546 (2000)
23. J.B. Krieger, G.J. Iafrate, Phys. Rev. B **33**, 5494 (1986)
24. P. Kwiat, H. Weinfurter, T. Herzog, A. Zeilinger, M. Kasevich, Phys. Rev. Lett. **74**, 4763 (1995)
25. R. Loudon, *The Quantum Theory of Light* (Clarendon, Oxford, 1983)
26. K.W. Madison, *Quantum Transport in Optical Lattices*, Ph.D. thesis, The University of Texas at Austin (1998)
27. K.W. Madison, C.F. Bharucha, P.R. Morrow, S.R. Wilkinson, Q. Niu, B. Sundaram, M.G. Raizen, Appl. Phys. B **65**, 693 (1997)
28. K.W. Madison, M.C. Fischer, R.B. Diener, Q. Niu, M.G. Raizen, Phys. Rev. Lett. **81**, 5093 (1998)
29. K.W. Madison, M.C. Fischer, M.G. Raizen, Phys. Rev. A **60**, R1767 (1999)
30. M.P. Marder, *Condensed Matter Physics* (Wiley and Sons, New York, 2000)
31. E.E. Mendez, G. Bastard, Phys. Today **46**(6), 34 (1993)
32. P.W. Milonni, J.H. Eberly, *Lasers* (Wiley and Sons, New York, 1988)
33. B. Misra, E.C.G. Sudarshan, J. Math. Phys. **18**, 756 (1977)
34. Q. Niu, M.G. Raizen, Phys. Rev. Lett. **80**, 3491 (1998)
35. Q. Niu, X.G. Zhao, G.A. Georgakis, M.G. Raizen, Phys. Rev. Lett. **76**, 4504 (1996)
36. E. Peik, M.B. Dahan, I. Bouchoule, Y. Castin, C. Salomon, Phys. Rev. A **55**, 2989 (1997)
37. E. Raab, M. Prentiss, A. Cable, S. Chu, D. Pritchard, Phys. Rev. Lett. **59**, 2631 (1987)
38. M.G. Raizen, C. Salomon, Q. Niu, Phys. Today **50**(7), 30 (1997)
39. J.J. Sakurai, *Modern Quantum Mechanics* (Addison-Wesley, New York, 1994)
40. D.A. Steck, *Sodium D Line Data (2001)*, available at http://steck.us/alkalidata
41. P. Valanju, E.C.G. Sudarshan, C.B. Chiu, Phys. Rev. D **21**, 1304 (1980)
42. S.R. Wilkinson, C.F. Bharucha, M.C. Fischer, K.W. Madison, P.R. Morrow, Q. Niu, B. Sundaram, M.G. Raizen, Nature **387**, 575 (1997)
43. R.G. Winter, Phys. Rev. **123**, 1503 (1961)
44. C. Zener, Proc. R. Soc. Lond. A **137**, 696 (1932)
45. C. Zener, Proc. R. Soc. Lond. A **145**, 523 (1934)

Chapter 9
Quantum Post-exponential Decay

Joan Martorell, J. Gonzalo Muga, and Donald W.L. Sprung

> *After a long time the rain let up, but the clouds stayed, and the*
> *lightning kept whimpering, and by and by a flash showed us a*
> *black thing ahead, floating, and we made for it.*
>
> Mark Twain

9.1 Introduction

Exponential decay is a very general phenomenon in the natural sciences. It occurs when a quantity N decreases at a rate proportional to its value,

$$\frac{dN}{dt} = -\frac{1}{\tau}N, \qquad N = N_0 e^{-t/\tau} . \tag{9.1}$$

It occurs in practically every field of physics, including unstable or excited elementary particles, nuclei, atoms, molecules, or quantum dots. Typical examples of microscopic decay processes are nuclear alpha-decay, atomic autoionization, spontaneous emission of a photon from an excited state, or escape from a potential trap. In the quantum world, however, the exponential law is not an obvious consequence of the basic dynamical law, which is Markovian (i.e., memoryless) and applies to amplitudes, not probabilities. Accordingly, the derivation of exponential decay from first principles has been the subject of much scrutiny and debate.

Quantum decay was first examined in nuclear physics: the decay of natural radioactive nuclei has been studied for more than a century [116]. Gamow [41] developed the first quantal theory for alpha-decay. He explained, at least qualitatively, the vast range of alpha-decay lifetimes as a tunneling phenomenon by

J. Martorell (✉)
Departament d'Estructura i Constituents de la Materia, Universitat de Barcelona, 08028 Barcelona, Spain, martorell@ecm.ub.es

J.G. Muga
Departamento de Química-Física, UPV-EHU, Apdo. 644, 48080 Bilbao, Spain, jg.muga@ehu.es

D.W.L. Sprung
Department of Physics and Astronomy, McMaster University, Hamilton, Ontario,
Canada L8S 4M1, dwsprung@mcmaster.ca

Martorell, J. et al.: *Quantum Post-exponential Decay.* Lect. Notes Phys. **789**, 239–275 (2009)
DOI 10.1007/978-3-642-03174-8_9 © Springer-Verlag Berlin Heidelberg 2009

imposing "outgoing wave boundary conditions" on the wave function at the outer edge of the confining barrier. Such an outgoing condition can be satisfied, in general, only for a discrete set of complex energies. The corresponding "Gamow states" are usually associated with "resonances," a concept that almost always arises in discussions of decay, especially exponential decay. Gamow states decay exponentially in time, with twice the imaginary part of the complex energy, divided by $\hbar$, giving the inverse lifetime. However, they are unphysical states because the wave function grows exponentially outside the interaction region in coordinate space. They can, however, be used successfully for performing "resonant state" or "pole" expansions of the physical wave function [45]. A simple explanation of why the Gamow state provides the correct decay rate, using a more conventional expansion in (real energy) stationary eigenstates, is given in [10].

Weisskopf and Wigner, in another influential paper [139, 140], worked out a more elaborate theory for spontaneous emission from an excited atom interacting with the quantized radiation field, as introduced by Dirac [23]. This approximate theory also predicts exponential decay and sets a standard paradigm for treating general decay problems in which an initial discrete bound state (or more generally a set of them) is coupled to ("embedded into") a perturbing continuum of states that destabilizes the discrete state, making it a decaying resonance. The relation between Fermi's golden rule and exponential decay was explored at length by Hillery [57] for radiative decay of a discrete atomic state, either in free space or in a cavity. The $2p \rightarrow 1s$ decay in hydrogen was examined in detail and estimates were made for the deviations at long and short times.

In summary, exponential decay is ubiquitous and a standard topic in quantum mechanics books where it is derived in first-order time-dependent perturbation theory using Fermi's golden rule. What is less well known, at least at textbook level, is that at both short and long times, quantum mechanics predicts deviations from the exponential law. The Zeno effect, a subject which has attracted much attention in recent years [11] and is treated in both the previous and present volumes of *Time in Quantum Mechanics*, is associated with the short time deviation. The deviations at long times are less often discussed and constitute the central topic of this chapter.

Figure 9.1 shows the survival probability for a particle initially confined inside a very simple spherically symmetric potential (s-wave decay). (Details of the potential will be given in Sect. 9.3.3.) Up to $t \simeq 50$ the decrease is exponential. After a transition near $t \simeq 100$, beyond $t \simeq 150$ the decay is again smooth and clearly follows a different law. This is the post-exponential decay at issue. We will discuss how it is determined by the corresponding Hamiltonian.

From another perspective, Figs. 9.2 and 9.3 show the absolute value of the radial s-wave function determined from a numerical solution of the time-dependent Schrödinger equation. The leftmost part of the x-axis corresponds to the inner well (see inset of Fig. 9.1). Figure 9.2 covers the range of times corresponding to exponential decay. As time progresses the amplitude in the well decreases exponentially and an outgoing wave packet expands beyond the barrier. The line for the earliest time shows the emission of several humps corresponding to higher energy wave packets. These are evidence that the chosen initial state is a superposition of several

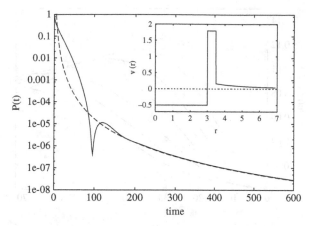

Fig. 9.1 Survival probability, $P(t)$, for an s-wave-confining potential consisting of a square well, a surrounding square barrier, and an outer inverse square potential. Units such that $\hbar = 2m = 1$. *Continuous line*: exact solution. *Dashed line*: prediction for asymptotic decay, Eq. (9.47). *Inset*: shape of the potential

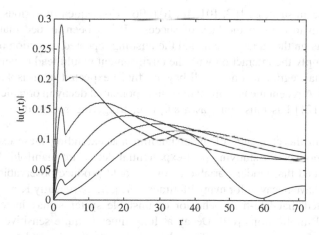

Fig. 9.2 Snapshots of $|u(r, t)|$, $0 < r < 72$, for times corresponding to exponential decay. From *top* to *bottom* at $r = 30$, the *continuous lines* correspond to times $t = 15, 30, 45, 60$, and 75

quasibound states. Figure 9.3 corresponds to post-exponential times and shows a much more regular trend: the wave function moves steadily outward and the part inside the inner well is a smooth continuation inward of the main outer part. It is this residual interior amplitude that shows the power law decrease.

From a theoretical viewpoint, this is an interesting regime, where the classical prediction of Eq. (9.1) differs from the quantal prediction. Quantum mechanics allows a very general argument leading to elegant results. On the experimental side, access to the post-exponential regime has proven to be difficult. There have been many attempts to provide convincing experimental evidence of post-exponential

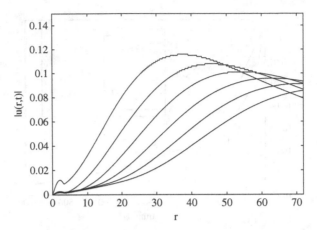

Fig. 9.3 Snapshots of $|u(r, t)|$, $0 < r < 72$, for times corresponding to intermediate and post-exponential decay. From *top* to *bottom* at the *left*, the *continuous lines* correspond to times $t = 75, 90, 105, 120, 135$, and 150

decay, with scant success [103, 101, 48, 104, 96]. Consequently, various ideas have been put forth to explain the lack of success. It has been argued that repetitive measurements on the same system, and the ensuing repeated reduction of the wave packet, or simply the interaction with the environment would lead to persistence of the exponential regime to times well beyond those expected in an isolated system [38, 71, 112]. A recent measurement of post-exponential decay in organic molecules in solution [121] has thus come as a surprise, triggering renewed interest in the subject.

Aside from the challenge of understanding and extending these experimental results, the motives for studying post-exponential decay are manifold. Winter, for example, argued that hidden variable theories could produce observable effects in samples that have decayed for many lifetimes [142]. More recently Krauss and Dent have suggested that late-time behavior of unstable systems may have important cosmological implications [69]. Decay at long times is quite sensitive to delicate measurement or environmental effects, becoming a testing ground for theories of these processes. At a fundamental level, the detailed form of long time deviations may help to distinguish between standard, Hermitian quantum mechanics [26], and modifications proposed to incorporate a microscopic arrow of time [98, 97]. From a practical point of view, Norman pointed out that post-exponential decay could set a limit to the validity of radioactive dating schemes [104]. In a recent paper we have also argued that the deviation could be a diagnostic tool for characteristics of certain cold atom traps [80]. Indeed, due to technological advances in lasers, semiconductors, nanoscience, and cold atoms, microscopic interactions are now relatively easy to manipulate, making decay controllable and post-exponential decay more accessible to experimental scrutiny and/or applications. For example, under appropriate conditions it could become the dominant regime and be used to implement an anti-Zeno effect, speeding up decay [73].

Finally, let us also mention that recent experiments on propagation of electric fields in periodic waveguide arrays provide a classical analog of a simple quantum system with exponential decay [74, 19]. These experiments may allow one to reach the post-exponential region in a particularly simple case, as we will discuss below.

9.1.1 What We Shall Not Discuss

This review focuses, as emphasized in the title chosen for the chapter, on systems whose decay is exponential over several lifetimes, i.e., systems where an isolated resonance is dominant, or whose decay is characteristic of a post-exponential regime, as for resonances very close to threshold [63, 42]. There are many phenomena where decay is not at all exponential: A trivial example is the survival probability of a freely moving quantum particle represented, say, by a Gaussian wave function. Overlapping resonances in nuclear physics can also make the decay quite complex [24], even for two degenerate discrete levels [115]. The level structure of polyatomic molecules leads to complicated coherence effects of interaction between levels and regular or irregular quantum beats [82]. Nonexponential decay signals may also result from the incoherent addition of purely exponential contributions from different members of the ensemble [27, 134, 102] or from temporal fluctuations of the environment on a timescale larger than the lifetime. Power law decay in the photoluminescence of conjugated polymers has also been reported but its mechanism involves incoherent processes [100, 16]. Another interesting phenomenon is the suppression of decay of dressed bound states, in or out of the continuum, formed by the interaction of the original discrete state(s) with the continuum states [127, 128, 78]. For other examples of nonexponential decay see [119, 126, 31]. Interesting as they are, these cases and systems are beyond the scope of this review, although we may briefly comment on some of them.

9.1.2 Previous Reviews

A number of monographs and review articles on the decay process in quantum mechanics, including post-exponential decay, exist. Among the textbooks, Sakurai [124] gives an excellent summary of the basic theory. The most frequently cited review is that of Fonda et al. [38]. It includes a section on the interpretation of repetitive experimental measurements. The monograph of Goldberger and Watson [47] uses the resolvent method to relate the asymptotic decay law to the threshold behavior of the energy density. Cohen-Tannoudji et al. [12] considered decay in the context of an initial discrete state coupled to a continuum of final states. In the same context Nakazato et al. [93] and Facchi et al. [32, 33] include a detailed discussion of Laplace transforms to extract asymptotic contributions to the survival probability. See also Longhi [76] for a recent extension to the more complex problem of coupling to a cavity with intrinsic gains and losses.

In this chapter we avoid unnecessary repetition of thoroughly reviewed material, although some well-known topics will be also briefly outlined for completeness. These include the Paley–Wiener theorem, see below, and a brief reminder of basic results of the discrete–continuum model in Sect. 9.2. We shall rather concentrate on some recent results and potential scattering. Decay of a single particle from a trap will be the main physical system under discussion in Sects. 9.2 (devoted to simple models) and 9.3 (on explicit decay laws depending on the long-range potential tail). Section 9.4 reviews different physical interpretations that have been given to post-exponential phenomena, and Sect. 9.5 the difficult route toward experimental observation. In the remainder of this section we introduce the basic and seminal work by Khalfin, with only a brief account of what preceded him.

9.1.3 Early Work and Paley–Wiener Theorem

Following the theories of Gamow, and Weisskopf and Wigner, Hellund [56] analyzed the decay of resonance radiation and showed that decay could not be exactly exponential. He also noted that it should be slower at long times. Later on, Höhler, going beyond the simplifications of the Weisskopf–Wigner model, obtained a power law for long times [58]; see also [114] for an early study of nonexponential alpha-decay.

Khalfin [65] discovered a very general result: the long-time decay for Hamiltonians with spectra bounded from below is slower than exponential. Here is the argument: Consider a system described by a time-independent Hamiltonian, H, initially in a normalized nonstationary state $|\Psi_0\rangle$. The survival amplitude of that state is defined to be the overlap of the initial state with the state at time t and is the expectation value of the time evolution operator

$$A(t) = \langle \Psi_0 | \exp(-iHt/\hbar) | \Psi_0 \rangle . \qquad (9.2)$$

The survival probability, sometimes called the "nondecay probability" [38], is[1]

---

[1] The survival probability is sometimes criticized as a measure of decay. For example, in a simple half-scattering problem with a particle located initially in a trapping potential, it may be difficult to measure. Some decay analyses therefore discuss other quantities, such as the nonescape probability from a region of space [43, 92], the probability density at chosen points of space [89, 88, 133], the flux [94, 95, 141], the arrival time [2]. For initially localized wave packets, there is no major discrepancy between survival probability and the nonescape probability. A claim to the contrary for long-time decay in [43] was criticized [89, 9, 136, 135] and confirmed later to be the result of a nonconverged computation [45]. Examination of densities, fluxes, or arrival time distributions may be interesting since a new variable is introduced but at the price of losing the simplicity and uniqueness of the survival probability.

A property of the survival amplitude not necessarily shared by other decay measures is its "stationarity." To formulate it we require a more precise notation: Let $P(t, t_0) = |\langle \psi(t_0) | \psi(t) \rangle|^2$ be the survival probability at time t of the state $\psi(t_0)$. It then follows from the hermiticity of H that $P(t, t_0) = P(t + t', t_0 + t')$, in other words, for a given wave packet the survival amplitude

$$P(t) = |A(t)|^2 .$$ (9.3)

The stationary states, $H|\Phi_{E,\lambda}\rangle = E|\Phi_{E,\lambda}\rangle$ (where λ denotes any other quantum numbers comprising a complete set of commuting observables), determine a basis. We assume that the Hamiltonian has only a continuous spectrum or at least that the discrete states of H are orthogonal to Ψ_0. Further, we assume that the spectrum is bounded from below: $E_m < E < \infty$. The completeness relation is then

$$\sum_\lambda \int_{E_m}^\infty dE \, |\Phi_{E,\lambda}\rangle\langle\Phi_{E,\lambda}| = \mathbf{1} ,$$ (9.4)

which gives

$$A(t) = \int_{E_m}^\infty dE \tilde{\omega}(E) \, e^{-iEt/\hbar} , \quad \text{where}$$

$$\tilde{\omega}(E) = \sum_\lambda |\langle\Phi_{E,\lambda}|\Psi_0\rangle|^2$$ (9.5)

is called the "energy density" of the quasi-stationary state. It is then natural to define

$$\omega(E) = \begin{cases} \tilde{\omega}(E) , & E \geq E_m \\ 0 , & E < E_m \end{cases}$$ (9.6)

and express the survival amplitude as the Fourier transform of the energy density [70],

$$A(t) = \int_{-\infty}^\infty dE \omega(E) e^{-iEt/\hbar} .$$ (9.7)

Khalfin showed that when $\omega(E)$ is so bounded from below, the Paley–Wiener theorem [111] requires that

$$\int_{-\infty}^\infty dt \frac{\ln |A(t)|}{1 + t^2} < \infty .$$ (9.8)

For this integral to converge as $t \to \infty$, the numerator must grow no faster than $\propto t^{2-a}$ with $a > 1$. This clearly rules out exponential decay at sufficiently long times. Assuming such a bound,

$$P(t) \geq \mathcal{A}e^{-\beta|t|^q} ,$$ (9.9)

$P(t_2, t_1)$ is a function only of the time difference $t_2 - t_1$. This may seem strange at first, but note that usually only the case in which the initial wave packet is localized in a region of space is of physical significance in a decay process.

with $\mathcal{A} > 0$, $\beta > 0$, and $q < 1$. An alternative proof, avoiding use of the Paley–Wiener theorem, has been given by Hack [54].

9.1.3.1 Decay at Post-exponential Times with Energy Integrals

Given an explicit expression for the energy density, to find analytic approximations for the post-exponential contribution to $P(t)$, one may replace the integration path in Eq. (9.5) by a contour integral in the complex plane [38, 67, 109]. For convenience, we place the origin of energy at threshold, $E_m = 0$, and choose a closed contour consisting of the positive real energy axis, a quarter circle of infinite radius running clockwise from the positive real axis to the negative imaginary axis, and continuing to the origin. Under very general conditions, integration along the circular arc gives a vanishing contribution. Then,

$$
\begin{aligned}
A(t) &= \int_0^\infty \mathrm{d}E \, \omega(E) \, \mathrm{e}^{-\mathrm{i}Et/\hbar} \\
&= \oint \mathrm{d}E \, \omega(E) \, \mathrm{e}^{-\mathrm{i}Et/\hbar} + \int_0^{-\mathrm{i}\infty} \mathrm{d}E \, \omega(E) \, \mathrm{e}^{-\mathrm{i}Et/\hbar} \\
&\equiv A_p(t) + A_v(t) ,
\end{aligned}
\tag{9.10}
$$

where we have analytically continued $\omega(E)$ into the lower half of the complex plane. Depending on the analytic structure of $\omega(E)$, this may or may not be a simple task. The contour integral $A_p(t)$ is $2\pi i$ times the sum of residues of poles in the fourth quadrant. These poles correspond to exponentially decaying terms.

The integral along the imaginary axis, $A_v(t)$, gives the dominant contribution to the decay at long times. We write the variable of integration as $\tilde{E} = iE$ to obtain

$$
A_v(t) = \mathrm{i} \int_0^\infty \mathrm{d}\tilde{E} \, \mathrm{e}^{-\tilde{E}t/\hbar} \, \omega(-i\tilde{E}) .
\tag{9.11}
$$

The exponential term becomes negligible after many lifetimes, say when $t \gg \tau = \hbar/\Gamma$, with Γ the width of the lowest lying resonance in $\omega(E)$. At such times t, the exponential in Eq. (9.11) restricts the range of significant contributions to the integral to values of $\tilde{E} \ll \Gamma$, which means to energies close to threshold.

A simple example is a pure Lorentzian energy density,

$$
\omega_L(E) = \frac{\Gamma/(2\pi)}{(E - E_R)^2 + \Gamma^2/4} .
\tag{9.12}
$$

In Sect. 9.2.2 we show how this appears, under suitable approximations, for a discrete state coupled to a continuum. (See also [38] pp. 604–606 for a similar but more elaborate model.) This energy density has a pole at $E = E_R - i\Gamma/2$, giving the exponential decay contribution

$$
A_p(t) = \mathrm{e}^{-iE_R t/\hbar} \, \mathrm{e}^{-\Gamma t/2\hbar} .
\tag{9.13}
$$

In addition, when $t \to \infty$,

$$
\begin{aligned}
A_v(t) &= i \frac{\Gamma}{2\pi} \int_0^\infty d\tilde{E} \frac{\exp(-\tilde{E}t/\hbar)}{(-i\tilde{E} - E_R)^2 + \Gamma^2/4} \\
&\simeq i \frac{\Gamma}{2\pi} \frac{\hbar}{E_R^2 + \Gamma^2/4} \frac{1}{t} + \mathcal{O}\left(\frac{1}{t^2}\right),
\end{aligned}
\tag{9.14}
$$

where in the second line we have set $\tilde{E} = 0$ in the denominator to evaluate the integral. This model therefore predicts that asymptotically, when the exponential term in Eq. (9.13) becomes negligible, $P(t) \simeq |A_v(t)|^2 \propto t^{-2}$.

Notice that due to the phase $-E_R t/\hbar$ in $A_p(t)$, in the range of times where the two contributions are of similar magnitude, interference oscillations are expected to occur. For a complete analysis of the decay for a more realistic truncated Lorentzian, including an exact expression for $A(t)$ and additional terms in the expansion of $A_v(t)$, see [130].

Suppose now that E_R decreases toward threshold at $E_m = 0$. As the pole moves toward the negative imaginary axis it increases the value of $A_v(t)$, eventually making it comparable to $A_p(t)$. In the limit $E_R = 0$ obviously the above decomposition is invalid, implying that the decay ceases to be exponential. This is known as small Q-value decay. For a more complete theoretical analysis of this situation and some particular model examples, see [63]. According to that analysis, small Q-value decay becomes noticeable when $Q \equiv E_R - E_m \lesssim \Gamma/2$.

The simplicity of the above model has made it popular in discussions of exponential decay. Jakobovits et al. [62] have shown that the steepest descent method applied directly to Eq. (9.5) also leads to exponential decay and shows that any corrections to it should usually be small.

Although useful as an illustration, Eq. (9.12) is far from being a realistic model for the energy density, in particular for the asymptotic law, since the Lorentzian form is generally not valid near threshold.

9.2 Simple Models and Examples

To fully appreciate post-exponential decay it is useful to understand first why exponential decay should be expected at all in a quantum system, even if only approximately, or over some limited time interval. The Gamow and Weisskopf–Wigner theories provide a clue, but a fresh look at the survival amplitude, Eq. (9.2), could raise doubts about the robustness of exponential decay, since a properly chosen initial state $\Psi(0)$ and energy density are amenable to an almost arbitrary decay law [46]. This puzzle is perhaps easier to understand in discrete–continuum decay models [113] (see below for an elementary illustration), but it has also been discussed in the context of scattering processes, e.g., in [47, 38]. If the initial preparation involves localization, the initial wave function will overlap prominently with scattering functions in certain regions of the energy spectrum, close to resonant poles, since they are more localized in the interaction region than ordinary, non-resonant waves. Contributions from broad resonances will decay faster, whereas for a narrow resonance it is the analytical structure of the pole rather than the complete state amplitude, which dominates the behavior, leading in practice to an "effective" Lorentzian energy density and through it to exponential decay.

9.2.1 Discrete State Embedded in a Continuum

Let us assume a simple Hamiltonian of the form

$$\mathcal{H} = E_\phi |\phi\rangle\langle\phi| + \int dE \; E \, |E\rangle\langle E| + \left\{ \int dE \; W(E) \, |\phi\rangle\langle E| + h.c. \right\} . \quad (9.15)$$

This is a typical nondegenerate single-channel Hamiltonian model which neglects continuum–continuum interactions. Other versions are due to Fano [34], Anderson [1], Lee [72], and Friedrichs [40] and have been successfully applied to study, for example, autoionization, photon emission, or cavities coupled to waveguides. The dynamics can be solved in several ways, using coupled differential equations for the time-dependent amplitudes and Laplace transforms, or finding the eigenstates with Feshbach's (P, Q) projector formalism [35], which allows separation of the inner (discrete) and outer (continuum) spaces and provides explicit expressions ready for exact calculation or phenomenological approaches. For modern treatments with emphasis on decay, see [24, 108]. Writing the eigenvector as [34, 24]

$$|\Phi_E\rangle = |\phi\rangle\langle\phi|\Phi_E\rangle + \int dE' \, |E'\rangle\langle E'|\Phi_E\rangle , \quad (9.16)$$

the coefficients for the discrete state are determined to be, using Laplace transform techniques or the projector P, Q formalism,

$$|\langle\phi|\Phi_E\rangle|^2 = \frac{|W(E)|^2}{[E - E_\phi - F(E)]^2 + \pi^2|W(E)|^4} , \quad (9.17)$$

where

$$F(E) = \mathcal{P} \int dE' |W(E')|^2/(E - E') . \quad (9.18)$$

If $W(E) \to W$ is energy independent, which is the essence of the Weisskopf–Wigner approximation, and if the range of integration is extended from $-\infty$ to ∞, then $F(E) = 0$ and $|\langle\phi|\Phi_E\rangle|^2$ becomes a Lorentzian,

$$|\langle\phi|\Phi_E\rangle|^2 = \frac{\Gamma/2\pi}{(E - E_\phi)^2 + \Gamma^2/4} , \quad (9.19)$$

with $\Gamma = 2\pi |W|^2$. Then, if the initial state is precisely ϕ, the energy density is a Lorentzian (Breit–Wigner form) and as we have shown in Sect. 9.1, its survival probability has an exponential component

$$P(t) = e^{-\Gamma t/\hbar} . \tag{9.20}$$

For weak coupling of bound state to continuum, the resonance is long-lived and its width is small, so that the constant-W approximation holds quite well. However, there is a fundamental physical limit to its validity, namely the existence of the energy threshold. If the resonance energy is close to threshold, this effect will be more noticeable, or even dominant, to the point of making the decay totally nonexponential [63, 42].

9.2.2 Simple Models Set in the Momentum Plane

Many decay models and in particular potential scattering models are treated in the complex momentum plane. The basic "mathematical" reason for exponential decay is easily seen to be the presence of a complex pole in the fourth quadrant of the momentum complex plane (second Riemann sheet of the energy plane), which, through its exponentially decaying residue, dominates the dynamics for some time. A simple analytical example of the deviation from exponentiality follows from the integral expression for the survival amplitude,

$$A(t) = \langle \Psi_0 | e^{-iHt/\hbar} | \Psi_0 \rangle = \frac{i}{2\pi} \int_C dq \, e^{-izt/\hbar} I(q) , \tag{9.21}$$

where

$$I(q) = \frac{q}{m} \langle \Psi_0 | \frac{1}{z - H} | \Psi_0 \rangle , \tag{9.22}$$

$z = q^2/2m$, and the contour C goes from $-\infty$ to ∞ passing above all singularities in the complex momentum plane q. One-dimensional motion of a particle of mass m is assumed. Consider now a pole expansion of the form $I(q) = \sum a_k/(q - q_k)$ [90, 45]. Since each term can be analyzed separately and combined linearly later, we concentrate on a single pole, $I(q) = a_r/(q - q_r)$, with q_r assumed to lie above the diagonal of the fourth quadrant and below the real axis. The integral is easily evaluated by deforming the contour to run diagonally across the second and fourth quadrants and picking up a small circle surrounding the pole. This gives

$$A(t) = \frac{1}{2} a_r w(-u_r) = a_r \left\{ \exp(-u_r^2) - \frac{1}{2} \text{sgn}[\text{Im}(u_r)] w[\text{sgn}[\text{Im}(u_r)] u_r] \right\} , \tag{9.23}$$

where

$$u_r = q_r/f, \quad f = (1 - i)(m\hbar/t)^{1/2} , \tag{9.24}$$

and $w(z) = \exp(-z^2)\mathrm{erfc}(-iz)$.[2] We have used the relation

$$w(-z) = 2e^{-z^2} - w(z) . \tag{9.25}$$

Equation (9.23) is particularly suitable for analyzing exponential decay, explicitly given by the pole, and its deviation, given by the line integral along the diagonal, evaluated as a w-function. The function $w(z)$ has the asymptotic expansion

$$w(z) \sim \frac{i}{\sqrt{\pi}\,z}\left(1 + \frac{1}{2z^2} + \dots\right) , \tag{9.26}$$

which, for long times, leads to a $1/t^{1/2}$ behavior.

However, an $I(q)$ with a single pole would be incompatible with time-reversal symmetry $A(t) = A(-t)^*$. The minimal model compatible with time-reversal must include, in addition to the resonance at q_r, the antiresonance pole at $-q_r^*$,

$$I(q) = \frac{a_r}{q - q_r} + \frac{a_r^*}{q + q_r^*} , \tag{9.27}$$

with $a_r = 1 + i\mathrm{Im}\,q_r/\mathrm{Re}\,q_r$ [44]. The contour integral along the diagonal which defines the u variable does not enclose this antiresonance pole, so it does not provide an exponentially decaying term but an additional contribution (the $w(u_r)$-function) that is significant only at short and long times [44]. In particular, it cancels exactly the asymptotic $t^{-1/2}$ decay from the resonance pole. However, the second terms in Eq. (9.26) do not cancel, resulting in a leading $t^{-3/2}$ behavior for $A(t)$.

So far this is a pure decay model rather than a Hamiltonian model; for an application see, e.g., [92]; but Hamiltonian realizations are possible. The separable potential considered in [90, 88, 89] leads to three core (state-independent) poles, two of them forming a resonance/antiresonance pair. Even closer to the minimal decay model is the delta-shell potential, discussed is the next section, when only the lowest resonance is excited and higher resonances can be neglected. Then it is exactly described by Eq. (9.27).

A general and useful result, independent of the assumed pole expansion, follows from noting that if the resolvent matrix element in Eq. (9.21) admits a series expansion, $c_0 + c_1 q + c_2 q^2 + \dots$, the leading term in the asymptotic formula, obtained by term by term integration, is

$$A(t) \sim \frac{1}{m\sqrt{2\pi i}}\,c_1\left(\frac{m\hbar}{t}\right)^{3/2} , \tag{9.28}$$

---

2 $\int_{-\infty}^{\infty} \frac{e^{-u^2}}{u-z} = i\pi\,\mathrm{sgn}[\mathrm{Im}(z)]\,w\{z\,\mathrm{sgn}[\mathrm{Im}(z)]\}$.

c_0 does not contribute by symmetry. For explicit examples of this, see [89], where a similar analysis is performed for the propagator and probability density. Further details are given in Sect. 9.4.2.

9.2.3 Exponential Decay as a Boundary Condition

Torrontegui et al. [133] have recently proposed a minimal, solvable 1D "source" model based on imposing an exponentially decaying amplitude for all times at one point in space ($x = 0$). This provides an economical approach to mimic analytically the wave function due to an exponentially decaying system with a long-lived resonance, while avoiding a detailed description of the interaction region where the decaying system is prepared. Deviations from exponential decay are observed in the probability density at $x > 0$.

9.2.4 One-Dimensional Well-Barrier Model of Confining Potential

One of the simplest 1D models of a decaying particle with a full description of the dynamics, including the preparation region, consists of a flat well surrounded by equal square barriers: in units $\hbar = 2m = 1$, we write $\mathcal{V}(x) = -V_0$ when $-a < x < a$; $\mathcal{V}(x) = V_b$ when $a < |x| < d$ and $\mathcal{V}(x) = 0$ when $|x| > d$. The initial state is chosen symmetric: $\Psi_0(x) = 1/\sqrt{a}\,\cos(\pi x/(2a))\,\Theta(a - |x|)$, and the continuum wave functions are

$$\Psi(x; E) = \begin{cases} \mathcal{N}\cos k_I x & , \quad |x| < a \\ A_b e^{\kappa x} + B_b e^{-\kappa x} & , \quad a < x < d \\ \frac{1}{\sqrt{2\pi k}}\cos[kx - \delta(k)] & , \quad d < x \end{cases} \tag{9.29}$$

for $x > 0$, while $\Psi(-x) = \Psi(x)$. Here $k^2 = E$, $k_I = \sqrt{k^2 + V_0}$, and $\kappa = \sqrt{V_b - k^2}$. The normalization of the outer part is determined by the condition

$$\int_{-\infty}^{\infty} dx\, \Psi(x; E)\, \Psi(x; E') = \delta(E - E'), \tag{9.30}$$

which leads to

$$\begin{aligned} \mathcal{N}^2 = \frac{1}{2\pi k}\Bigg[& \left(\cosh^2(\kappa b) + \frac{\kappa^2}{k_I^2}\sinh^2(\kappa b)\right)\cos^2(kd - \delta(k)) \\ & + \frac{k^2}{\kappa^2}\left(\sinh^2(\kappa b) + \frac{\kappa^2}{k_I^2}\cosh^2(\kappa b)\right)\sin^2(kd - \delta(k)) \\ & + 2\frac{k}{\kappa}\sinh(\kappa b)\cosh(\kappa b)\sin(kd - \delta(k))\cos(kd - \delta(k))\left(1 + \frac{\kappa^2}{k_I^2}\right)\Bigg]. \end{aligned} \tag{9.31}$$

In terms of these, the energy density is

$$\omega(E) \equiv |\langle \Psi_0 | \Psi(E)\rangle|^2 = \left| \mathcal{N} \frac{1}{\sqrt{a}} \int_{-a}^{a} dx \ \cos k_I x \ \cos\left(\pi \frac{x}{2a}\right)\right|^2$$

$$= \frac{\mathcal{N}^2}{a} \left[\frac{\pi/a \ \cos k_I a}{k_I^2 - \pi^2/(4a^2)}\right]^2 . \tag{9.32}$$

The asymptotic behavior, when $E \to 0$, is easily found to be $\omega(E) \propto k^{-1}$. Applying Eq. (9.11) one finds

$$A_v(t) \propto t^{-3/2} , \quad P(t) \propto t^{-3} . \tag{9.33}$$

Figure 9.4 is drawn for the parameters: $V_0 = 0.5$, $V_b = 1.8$, $a = 1.0$, and $d = 1.4$. The exact survival probability is compared to some alternative asymptotic forms of $P(t)$ and shows that indeed the asymptotic time dependence is that of Eq. (9.33).

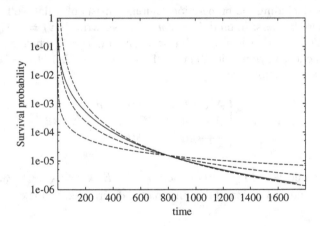

Fig. 9.4 Survival probability, $P(t)$ vs. time for the 1D-confining potential described in the text. *Continuous line*: exact solution of the TDSE. The three *dashed lines* correspond to asymptotic decays proportional to $1/t$, $1/t^2$, and $1/t^3$. The proportionality constants have been adjusted to reproduce the exact $P(t = 800)$

9.3 Three-Dimensional Models of a Particle Escaping from a Confining Potential

In this section, we directly obtain the energy densities and asymptotic decay profile for specific and increasingly realistic potential forms.

9.3.1 The Delta-Shell Model

Winter [141] gave the first *explicit* example of the validity of Khalfin's prediction. He studied 1D motion on the half-line $x \geq 0$, with a delta barrier at $x = a$ and zero potential elsewhere. He took the initial wave function to be the ground state of an infinite well, $\Psi_0(x, t = 0) = \sqrt{2/a}\sin(\pi x/a)\,\Theta(a - x)$, and showed that the mean momentum and the current at long times deviated from those of pure exponential decay, making a few analytical approximations. Recently the same problem has been re-analyzed by Dicus et al. [22] and indeed they find that the survival probability deviates from exponential at long times.

Winter's model and its variants have been applied on many subsequent occasions, for example, to study the effect of a distant detector (by adding an absorptive potential) [18], anomalous decay from a flat initial state [31], resonant state expansions [45], initial state reconstruction [92], or the relevance of the non-Hermitian Hamiltonian concept (associated with a projector formalism for internal and external regions of space) in potential scattering [125]. In [125] the model was extended to a chain of delta functions to study overlapping resonances.

Del Campo et al. [17] have also generalized the model for a Tonks–Girardeau gas of N bosons with strongly repulsive contact interactions as well as spinless fermions with strongly attractive contact interactions, and studied the long-time asymptotics and some new effects in the few-body decay of these systems.

9.3.2 Well-Behaved Short-Range Potentials

Spherically symmetric potentials $V(r)$ in three dimensions conserve angular momentum. One can separate the wave function into partial waves or channels of given angular momentum ℓ. For each partial wave, the radial Schrödinger equation is

$$\frac{d^2 w_\ell}{dr^2} - \frac{\ell(\ell + 1)}{r^2} w_\ell + [k^2 - \mathcal{V}(r)]w_\ell = 0 , \qquad (9.34)$$

where $\mathcal{V}(r) = (2m/\hbar^2)V(r)$, $k^2 = (2m/\hbar^2)E$, and $w_\ell(k, r)/r$ is the radial wave function in channel ℓ. From the low-energy behavior of w_ℓ one can extract the threshold behavior of the energy density $\omega(E)$ and from there, the time dependence of the post-exponential survival probability. For so-called well-behaved potentials this is a standard exercise in potential scattering theory [131]. These potentials decrease as $1/r^3$ or faster when $r \to \infty$ and are less singular than $1/r^{3/2}$ when $r \to 0+$. We will now summarize the derivation given in [80]. We work with solutions normalized as

$$\int_0^\infty dr\, w_\ell(k', r)^* w_\ell(k, r) = \delta(k' - k) \qquad (9.35)$$

and obeying the boundary condition $w_\ell(k, 0) = 0$. At large distance

$$\lim_{r \to \infty} w_\ell(k, r) = (2/\pi)^{1/2} \sin(kr - \pi\ell/2 + \delta_\ell) , \qquad (9.36)$$

where δ_ℓ is the phase shift for partial wave ℓ. The corresponding solutions normalized to energy, rather than momentum, are

$$w_\ell(E, r) = \sqrt{\frac{m}{\hbar^2 k}} \, w_\ell(k, r) . \qquad (9.37)$$

In turn, these solutions are related to the regular solutions $\hat{\phi}_\ell$ (defined by their behavior as Riccati–Bessel functions $\hat{j}_\ell$, $\hat{\phi}_\ell(r) \sim \hat{j}_\ell(kr)$ when $r \to 0$) by

$$w_\ell(k, r) = \sqrt{\frac{2}{\pi}} \frac{\hat{\phi}_\ell}{|f_\ell(k)|} , \qquad (9.38)$$

where $f_\ell(k)$ is the Jost function defined as in [131]. From the definition, Eq. (9.5), and particularizing to a single-channel ℓ,

$$\omega_\ell(E) = |\langle u_i | w_\ell(E) \rangle|^2 = \frac{2m}{\pi\hbar^2} \frac{1}{k} \frac{|\langle u_i | \hat{\phi}_\ell \rangle|^2}{|f_\ell(k)|^2} , \qquad (9.39)$$

where $u_i(r)$ is the radial wave function of the initial state, normalized to unity, and it is assumed that $u_i(r) = 0$ beyond some fixed distance $r = r_a$. Aside from the exceptional case where there is a bound state at zero energy, the Jost function, $f_\ell(k)$, tends to a nonvanishing constant $f_\ell(0)$ as $k \to 0$, and the threshold dependence of the energy density follows from that of $\hat{\phi}_\ell(k, r)$. It can be shown that [131]

$$|\hat{\phi}_\ell(k, r)| \le \gamma_\ell \left(\frac{|kr|}{1 + |kr|} \right)^{\ell+1} e^{|\mathrm{Im} kr|} \quad \text{as } k \to 0 , \qquad (9.40)$$

where γ_ℓ is some constant. This guarantees the desired analytic properties and allows the use of Eq. (9.10). Since the limiting behavior at small k is that of the Ricatti–Bessel functions, it is clear that when $k \to 0$,

$$|\langle u_i | \hat{\phi}_\ell \rangle|^2 \propto k^{2l+2} \Rightarrow \omega(E) \simeq \zeta \, k^{2l+1} , \qquad (9.41)$$

where ζ is a constant.

To evaluate $A_v(t)$, we have to continue this function to small energies on the negative imaginary axis, $\omega(-i\tilde{E}) \simeq \zeta(-i)^{l+1/2} \tilde{E}^{l+1/2}$.

Inserting this into Eq. (9.11) one immediately finds

$$A_v(t) \simeq -(-i)^{l+3/2} \zeta \, \Gamma(l + 3/2) \, t^{-(l+3/2)} . \qquad (9.42)$$

This is the well-known result that the asymptotic decay in a channel with angular momentum ℓ in a well-behaved exterior potential follows a power law, $P(t) \propto 1/t^{2l+3}$, with an integer exponent.

9.3.3 The Long-Range Part of the Potential

The above short-range potentials include as a special case potentials that are strictly constant beyond a certain distance. We are now going to discuss the opposite situation: how the decay law is modified when the outer tail of the potential decreases more slowly than $1/r^3$ and find cases where the decay is no longer a power law. For simplicity we will consider only s-waves ($\ell = 0$).

9.3.3.1 Inverse Square Potentials

The case of potentials with nonsingular inner part, and smoothly decreasing as $1/r^2$ when $r \to \infty$, is particularly interesting for its simplicity and for the range of behaviors exhibited [80]. Moreover, attractive inverse square potentials occur physically as effective radial potentials between a charged wire and a polarizable neutral atom [20], the strength factor being proportional to the square of the linear charge density of the wire and thus controllable [20, 21, 55]. Combined with a repulsive centrifugal term, an arbitrary α/r^2 potential can be implemented. It is also possible to modify the inner region and implement a potential minimum by a time-varying sinusoidal voltage in the high frequency limit [55] or by replacing the wire by a charged optical fiber with blue detuned light propagating along the fiber with the cladding removed [21]. Decay experiments with cold atoms showing exponential laws have been performed [20], and the ability to modify the potential parameters makes the observation and study of the long-time power law in these systems a realistic prospect.[3]

For s-waves, when $E \to 0$, one can prove that

$$w(E, r) \simeq \sqrt{\frac{2mk}{\pi \hbar^2} \frac{1}{|f(k)|}} \, \phi_0(r) \,, \qquad (9.43)$$

where for brevity we have omitted the subindex $\ell = 0$. We denote by $\phi_0(r)$ the zero-energy solution of the Schrödinger equation that satisfies the boundary conditions: $\phi(r = 0) = 0$, $(\partial\phi(r)/\partial r)_{r=0} = 1$. It follows that near threshold

$$\langle w(E)|u_i \rangle \propto \frac{\sqrt{k}}{|f(k)|} \,, \qquad \omega(E) \propto \frac{k}{|f(k)|^2} \,. \qquad (9.44)$$

Earlier we used the property that for short-range potentials $f(0) \neq 0$, giving $\omega(E) \propto k$ when $\ell = 0$. For a potential which behaves as $1/r^2$ as $r \to \infty$, this is no longer true, and several cases must be considered: We write the asymptotic ($r \to \infty$) form of the potential as

---

[3] Dipole (r^{-2}) potentials are also the effective interaction between an electron and an excited hydrogen atom [107] or a polar molecule such as HCl [25, 30]. Control of the interactions in this case is, however, much reduced.

$$\mathcal{V}(r) \simeq \frac{\alpha}{r^2} \, . \tag{9.45}$$

Referring only to the tail and implying nothing about the inner part, we will call potentials with $\alpha > 0$ (resp. $\alpha < 0$) repulsive (attractive). In addition we will distinguish between weakly attractive, $-1/4 < \alpha < 0$, and strongly attractive potentials, $\alpha < -1/4$.

It is convenient to introduce β via $\beta(\beta+1) = \alpha$. Then weakly attractive potentials correspond to $-1/2 < \beta < 0$ and repulsive potentials to $\beta > 0$. In both cases, the solutions of the Schrödinger equation at large r can be written as linear combinations of the Ricatti–Bessel functions $\hat{j}_\beta(kr)$ and $\hat{n}_\beta(kr)$, and the order of the related cylinder functions $\nu = \beta + 1/2$ remains positive. Using the analytic properties of the Bessel functions as $kr \to 0$, to lowest order in k,

$$|f(k)| \simeq \mathcal{D} \, k^{-\beta} \, ; \qquad\qquad \omega(E) \simeq \zeta \, k^{2\beta+1} \, , \tag{9.46}$$

where ζ and $\mathcal{D}$ are constants. The formal similarity to Eq. (9.41) is evident and corresponds to the analogy between the form of $\mathcal{V}(r)$ in Eq. (9.45) written in terms of β: $\mathcal{V}(r) = \beta(\beta + 1)/r^2$ and the centrifugal term in Eq. (9.34). It follows that

$$A_v(t) \simeq -(-\mathrm{i})^{\beta+3/2} \, \zeta \, \Gamma(\beta + 3/2) \, t^{-(\beta+3/2)} \, . \tag{9.47}$$

It should be emphasized that here β can take any real value greater than $-1/2$. The previously considered well-behaved potentials always lead to power laws with integer exponents $2\ell + 3$ for the asymptotic survival probability. In contrast to them, for repulsive or weakly attractive inverse square tail potentials the asymptotic decay law is still a power law but with an exponent that can be arbitrarily adjusted by varying the strength β.

To illustrate this result, we used a very simple model of confining potential (shown as an inset of Fig. 9.1):

$$\mathcal{V}(r) = \begin{cases} -\mathcal{V}_0 \, , & 0 < r \leq r_a \\ \mathcal{V}_b \, , & r_a < r \leq r_d \\ \frac{\alpha}{r^2} \, , & r_d < r \end{cases} \, . \tag{9.48}$$

Taking units where $\hbar = 2m = 1$, the parameters of the specific example are as follows: $\mathcal{V}_0 = 0.5, r_a = 3.0, \mathcal{V}_b = 1.8, r_d = 3.4$, and $\alpha = 0.39$ ($\beta = 0.3$). The radial wave function for the initial state is again chosen as $u_i(r) = \sqrt{2/r_a} \, \sin(\pi r/r_a) \, \Theta(r_a - r)$. In terms of these one can work out an explicit expression for the constant ζ appearing in Eq. (9.47). Figure 9.1 compares the exact survival probability to the excellent prediction given by Eq. (9.47), for sufficiently long times.

9.3.3.2 Post-exponential but Not Yet Asymptotic

In all the above discussion we have assumed that there was no intermediate range of times for which the decay was no longer exponential or mixed, and yet it was not fully asymptotic. In fact, for attractive potentials this range exists and can be misleading: Explicit calculations show a deviation from the exponential decay law that can be well fitted over a sizeable range of times with a power law, but neither the exponent nor the absolute value is well predicted by the above expressions. A more careful examination of the $k \to 0$ behavior of the Jost function shows that at least two more terms are needed in the truncated expansion and that it has to be rewritten as

$$|f(k)|^2 \simeq k(\lambda_- k^{-2\nu} + \lambda_0 + \lambda_+ k^{2\nu}) , \qquad (9.49)$$

with $\nu = \beta + 1/2$ as defined earlier and the λ's are constants. (Note that when $\beta \to -1/2$, $\nu \to 0$, and $k^{2\nu} \simeq 1$.) Further details may be found in [80].

9.3.3.3 Strongly Attractive Outer Potentials

When $\alpha < 1/4$, then ν, the order of the cylinder functions, becomes imaginary, $\nu \to i\nu' = i\sqrt{1/4 - \alpha}$, and the asymptotic form of the Jost function behaves very differently [107, 25, 80]:

$$\frac{|f(k)|^2}{k} \simeq \mathcal{N} \left[\sinh^2(\nu'\pi/2) + \cos^2(\nu' \ln k + \psi)\right] . \qquad (9.50)$$

The expression for ψ is given in [80]. Obviously, from this expression one cannot derive any simple form of power law decay for the asymptotic survival probability. Still, using the method of steepest descents, one can write an approximate form for $A_\nu(t)$ and argue that the decay must be close to $1/t$, as found in explicit numerical calculations.

9.3.4 Potentials with a α/r^p Asymptotic Decrease, $2 < p < 3$

For these potentials, the threshold dependence of the energy density is still given by Eq. (9.44), so that the small k behavior is again determined by that of the Jost function. Klaus [66] has given the required expressions for the Jost function. (He wrote the potential at large $r \to \infty$ as $\mathcal{V}(r) \simeq q_0/r^{2+\varepsilon}$, so his q_0 is our α and $2 + \varepsilon$ is our p.) He discussed the case $0 < \varepsilon \leq 1$ and considered two options: either (i) $f(0) \neq 0$ or (ii) $f(0) = 0$. Since the second case is exceptional, corresponding to a bound state at zero energy, we will discuss only case (i), where the Jost function is nonvanishing at zero energy,

$$f(k) = f(0) \left(1 + a_0\, e^{-i\pi\varepsilon/2}\, k^\varepsilon\right) + \mathcal{O}(k^\varepsilon)\,,$$

$$a_0 \equiv -\frac{q_0\, 2^\varepsilon}{\varepsilon(\varepsilon + 1)}\, \Gamma(1 - \varepsilon)\,. \tag{9.51}$$

Inserting this into Eq. (9.44) and keeping only terms up to order k^ε, we have

$$\omega(E) \simeq \Lambda \frac{k}{|f(0)|^2}(1 - \lambda k^\varepsilon)\,,$$

$$\lambda \equiv 2a_0\, \cos(\pi\varepsilon/2)\,, \tag{9.52}$$

where Λ is a constant. We can now determine the corresponding $A_v(t)$:

$$
\begin{aligned}
A_v(t) &= \int_0^{-i\infty} dE\, \omega(E) e^{-iEt/\hbar} \\
&= i \int_0^\infty dx\, e^{-xt}\, \Lambda \frac{e^{-i\pi/4}}{|f(0)|^2} x^{1/2} \left(1 - \lambda e^{-i\pi\varepsilon/4} x^{\varepsilon/2}\right) \\
&= \frac{i\Lambda e^{-i\pi/4}}{|f(0)|^2} \left(\int_0^\infty dx\, e^{-xt} x^{1/2} - \lambda e^{-i\pi\varepsilon/4} \int_0^\infty dx\, e^{-xt} x^{(1+\varepsilon)/2}\right) \\
&= \frac{i\Lambda e^{-i\pi/4}}{|f(0)|^2} \left(\Gamma(3/2)t^{-3/2} - \lambda e^{-i\pi\varepsilon/4} \Gamma((3+\varepsilon)/2)t^{-3/2-\varepsilon/2}\right)\,. \tag{9.53}
\end{aligned}
$$

In the above we wrote $E = -ix$ and therefore $k = e^{-i\pi/4}\sqrt{x}$. A point worth remarking is that $|f(0)|$ plays the role of a normalization factor, so that if we are primarily interested in the "slope" of the post-exponential decay, we can ignore $|f(0)|$, and it is the values of ε and q_0 that fix the slope. When t is very large, the second term is negligible, and the decay again follows a simple power law with $P(t) \propto 1/t^3$. At intermediate times, the second term is sizeable and, depending on the value of ε, the effective power law will deviate significantly from inverse cubic. (Note that when $\varepsilon \to 0$, the truncated expansion in Eq. (9.52) is no longer adequate.)

9.4 Physical Interpretation of Post-exponential Decay

The argument based on the Paley–Wiener theorem is sound and convincing but perhaps not very illuminating from a physical perspective. We have also seen that post-exponential decay can be mathematically attributed to the fact that the pole contribution decreases until eventually it is comparable to or smaller than a line integral, whose value arises predominantly from a saddle point at threshold, associated with slow particles. This is surely more intuitive, yet not fully satisfying for those seeking a more pictorial, rather than a complex plane, understanding of the phenomenon. In this vein, Hellund proposed an electrostatic analog [56] which relates quantum emission of radiation to the damped oscillation of a charge describable in purely classical (stochastic) terms. He thus interpreted the deviation from exponential decay as a "straggling phenomenon" characteristic of diffusion processes. Since quantum dynamics cannot be generally reduced to a classical diffusion process it would be interesting to explore the general applicability of this concept in other decay models. Other interpretational efforts are reviewed next.

9.4.1 Initial State Reconstruction

A powerful, general argument to explain nonexponential decay is the so-called "initial state reconstruction" [29, 37]. Following Feshbach, define projection operators $P = |\Psi_0\rangle\langle\Psi_0|$ and $Q = 1 - P$. Then

$$
\begin{aligned}
A(t) &= \langle\Psi_0|e^{-iH(t-t')/\hbar}(P + Q)e^{-iHt'/\hbar}|\Psi_0\rangle \\
&= A(t - t')A(t') + \langle\Psi_0|e^{-iH(t-t')/\hbar}\, Q\, e^{-iHt'/\hbar}|\Psi_0\rangle
\end{aligned}
\tag{9.54}
$$

for any intermediate time t'. When the second term can be neglected, the relation reduces to $A(t) = A(t - t')A(t')$, whose solution $A_e(t) = \exp(-\gamma t)$ is the exponential decay law. The neglected term corresponds to "initial state reconstruction": at intermediate times, t', the system is in states orthogonal to $|\Psi_0\rangle$ (in the subspace Q) but at time t it is found again in the "reconstructed" initial state. Since A_e decreases exponentially, we expect that at long enough times, the second term in Eq. (9.54) cannot in general be neglected and the exponential regime ceases to be valid. The second term in Eq. (9.54) is sometimes called the "memory" term, as it implies that quantum mechanics allows one to determine the absolute age of the decaying system [48]. We introduce the following notation:

$$
A_P(t) \equiv A(t - t')A(t'), \quad A_Q(t) \equiv A(t) - A_P(t) .
\tag{9.55}
$$

Note that these amplitudes depend on t and also on t', but we have not written this second dependence explicitly. The survival probability can be decomposed as

$$
P(t) = |A_P(t)|^2 + |A_Q(t)|^2 + 2\mathrm{Re}\left[A_P^*(t)A_Q(t)\right]
\tag{9.56}
$$

and again each term depends on t'. The third contribution will be called the interference or mixed term. Muga et al. [92] investigated the relative importance of the three terms. If the interference is negligible, the reconstruction process is similar to a consistent "classical" history in which we can assign probabilities to alternative paths, and reconstructed states can indeed be located elsewhere at an intermediate time. In simple terms, when the histories are consistent we may plainly say that events have happened in one or the other order with certain probabilities, without the need to invoke virtual paths and complex amplitudes. If, on the contrary, interference matters, the reconstruction is not a consistent history [50]. There remains of course an arbitrariness in the definition of "negligible," since the interference term is often small compared to the others, but rarely 0. One should accept, in other words, that the "consistency" or "classicality" of the histories is not absolute and sharply defined, but a contingent quality that may, nonetheless, be precisely quantified. The calculations, done for Winter's model and for the 2-pole model of Sect. 9.2, showed a significant difference between short- and long-time deviations from exponential decay as far as the role of state reconstruction is concerned. It becomes a consistent history for long times but not for short times. Reconstruction, and long-time decay, was hindered by placing an absorber outside the interaction region; more on this later.

9.4.2 Is Free Motion the Origin of Post-exponential Decay?

Jacob and Sach [61], in their field theoretical analysis of a scalar particle coupled to two pions, found nonexponential terms decaying like $t^{-3/2}$ in the amplitude. Their explanation was geometrical: If a particle is produced at point x having velocity between v and dv, it will appear after a time t within a spherical shell of radius vt centered on x, and the thickness of the shell will be tdv. The probability that it will be found within a small element of volume within the shell is inversely proportional to the volume $4\pi t^3 v^2 dv$ of the shell. Hence the probability amplitude is proportional to $t^{-3/2}$. It is a simple picture but unfortunately cannot be a universal explanation, since, for example, the decay amplitude in 1D scattering is generically $t^{-3/2}$, see, for instance, the example given in Sect. 9.2.4, Eq. (9.33), and Fig. 9.4, whereas the above argument translated to 1D would give only $t^{-1/2}$. The post-exponential power law behavior is sometimes interpreted as expressing the dominance of free motion [11, 106], but explicit calculations of the long-time propagator for specific potential scattering models in one dimension show that this is not the case in general. Muga, Delgado, and Snider [89] expressed the propagator as an integral in the momentum q-plane

$$\langle x | e^{-iHt/\hbar} | x' \rangle = \frac{i}{2\pi} \int_c dq \, I(q) \exp(-izt/\hbar) , \qquad (9.57)$$

$$I(q) = \frac{q}{m} \langle x | \frac{1}{z - H} | x' \rangle , \qquad (9.58)$$

where $z = q^2/2m$ and the contour C runs from $-\infty$ to ∞ passing above the singularities of the resolvent. C is then deformed to pass along the diagonal of the second and fourth quadrants so that the long-time dependence is explicitly extracted from the behavior of $I(q)$ at the origin. By decomposing the resolvent according to

$$\frac{1}{z - H} = \frac{1}{z - H_0} + \frac{1}{z - H_0} V \frac{1}{z - H}, \qquad (9.59)$$

(with H_0 the kinetic energy) $I(q)$ separates into "free" and "scattered" parts, $I = I_f + I_s$, which can be calculated for specific models. For 1D motion on the full line, the free motion part gives $I_f = -i/\hbar$, which implies $\langle x | e^{-iH_0t/\hbar} | x' \rangle \sim t^{-1/2}$

at long times, differing from the generic behavior of the propagator (except for the exceptional case of a zero-energy pole of the resolvent matrix element). Indeed, one obtains $\langle x|e^{-iHt/\hbar}|x'\rangle \sim t^{-3/2}$ when the interaction is taken into account. This comes about because of a cancellation between free motion and scattered contributions, i.e., $I_s(0) = -I_f(0) \neq 0$, which can be checked in specific potentials. For motion restricted to the half-line (or 3D partial waves) both terms vanish, $I_s(0) = I_f(0) = 0$, so both free and scattering components provide generically terms of the same order, $t^{-3/2}$, to the propagator. (Exceptions due to zero-energy resonances or specially chosen states have been considered by Miyamoto [84, 86, 87].)

Free motion is also implicit in an argument by Newton which makes use of classical mechanics [94, 95]. Provided that a point source emits particles with an exponential decay law and with a certain velocity distribution, their current density at a distant point would eventually depend on time according to an inverse power law. This is a suggestive observation, although it does not explain why the source itself, i.e., the survival probability, ceases to decay exponentially. In this respect Winter [141] provides an interesting result from the analysis of its model for a strong delta case. At the location of the delta, $x = a$, he defines an average local velocity as the ratio between the flux and the density and finds, for the exponential regime, the average velocity associated with the resonance, whereas in the post-exponential regime he has a/t, again suggesting a simple classical-like and free motion type of explanation.

9.5 Toward Experimental Observation

Already in 1911, Rutherford looked for experimental deviations from the exponential law in the alpha decay of ^{222}Ra, but found no deviations up to 27 half-lives. Similar searches have been made over the years, in experiments on the decay of radioactive nuclei and unstable particles, never finding any clear evidence. Particularly interesting for its accuracy is the experiment of Norman et al. [103] in 1988. They observed the decay of a sample of ^{56}Mn, with a half-life of $2\frac{1}{2}$ hours, for a total of 45 half-lives. That corresponds to a reduction of the initial activity by an impressive factor of $\simeq 3 \times 10^{14}$; still no significant deviation from the exponential law was found. (This paper includes a list of prior measurements and their corresponding number of half-lives.) It was pointed out, however, by Avignone [3] that an estimate by Winter implies that deviations from the exponential law would be expected to occur only at times of the order of 200 lifetimes in that experiment. Clearly, that is well beyond feasibility.

Already at that time, other possible reasons for the persistence of exponential decay were summarized by Greenland [48]. As explained earlier, any process suppressing initial state reconstruction will extend exponential decay to longer times than in an isolated system. In the case of radioactive decay, any fragments leaving the nucleus might interact with the surrounding electrons or with other atoms, irreversibly suppressing the reconstruction of the undecayed state. For the ^{56}Mn experiment it means that the decaying nucleus must not interact with its surroundings for the duration of the experiment.

Deviations from exponential decay have also been sought in particle physics. For a review of experiments up to 1968, see Nikolaev [101]. More recent attempts have similarly failed to detect any deviation. Tomono et al. [132] made precise measurements of the muon lifetime. Their experiment stopped at $17\,\mu s$, around 8 muon lifetimes, whereas theoretical models predict deviations to occur only at times beyond $200\,\mu s$. Novkovic et al. [105] have measured the decay of ^{198}Au up to 25 lifetimes, also finding no evidence for nonexponential decay.

In atomic physics, Robiscoe [120] calculated the long-time correction for the $2P \rightarrow 1S$ transition in a nonrelativistic two-level hydrogenic atom, finding that it would dominate over the exponential term only after 125 lifetimes ($Z = 1$). His calculated modifications to the Lorentzian line-shape were also minute. Assuming a small Γ/E_R ratio, and an atom–field coupling linear in energy (which hold generally for all spontaneous electric dipole processes), he concluded that the deviations are undetectably small for all of the most prominent spontaneous atomic transitions. Nicolaides and Mercouris [99] arrived at a more optimistic conclusion in studying decay of an autoionizing state close to threshold, the $\mathrm{He^-}1s2p^2$ ^{4}P core-excited shape resonance, with time-dependent methods. For that case the post-exponential decay sets in after 12 lifetimes.

9.5.1 Effects of Measurement and/or Environment

9.5.1.1 Collapse Models

The effect of the environment on the decaying particle, understood in a broad sense that could include the measurement apparatus, has been modeled by assuming randomly distributed collapsing interactions [28, 38]. It is argued, for example, that as an unstable elementary particle decays in a bubble chamber, each bubble is a measurement indicating that the particle has not yet decayed (has survived), so that a reduction takes place, resetting the system into the initial undecayed state. Therefore, the decay law that should be observed in experiment will be an environment-affected $F(t)$ rather than $P(t)$. The probability that the system is not subjected to any measurement in a time interval δt is taken to be $\exp(-\lambda\delta t)$. As detailed in [38], the survival probability $F(t)$ resulting from these measurements satisfies

$$F(t) = \mathrm{e}^{-\lambda t} P(t) + \lambda \int_0^t \mathrm{d}t'\, \mathrm{e}^{-\lambda t'} P(t') F(t - t') , \qquad (9.60)$$

where $P(t)$ is the survival probability without measurements. The integral is a convolution. Using Laplace transforms one easily finds that $f(s) = \mathcal{L}[F(t)]$ is given by

$$f(s) = \frac{p(s+\lambda)}{1 - \lambda p(s+\lambda)}, \tag{9.61}$$

where $p(s) = \mathcal{L}[P(t)]$. There is a particularly simple case: $P(t) = \exp(-\gamma t)$, for which $F(t) = P(t)$. More generally, a zero of the denominator of Eq. (9.61) on the negative real axis gives a contribution to $F(t)$ of exponential form. If the parameter λ is sufficiently large, this term will dominate and the corrections to the original lifetime are small. Moreover, the post-exponential regime will be suppressed.

Benatti and Floreani [7] have worked out a density operator model for elementary particle decays, taking into account incoherent interactions with the environment, and reached a similar conclusion.

9.5.1.2 Unitary Model for System–Environment Coupling

Lawrence [71, 112] studied the effect of interactions with the environment, using a unitary 3D $\ell = 0$ model of a particle confined by a delta-shell potential at $r = a$. Once outside the barrier, the particle interacts with N continuous spins. The interaction Hamiltonian is: $H = \eta \Theta(r-a) \sum_{i=1}^{N} S_i$, where S_i are operators that correspond to projections on a chosen axis of the ith spin and have a continuous spectrum, $S_i|\mu_i\rangle = \mu_i|\mu_i\rangle$, with $\mu_i \in [-1, 1]$. This leads to very simple expressions for the resulting energy density and analytic results for the asymptotic survival probability. Whereas in the absence of environmental coupling the latter shows a t^{-3} dependence, the coupling to N spins leads to $P(t) \sim t^{-2N-3}$, and the exponential regime is extended by a time that increases faster than linearly with N. Although the model is very simple, it provides a convincing demonstration of the relevance of the environment effects and the substantial modifications to the survival probability they entail. It is interesting also that for weak coupling, the dramatic effect at late times is accompanied by a negligible effect at early and intermediate times.

9.5.1.3 Effects of Adiabatic Switching and of Fluctuations of the Interaction

Greenland and Lane [49], using the model of Eq. (9.15) for photoionization, studied the effect of laser fluctuations (i.e., fluctuations in discrete to continuum matrix elements) and obtained for the decay rate the usual result, $2\pi W^2$, averaged over the bandwidth. They also argued, based on arguments similar to those in [38], that laser fluctuations eliminate the post-exponential region unless the transition is fast on the timescale of the fluctuations.

Mittelman and Tip [83], and Robinson [118] examined the effect of an adiabatic switching-on of the discrete–continuum coupling. The treatment of Robinson is more general and shows, in agreement with Mittelman and Tip, that adiabatic switching can attenuate the correction to exponential decay at long times. However, it differs from [83] in that the reduction depends on the rise time of the coupling potential, instead of the observation time. It thus leaves open the possibility of observing post-exponential decay, in particular for photoionization near threshold. Memory effects due to finite-time switching conditions for the release of the initial state have been recently revisited by Martorell et al. with the analytically solvable model of Sect. 9.5.3 [81].

9.5.1.4 Ways to Enhance Post-exponential Decay

Most of the proposals to enhance post-exponential decay to make it visible, or to advance the time at which it dominates, are based on the idea of placing a resonance close to threshold [98, 63, 42, 123, 143, 73]. In this respect, systems with controllable parameters are preferred, for example, photodetachment of electrons from negative ions, spontaneous emission in a cavity close to cutoff frequency or in a photonic band-gap material, see the chapter by Alonso and de Vega in this volume. Lewenstein and Rzazewski [73] proposed to manipulate the decay making the long-time decay dominant even at early stages so that repeated measurements could lead to an acceleration of decay (anti-Zeno effect). A precedent for this idea was included in the well-known article on the Zeno effect by Chiu, Sudarshan and Misra [11]. They pointed out that if the decay is measured periodically at very long-time intervals T (in the post-exponential range), the decay law at time $t = NT$ will be $F(t) = P(T)^N \propto (t/N)^{-\kappa N}$ if the survival probability for the isolated system is $P(t) \propto 1/t^\kappa$.

Another idea is to control the initial state so as to maximize the post-exponential term. This has been explored by Miyamoto [85] with an N-level Friedrichs model. One further option would be to use scattering singularities, such as a zero-energy resonance, naturally or artificially, to provide slower long-time decay [89, 87, 80]. The escape of interacting cold atoms from a trap, in the strongly interacting Tonks–Girardeau regime, has been also proposed to improve the visibility of the long-time decay, because the signal would be enhanced proportionally to the number of atoms [17].

Finally, Torrontegui et al. have recently found that the probability density at the exponential to post-exponential transition time, and thus its observability, increases with the distance of the detector from the source, up to a critical distance beyond which exponential decay is no longer observed [133]. This result is inspired by an earlier classical model by Newton [94, 95]. Quantum solvable models provide explicit expressions for the dependence of the transition on resonance and observational parameters, facilitating the choice of optimal conditions.

9.5.1.5 Complex Potentials

The detector model proposed in [92, 18] is formally Winter's model complemented by an imaginary absorbing potential,

$$H = -\frac{\partial^2}{\partial x^2} + \eta\delta(x - 1) - iV_c\Theta(x - X_c), \qquad x \geq 0, \qquad (9.62)$$

and the initial state is as usual the ground state of an infinite well between 0 and 1. This is an "optical model" where the absorbing potential represents the effect of detection of the escaping atom by laser excitation and photon emission in the outer region; see Chap. 4.

This Hamiltonian may hold N_{loc} discrete, localized eigenstates of complex energy with purely exponential decay. The detector is placed at the barrier edge at $X_c = 1$, or further out. The survival amplitude may be written in terms of eigenstates forming a biorthogonal basis [91]:

$$A(t) = \sum_{l=1}^{N_{loc}} C_l \hat{C}_l \, e^{-iE_l t} + \int_0^\infty f(q) \, e^{-i(q^2 - iV_c)t} \, dq \, , \qquad (9.63)$$

where $C_l = \langle \psi_0 | u_l \rangle$, $\hat{C}_l = \langle \hat{u}_l | \psi_0 \rangle$, $f(q) = \langle \psi_0 | \phi_q \rangle \langle \hat{\phi}_q | \psi_0 \rangle$, and $|u_l\rangle$ and $|\hat{u}_l\rangle$ are, respectively, right and left localized eigenstates. The continuum eigenstates appearing above, $|\phi_q\rangle$ and $|\hat{\phi}_q\rangle$, satisfy

$$H|\phi_q\rangle = E_q|\phi_q\rangle = (q^2 - iV_c)|\phi_q\rangle \, , \qquad (9.64)$$

$$\langle \hat{\phi}_q | H = E_q \langle \hat{\phi}_q | = (q^2 - iV_c)\langle \hat{\phi}_q | \, , \qquad (9.65)$$

$$\langle \phi_q | \hat{\phi}_{q'} \rangle = \delta(q - q') \, . \qquad (9.66)$$

Note that $|\phi_q\rangle$ and its corresponding biorthogonal partner are not usual scattering states because the exterior region is not free from interaction ($V(x) \neq 0$ when $x \to \infty$). However, the potential is constant there and this enables us to write the solution in the external region in terms of an S matrix,

$$\phi_q(x) = \frac{1}{(2\pi)^{1/2}} \begin{cases} C_1 \sin kx, & 0 \leq x \leq 1 \\ A e^{ikx} + B e^{-ikx}, & 1 \leq x \leq X_c \, , \\ e^{-iqx} - S(q) e^{iqx}, & x \geq X_c \end{cases} \qquad (9.67)$$

where $k = (q^2 - iV_c)^{1/2}$ is the wavenumber inside, q the wavenumber outside, and C_1, A, B, and S are obtained from the matching conditions at $x = 1$ and $x = X_c$. For scattering-like solutions, q is positive. There are two branch points of k in the complex q-plane; we take the branch cut to connect these points. Similarly, the root in $q = (k^2 + iV_c)^{1/2}$ is defined with a branch cut joining the two branch points in the k-plane. In contrast to scattering-like states of the continuum, localized states are characterized by a complex q with positive imaginary part.

The main effect of increasing V_c, at weak intensities, is the progressive suppression of long-time deviations, as illustrated in Fig. 9.5. Beyond some threshold strength, even the exponential decay is affected. A quantitative approximation to the survival probability helps to understand these effects: Let q_r be the resonance with the longest lifetime and let us assume that it is narrow and isolated. If all other resonances have already decayed, for weak enough absorption, i.e., $N_{loc} = 0$, the integral of Eq. (9.63) can be approximated, using contour deformation in the

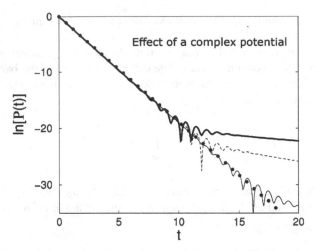

Fig. 9.5 $\ln[P(t)]$ for different absorptive step potentials, see Eq. (9.62). $V_c = 0$ (*thick solid line*), 0.1 (*dashed line*), 0.3 (*thin solid line*), and 0.5 (*dots*). $X_c = 1$ and $\eta = 5$

complex q-plane, by the residue corresponding to the first resonance plus a saddle contribution,

$$A(t) \approx -2\pi\mathrm{i}\,\mathrm{Res}\,[f(q)]_{q=q_r}\,\mathrm{e}^{-\mathrm{i}\mathcal{E}_r t}\mathrm{e}^{-\Gamma_r t/2} - \frac{\sqrt{\pi}\mathrm{i}}{8}\ddot{f}(0)\mathrm{e}^{-V_c t}\frac{1}{t^{3/2}}, \qquad (9.68)$$

where $\mathcal{E}_r$ represents the energy of the decaying particle, Γ_r the corresponding decaying width, and $\ddot{f}(0) = [\mathrm{d}^2 f(q)/\mathrm{d}q^2]_{q=0}$. The second term is responsible for the deviation from the exponential decay in the survival $P(t) = |A(t)|^2$. The novelty with respect to the nonabsorption case, $V_c = 0$, is that the deviation is not given by a purely algebraic term: The usual algebraic dependence is multiplied by the exponentially decaying factor $\exp(-V_c t)$. By increasing V_c, the deviation term decays more and more rapidly until, at threshold, i.e., $\Gamma_r = V_c^{\mathrm{th}}$, the deviation decays faster than the residue term. This threshold value corresponds exactly to the passage from a resonance to a localized, normalizable state with purely exponential decay. While for $V_c < V_c^{\mathrm{th}}$, the dominant term at long times is the saddle contribution (proportional to $\exp(-V_c t) t^{-3/2}$), in the opposite case, $V_c > V_c^{\mathrm{th}}$, the decay is purely exponential, and the dominant contribution comes from the discrete part of the spectrum.

9.5.1.6 Effect of a Distant Detector

A still controversial and rather crucial question is how is the decay affected by the distance between detector and system in indirect measurements [13–15, 59, 60, 79, 68, 137, 138, 110, 122]? In their conceptual analysis of the Zeno effect, Home and Whitaker [59] stated that the only real paradox is that the system is predicted to have its decay affected by a detector at a macroscopic distance. Indeed, a common sense expectation is that a greater separation of the detector from the initial location of the system ought to reduce the perturbing effects of measurement, but theories confirming this expectation have been disputed [13–15, 137]. The need for more work to arrive at a definite conclusion is clear [68].

One way to proceed is by study of explicit models. In [18], with the Hamiltonian model of Eq. (9.62), it was shown that the disturbance of the measurement (represented by the complex absorbing potential) on the survival amplitude disappears with increasing distance X_c between the initial state and the detector, as well as by improving its efficiency.

9.5.1.7 Observation of Power Laws in Organic Molecules, Quantum Dots, or Nanocrystals

In view of the above arguments favoring persistence of exponential decay at long times and of the distorting environmental effects on the power law decay, it came as a surprise that the experiments of Rothe et al. [121] showed a clean power law decrease after the cross over from the exponential regime. In those experiments, the luminescence decays of several species of dissolved organic molecules were measured, most with lifetimes of the order of nanoseconds, but some up to a tenth of a millisecond. Power law decays with algebraic exponents between $\simeq 2$ and $\simeq 4$ were deduced. The strategy followed was to look for resonant decays with large Γ and not too small E_R. As shown in Eq. (9.14), larger Γ increases the asymptotic component and not too small E_R avoids small Q nonexponential decay. Having the molecules in solution increases the widths, and for large molecules, intramolecular structure also favors increased broadening. In this case, environmental effects increase rather than decrease the asymptotic terms.

Unfortunately, the theoretical description of luminescence decay in such organic molecules is difficult [6, 4, 5], and therefore a detailed comparison with theory is missing. The possibility of other origins for the observed power law decrease in fluorescence has been ruled out by Rothe et al. [121]. Still, very recently Sher et al. [129] have also measured power law decay in the fluorescence blinking of various semiconductor nanocrystals. They reproduced those with Monte Carlo simulations on a three-level model. These authors claim that their approach can be also useful in the analysis of nonexponential fluorescence decays of molecular systems. In fact colloidal semiconductor quantum dots, nanorods, nanowires, and some organic dyes exhibit power law distributions of on- and off-times of emission intermittency [39]. They can be explained with models in which an electron jumps into one of the multiple traps and returns, but not all facets of the experiment are well understood. In condensed molecular solids composed of conjugated polymers [100], the power law luminescence found experimentally is attributed to electron–hole pair recombination and explained using inhomogeneous, statistical theories. Therefore, the field is still open, both theoretically and experimentally, and more conclusive work on these systems would be highly desirable.

9.5.2 Indirect Measurement

Rather recently, Kelkar et al. [64] made an interesting proposal for extracting the long-time survival probability from the low-energy $\ell = 0$ phase shift of α–α scattering, where it is affected by a 2^+ resonant state (virtual formation of ^8Be). By fitting the experimental phase shift to a parameterized form, they extracted a power law behavior $1/t^{6.36}$ for $P(t)$. The explanation for this value is murky, since the long-range part of the α–α potential is dominated by Coulomb repulsion, and the post-exponential decay law should be modified by that long-range potential tail. Applying the recipe for well-behaved potential tails with $\ell = 2$, the exponent would be 7. The result is nevertheless interesting, as it is a different way to approach the problem, which may have application to other systems. Their result corresponds to an estimated changeover time of 30 lifetimes, which makes a direct observation not feasible.

9.5.3 A Classical Analog of a Decaying Quantum System

Recently Longhi [74, 75] proposed to use a system of identical parallel wave guides as an optical analog of a decaying quantum system. A laser beam injected in the first guide leaks amplitude to the second as it travels down the guide [77]. The second guide similarly leaks to the first and third, and so on. Distance along the guide plays the role of time. Let $c_n(t)$ be the amplitude in the guide labeled n at distance t from the point of injection. If the guides have identical cross section and are equidistant, the transverse motion is in a periodic system with nearest neighbor interactions. The system is equivalent to a particle in a tight-binding model with Schrödinger equation

$$i\hbar \frac{d|\psi\rangle}{d(gt)} = \frac{1}{g} H_{TB}|\psi\rangle = \hat{H}_{TB}|\psi\rangle \qquad \text{with}$$

$$\hat{H}_{TB} = \begin{pmatrix} \varepsilon_1/g & -\Delta & 0 & \cdots 0 & 0 \\ -\Delta & \varepsilon_0/g & -1 & \cdots 0 & 0 \\ \vdots & \vdots & \vdots & \cdots & \vdots \\ 0 & 0 & 0 & \cdots \varepsilon_0/g & -1 \\ 0 & 0 & 0 & \cdots -1 & \varepsilon_0/g \end{pmatrix}. \qquad (9.69)$$

Here g is the coupling parameter between adjacent wave guides and ε_0 is the common site energy for all sites in the periodic portion of the system. Site 1 is special; its site energy ε_1 may differ from ε_0 and its coupling parameter is $g\Delta$, with $0 \leq \Delta \leq 1$. For example, if the first wave guide is placed farther apart from its neighbor, we should expect $|\Delta| < 1$, and if its profile is wider, $\varepsilon_1 < \varepsilon_0$.

From here on we will measure distance in units of $\hbar/g$ and suppose that $\varepsilon_1 = \varepsilon_0$ by making the guides to have identical cross-sections. Then $\hat{H}_{TB}$ is a dimensionless Hamiltonian of a "particle" in a semi-infinite discrete space. Initially, only $c_1(t = 0) = 1$ is nonzero. If Δ is significantly less than 1, the beam is trapped behind a barrier and will only slowly leak out to site 2. The exact time-dependent solution is

$$c_1(t) = \sum_{s=0} \alpha^{2s} \frac{2s+1}{t} J_{2s+1}(2t) .$$ (9.70)

Here, $\alpha^2 \equiv 1 - \Delta^2$ is the "decoupling" parameter. If $\Delta = 0$ (by placing the first guide far from the rest of the system), a well-known sum rule satisfied by the Bessel functions gives $c_1(t) = 1$ at all times. At the other extreme, for $\Delta \sim 1$, small α^2 makes the series converge rapidly. $c_1(t)$ plays the role of the survival amplitude of a decaying quantum analog.

For the other sites, the solution is

$$c_n(t) = i^{n-1} \Delta \sum_{s=0} \alpha^{2s} \frac{2s+n}{t} J_{2s+n}(2t) .$$ (9.71)

Because $J_N(2t) \sim t^N/N!$ at small t, it is clear that the initial conditions at $t = 0$ have been satisfied.

Longhi showed that there are three distinct regions of t-dependence. At small t, from the properties of $J_1(2t)$,

$$c_1(t) \approx 1 - t^2/2 .$$ (9.72)

Next there is an intermediate region where exponential decay holds, with

$$c_1(t) \approx \frac{1+\alpha^2}{2\alpha^2} \exp[-\gamma_0 t/2] , \quad \text{where}$$

$$\gamma_0 = \Delta^2/\alpha$$ (9.73)

is the decay constant of the Gamow state. Weak coupling gives very slow decay and strong coupling $\Delta \sim 1$ very rapid decay.

At still longer t, the exponential has fallen so far that it is smaller than the asymptotic value of the sum of Bessel functions. In this limit we can approximate each term and perform the sum to obtain the asymptotic value for $c_n(t)$.

Use of the asymptotic form

$$J_{n+2s}(2t) \sim \frac{(-)^s}{\sqrt{\pi t}} \left[\cos \Phi_n - \frac{4(n+2s)^2 - 1}{16t} \sin \Phi_n \right] ,$$

$$\Phi_{n+2s} \equiv 2t - (n + \frac{1}{2})\frac{\pi}{2} - s\pi$$ (9.74)

leads, in lowest order, to

$$c_n(t) \sim \Delta \frac{i^{n-1}}{\sqrt{\pi t^3}} \frac{n + (n-2)\alpha^2}{(1+\alpha^2)^2} \cos(2t - (2n+1)\frac{\pi}{4}) ,$$

$$c_1(t) \sim \frac{1}{\sqrt{\pi t^3}} \frac{1-\alpha^2}{(1+\alpha^2)^2} \cos(2t - 3\pi/4) .$$ (9.75)

These can be further embellished by adding the next to leading order term in the asymptotic form. In any case it is evident that the post-exponential decay is a power law with exponent $-3/2$.[4]

It turns out to be an excellent approximation to write $c_1(t)$ as the sum of the exponential plus the asymptotic approximation, for all but the very shortest t. In the cross-over region their phases are correct so that the interference oscillations are nicely reproduced. For example,

$$c_1(t) \approx \frac{1+\alpha^2}{2\alpha^2}\, e^{-\gamma_0 t/2} + \frac{S_1}{t}\, J_1\left(2t - \frac{a_1}{t}\right), \qquad (9.76)$$

for $t >> a_1$, where

$$S_1 = \frac{1-\alpha^2}{(1+\alpha^2)^2}, \qquad a_1 = \frac{6\alpha^2}{(1+\alpha^2)^2} + \frac{3}{16}. \qquad (9.77)$$

When α is small ($\Delta \to 1$), there is almost no exponential region, whereas when $\alpha \to 1$, the exponential region extends to very large times. This is illustrated in Fig. 9.6: It can be seen that when $\Delta = 0.4$ in the time range shown the decay is exponential. When $\Delta = 0.95$ the post-exponential decay is reached very soon, and the survival probability is still quite sizeable. The figure also shows the effect of disorder on the post-exponential decay: Instead of assuming a constant g for the tunneling matrix elements, we have added a random fluctuation of $\pm 1\,\%$ to

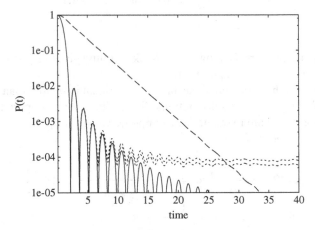

Fig. 9.6 Survival probability, $P(t) = |c_1(t)|^2$. *Continuous line:* $\Delta = 0.95$; *long dashed line:* $\Delta = 0.4$. The *dotted lines* defining a band of values correspond to the simulations with a random fluctuation of $\pm 1\,\%$ in the tunneling matrix elements

---

[4] The long-time regime has been studied also for finite-time switching in [81] and to examine its observability as a function of distance in [133].

each of the tunneling matrix elements between consecutive sites. By running a large number of cases we determine an average and a mean square deviation. The band corresponds to the average ± 3 standard deviations. As can be seen the effect on the post-exponential decay is significant and suggests that localization effects are taking place at long times. Note that if that random fluctuation could be attributed to "environmental" effects, our model does not predict an extension of the range of exponential decay.

We have recently investigated initial switching effects in this model. It was assumed that the coupling $\Delta(t)$ was increased from 0 to its final value Δ over a time T rather than instantaneously at time 0. The result is an additional phase shift T added to a_1 in Eq. (9.77), and S_1 is multiplied by $\sin T/T$. Together these changes alter the interference between exponential and asymptotic decay contributions in a way which ought to be observable [81].

The group in Milan have carried out two experiments [19, 8] which verify Longhi's theory. In the first, they arranged $\Delta = 1$ ($\alpha = 0$) and found that the measured amplitude in guide n followed the expected form of the Bessel function $c_n(t) = J_n(2t)/t$ [117].

In the second paper [8], Longhi and collaborators verified the optical analog of the Zeno effect. To do this, the wave guides were broken into longitudinal segments of length $t = L$. The beam in guide $n = 1$ entering a new segment effectively resets the initial condition (albeit with a reduced amplitude less than unity). The segment length L was chosen to be short, to correspond to the short-time behavior, Eq. (9.72). What they found was that the decay of $c_1(t)$ in each short segment was reset to the quadratic behavior, so that it never reached the exponential decay regime. On the contrary, without segmentation they did see a switch-over to Eq. (9.73). It will be interesting to see whether their experiment can be extended to longer length scales to verify the changeover to post-exponential decay, described in Eq. (9.76).

9.6 Final Comments

We have provided an overview of quantum post-exponential decay, from the early works to the most recent analyses and experiments. Several simple models were discussed, which help to understand the fundamental properties of the long-time deviation from exponential decay, and provide clues for its control and for setting new experiments, which is indeed one of the major challenges. Post-exponential decay is a subtle phenomenon, involving weak and low-energy signals, easy to perturb and hard to detect. However, recent progress in artificial mesoscopic semiconductor structures or trap design and detection of ultra-cold atoms may offer access to this regime for simple quantum systems. Further challenges are to gain a full understanding of post-exponential decay in complex systems, such as organic molecules in solution, as well as a more intuitive, physical interpretation of the effect.

Acknowledgments We thank A. del Campo, F. Delgado, W. Van Dijk, L. Schulman, and E. Torrontegui for useful discussions. We are grateful to DGES-Spain for support through grant FIS2006-10268-C03-01; to UPV-EHU (GIU07/40); to NSERC-Canada for Discovery grant RGPIN-3198; and to the Max Planck Institute for the Physics of Complex Systems in Dresden.

References

1. P.W. Anderson, Phys. Rev. **124**, 41 (1961)
2. C. Anastopoulos, J. Math. Phys. **49**, 022103 (2008)
3. F.T. Avignone, Phys. Rev. Lett. **62**, 2664 (1988)
4. W. Barford, Phys. Rev. B **65**, 205118 (2002)
5. W. Barford, Phys. Rev. B **70**, 205204 (2004)
6. W. Barford, R.J. Bursill, R.W. Smith, Phys. Rev. B **66**, 115205 (2002)
7. F. Benatti, R. Floreanini, Phys. Lett. B **428**, 149 (1998)
8. P. Biagoni, G. Della Valle, M. Ornigotti, M. Finazzi, L. Duò, P. Laporta, S. Longhi, Opt. Express **16**, 3762 (2008)
9. R.M. Cavalcanti, Phys. Rev. Lett. **80**, 4353 (1998)
10. R.M. Cavalcanti, C.A.A. de Carvalho, Rev. Bras. Ens. Fis. **21**, 464 (1999)
11. C.B. Chiu, E.C.G. Sudarshan, B. Misra, Phys. Rev. D **17**, 520 (1977)
12. C. Cohen-Tannoudji, J. Dupont-Roc, G. Grynberg, *Atom–Photon Interactions: Basic Processes and Applications* (Wiley, New York, 1992), pp. 220–221
13. B. Crosigni, D. di Porto, Nuovo Cimento B **109**, 555 (1994)
14. B. Crosigni, D. di Porto, Europhys. Lett. **35**, 165 (1996)
15. B. Crosigni, D. di Porto, Europhys. Lett. **39**, 233 (1997)
16. C.M. Cuppoletti, L.J. Rothberg, Synth. Met. **139**, 867 (2003)
17. A. del Campo, F. Delgado, G. García-Calderón, J.G. Muga, Phys. Rev. A **74**, 013605 (2006)
18. F. Delgado, J.G. Muga, G. García-Calderón, Phys. Rev. A **74**, 062102 (2006)
19. G. Della Valle, S. Longhi, P. Laporta, P. Biagioni, L. Dou, M. Finazzi, Appl. Phys. Lett. **90**, 261118 (2007)
20. J. Denschlag, O. Umshaus, J. Schmiedmayer, Phys. Rev. Lett. **81**, 737 (1998)
21. J. Denschlag, D. Cassettari, A. Chenet, S. Schneider, J. Schmiedmayer, Appl. Phys. B **69**, 291 (1999)
22. D.A. Dicus, W.W. Repko, R.F. Schwitters, T.M. Tinsley, Phys. Rev. A **65**, 032116 (2002)
23. P.A.M. Dirac, Proc. R. Soc. Lond. (A) **114**, 243, 710 (1927)
24. F.M. Dittes, Phys. Rep. **339**, 215 (2000)
25. W. Domcke, L.S. Cederbaum, J. Phys. B (At. Mol.) **14**, 149 (1980)
26. S.D. Druger, M.A. Samuel, Phys. Rev. A **30**, 640 (1984)
27. M. Dworzak et al., Superlatt. Micro. **36**, 763 (2004)
28. H. Ekstein, A.J.F. Siegert, Ann. Phys. (NY) **68**, 509 (1971)
29. L. Ersak, Yad. Fiz. **9**, 458 (1969); English translation: Sov. J. Nucl. Phys. **9**, 263 (1969)
30. H. Estrada, W. Domcke, J. Phys. B: At. Mol. Phys. **17**, 279 (1984)
31. P. Exner, Rep. Math. Phys. **59**, 351 (2007)
32. P. Facchi, S. Pascazio, *La Regola d'oro di Fermi* (Bibliopolis, Napoli, 1999)
33. P. Facchi, S. Pascazio, J. Phys. A: Math. Gen. **41**, 493001 (2008)
34. U. Fano, Phys. Rev. **124**, 1866 (1961)
35. H. Feshbach, Ann. Phys. **5**, 357 (1958)
36. H. Feshbach, Ann. Phys. **19**, 287 (1962)
37. L. Fonda, G.C. Ghirardi, Il Nuovo Cimento **7A**, 180 (1972)
38. L. Fonda, G.C. Ghirardi, A. Rimini, Rep. Prog. Phys. **41**, 587 (1978)
39. P. Frantsuzov, M. Kuno, B. Janko, R.A. Marcus, Nat. Phys. **4**, 519 (2008)
40. K.O. Friedrichs, Commun. Pure Appl. Math. **1**, 361 (1948)

41. G. Gamow, Zeit. Phys. **51**, 204 (1928)
42. G. García-Calderón, J. Villavicencio, Phys. Rev. A **73**, 062115 (2006)
43. G. García-Calderón, J.L. Mateos, M. Moshinsky, Phys. Rev. Lett. **74**, 337 (1995)
44. G. García-Calderón, V. Riquer, R. Romo, J. Phys. A **34**, 4155 (2001)
45. G. García-Calderón, I. Maldonado, J. Villavicencio, Phys. Rev. **76**, 012103 (2007)
46. M.L. Goldberger, K.M. Watson, Phys. Rev. **136**, B1472 (1964)
47. M.L. Goldberger, K.M. Watson, *Collison Theory* (John Wiley and Sons, New York, 1964), Ch. 8, pp. 448–450
48. P.T. Greenland, Nature **335**, 298 (1988)
49. P.T. Greenland, A.M. Lane, Phys. Lett. A **117**, 181 (1986)
50. R.B. Griffiths, J. Stat. Phys. **36**, 219 (1984)
51. R. Omnès, *The Interpretation of Quantum Mechanics* (Princeton University Press, Princeton, 1994)
52. M. Gell-Mann, J.B. Hartle, Phys. Rev. D **47**, 3345 (1993)
53. J.J. Halliwell, in *Time in Quantum Mechanics*, J.G. Muga, R. Sala, I.L. Egusquiza (eds.) (Springer, Berlin, 2002)
54. M.N. Hack, Phys. Lett. A **90**, 220 (1982)
55. L.V. Hau, M.M. Burns, J.A. Golovchenko, Phys. Rev. A **45**, 6468 (1992)
56. E.J. Hellund, Phys. Rev. **89**, 919 (1953)
57. M. Hillery, Phys. Rev. A **24**, 933 (1981)
58. G. Höhler, Zeit. Phys. **152**, 546 (1958)
59. D. Home, M.A.B. Whitaker, Ann. Phys. (NY) **258**, 237 (1997)
60. M. Hotta M. Morikawa, Phys. Rev. A **69**, 052114 (2004)
61. R. Jacob, R.G. Sachs, Phys. Rev. **121**, 350 (1961)
62. H. Jakobovits, Y. Rotshchild, J. Levitan, Am. J. Phys. **63**, 439 (1995)
63. T. Jittoh, S. Matsumoto, J. Sato, Y. Sato, K. Takeda, Phys. Rev. A **71**, 012109 (2005)
64. N.G. Kelkar, M. Nowakowski, P. Khemchandani, Phys. Rev. C **70**, 024601 (2004)
65. L.A. Khalfin, Sov. Phys. JETP **6**, 1053 (1958)
66. M. Klaus, J. Math. Phys. **29**, 148 (1988)
67. P.L. Knight, Phys. Lett. A **61**, 25 (1977)
68. K. Koshino, A. Shimizu, Phys. Rep. **412**, 191 (2005)
69. L.M. Krauss, J. Dent, Phys. Rev. Lett. **100**, 171301 (2008)
70. N.S. Krylov, V.A. Fock, J. Exptl. Theor. Phys. (U.S.S.R.) **17**, 93 (1947)
71. J. Lawrence, J. Opt. B: Quantum Semicl. Opt. **4**, S446 (2002)
72. T.D. Lee, Phys. Rev. **95**, 1329 (1954)
73. M. Lewenstein, K. Rzazewski, Phys. Rev. A **61**, 022105 (2000)
74. S. Longhi, Phys. Rev. Lett. **97**, 110402 (2006)
75. S. Longhi, Phys. Rev. E **74**, 026602 (2006)
76. S. Longhi, Phys. Rev. A **74**, 063826 (2006)
77. S. Longhi, G. Della Valle, M. Ornigotti, F. Laporta, Phys. Rev. B **76**, 201101(R) (2007)
78. S. Longhi, Eur. Phys. J. B **57**, 45 (2007)
79. M.G. Makris, P. Lambropoulos, Phys. Rev. A **70**, 044101 (2004)
80. J. Martorell, J.G. Muga, D.W.L. Sprung, Phys. Rev. A **77**, 042719 (2008)
81. J. Martorell, D.W.L. Sprung, W. Van Dijk, J.G. Muga, Phys. Rev. A **79**, 062104 (2009); arXiv:0905.3970
82. E.S. Medvedev, Sov. Phys. Usp. **34**, 16 (1991)
83. M.H. Mittelman, A. Tip, J. Phys. B: At. Mol. Phys. **17**, 571 (1984)
84. M. Miyamoto, Phys. Rev. **68**, 022702 (2003)
85. M. Miyamoto, Phys. Rev. A **70**, 032108 (2004)
86. M. Miyamoto, Phys. Rev. A **69**, 042704 (2004)
87. M. Miyamoto, Open Sys. Inform. Dyn. **13**, 291 (2006)
88. J.G. Muga J.P. Palao, J. Phys. A: Math. Gen. **31**, 9519 (1998)
89. J.G. Muga, V. Delgado, R.F. Snider, Phys. Rev. B **52**, 16381 (1995)
90. J.G. Muga, G.W. Wei, R.F. Snider, Ann. Phys. (NY) **252**, 336 (1996)

91. J.G. Muga, J.P. Palao, B. Navarro, I.L. Egusquiza, Phys. Rep. **395**, 357 (2004)
92. J.G. Muga, F. Delgado, A. del Campo, G. García-Calderón, Phys. Rev. A 73, 052112 (2006)
93. H. Nakazato, M. Namiki, S. Pascazio, Int. J. Mod. Phys. **B10**, 247 (1996)
94. R.G. Newton, Ann. Phys. (NY) **14**, 333 (1961)
95. R.G. Newton, *Scattering Theory of Waves and Particles*, 2nd edn (Dover, Mineola, 2002), Chap. 19
96. T.D. Nghiep, V.T. Hahn, N.N. Son, Nucl. Phys. B (Proc. Suppl.) **66**, 533 (1998)
97. C.A. Nicolaides, Phys. Rev. A **66**, 022118 (2002)
98. C.A. Nicolaides, D.R. Beck, Phys. Rev. Lett. **38**, 683, 1037 (1977)
99. C.A. Nicolaides, T. Mercouris, J. Phys. B **29**, 1151 (1996)
100. V.R. Nikitenko, D. Hertel, H. Bässler, Chem. Phys. Lett. **348**, 89 (2001)
101. N.N. Nikolaev, Sov. Phys. Usp. **11**, 522 (1968)
102. I.S. Nikolaev, P. Lodahl, A.F. Van Driel, A.F. Koenderink, W.L. Vos, Phys. Rev. B **75**, 115302 (2007)
103. E.B. Norman, S.B. Gazes, S.G. Crane, D.A. Bennett, Phys. Rev. Lett. **60**, 2246 (1988)
104. E.B. Norman, B. Sur, K.T. Lesko, R.M. Larimer, D.J. DePaolo, T.L. Owens, Phys. Lett. B **357**, 521 (1995)
105. D. Novkovic et al., Nucl. Inst. Methods A **566**, 477 (2006)
106. H.M. Nussenzveig, *Causality and Dispersion Relations* (Academic Press, New York, 1972)
107. T.F. O'Malley, Phys. Rev. **137**, A1668 (1965)
108. J. Okolowicz, M. Ploszajczak, I. Rotter, Phys. Rep. **374**, 271 (2003)
109. D.S. Onley, A. Kumar, Am. J. Phys. **60**, 432 (1992)
110. M. Ozawa, Phys. Lett. A **356**, 411 (2006)
111. R.E. Paley, N. Wiener, *Fourier Transforms in the Complex Domain* (American Mathematical Society Colloquium Publication **19**, 1934)
112. R.E. Parrott, J. Lawrence, Europhys. Lett. **57**, 632 (2002)
113. A. Peres, Ann. Phys. (NY) **129**, 33 (1980)
114. J. Petzold, Zeit. Phys. **155**, 422 (1959)
115. P.M. Radmore, P.L. Knight, Phys. Lett. A **102**, 180 (1984)
116. M. Razavy, *Quantum Theory of Tunnelling* (World Scientific, Singapore, 2003), p. 18.
117. M. Razavy, Can. J. Phys. **57**, 1731 (1979)
118. E.J. Robinson, Phys. Rev. A **33**, 1461 (1986)
119. E.J. Robinson, J. Phys. B: At. Mol. Opt. Phys. **27**, L305 (1994)
120. R.T. Robiscoe, Phys. Lett. A **100**, 407 (1984)
121. C. Rothe, S.I. Hintschich, A.P. Monkman, Phys. Rev. Lett. **96**, 163601 (2006)
122. L.C. Ryff, Eurphys. Lett. **39**, 231 (1997)
123. K. Rzazewski, M. Lewenstein, J.H. Eberly, J. Phys. B: At. Mol. Phys. 15, L661 (1982)
124. J.J. Sakurai, *Modern Quantum Mechanics. Rev. Ed.* (Addison-Wesley, New York, 1994), Supplement II, pp. 481–486
125. D.V. Savin, V.V. Sokolov, H.J. Sommers, Phys. Rev. E **67**, 026215 (2003)
126. L.S. Schulman, B. Gaveau, J. Phys. A **28**, 7359 (1995)
127. L.S. Schulman, C.R. Doering, B. Gaveau, J. Phys. A **91**, 2053 (1991)
128. L.S. Schulman, D. Tolkunov, E. Mihokova, Phys. Rev. Lett. **96**, 065501 (2006)
129. P.H. Sher, J.M. Smith, P.A. Dalgarno, R.J. Warburton, X. Chen, P.J. Dobson, S.M. Daniels, N.L. Pickett, P. O'Brien, Appl. Phys. Lett. **92**, 101111 (2008)
130. K.M. Sluis, E.A. Gislason, Phys. Rev. A **43**, 4581 (1991)
131. J.R. Taylor, *Scattering Theory* (J. Wiley, New York, 1972)
132. D. Tomono et al., Nucl. Inst. Methods A **503**, 283 (2003)
133. E. Torrontegui, J.G. Muga, J. Martorell, D.W.L. Sprung, arXiv:0903.4156
134. R.A. Valleé, K. Baert, B. Kolaric, M. Van der Auweraer, K. Clays, Phys. Rev. B **76**, 045113 (2007)
135. W. Van Dijk, Y. Nogami, Phys. Rev. Lett. **90**, 028901 (2003)
136. W. Van Dijk, Y. Nogami, Phys. Rev. C **65**, 024608 (2002)

137. S. Wallentowitz, P.E. Toschek, Phys. Rev. A **72**, 046101 (2005)
138. S. Wallentowitz, P.E. Toschek, Phys. Lett. A **367**, 420 (2007)
139. V.F. Weisskopf, E. Wigner, Zeit. Phys. **63**, 54 (1930)
140. V.F. Weisskopf, E. Wigner, Zeit. Phys. **65**, 18 (1930)
141. R.G. Winter, Phys. Rev. **123**, 1503 (1961)
142. R.G. Winter, Phys. Rev. **126**, 1152 (1962)
143. J. Zakrzewski, K. Rzazewski, M. Lewenstein, J. Phys. B: At. Mol. Phys. **17**, 729 (1984)

Chapter 10
Timescales in Quantum Open Systems: Dynamics of Time Correlation Functions and Stochastic Quantum Trajectory Methods in Non-Markovian Systems

Daniel Alonso and Inés de Vega

> *Past experience, if not forgotten, is a guide for the future*
> Chinese proverb

10.1 Introduction

The dynamics of a system in interaction with another system, the later considered as a reservoir, is studied in many different domains in physics. This approach is useful not only to address fundamental questions like quantum decoherence and the measurement problem [1] but also to deal with practical and theoretical problems appearing in the emerging fields of nanotechnology [2, 3] and quantum computing [4–6], as well as in systems of ultra-cold atoms [7]. In many of these cases, the basic approximation is the Markov assumption in which there is a clear separation of the typical timescales associated with the system and the reservoir or environment. This separation of timescales, together with other assumptions like the weak coupling between the system and the reservoir, has been central in the development of several fields, in particular in quantum optics [8, 9]. However, in the last few years, new systems where the Markov approximation is no longer valid have received increasing attention. This has produced a growing interest in developing a theory of non-Markovian quantum open systems.

The interaction of systems with structured environments, e.g., atoms coupled to the radiation field within a photonic crystal or a cavity [10], the interaction of systems with spin baths [11] (e.g., quantum dots in a bath of atomic nuclei) or with

D. Alonso (✉)
Departamento de Física Fundamental y Experimental, Electrónica y Sistemas, Instituto Universitario de Estudios Avanzados (*IUdEA*) en Física Atómica, Molecular y Fotónica, Universidad de La Laguna, La Laguna 38203, Tenerife, Spain, dalonso@ull.es

I. de Vega
Max Planck Institute of Quantum Optics, Garching 85748, Germany, ines.devega@mpq.mpg.de

Alonso, D., de Vega, I.: *Timescales in Quantum Open Systems: Dynamics of Time Correlation Functions and Stochastic Quantum Trajectory Methods in Non-Markovian Systems.* Lect. Notes Phys. **789**, 277–301 (2009)
DOI 10.1007/978-3-642-03174-8_10   © Springer-Verlag Berlin Heidelberg 2009

environments at low temperatures, the non-Markovian dynamics of entanglement [12], or the dynamics of a continuous atom laser based on focusing to an independently formed atomic condensate [13], are just examples of current problems where understanding non-Markovian features is crucial.

10.2 Atoms in a Structured Environment, an Example of Non-Markovian Interaction

One of the most basic concepts in physics is the spontaneous emission of an atom. It is known that spontaneous emission from an excited atom can be modified by placing the atom in a structured environment [14, 15]. Such environments are physically realized by placing the atoms in cavities or within photonic crystals. In both cases, the photonic density of states of the radiation field is severely modified. On the one hand, the boundary conditions that a cavity imposes to the radiation field modify its photonic density of states, which is no longer a growing function of the frequency, but becomes a Lorentzian centered in the cavity resonant frequency. This resonant frequency is related to the cavity length. Following the well-known Fermi Golden rule, the spontaneous emission rate of an atom is proportional to the photonic density of states of the field in the central emission frequency. Consequently, the emission of an atom within a cavity can be suppressed if the central emission frequency is off-resonant with respect to the cavity resonant frequency, i.e., with respect to the center of the Lorentzian. On the other hand, photonic crystals present a periodicity in the refraction index, which produces strong Bragg scattering for photons within a certain frequency range. Hence, these photons are not present within the crystal, and therefore the photonic density of states is 0 for the corresponding frequency range. As a consequence, the photonic density of states typically presents a band-gap profile consisting of regions where it varies abruptly (the bands) and regions where it is 0 (the gaps). The spontaneous emission of an atom near a band-gap edge [16, 17] differs from the exponential decay and exhibits oscillations, see Chap. 9 in this volume. Furthermore, even at zero temperature the steady-state population may remain finite for the excited state [18]. A similar phenomenon, the so-called limited quantum decay, is discussed in [19] for the case of a single level coupled to a continuum which is bounded in energy.

As noted before, a key element in the interaction of an atom with the electromagnetic field is the density of electromagnetic modes of such field for the atomic transition frequency. However, it is also important to consider the profile of the density of states around such a frequency. The electromagnetic field may be viewed as a reservoir or environment in interaction with the atom. If the density of photons around the atomic transition frequency is smooth then the interaction between the atom and the reservoir is, to a good approximation, Markovian. This means that there is a timescale associated with the dynamics of the atom that is slow and also a fast timescale associated with the correlation function of the environment. This in turn leads to an atomic dynamics that does not depend on its previous history.

The spontaneous decay in this case is almost exponential with a characteristic decay rate and the spectrum of the photons emitted exhibits a Lorentzian profile. A typical example is the interaction of an atom with a free electromagnetic field.

On the other hand, if the photonic density of states varies on a frequency range that is comparable with the spontaneous emission rate, the dynamics of the atom in contact with the reservoir may show noticeable non-Markovian effects. In such case the decay is nonexponential and the spectrum of the emitted light does not present a Lorentzian profile [20]. For non-Markovian interactions, the timescale associated with the environmental correlation function is not fast enough in comparison to that of the atomic dynamics, and the atomic dynamics depends on its previous history.

We may find other examples where non-Markovian effects may be important, as in the dynamics of qubits [21]. In this chapter we shall focus on atoms in non-Markovian environments and will particularize to atoms in photonic crystals.

10.3 Two Complementary Descriptions of the Dynamics of a Quantum Open System

The description of the dynamics of a quantum open system is usually done in a reduced scheme in which the degrees of freedom of the environment are averaged out. There are different ways of tackling the problem: we may use the reduced density matrix of the system that evolves according to some master equation or we may consider some Schrödinger type of equation that evolves vectors in the Hilbert space of the system conditioned to the state of the environment. In the last context, a frequent approach is based on the so-called Stochastic Schrödinger equations. Alternatively, other Monte Carlo methods have been developed, as the quantum jump approach (see Chap. 6). Along the chapter, we will consider a large system described by a Hamiltonian H, which consists of the addition of a system Hamiltonian H_S, an environment Hamiltonian H_E, and an interaction Hamiltonian between the system and the environment H_I. In the examples discussed here, the system is an atom, the environment is an electromagnetic field, and their interaction is described by assuming the dipolar approximation.

10.3.1 Reduced Density Matrix Approach

The state of the system at time t can be characterized by its reduced density matrix, $\rho_S(t)$, that is obtained by tracing out the degrees of freedom of the environment from the density matrix of the whole system $\rho(t)$, i.e., $\rho_S(t) = Tr_E(\rho(t))$, with Tr_E being the trace over the environmental degrees of freedom. The reduced density matrix depends on time and evolves according to some mapping $\mathcal{L}_t$ that when applied to and initial state $\rho_S(0)$ gives the state at later time $\rho_S(t)$. In the Markov case such mapping reduces to a master equation of Lindblad form ($\hbar = 1$) [22, 23]:

$$\frac{d\rho_S(t)}{dt} = -i[H_s, \rho_S(t)] + \frac{1}{2}\sum_n([L_n\rho_S(t), L_n^\dagger] + [L_n, \rho_S(t)L_n^\dagger]) \equiv \mathcal{L}_t\rho_S(t),$$

$$(10.1)$$

where L_n are operators in the Hilbert space of the system that model the coupling of the system to its environment.

In the non-Markovian case, the Lindblad equation does not rule the dynamics of ρ_S, and a more general master equation is needed. An example is the Redfield equation that has been proven to be valid for long times and within the weak coupling limit [24, 25] by considering a slippage or controlled modification of the initial condition, $\rho_S(0)$ [26, 27]. In general, non-Markovian master equations contain some memory effects in the system–environment interaction which are absent in the Lindblad equation. Apart from the Redfield equation, some other non-Markovian master equation approaches are discussed in [28–30] where memory effects are taken into account. Another set of time-convolutionless master equations is found in [31, 32]. Other useful methods rely on the computation of the evolution of the Heisenberg equations of motion [33, 34].

With the solution of the master equation it is then possible to compute the expectation values of system observables. If A is one observable of the system its expectation value is then given by

$$\langle A(t) \rangle = Tr_S(\rho_S(t)A).$$

$$(10.2)$$

In this particular sense the dynamics of the atom is solved within the approximations made.

10.3.2 Stochastic Schrödinger Equation Approach

Complementary to the master equation approach, different methods have been developed in many fields that consider the evolution of vectors in the Hilbert space of the system conditioned to the dynamics of the environment. In such schemes the environment acts as a source of quantum noise, z_t [8, 35–38], affecting *stochastically* the evolution of the system. If $\psi(z, t)$ is the time-dependent system vector, and its evolution is driven by a quantity z_t that depends on the coordinates of the environment z, one can obtain the reduced density matrix of the system as

$$\rho_S(t) = M\left[|\psi_t(z^*)\rangle\langle\psi_t(z)|\right],$$

$$(10.3)$$

where M is the average over the environmental noise. Such stochastic equations are said to give an unraveling of the reduced density matrix $\rho_S(t)$. In the Markov case there are several unravelings for the Lindblad master equation. Each unraveling corresponds to a particular way of making decomposition (10.3) of the reduced density matrix. We may found in this context the quantum jump or Monte Carlo wave function approach [8, 39–43] that involves jumps occurring at random times.

Alternatively, the quantum state diffusion approach considers a continuous time evolution [44–47]. These two approaches or unravelings have provided powerful methods to numerically compute the evolution of the system quantum mean values. Moreover, some of them provide us with an interpretation of a particular measuring scheme. In particular, different measurement schemes of the emitted light in a homodyne or heterodyne detection lead to different stochastic Schrödinger equations with distinct diffusive terms [48, 49]. A particular example is the continuous measurements on atoms described by diffusive type of Schrödinger equation [50]. The relation of continuous measurements and the stochastic schemes for state vectors in the Markovian case have been extensively discussed, see for instance [51, 36].

For non-Markovian interactions, several groups have developed a theory of stochastic Schrödinger equations for state vectors [52–54, 26, 27, 55–58, 33, 1]. A different method, which is based on using a pair of state vectors rather than a single one, was proposed in [59, 36] and further developed in [60]. Nonetheless, the physical interpretation of non-Markovian Schrödinger equations is still on debate [61–63].

10.3.3 Derivation of a Non-Markovian Stochastic Schrödinger Equation (NMSSE)

In this section we shall derive an NMSSE, as the one obtained in [52–54]. To that end we shall consider a Hamiltonian of the following form:

$$H = H_S + H_I + H_E = H_S + \sum_\lambda (g_\lambda L a_\lambda^\dagger + g_\lambda^* L^\dagger a_\lambda) + \sum_\lambda \omega_\lambda a_\lambda^\dagger a_\lambda , \qquad (10.4)$$

where $\hbar = 1$, H_E is a set of harmonic oscillators described by their creation and annihilation operators $a_\lambda^\dagger$, a_λ, and with frequencies ω_λ. In the interaction Hamiltonian, L is a system operator that describes the coupling to the environment and g_λ are the coupling constants of the λth mode to the system. This Hamiltonian is well suited to describe different systems, in particular the interaction between an atom and an electromagnetic field when one-photon processes are dominant.

Let $|\Phi_t\rangle$ be the state of the whole system in the interaction picture with respect to the environment. In this representation the Schrödinger equation reads

$$i\partial_t |\Phi_t\rangle = (H_S + H_I(t))|\Phi_t\rangle$$
$$= H_S|\Phi_t\rangle + \sum_\lambda (g_\lambda L e^{i\omega_\lambda t} a_\lambda^\dagger + g_\lambda^* L^\dagger e^{-i\omega_\lambda t} a_\lambda)|\Phi_t\rangle . \qquad (10.5)$$

Since the environment is a set of harmonic oscillators, it is convenient to describe it by using a coherent state basis in the Bargmann representation. A coherent state of the oscillators $|z\rangle = |z_1 z_2 \cdots z_\lambda \cdots\rangle$ satisfies the relations $a_\lambda|z\rangle = z_\lambda|z\rangle$ and

$\langle z|a_\lambda = \frac{\partial}{\partial z_\lambda^*} \langle z|$. The coherent states fulfill a resolution of the identity of the form $1 = \int d\mu(z)|z\rangle\langle z|$, where the measure $d\mu(z)$ is given by

$$
\begin{aligned}
d\mu(z) &= \frac{d^2 z_1 \, e^{-|z_1|^2}}{\pi} \frac{d^2 z_2 \, e^{-|z_2|^2}}{\pi} \cdots \frac{d^2 z_\lambda \, e^{-|z_\lambda|^2}}{\pi} \cdots \\
&= \prod_m \frac{d^2 z_m}{\pi} e^{-|z_m|^2} .
\end{aligned}
\tag{10.6}
$$

The state $|\Phi_t\rangle$ of the total system may be expressed, using the identity in the coherent state basis, as

$$
|\Phi_t\rangle = \int d\mu(z)|z\rangle|\psi_t(z^*)\rangle ,
\tag{10.7}
$$

where $|\psi_t(z^*)\rangle = \langle z|\Phi_t\rangle$ is a vector in the Hilbert space of the system and it depends on the state of the environment $|z\rangle$. The solution of the Schrödinger equation (10.5) can be obtained if one finds the system state vector $|\psi_t(z^*)\rangle$. The dynamical equation for $|\psi_t(z^*)\rangle$, which follows from Eq. (10.5), is

$$
i\partial_t|\psi_t(z^*)\rangle = H_S|\psi_t(z^*)\rangle + Lz_t|\psi_t(z^*)\rangle + L^\dagger \int_0^t ds\alpha(t-s)\frac{\delta|\psi_t(z^*)\rangle}{\delta z_s} ,
\tag{10.8}
$$

where

$$
z_t = \sum_\lambda g_\lambda z_\lambda^* e^{i\omega_\lambda t}
\tag{10.9}
$$

is a time-dependent function that takes into account the rotation of the environment oscillators weighted by their coupling to the system. This function acts as a driving in (10.8). The larger is the coupling of a particular oscillator to the system, the greater is its contribution to the driving of the system. In the third term of the RHS of (10.8), a dissipative term containing a memory kernel $\alpha(t-s) = \sum |g_\lambda|^2 e^{-i\omega_\lambda(t-s)}$ appears. The functional derivative $\frac{\delta|\psi_t(z^*)\rangle}{\delta z_s}$ still has to be handled in order to use Eq. (10.8). In some simple cases, this functional derivative may be computed explicitly, but in general some approximations are needed. On the other hand, an important set of relations that has been used along the derivation of (10.8) is

$$
\int d\mu(z)z_t = M[z_t] = 0 ,
$$
$$
M[z_t z_s] = 0, \quad \text{and} \quad M[z_t^* z_s] = \alpha(t-s) .
\tag{10.10}
$$

This set of relations allows a stochastic interpretation of (10.8). Indeed, we may think on z_t as a colored complex Gaussian noise with mean and correlations given by (10.10), where $\alpha(t-s)$ is the so-called *environment correlation function*. The non-Markovian character of (10.8) is contained in the fact that $\alpha(t-s)$ is not of delta

distribution. Let us point out that Eq. (10.8), together with the representation of state
(10.7), corresponds to an equivalent representation of original problem (10.5).

At this point we may proceed in two ways. If we have all the information of
the coupling constants and the frequencies of the harmonic oscillators of the environment, we may sample at random and according to the measure $d\mu(z)$, a set of
coherent state coordinates z. With such set we construct z_t and the environment
correlation function $\alpha(t - s)$. For a given initial state of the system we may then
try to solve (10.8). Such solution $|\psi_t(z^*)\rangle$ is associated with a single realization of
the coordinates z and represents a single *quantum trajectory* of the system. If we
sample again the coherent state measure $d\mu(z)$ we will obtain different trajectories.
Alternatively we may proceed phenomenologically, i.e., we assume that we know
the correlation function of the environment $\alpha(t - s)$, then we may synthesize a
colored complex Gaussian noise [26, 64] such that it has a correlation $\alpha(t-s)$ and so
properties (10.10) are fulfilled. Again, every different realization of the synthesized
noise would give a different trajectory.

With the set of trajectories $|\psi_t(z^*)\rangle$ obtained we may in principle construct the
solution $|\Phi_t\rangle$ of (10.5). More of our interest is the reduced density matrix of the
system that will be given by (10.3). In particular, if A is an observable of the system,
its expectation value $\langle A(t)\rangle = \langle \Phi_t|A|\Phi_t\rangle$ is

$$\langle A(t)\rangle = Tr_S(\rho_S(t)A) = M[\langle \psi_t|A|\psi_t\rangle] \,. \tag{10.11}$$

Therefore, the full description of the reduced dynamics can be tackled through
the solution of (10.8).

As we mentioned above the functional derivative $\frac{\delta|\psi_t(z^*)\rangle}{\delta z_s}$ is difficult to manipulate
in most of the cases. Nonetheless, it is possible to construct some particular forms
of it. One of them consists on writing $\frac{\delta|\psi_t(z^*)\rangle}{\delta z_s} = \mathcal{O}(t, s, z)|\psi_t(z^*)\rangle$ and constructing
the operator $\mathcal{O}(t, s, z)$ [55, 65]. For instance, within the weak coupling limit, the
operator $\mathcal{O}(t, s, z)$ is given by a perturbative expansion on the coupling constant
that leads to an approximation, up to second order, of (10.8) of the form

$$i\partial_t|\psi_t(z^*)\rangle = H_S|\psi_t(z^*)\rangle + Lz_t|\psi_t(z^*)\rangle + L^\dagger \int_0^t ds\alpha(t - s)V_{t-s}L|\psi_t(z^*)\rangle$$
$$+\text{higher orders} \,. \tag{10.12}$$

Here, we have defined the operator V_t that acts on system operators $A, B, C, ...$
as $V_t ABC... = e^{-iH_St}Ae^{iH_St}BC....$ Let us point out that higher order terms in the
expansion of $\mathcal{O}(t, s, z)$ in terms of the coupling constant may involve the noise z_t
as shown in [66]. Here the time under which the perturbative expansion is valid
decreases as the square of the coupling constant, as it is stated by the Van Hove
limit [67, 68]. A discussion in the context of a two-level system immersed in a
photonic crystal may be found in [69].

10.4 Dynamics of Multiple Time Correlation Functions

Many properties of physical systems are described by correlation functions. Among others, spectral properties of light in interaction with matter rely on the knowledge of multiple time correlation functions (MTCF). In the case of a system in interaction with an environment and within the Born–Markov approximation, it is known that the dynamics of multiple time correlation functions are ruled by the so-called quantum regression theorem [70–72, 9]. Basically, the statement is that the dynamical equations for the expectation values of system observables (one-time functions) are the same as those of two-time correlation functions. This principle is also valid to derive the dynamical equations of higher order correlations.

10.4.1 Two-Time Correlation Functions in the Markov Case: Quantum Regression Formula

Let us discuss in more detail the quantum regression theorem. Our aim is to compute the expectation value of a set of system observables $A_1, A_2, ...$ According to the discussion in Sect. 10.3 this may be done through the knowledge of the reduced density matrix of the systems, $\rho_S(t)$. Alternatively we may try to compute the expectation $\langle A_1(t) \rangle$ at any time studying its dynamics, i.e., the dynamical equation it satisfies. This equation may be obtained from Eq. (10.1) and integrated along with the initial condition $\langle A_1(0) \rangle$. The resulting linear equation is

$$\frac{d \langle A_i(t) \rangle}{dt} = \sum_j L_{ij}(t) \langle A_j(t) \rangle . \tag{10.13}$$

From the quantum regression theorem it follows that the two-time correlation functions have the same equations of motion, i.e.,

$$\frac{d \langle A_i(t + \tau) A_k(t) \rangle}{d\tau} = \sum_j L_{ij}(t) \langle A_j(t + \tau) A_k(t) \rangle . \tag{10.14}$$

This set of equations is rather convenient to compute correlations. The conditions of its validity are discussed in [73] and also in [72, 74].

For non-Markovian interactions, contrary to the Markov case, correlations verify a different set of equations than quantum mean values. In addition, these equations contain the memory kernel associated with the environment (the environment correlation function).

In the next section we shall focus on the dynamics of two-time correlation functions of system observables. While higher order correlations can also be evaluated, we find that to illustrate the theory it is enough to consider only two-time correlations. We will use the system of equations derived in [34, 75] in the weak coupling limit.

10.4.2 Two-Time Correlation Function of System Observables in the Non-Markovian Case

Let us take a set of N system observables $A_1(t_1), A_2(t_2), \ldots, A_N(t_N)$ in Heisenberg representation, such that $t_1 > t_2 > \cdots > t_N$. If Ψ_0 is the initial state of the total system, $C_A(t) = \langle \Psi_0 | A_1(t_1) A_2(t_2) \cdots A_N(t_N) | \Psi_0 \rangle$ is an N-time correlation function of the system. This is the object of our interest.

To begin let us consider the Heisenberg evolution equation for a system observable $A(t) = \mathcal{U}^{-1}(t, 0) A \mathcal{U}(t, 0)$, where $\mathcal{U}(t, 0)$ is the evolution operator with the total Hamiltonian,

$$\frac{dA(t_1)}{dt_1} = i\mathcal{U}^{-1}(t_1, 0)[H_T, A]\mathcal{U}(t_1, 0) = -i[H_S(t_1), A(t_1)]$$
$$+ i \sum_\lambda g_\lambda \left(a_\lambda^\dagger(t_1, 0)[L(t_1), A(t_1)] + [L^\dagger(t_1), A(t_1)] a_\lambda(t_1, 0) \right). \quad (10.15)$$

We can replace in (10.15) the formal solution of the evolution equation of the environmental operators, $da_\lambda(t_1, 0)/dt_1 = i[H_T(t_1), a_\lambda(t_1, 0)] = -i\omega_\lambda a_\lambda(t_1, 0) - ig_\lambda L(t_1)$,

$$a_\lambda(t_1, 0) = e^{-i\omega_\lambda t_1} a(0, 0) - ig_\lambda \int_0^{t_1} d\tau e^{-i\omega_\lambda(t_1 - \tau)} L(\tau). \quad (10.16)$$

The single evolution equation (10.15) becomes

$$\frac{dA(t_1)}{dt_1} = i[H_S(t_1), A(t_1)] - v^\dagger(t_1)[L(t_1), A(t_1)]$$
$$+ \int_0^{t_1} d\tau \alpha^*(t_1 - \tau) L^\dagger(\tau)[A(t_1), L(t_1)] + [L^\dagger(t_1), A(t_1)] v(t_1)$$
$$+ \int_0^{t_1} d\tau \alpha(t_1 - \tau)[L^\dagger(t_1), A(t_1)] L(\tau), \quad (10.17)$$

with $\alpha(t) = \sum_\lambda |g_\lambda|^2 e^{-i\omega_\lambda t}$ being the *environment correlation function*. Generally, for an environment with a large number of degrees of freedoms this function decays in a typical timescale τ_B *environment correlation time*. We have also defined the bath operators

$$v^\dagger(t_1) = -i \sum_\lambda g_\lambda a_\lambda^\dagger(0, 0) e^{i\omega_\lambda t_1},$$
$$v(t_1) = i \sum_\lambda g_\lambda a_\lambda(0, 0) e^{-i\omega_\lambda t_1}. \quad (10.18)$$

From (10.17) the evolution equation of the quantum mean value of A for an initial state of the form $| \Psi_0 \rangle = | \psi_0 \rangle | 0 \rangle$, with $|0\rangle$ the vacuum state for the environment, is equal to

$$\frac{d}{dt_1}\langle \Psi_0 \mid A(t_1) \mid \Psi_0 \rangle = i\langle \Psi_0 \mid [H_S(t_1), A(t_1)] \mid \Psi_0 \rangle$$

$$+ \int_0^{t_1} d\tau \alpha(t_1 - \tau)\langle \Psi_0 \mid [L^\dagger(t_1), A(t_1)]L(\tau) \mid \Psi_0 \rangle$$

$$+ \int_0^{t_1} d\tau \alpha^*(t_1 - \tau)\langle \Psi_0 \mid L^\dagger(\tau)[A(t_1), L(t_1)] \mid \Psi_0 \rangle .$$

$$(10.19)$$

It is important to note that in dynamical equation (10.17) the environment operators $v(t)$ and $v^\dagger(t)$ represent an external driving acting on the system due to its interaction with the environment. These *forces* are related to the correlation function $\alpha(t - t')$, in fact it can be shown that $\langle 0|v(t)v^\dagger(t')|0\rangle = \alpha(t - t')$, so that the environment correlation function is the autocorrelation function of the environment *forces* acting on the system. Furthermore, $\langle 0|v^\dagger(t)|0\rangle = \langle 0|v(t)|0\rangle = 0$.

Let us now calculate the following evolution equation:

$$\frac{dA(t_1)B(t_2)}{dt_1} = i\mathcal{U}^{-1}(t_1)[H_T, A]\mathcal{U}(t_1)B(t) = i[H_S(t_1), A(t_1)]B(t_2)$$

$$+i\sum_\lambda g_\lambda \left(a_\lambda^\dagger(t_1, 0)[L(t_1), A(t_1)]B(t_2) \right.$$

$$\left. +[L^\dagger(t_1), A(t_1)]a_\lambda(t_1, 0)B(t_2) \right) .$$

$$(10.20)$$

The idea again is to eliminate the dependence on the environmental operators once the average over the total system state is performed. First, we replace the analytical solution of the creation operator $a_\lambda^\dagger(t_1, 0)$, so that the term $a_\lambda^\dagger(0, 0)$ appears in the left hand side of the expression and can be eliminated when applying the vacuum initial state. Second, we move the annihilation operator to the right-hand side by doing the following:

$$a_\lambda(t_1, 0)B(t_2) = \mathcal{U}^{-1}(t_2)a_\lambda(t_1, t_2)B\mathcal{U}(t_2)$$

$$= \mathcal{U}^{-1}(t_2)e^{-i\omega_\lambda(t_1 - t_2)}a_\lambda(0, 0)B\mathcal{U}(t_2) - ig_\lambda \int_{t_2}^{t_1} d\tau e^{-i\omega_\lambda(t_1 - \tau)}L(\tau)B(t_2)$$

$$= B(t_2)a_\lambda(t_2, 0) - ig_\lambda \int_{t_2}^{t_1} d\tau e^{-i\omega_\lambda(t_1 - \tau)}L(\tau)B(t_2) ,$$

$$(10.21)$$

where we have used

$$a_\lambda(t_1, t_2) = e^{-i\omega_\lambda(t_1 - t_2)}a_\lambda(t_2, t_2) - ig_\lambda \int_{t_2}^{t_1} d\tau e^{-i\omega_\lambda(t_1 - \tau)}L(\tau, t_2), \quad (10.22)$$

with $a_\lambda(t_2, t_2) = a_\lambda(0, 0) \equiv a_\lambda$ and $[B, a_\lambda(0, 0)] = 0$. We now insert in the former expression the solution of $a_\lambda(t_2, 0)$, which is of the form (10.16), and obtain

$$a_\lambda(t_1, 0)B(t_2) = e^{-i\omega_\lambda t_1} B(t_2)a_\lambda(0, 0) - ig_\lambda \int_0^{t_2} d\tau e^{-i\omega_\lambda(t_1-\tau)} B(t_2)L(\tau)$$

$$-ig_\lambda \int_{t_2}^{t_1} d\tau e^{-i\omega_\lambda(t_1-\tau)} L(\tau)B(t_2) . \tag{10.23}$$

Replacing (10.23) by (10.20) and considering the solution of $a_\lambda^\dagger(t_1, 0)$, we obtain

$$\frac{dA(t_1)B(t_2)}{dt_1} = i[H_S(t_1), A(t_1)]B(t_2) - v^\dagger(t_1)[L(t_1), A(t_1)]B(t_2)$$

$$-\int_0^{t_1} d\tau \alpha^*(t_1-\tau)L^\dagger(\tau)[L(t_1), A(t_1)]B(t_2) + [L^\dagger(t_1), A(t_1)]B(t_2)v(t_1)$$

$$+\int_{t_2}^{t_1} d\tau \alpha(t_1 - \tau)[L^\dagger(t_1), A(t_1)]L(\tau)B(t_2)$$

$$+\int_0^{t_2} d\tau \alpha(t_1 - \tau)[L^\dagger(t_1), A(t_1)]B(t_2)L(\tau) . \tag{10.24}$$

The evolution of the quantum mean value $\langle A(t_1)B(t_2)\rangle$ is again obtained by applying the total initial state on both sides of the former expression. When such initial state is $| \psi_0\rangle | 0\rangle$, we obtain the following:

$$\frac{d\langle \Psi_0 | A(t_1)B(t_2) | \Psi_0\rangle}{dt_1} = i\langle \Psi_0 | [H_S(t_1), A(t_1)]B(t_2) | \Psi_0\rangle$$

$$+\int_0^{t_1} d\tau \alpha^*(t_1 - \tau)\langle \Psi_0 | L^\dagger(\tau)[A(t_1), L(t_1)]B(t_2) | \Psi_0\rangle$$

$$+\int_{t_2}^{t_1} d\tau \alpha(t_1 - \tau)\langle \Psi_0 | [L^\dagger(t_1), A(t_1)]L(\tau)B(t_2) | \Psi_0\rangle$$

$$+\int_0^{t_2} d\tau \alpha(t_1 - \tau)\langle \Psi_0 | [L^\dagger(t_1), A(t_1)]B(t_2)L(\tau) | \Psi_0\rangle . \tag{10.25}$$

Equations (10.19) and (10.25) represent the evolution of quantum mean values and two-time correlations, respectively, obtained *without the use of any approximation*. However, it is clear that these equations are open, in the sense that quantum mean values depend on two-time correlations, while two-time correlations depend on three-time correlations. In general, when no approximations are made, N-time correlation depends on $(N + 1)$-time correlations, which gives rise to a hierarchy structure of MTCF as described in [76].

At this stage we consider that the system and the environment are weakly coupled. If we define $V_t A B = e^{iH_S t} A e^{-iH_S t} B$ and $A(t) = e^{iH_T t} A e^{-iH_T t}$, we can write a weak coupling approximations of Eqs. (10.19) and (10.25) up to *second order* in the coupling constant (see [34, 75] for details).

Then we obtain the following equation for quantum mean values:

$$
\frac{d}{dt_1} \langle \Psi_0 \mid A(t_1) \mid \Psi_0 \rangle = i \langle \Psi_0 \mid \{[H_S, A]\}(t_1) \mid \Psi_0 \rangle
$$
$$
+ \int_0^{t_1} d\tau \alpha^*(t_1 - \tau) \langle \Psi_0 \mid \{V_{\tau-t_1} L^\dagger [A, L]\}(t_1) \mid \Psi_0 \rangle
$$
$$
+ \int_0^{t_1} d\tau \alpha(t_1 - \tau) \langle \Psi_0 \mid \{[L^\dagger, A] V_{\tau-t_1} L\}(t_1) \mid \psi_0 \rangle
$$
(10.26)

and for two-time correlations

$$
\frac{d}{dt_1} \langle \Psi_0 \mid A(t_1) B(t_2) \mid \Psi_0 \rangle = i \langle \Psi_0 \mid \{[H_S, A]\}(t_1) B(t_2) \mid \Psi_0 \rangle
$$
$$
+ \int_0^{t_1} d\tau \alpha^*(t_1 - \tau) \langle \Psi_0 \mid \{V_{\tau-t_1} L^\dagger [A, L]\}(t_1) B(t_2) \mid \Psi_0 \rangle
$$
$$
+ \int_0^{t_1} d\tau \alpha(t_1 - \tau) \langle \Psi_0 \mid \{[L^\dagger, A] V_{\tau-t_1} L\}(t_1) B(t_2) \mid \psi_0 \rangle
$$
$$
+ \int_0^{t_2} d\tau \alpha(t_1 - \tau) \langle \Psi_0 \mid \{[L^\dagger, A]\}(t_1) \{[B, V_{\tau-t_2} L]\}(t_2) \mid \Psi_0 \rangle .
$$
(10.27)

As noted in [34, 75], while the first two terms of (10.27) are analogous to those of (10.26), the equation for two-time correlations contains an additional term that does not appear in the evolution of quantum mean values. Note that this term vanishes for Markovian interactions, since then the correlation $\alpha(t_1 - \tau) = \Gamma \delta(t_1 - \tau)$ is 0 in the domain of integration from 0 to t_2. This result is consistent with the quantum regression theorem discussed in Sect. 10.4.1.

In conclusion, Eq. (10.27) shall be used in general to evaluate the evolution of non-Markovian two-time correlations. Moreover, as it is shown in [76], Eqs. (10.26) and (10.27) are just the first two equations of a full hierarchy of equations for multiple time correlations. This hierarchy is closed in the sense that an N-time correlation function depends at most on other N-time correlations.

Let us remark that it is possible to show that under certain conditions, multiple time correlation functions evolve as expectation values in the stationary limit, even in the non-Markovian case [77].

In the same vein that for the expectation values, it is possible to compute a particular multiple time correlation function with a stochastic scheme. We shall discuss this issue in the next section.

10.4.3 Non-Markovian Stochastic Trajectory Methods for MTCF

It is possible to develop a stochastic scheme to compute multiple time correlation functions. The advantage of this method is that it allows the evaluation of a specific correlation function, in contrast to the equations discussed in the previous sections where the evolution of a certain correlation is coupled to some other correlations.

If we take the partial interaction picture with respect to the environment, the N-time correlation function is defined as $C_A(\mathbf{t}|\Psi_0) = \Psi_0| \prod_{i=1}^{N} \mathcal{U}_I^{-1}(t_i, 0) A_i \mathcal{U}_I(t_i, 0)|\Psi_0\rangle$, where $\mathcal{U}_I$ is the evolution operator of the system in the interaction picture. Within the Bargmann representation [9, 66], we write

$$C_A(\mathbf{t}|\Psi_0) = \int d\mu(z)\langle\psi_0|G^{-1}(0, 1) \prod_{i=1}^{N} A_i G(i, i+1)|\psi_0\rangle , \qquad (10.28)$$

with $t_0 = 0$, $t_{N+1} = 0$, and $z_{N+1} = z_0$. We have introduced the *reduced propagators* $G(i, i+1) \equiv G(z_i^* z_{i+1}|t_i t_{i+1}) = \langle z_i|\mathcal{U}_I(t_i, t_{i+1})|z_{i+1}\rangle$, which act on the system Hilbert space and give the evolution of system state vectors from t_{i+1} to t_i, given that in the same time interval the environment coordinates go from z_{i+1} to z_i. It is clear then that once their time evolution is solved, the time correlation function (10.28) can be obtained. It can be shown that the reduced propagator satisfies the evolution equation [34]

$$\frac{\partial G(i, i+1)}{\partial t_i} = \left(-iH_S + Lz_{i,t_i}^* - L^\dagger z_{i+1,t_i}\right)G(i, i+1)$$

$$-L^\dagger \int_{t_{i+1}}^{t_i} d\tau \alpha(t_i - \tau)\langle z_i|\mathcal{U}_I(t_i, t_{i+1})L(\tau, t_{i+1})|z_{i+1}\rangle ,(10.29)$$

with $L(t', t) = e^{iH_B t}e^{-iH(t-t')}Le^{iH(t-t')}e^{-iH_B t'}$. Also, $z_{i,t} = i\sum_\lambda g_\lambda z_{i,n}e^{i\omega_\lambda t}$ is a time-dependent function, $\alpha(t - \tau) = \sum_\lambda |g_\lambda|^2 e^{-i\omega_\lambda(t-\tau)}$, and the initial condition $G(i, i+1) = \exp(z_i^* z_{i+1})$. Thus the function $z_{i,t}$ is a sum of time-dependent coherent states and $\alpha(t - \tau)$ is its time autocorrelation function, as it can be verified by computing the average $M[z_{i,t}z_{i,\tau}^*]$ regarding the measure $d\mu(z)$ as shown in the previous section:

$$\langle z_i|\mathcal{U}_I(t_i, \tau)L\mathcal{U}_I(\tau, t_{i+1})|z_{i+1}\rangle = \mathcal{M}_l\left[\langle z_i|\mathcal{U}_I(t_i, \tau)|z_l\rangle L\langle z_l|\mathcal{U}_I(\tau, t_{i+1})|z_{i+1}\rangle\right]$$

$$= \mathcal{M}_l\left[G(z_i^* z_l|t_i\tau)LG(z_l^* z_{i+1}|\tau t_{i+1})\right] , \qquad (10.30)$$

where in the second line we have inserted $1 = \int \frac{d^2 z}{\pi}e^{-|z|^2}|z\rangle\langle z|$, and we have defined

$$\mathcal{M}_l[\cdots] = \int d\mu(z_l) \cdots . \tag{10.31}$$

With this notation, Eq. (10.29) can be rewritten as

$$\frac{\partial G(z_i^* z_{i+1} | t_i t_{i+1})}{\partial t_i} = \left(- iH_S + Lz_{i,t_i}^* - L^\dagger z_{i+1,t_i} \right) G(z_i^* z_{i+1} | t_i t_{i+1})$$

$$-L^\dagger \int_{t_{i+1}}^{t_i} d\tau \alpha(t_i - \tau) \mathcal{M}_l \left[G(z_i^* z_l | t_i \tau) L G(z_l^* z_{i+1} | \tau t_{i+1}) \right] .$$

$$\tag{10.32}$$

In this equation, the last term expresses how the dissipation at time t depends on previous trajectories of other system propagators [78]. For that reason, Eq. (10.29) cannot in general be expressed in terms of the particular propagator evolved $G(i, i + 1)$, and hence it is not a closed equation for this propagator. Only in very exceptional cases this can be done in an exact way, while in most of the systems it is necessary to perform some approximations to close the equation.

One possible approximation is to assume that

$$\langle z_i | \mathcal{U}_I(t_i, t_{i+1}) L(\tau, t_{i+1}) | z_{i+1} \rangle = O(z_{i+1} z_i, t_{i+1}, \tau) G(i, i + 1), \tag{10.33}$$

where the operator O has to be constructed [55], for instance, by treating $L(\tau, t_{i+1})$ in the weak coupling limit. In terms of $O(z_{i+1} z_i, t_{i+1}, \tau)$, Eq. (10.29) reads

$$\frac{\partial G(i, i + 1)}{\partial t_i} = \left(- iH_S + Lz_{i,t_i}^* - L^\dagger z_{i+1,t_i} \right.$$

$$\left. -L^\dagger \int_{t_{i+1}}^{t_i} d\tau \alpha(t_i - \tau) O(z_{i+1} z_i, t_{i+1}, \tau) \right) G(i, i + 1) . \tag{10.34}$$

Equations (10.29) or (10.34) depend on two time-dependent functions, z_{i,t_i}^* and z_{i+1,t_i}, which take into account the "history" of the environment and lead to a conditioned dynamics of the system with respect to the environment dynamics. They constitute the starting point to compute the non-Markovian MTCFs within a Monte Carlo method by choosing the variables z_i randomly according to the distribution $d\mu(z)$. For a single realization, a value of the integrand appearing in (10.28) can be obtained; first, evolving $|\psi_0\rangle$ from $(t_{N+1} = 0, z_{N+1} = z_0)$ to (t_N, z_N) so that a vector $|\phi_N\rangle = G(N, N + 1)|\psi_0\rangle$, second, applying A_N to $|\phi_N\rangle$ so that we get $|\tilde{\phi}_N\rangle = A_N|\phi_N\rangle$, third, evolving $|\tilde{\phi}_N\rangle$ with $G(N - 1, N)$, and so on. The process continues until the vector $|\phi_1\rangle = G(1, 2)|\tilde{\phi}_2\rangle$ is obtained and finally it is used to compute $\langle \psi_1 | A_1 | \phi_1 \rangle$, with $|\psi_1\rangle = G(0, 1)|\psi_0\rangle$. In the end, the sum over many of these "histories" with respect to the measure $d\mu(z)$ leads to the MTCFs defined in (10.28).

Notice that since the equation for the reduced propagator (10.29) is made for an initial state of the environment different from the vacuum, it can be used to compute the expectation values and correlation functions of system observables with more general initial conditions than the one usually taken, i.e., $|\Psi_0\rangle = |\psi_0\rangle|0\rangle$ [64].

The choice between using the stochastic method and the system of equations for computing the MTCFs has to be made according to the particular problem. When an N-time correlation function has to be computed with the first method, the system of equations will contain all possible correlations of the matrices $\mathcal{Y}$ that form a basis for the QOS. The correlation of other system observables can be computed by combining correlations of this basic set of observables. In turn, the stochastic method allows us to compute only the particular correlation function that is needed, and not the whole set of $\mathcal{Y}^N$ correlations that appears interrelated in the set of differential equations. Hence, if the system has a large number of degrees of freedom, so that $\mathcal{Y}$ is a large set, the stochastic method is in general more convenient.

10.5 Examples

To illustrate the theory let us discuss the particular examples of the spontaneous emission of an atom and the fluorescence in a structured environment.

10.5.1 Atomic Emission Spectra

In this section, we derive the formula necessary to obtain the emission spectra of a two-level atom in non-Markovian interaction with the surrounding radiation field. We follow a well-known photodetection model of experiment, the *gedanken spectrum analyzer*, that provides an operational definition of the spectral profile [79]. The Hamiltonian of the emitting atom (with levels $|1\rangle$ and $|2\rangle$) is given by

$$H_S = -\frac{\omega_{12}}{2}(\sigma_{22} - \sigma_{11}) = \frac{\omega_{12}}{2}\sigma_z , \qquad (10.35)$$

where $\sigma_{i,j} = |i\rangle\langle j|$, with $\{i, j\} = 1, 2$, are the atomic pseudospin operators in the atomic basis, and the total Hamiltonian of emitting atom and radiation field is described by a Hamiltonian H_R, given by $H_R = H_S + H_B + \sum_\lambda g_\lambda(L^\dagger a_\lambda + a_\lambda^\dagger L)$. In order to detect the emitted radiation, suppose that we have a detecting atom placed in $\mathbf{r}$ with Hamiltonian $H_D = \omega\sigma_z/2$, where ω is its rotating frequency. The Hamiltonian of the total system (detector atom, emitting atom, and radiation field) is

$$H = H_D + H_R + W . \qquad (10.36)$$

Here the coupling between the detecting atom H_D with H_B is dipolar and given by a Hamiltonian W, which in the interaction picture with respect to the detector is given by

$$\tilde{W}(t) = \left[\sigma_{21}\mathbf{d}^D \cdot \mathbf{E}^{(+)}(\mathbf{r}, t)e^{i\omega t} + \sigma_{12}\mathbf{d}^D \cdot \mathbf{E}^{(-)}(\mathbf{r}, t)e^{-i\omega t}\right] . \qquad (10.37)$$

Here, we have considered $d_{21}^{\mathcal{D}}\hat{\mathbf{d}}^{\mathcal{D}} = d_{12}^{\mathcal{D}}\hat{\mathbf{d}}^{\mathcal{D}} = \langle 1 \mid \mathbf{D} \mid 2 \rangle = \mathbf{d}^{\mathcal{D}}$. The superindex $\mathcal{D}$ reminds that these are the components of the detector's dipole. It is important to note here that the field operators $\mathbf{E}^{(+)}$ and $\mathbf{E}^{(-)}$ correspond to the field emitted by the atoms and the background radiation field. The positive part of the field at the position $\mathbf{r}$ is defined as

$$\mathbf{E}^{(+)}(\mathbf{r}, \mathbf{r}_a, t) = \sum_{\lambda} \varepsilon_{\lambda} A_{\lambda}(\mathbf{r}) a_{\lambda}(\mathbf{r}_a, t) \mathbf{e}_{\lambda} \tag{10.38}$$

and $\mathbf{E}^{(-)}(\mathbf{r}, \mathbf{r}_a, t) = [\mathbf{E}^{(+)}(\mathbf{r}, \mathbf{r}_a, t)]^{\dagger}$ [80]. In the last expression (and from now on) we have added explicitly the dependence on the position $\mathbf{r}_a$ of the source dipole (or emitting atom) that originates the field. The quantity $\varepsilon_{\lambda} = \sqrt{\frac{\omega_{\lambda}}{2\varepsilon_0}}$, with υ the quantization volume. In terms of the coupling strengths we find that $g_{\lambda} \equiv g_{\lambda}(\mathbf{r}) = \varepsilon_{\lambda} A_{\lambda}(\mathbf{r})\mathbf{d} \cdot \mathbf{e}_{\lambda}$.

A shutter is placed between the radiating atom and the detector. In that way, the radiation illuminates the detector only for the time T in which the shutter is open. In order to excite the detector, the time of observation T needs to be much larger than the inverse of the natural width γ of the detecting atom's excited level. In addition, T should be larger than the reciprocal of the spectral width $1/\Gamma$ of the emitting atom. With this setup, the spectral distribution of the fluorescence light, $P(\omega, T)$, is defined as the probability of excitation of the detecting atom at the time of observation T, i.e.,

$$P(\omega, T) = Tr_{R,\mathcal{D}}(\mid 2\rangle\langle 2 \mid \rho(T)) , \tag{10.39}$$

where $\rho(T)$ is the density matrix of the total system at time T. Replacing the Taylor expansion of the density matrix $\rho(T)$ for $\rho(T) \approx \rho(0)$, and after some manipulations, $P(\omega, T)$ is obtained as

$$P(\omega, T) = \int_0^T dt \int_0^T dt' e^{i\omega(t-t')} g^{(1)}(\mathbf{r}, \mathbf{r}_a; t, t') . \tag{10.40}$$

Here, the average $\langle \cdots \rangle = Tr_R (\rho_R \cdots)$, and we have defined

$$g^{(1)}(\mathbf{r}, \mathbf{r}_a; t, t') = \langle \mathbf{d}^{\mathcal{D}} \cdot \mathbf{E}^{(-)}(\mathbf{r}, \mathbf{r}_a, t)\mathbf{d}^{\mathcal{D}} \cdot \mathbf{E}^{(+)}(\mathbf{r}, \mathbf{r}_a, t') \rangle \tag{10.41}$$

as the first-order correlation of the projection of the emitted field in the direction of the dipole. In the last expression, the operators $\mathbf{d}^{\mathcal{D}} \cdot \mathbf{E}^{(-)}(\mathbf{r}, \mathbf{r}_a, t)$ and $\mathbf{d}^{\mathcal{D}} \cdot \mathbf{E}^{(+)}(\mathbf{r}, \mathbf{r}_a, t')$ should be replaced by their expression in terms of the system operators $L^{\dagger}$ and L, respectively. This is done by inserting in (10.38), and in its complex conjugated, solution (10.16) for $a_{\lambda}^{\dagger}(t, 0)$ and $a_{\lambda}(t', 0)$, respectively. Taking into account that the term proportional to $a_\lambda^\dagger(0, 0)\, a_\lambda(0, 0)$ does not contribute to photodetection signals since the field is in the vacuum state $|0\rangle$, then

$$P(\omega, T) = \int_0^T dt \int_0^T dt' e^{i\omega(t-t')}$$

$$\times \left\{ \int_0^t d\tau \int_0^{t'} d\tau' \alpha^*(t-\tau)\alpha(t'-\tau')\langle L^\dagger(\tau)L(\tau')\rangle \right\}. \quad (10.42)$$

This formula emphasizes the role of the system fluctuations $\langle L^\dagger(\tau)L(\tau')\rangle$ in measurable quantities like the power spectrum of emitted light.

Here it has been assumed that there is no spatial dependence of the environment correlation function. More details of the derivation can be found in [81].

In the Markov case, the environmental correlation is a delta function, $\alpha(t-\tau) = \Gamma\delta(t-\tau)$, and the last formula is just

$$P(\omega, T) = \Gamma^2 \int_0^T dt \int_0^T dt' e^{i\omega(t-t')}\langle L^\dagger(t)L(t')\rangle, \quad (10.43)$$

which in the stationary limit, i.e., with an observation time $T \to \infty$, leads to the usual expression for the power spectra [9]. In addition, within the Markov approximation the system correlations $\langle L^\dagger(0)L(\tau)\rangle$ can be computed with the quantum regression theorem.

In the non-Markovian case, we cannot assume that the correlation function is a delta, and it is necessary to use the original formula, (10.42) for the spectra, and the system of equations (10.27) in order to compute the system correlations.

Let us now use formulas (10.42) and (10.43) to compute the non-Markovian and Markovian spectra, respectively [82], see Fig. 10.1. The non-Markovian case corresponds to choosing γ small enough so that the correlation function decays within a nonzero correlation time. Since we are dealing with spontaneous emission processes, in which the correlation functions $\langle L^\dagger(t)L(t')\rangle$ relax to a zero value, we choose the observing time of the detector $T > T_{CA}$, where T_{CA} is the relaxation time of the two-time correlation. Notice that when a laser is tuned to the atomic rotating frequency, then the two-time correlations do not decay to a zero value, so that the condition $T > T_{CA}$ is not sufficient to obtain a stationary spectrum. In this case it is necessary to define the spectra in a stationary limit $T \to \infty$.

In the derivation of Eq. (10.42) we have assumed that (10.41) does not depend on the spatial coordinates and therefore no spatial dependence is considered in the correlation function of the environment. However, there are systems in which it is crucial to consider this spatial dependence, for instance, where the evanescent components of the emitted field are relevant [83]. We shall see in the next section an example in which such spatial dependence has to be taken into account explicitly. The emission spectra (10.42) are then replaced by a more general expression which includes the relative position of the detector with respect to the emitting atom:

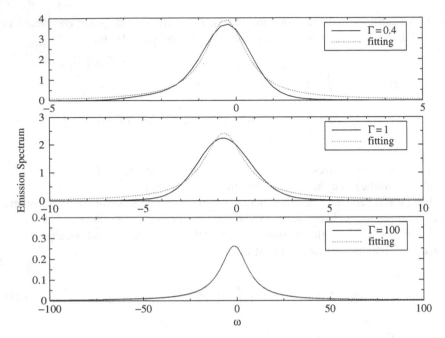

Fig. 10.1 (Color online) The spontaneous emission spectra are computed with formula (10.42) for several values of γ, and by choosing $T > T_A$, where T_A is the atomic relaxation time. In order to observe the departure from the Lorentzian profile typical of Markovian interactions, the result is numerically fitted with a Lorentzian function. When γ is small, so that non-Markovian effects are important in the atomic dynamics, the Lorentzian fitting is not appropriate, which means that the use of formula (10.42) is necessary to compute the spectra. For $\gamma = 100$ the interaction is practically Markovian, and the spectra correspond perfectly to a Lorentzian

$$
P(\omega, T) = \int_0^T dt \int_0^T dt' e^{i\omega(t-t')} g^{(1)}(\mathbf{r}, \mathbf{r}_a, t, t')
$$

$$
= \int_0^T dt \int_0^T dt' e^{i\omega(t-t')} \left\{ \int_0^t d\tau \int_0^{t'} d\tau' S^*(\mathbf{r}, \mathbf{r}_a, t, \tau) S(\mathbf{r}, \mathbf{r}_a, t', \tau') \right.
$$

$$
\left. \times \langle L^\dagger(\tau) L(\tau') \rangle \right\} ,
\tag{10.44}
$$

where

$$
S(\mathbf{r}, \mathbf{r}_a, t, \tau) = \sum_\lambda g_\lambda^{\mathcal{D}} g_\lambda e^{-i\omega_\lambda(t-\tau)} e^{i(\mathbf{r}-\mathbf{r}_a)\mathbf{k}}
\tag{10.45}
$$

is the spatially dependent correlation function. In order to compute the fluorescence spectra, the limit of $T \to \infty$ has to be taken, so that the signal is observed in the stationary limit. In that case, the above formula corresponds to a double Laplace transform of a convolution,

$$
P(\omega) = S^*(\mathbf{r}, \mathbf{r}_a, -\omega) S(\mathbf{r}, \mathbf{r}_a, \omega) \langle L^\dagger(-\omega) L(\omega) \rangle .
\tag{10.46}
$$

10.5.2 Fluorescence in a Structured Environment

We shall focus in this section on the fluorescence spectra of a two-level atom immersed in a structured environment. We shall use the results of previous sections, which are valid for an arbitrary system, provided that it is linearly and weakly coupled to its environment. If the atom is additionally driven by a laser, the coupling operator L appearing in the Hamiltonian (10.4) has a particular form, which we shall derive in the following.

Let us first consider the interaction Hamiltonian of the atom with a classical laser field, which in the rotating wave approximation can be written as follows [79, 18]:

$$H_{SL} = \varepsilon(\sigma_{21}e^{-i(\omega_L t + \phi_T)} + \sigma_{12}e^{i(\omega_L t + \phi_T)}) , \tag{10.47}$$

where ω_L is the frequency of the laser, ε its Rabi frequency, and ϕ_T its phase. Because of the magnitude of the laser field, the Hamiltonian H_{SL} should be considered as part of the noninteracting Hamiltonian H_0, so that now $H_0 = H_S + H_B + H_{SL}$. The explicit time dependence of the Hamiltonian can be eliminated by applying the following unitary operator:

$$U_t = e^{[i\omega_L t + i\phi_T]\left[\sum_\lambda a_\lambda^\dagger a_\lambda + (\sigma_{22} - \sigma_{11})\right]} . \tag{10.48}$$

This operation transforms the Hamiltonian into the rotating frame of the laser, where it can be written as $H' = H_0' + H_I'$. Here,

$$H_0' = \sum_\lambda \Delta_\lambda a_\lambda^\dagger a_\lambda + \frac{1}{2}\omega_S\sigma_3 + \varepsilon[\sigma_{21} + \sigma_{12}] , \tag{10.49}$$

with $\Delta_\lambda = \omega_\lambda - \omega_L$, and

$$H_I' = i \sum_\lambda g_\lambda(\sigma_{12}a_\lambda^\dagger - a_\lambda\sigma_{21}) . \tag{10.50}$$

A new unitary operation V,

$$V = \begin{pmatrix} c & -s \\ s & c \end{pmatrix} \tag{10.51}$$

may transform the system into a dressed picture, where the Hamiltonian $\tilde{H} = V^{-1}H'V$ has the form (10.4). The constants appearing in the transformation matrix V are $c = \cos\phi$ and $s = \sin\phi$, where the angle, ϕ, is given by $\sin^2\phi = \frac{1}{2}[1 - sgn(\Delta_{SL})/\sqrt{\varepsilon^2/\Delta_{SL}^2) + 1}]$, with $\Delta_{SL} = \omega_S - \omega_L$. The noninteracting dressed state Hamiltonian $\tilde{H}_0 = \tilde{H}_S + \tilde{H}_B + \tilde{H}_{SL}$ is equal to

$$\tilde{H}_0 = \Omega R_3 + \sum_\lambda \Delta_\lambda a_\lambda^\dagger a_\lambda \qquad (10.52)$$

and the interaction Hamiltonian $\tilde{H}_I$ has the same form as in (10.4) once the interaction operator L is defined as

$$L = \mathbf{cs}R_3 + \mathbf{c}^2 R_{12} - \mathbf{s}^2 R_{21} . \qquad (10.53)$$

Here, $R_{ij} = |\tilde{i}\rangle\langle\tilde{j}|$ are the atomic operators defined in the dressed state basis $\{|\tilde{1}\rangle, |\tilde{2}\rangle\}$, and $R_3 = R_{22} - R_{11}$. The quantity $\Omega = [\varepsilon^2 + \Delta_{SL}^2/4]^{1/2}$ is the so-called generalized Rabi frequency.

Once the coupling operator L is known, the two-time correlation $\langle L^\dagger(\tau)L(\tau')\rangle$ can be easily computed with Eq. (10.27). Note that the form of L is such that the last term of (10.27) is nonzero, and therefore the quantum regression theorem is not fulfilled.

In order to obtain the emission spectra with (10.46), it is necessary to derive the Laplace transform of the spatial-dependent correlation function, $S^*(\mathbf{r}, \mathbf{r}_a, \omega)$.

We should then consider the explicit form of the coupling constants appearing in (10.4) which for a dipolar coupling is

$$g_\lambda = -\mathrm{i}\sqrt{\frac{1}{2\hbar\varepsilon_0\omega_\lambda}}\omega_{12}\hat{\mathbf{e}}_{\mathbf{k}\sigma} \cdot \mathbf{d}_{12}\mathrm{e}^{\mathrm{i}\mathbf{k}\cdot\mathbf{r}} , \qquad (10.54)$$

where $\mathbf{d}_{12}$ is the atom dipolar moment, $\hat{\mathbf{e}}_{\mathbf{k}\sigma}$ is the unit vector in the direction of the polarization σ for a given wave vector $\mathbf{k}$, and ε_0 is the electric vacuum permittivity.

By using this definition, and following Eq. (10.45), we shall now write

$$S(\mathbf{r}, \mathbf{r}_a, \tau) = \gamma(\frac{a}{2\pi})^3 \sum_\sigma \int_{1BZ} \mathrm{d}\mathbf{k}\frac{|\hat{e}_{\mathbf{k},\sigma} \cdot \hat{u}_d||\hat{e}_{\mathbf{k},\sigma} \cdot \hat{u}_d^{\mathcal{D}}|}{\omega(\mathbf{k})}\mathrm{e}^{-\mathrm{i}\omega(\mathbf{k})\tau}\mathrm{e}^{\mathrm{i}(\mathbf{r}-\mathbf{r}_a)\cdot\mathbf{k}} , \quad (10.55)$$

where $\hat{u}_d^{\mathcal{D}}$ is the unitary vector corresponding to the dipolar moment of the detector, $\gamma = \omega_{12}^2 d_{12}^2/(2\varepsilon_0\hbar)$, and the integrals are performed over the first Brillouin zones of the crystal. In order to calculate the integrals appearing in (10.55), it is clear that the dispersion relation $\omega(\mathbf{k})$ is needed.

Analogously to the tight binding model in solid state physics, the dynamics near the edge of the band is often described through an effective mass approximation. This approximation is based on an expansion of the full dispersion relation in the vicinity of the band edge (see for instance [84–87]). Hence, considering the simplest case of a cubic lattice, the dispersion relation has the parabolic form $\omega(\mathbf{k}) = \omega_c + \mathcal{A}(\mathbf{k} - \mathbf{k_0})^2$, where $\mathbf{k_0}$ is the origin of the first Brillouin zone of the crystal (which is the unitary cell in $\mathbf{k}$ space) about which we perform the expansion in each direction, ω_c is the frequency of the band edge, and $\mathcal{A}$ is a constant that depends on the specific photonic crystal considered.

In order to calculate the spatial-dependent correlation function, $S(\mathbf{r}, \mathbf{r}_a, t, \tau)$, we assume that the quantity $\frac{|\hat{e}_{\mathbf{k},\sigma} \cdot \hat{u}_d||\hat{e}_{\mathbf{k},\sigma} \cdot \hat{u}_d^{\mathcal{D}}|}{\omega(\mathbf{k})}$ is a slowly varying function in the reciprocal space (particularly in the region nearby the symmetric point $\mathbf{k}_0$). In that case, we can express function (10.55) as

$$S(\mathbf{r}, \mathbf{r}_a, \tau) = \gamma \left(\frac{a}{2\pi}\right)^3 \sum_\sigma \frac{|\hat{e}_{\mathbf{k}_0,\sigma} \cdot \hat{u}_d||\hat{e}_{\mathbf{k}_0,\sigma} \cdot \hat{u}_d^{\mathcal{D}}|}{\omega(\mathbf{k}_0)} \int_{1BZ} d\mathbf{k} e^{-i\omega(\mathbf{k})\tau} e^{i(\mathbf{r}-\mathbf{r}_a)\cdot\mathbf{k}}. \quad (10.56)$$

Its Laplace transform is then given by

$$S(\mathbf{r}, \mathbf{r}_a, \omega) = \gamma \left(\frac{a}{2\pi}\right)^3 \sum_\sigma \frac{|\hat{e}_{\mathbf{k}_0,\sigma} \cdot \hat{u}_d||\hat{e}_{\mathbf{k}_0,\sigma} \cdot \hat{u}_d^{\mathcal{D}}|}{\omega(\mathbf{k}_0)} \int_{1BZ} d\mathbf{k} \frac{e^{i(\mathbf{r}-\mathbf{r}_a)\cdot\mathbf{k}}}{i(\omega(\mathbf{k}) - \omega)}. \quad (10.57)$$

In order to perform the last integrals analytically, we consider the parabolic dispersion relation and the limit in which the detector and the emitting atom are far away from each other. In that way, we can make the change of variable $\mathbf{q} = \mathbf{k} - \mathbf{k}_0$, so that we have $\omega(\mathbf{k}) = \omega_c + \mathcal{A}\mathbf{q}^2$ and extend the limits of integration to infinity. Then

$$S(\mathbf{r}, \mathbf{r}_a, \omega) = -Q \frac{2\pi}{id} \int_0^\infty dq\, q \frac{e^{ikd} - e^{-ikd}}{\omega(q) - \omega}, \quad (10.58)$$

here we have assumed that $\mathbf{q} = q(\sin\theta\cos\phi, \sin\theta\sin\phi, \cos\theta)$ and $\mathbf{d} = \mathbf{r} - \mathbf{r}_a$ being parallel to the z-axis. The last consideration is not too restrictive. Any arbitrary rotation of the detector (i.e., of $\mathbf{d}$) may give rise to the same result of the integral, which is performed along the whole reciprocal space. The constant

$$Q = \gamma \left(\frac{a}{2\pi}\right)^3 e^{i\mathbf{k}_0\cdot\mathbf{d}}\left[1 - \frac{(\mathbf{k}_0 \cdot \mathbf{u}_d)^2}{k_0^2}\right]\left[1 - \frac{(\mathbf{k}_0 \cdot \mathbf{u}_d^{\mathcal{D}})^2}{k_0^2}\right], \quad (10.59)$$

provided that the $d_{12}^{\mathcal{D}} = d_{12}$. Considering that θ and $\theta^{\mathcal{D}}$ are the angles between $\mathbf{k}_0$ and $\mathbf{u}_d$ and $\mathbf{u}_d^{\mathcal{D}}$, respectively, one finds

$$Q = \gamma \left(\frac{a}{2\pi}\right)^3 e^{ik_0d} \sin^2\theta \sin^2\theta^{\mathcal{D}}. \quad (10.60)$$

Performing the last integral of (10.58), we just obtain

$$S(\mathbf{r}, \mathbf{r}_a, \omega) = \frac{Q2\pi^2}{id\mathcal{A}} e^{-d/l}, \quad (10.61)$$

where $l = 1/\sqrt{\frac{\omega - \omega_c}{\mathcal{A}}}$ is the so-called localization length. Note that there are two possibilities: if $\omega_c < \omega$ then S is proportional to $\exp(-id/l)$ and following Eq. (10.46) the power spectrum decays as d^{-2}; if $\omega_c > \omega$, S is proportional to $\exp(-d/|l|)$, and the power spectrum decays exponentially. Hence, emitted photons with frequencies within the gap will not contribute to the power spectra if the detector is placed far away from the emitter. The reason is that in the long-time limit, all photons within the gap are emitted in the form of evanescent modes and remain spatially localized nearby the emitting atom (see Fig. 10.2). In addition, increasing the frequency detuning with respect to the band edge, which we assume placed at $\omega_c = 0$, gives rise to a decreasing localization length l, and therefore to a faster spatial decaying of the function $\mathcal{S}$.

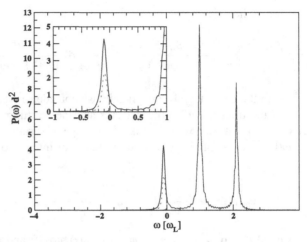

Fig. 10.2 (Color online) The power spectrum $P(\omega)$ times d^2 is considered for different values of the detector distance d, which is measured in units of $\sqrt{\mathcal{A}/\omega_L}$. *Solid line*: $d = 0.1$, *dotted line*: $d = 10$, and *dashed line*: $d = 100$. The Rabi frequency is $\Omega = 0.55$, so that the left sideband of the Mollow triplet is within the gap, while the right sideband lies within the band. When the detector is placed sufficiently far away from the emitting atom, the left sideband is no longer detected. More specifically, no emission is detected within the gap region

10.6 Discussion and Conclusions

In the last few decades the development of new systems, like quantum dots and atom lasers, and the engineering of new materials, like photonic crystals, have given rise to phenomena which cannot be described by considering the Markov approximation. This has produced a growing interest in developing a theory of non-Markovian quantum open systems, where a finite relaxation time for the surrounding environment is considered. Such finite relaxation time turns out to play an important role, giving rise to some important memory effects in the evolution of the quantum open system.

Memory effects are observed in all the system dynamical quantities, particularly in its expectation values and its fluctuations, which are encoded by multiple time correlation functions. In this chapter, we have discussed the different methods that exist to compute such quantities. Using a stochastic Schrödinger equation approach, and developing the dynamical equations that quantum observables and correlations of the system satisfy. Both approaches are complementary to each other, and the choice between them should be made according to the specific needs we have.

Apart from the theoretical interest, the importance of multiple time correlation functions relies on the fact that many measurable quantities are related to them. For instance, an atom interacting with the electromagnetic field emits some radiation that can be detected in the laboratory. In many experiments, the measured quantity is the number of photons, a quantity that turns out to be proportional to a certain two-time correlation function of system observables. Henceforth, predicting this experimental result requires a theory to describe not only quantum mean values but also multiple time correlation functions. For the Markov case there are well-developed tools to compute multiple time correlation functions, the most relevant being the so-called quantum regression theorem. For the non-Markovian case, the structure of the evolution equations of system fluctuations is much more complicated, since some memory effects should be taken into account. In general, non-Markovian fluctuations do not follow the predictions of the quantum regression theorem, an issue that has been widely treated in the literature and we have discussed in this chapter.

Acknowledgments We would like to thank J.G. Muga for his kind invitation to contribute to this volume and its invaluable support. We thank H. Carmichael, G.C. Hegerfeldt, A. Ruíz, and L.S. Schulman for their comments at different stages of this work and G. Nicolis, P. Gaspard, J.I. Cirac, and W.T. Strunz for support and encouragement. This work has been supported by Ministerio de Ciencia y Tecnología of Spain (FIS2007-64018) and by the EU projects CONQUEST and SCALA.

References

1. J. Gambetta, H. Wiseman, Phys. Rev. A **66**, 012108 (2002)
2. M. Esposito, P. Gaspard, Phys. Rev. E **76**, 041134 (2007)
3. C. Flindt, T. Novotny, A. Braggio, M. Sassetti, A.-Pekka Jauho, Phys. Rev. Lett. **100**, 150601 (2008)
4. D. Lidar, Phys. Rev. Lett. **100**, 160506 (2008)
5. F. Verstraete, M.M. Wolf, J.I. Cirac, arXiv:0803.1447 (2008)
6. B. Kraus, H.P. Büchler, S. Diehl, A. Kantian, A. Micheli, P. Zoller, Phys. Rev. A **78**, 042307 (2008)
7. S. Diehl, A. Micheli, A. Kantian, B. Kraus, H.P. Büchler, P. Zoller, arXiv:0803.1482 (2008)
8. H.J. Carmichael, *An Open Systems Approach to Quantum Optics*, Lecture Notes in Physics, Monographs Series, vol. 18 (Springer-Verlag, Berlin, 1993)
9. H.J. Carmichael, *Statistical Methods in Quantum Optics 1*. Texts and Monographs in Physics (Springer, Berlin, 1999)
10. B.M. Garraway, B.J. Dalton, J. Phys. B **39**, S767 (2006)
11. H.P. Breuer, F. Petruccione, Phys. Rev. E **76**, 016701 (2007)
12. B. Bellomo, R.L. Franco, G. Compagno, Phys. Rev. Lett. **99**, 160502 (2007)
13. C. Lazarou, G.M. Nikolopoulos, P. Lambropoulos, J. Phys. B: At. Mol. Opt. Phys. **40**, 2511 (2007)
14. E. Purcell, Phys. Rev. **69**, 681 (1946)
15. D. Kleppner, Phys. Rev. Lett. **47**, 233 (1981)
16. E. Yablonovitch, Phys. Rev. Lett. **58**, 2059 (1987)
17. S. John, Phys. Rev. Lett. **58**, 2486 (1987)

18. S. John, T. Quang, Phys. Rev. A **50**, 1764 (1994)
19. B. Gaveau, L.S. Schulman, J. Phys. A: Math. Gen. **28**, 7359 (1995)
20. M. Lewenstein, J. Zakrzewski, T.W. Mossberg, Phys. Rev. A **38**, 808 (1988)
21. S. Maniscalco, F. Petruccione, Phys. Rev. A **73**, 12111 (2006)
22. G. Lindblad, Commun. Math. Phys. **48**, 119 (1976)
23. G.V. Gorini, A. Kossakowski, E.C.G. Sudarshan, J. Math. Phys. **17**, 821 (1976)
24. A.G. Redfield, IBM J. Res. Dev. **1**, 19 (1957)
25. A.G. Redfield, Adv. Magn. Reson. **1**, 1 (1965)
26. P. Gaspard, M. Nagaoka, J. Chem. Phys. **111**, 5676 (1999)
27. P. Gaspard, M. Nagaoka, J. Chem. Phys. **111**, 5668 (1999)
28. R. Zwanzig, J. Chem. Phys. **33**, 1338 (1960)
29. S. Nakajima, Prog. Theor. Phys. **20**, 948 (1958)
30. I. Prigogine, *Non-Equilibrium Statistical Mechanics* (John Wiley & Sons Inc., New York, 1962)
31. W.T. Strunz, T. Yu, Phys. Rev. A **69**, 052115 (2004)
32. H.P. Breuer, B. Kappler, F. Petruccione, Ann. Phys. **291**, 36 (2001)
33. J.D. Cresser, Laser Phys. **10**, 337 (2000)
34. D. Alonso, I. de Vega, Phys. Rev. Lett. **94**, 200403 (2005)
35. C. Gardiner, P. Zoller, *Quantum Noise*. Springer Series in Synergetics (Springer-Verlag, Berlin, 2004)
36. H. Breuer, F. Petruccione, *The Theory of Open Quantum Systems* (Oxford University Press, New York, 2002)
37. M. Orszag, *Quantum Optics: Including Noise Reduction, Trapped Ions, Quantum Trajectories, and Decoherence*, 2nd edn (Springer, Berlin, 2008)
38. P. Lambropoulos, D. Petrosyan, *Fundamentals of Quantum Optics and Quantum Information* (Springer, Berlin, 2006)
39. A. Barchielli, V. Belavkin, J. Phys. A: Math. Gen. **24**, 1495 (1991)
40. J. Dalibard, Y. Castin, K. Molmer, Phys. Rev. Lett. **68**, 580 (1992)
41. R. Dum, P. Zoller, H. Ritsch, Phys. Rev. A **45**, 1879 (1992)
42. G.C. Hegerfeldt, M.B. Plenio, Phys. Rev. A **56**, 2334 (1997)
43. M.B. Plenio, P.L. Knight, Rev. Mod. Phys. **70**, 101 (1998)
44. N. Gisin, I.C. Percival, J. Phys. A: Math. Gen. **25**, 5677 (1992)
45. N. Gisin, I.C. Pecival, J. Phys. A: Math. Gen. **26**, 2233 (1993)
46. N. Gisin, I.C. Percival, J. Phys. A: Math. Gen. **26**, 2245 (1993)
47. N. Gisin, P. Knight, I. Percival, R. Thompson, D. Wilson, J. Mod. Opt. **40**, 1663 (1993)
48. H.M. Wiseman, G.J. Milburn, Phys. Rev. A **47**, 1652 (1993)
49. H.M. Wiseman, G.J. Milburn, Phys. Rev. A **47**, 642 (1993)
50. T.B.L. Kist, M. Orszag, T.A. Brun, L. Davidovich, J. Opt. B: Quantum Semicl. Opt. **1**, 251 (1999)
51. H. Breuer, F. Petruccione, Fortschr. Phys./Prog. Phys. **45**, 39 (1997)
52. W. Strunz, Phys. Lett. A **224**, 25 (1996)
53. L. Diósi, W.T. Strunz, Phys. Lett. A **235**, 569 (1997)
54. L. Diósi, N. Gisin, W. Strunz, Phys. Rev. A **58**, 1699 (1998)
55. T. Yu, L. Diósi, N. Gisin, W.T. Strunz, Phys. Rev. A **60**, 91 (1999)
56. W.T. Strunz, L. Diósi, N. Gisin, Phys. Rev. Lett. **82**, 1801 (1999)
57. W.T. Struntz, L. Diósi, N. Gisin, T. Yu, Phys. Rev. Lett. **83**, 4909 (1999)
58. M.W. Jack, M.J. Collet, Phys. Rev. A **61**, 062106 (2000)
59. H.P. Breuer, B. Kappler, F. Petruccione, Phys. Rev. A **59**, 1633 (1999)
60. H.P. Breuer, Eur. Phys. J. D **29**, 105 (2004)
61. L. Diósi, Phys. Rev. Lett. **100**, 080401 (2008)
62. H.M. Wiseman, J.M. Gambetta, Phys. Rev. Lett. **101**, 140401 (2008)
63. L. Diósi, Phys. Rev. Lett. Erratum **101**, 149902 (2008)
64. I. de Vega, D. Alonso, P. Gaspard, W.T. Strunz, J. Chem. Phys. **122**, 124106 (2005)

65. W. Struntz, unpublished (2001)
66. W.T. Strunz, Chem. Phys. **268**, 237 (2001)
67. L. Van Hove, Physica **21**, 517 (1955)
68. R. Zwanzig, *Nonequilibrium Statistical Mechanics* (Oxford University Press, Oxford, 2001)
69. I. de Vega, D. Alonso, P. Gaspard, Phys. Rev. A **71**, 23812 (2005)
70. M. Lax, Phys. Rev. **129**, 2342 (1963)
71. M. Lax, Phys. Rev. A **172**, 350 (1968)
72. M. Lax, Opt. Comm. **179**, 463 (2000)
73. G.W. Ford, R.F. O'Connell, Phys. Rev. Lett. **77**, 798 (1996)
74. G.W. Ford, R.F. O'Connell, Opt. Comm. **179**, 451 (2000)
75. I. de Vega, D. Alonso, Phys. Rev. A **73**, 22102 (2006)
76. D. Alonso, I. de Vega, Phys. Rev. A **75**, 52108 (2007)
77. A.A. Budini, J. Stat. Phys. **131**, 51 (2008)
78. A.A. Budini, Phys. Rev. A **63**, 012106 (2001)
79. M.O. Scully, M.S. Zubairy, *Quantum Optics* (Cambridge University Press, Cambridge, 1997)
80. C. Cohen-Tannoudji, J. Dupont-Roc, G. Grynberg, *Atom-Photon Interactions. Basic Processes and Applications* (Willey Interscience, New York, 1992)
81. I. de Vega, D. Alonso, Phys. Rev. A **77**, 043836 (2008)
82. D. Alonso, I. de Vega, E. Hernández-Concepción, Comptes Rendus-Physique **8**, 684 (2007)
83. Y. Yang, S.Y. Zhu, Phys. Rev. A **62**, 013805 (2000)
84. M. Florescu, S. John, Phys. Rev. A **64**, 033801 (2001)
85. S. John, T. Quang, Phys. Rev. Lett. **74**, 3419 (1995)
86. S. John, J. Wang, Phys. Rev. Lett. **64**(20), 2418 (1990)
87. S. John, J. Wang, Phys. Rev. B **43**, 12772 (1991)

Chapter 11
Double-Slit Experiments in the Time Domain

Gerhard G. Paulus and Dieter Bauer

11.1 Introduction

Writing about "double-slits in time" bears the risk of entering mined area. It is thus advisable to first set the stage by describing the kind of experiments, either real ones or gedankenexperiments, we are going to discuss.

Let us consider some massive particle(s) fulfilling a nonrelativistic, time-dependent Schrödinger equation. Initially, the particle(s) are confined inside some region, e.g., a particle beam impinging on a shutter, atoms in a magneto-optical trap (MOT), or electrons in an atom. At later times particle(s) are allowed to leave the region upon passing through externally controlled "time slits." Such time slits can be realized by chopping or modulating a particle beam [62, 4, 23, 29], using laser light as an ultrafast atomic mirror [63, 3], or ionizing atoms using intense laser pulses [41], respectively.

Moshinsky studied in the seminal 1952 paper entitled "Diffraction in Time" [49] the case of a perfectly absorbing shutter at $z = 0$, blocking a particle beam of wave vector k_z along the z-axis $\langle z|\Psi(t = 0)\rangle = \exp[ik_z z]\Theta(-z)$ (with Θ the Heaviside step function) for times $t < 0$ but removed suddenly at $t = 0$. Moshinsky showed that for $t > 0$

$$|\langle z|\Psi(t)\rangle|^2 = \frac{1}{2}\left\{\left[\frac{1}{2} + C(\xi)\right]^2 + \left[\frac{1}{2} + S(\xi)\right]^2\right\} \qquad (11.1)$$

holds, where $\xi = (k_z t - z)/(\sqrt{\pi}t)$ and C and S are the Fresnel integrals (units are used where $\hbar$ and the mass of the particle are unity). Plotting the probability

G.G. Paulus (✉)
Institute of Optics and Quantum Electronics, Friedrich Schiller University of Jena, Jena 07743, Germany; Department of Physics, Texas A&M University, College Station, TX 77843-4242, USA, ggpaulus@ioq.uni-jena.de

D. Bauer
Max-Planck-Institut für Kernphysik, Postfach 103980, Heidelberg 69029, Germany, dbauer@mpi-k.de

Paulus, G.G., Bauer, D.: *Double-Slit Experiments in the Time Domain*. Lect. Notes Phys. **789**, 303–339 (2009)
DOI 10.1007/978-3-642-03174-8_11

density (11.1) at a given spatial point z' as a function of *time* one obtains the diffraction pattern well known from *spatial* diffraction at a straight edge (see, e.g., [8]). Times $t < z'/k_z$, i.e., times less than the "time of flight" to the chosen position z', correspond to the "shadow" region where the probability that a particle yet arrived is rather small (although nonzero). However, for times $t > z'/k_z$, the probability density does not simply assume the constant value it has for $z < 0$ but oscillates around this value, exactly as the light intensity does in the illuminated region behind a straight edge. This diffraction in time has been revisited several times in the literature. In [45] it is analyzed in terms of Wigner distributions and tomographic probabilities. The effect of the aperture function on the temporal diffraction pattern was studied in [19], and in the very recent reference [20], Moshinsky's shutter was replaced by a moving mirror, which offers the possibility to slow down or accelerate atom beams and may be useful for atom interferometry in the time domain. A valuable review on exact solutions for time-dependent phenomena of such kind including concise introductions into the relevant theoretical methods can be found in [39].

While spatial diffraction also occurs for massive particles, diffraction in time does not occur for light. This – perhaps at first sight puzzling – asymmetry is due to the fact that only for stationary solutions both the Schrödinger and the wave equations reduce to the Helmholtz equation and thus have formally equivalent solutions. For time-dependent phenomena the different dispersion of matter and light waves yields different dynamics[1] [10, 11].

When talking or writing about "interference in time" a common objection one encounters is that this cannot be because time is not an operator, at least in "mainstream" quantum theory [30, 31]. As a consequence, the "occurrence time" of physical events is not an observable and thus does not fulfill an uncertainty relation with the energy (i.e., usually the Hamiltonian) of the system. However, aside from the fact that interference per se does not require operators, uncertainty relations between times and energies emerge for reasons different from noncommuting operators in many quantum mechanical calculations. In fact, in the theoretical analysis of Sects. 11.2 and 11.4, we shall also meet (at least) two variants of the time–energy uncertainty relation: peaks in energy spectra become narrower if the interaction time increases and interference structures in energy depend on the time delay which characterizes the preparation of the quantum state, in our case the delay between two or more "slits" in time. An excellent overview of the multifaceted aspects of the time–energy uncertainty relation, in general, is given in Chap. 3 of the first volume of this book series [12] (see also [28], the classic papers [2, 44], and, e.g., [50, 38]).

11.2 Wave Packet Interference in Position and Momentum Space

The theoretical analysis in this section and later on in Sect. 11.4 is performed for electronic wave packets using atomic units. The generalization to any kind of matter waves is straightforward (see also [11]).

---

[1] More mathematically, the first time derivative in the Schrödinger equation and the second time derivative in the wave equation yield different dynamics.

The wave function of a free electronic Gaussian wave packet fulfilling the time-dependent Schrödinger equation (TDSE)

$$i\frac{\partial}{\partial t}|\psi_{\mathbf{k}}(t)\rangle = \frac{1}{2}\hat{\mathbf{p}}^2|\psi_{\mathbf{k}}(t)\rangle \tag{11.2}$$

reads in position space

$$\langle \mathbf{r}|\psi_{\mathbf{k}}(t)\rangle = \frac{A}{\left(1+it/a^2\right)^{3/2}} \exp\left[-\frac{r^2 - 2ia^2\mathbf{k}\cdot\mathbf{r} + ia^2k^2t}{2a^2\left(1+it/a^2\right)}\right] \tag{11.3}$$

and in momentum space[2]

$$\langle \mathbf{p}|\psi_{\mathbf{k}}(t)\rangle = Aa^3 \exp\left[-\frac{1}{2}a^2(\mathbf{k}-\mathbf{p})^2 - \frac{1}{2}ip^2t\right]. \tag{11.4}$$

Here, A is a normalization constant, $\mathbf{k}$ is the center-of-mass momentum of the wave packet (i.e., equal to the group velocity in atomic units), and a determines the width of the wave packet.

11.2.1 Double-Slit in Space

Let us first analyze the "standard" double-slit in space with the help of Gaussian wave packets. We assume that two wave packets of equal width and amplitudes start at $t = 0$ with equal center-of-mass momentum $\mathbf{k} = k_z\mathbf{e}_z$, $k_z > 0$, at the two spatial positions $\mathbf{r} = \pm d\mathbf{e}_x$. The wave function at a later time $t > 0$ reads in position space, according to (11.3),

$$\langle \mathbf{r}|\psi_{k_z}(t)\rangle = \frac{A}{\left(1+it/a^2\right)^{3/2}} \exp\left[-\frac{y^2 + z^2 - 2ia^2k_z z + ia^2k_z^2 t}{2a^2\left(1+it/a^2\right)}\right]$$

$$\times \left\{\exp\left[-\frac{(x-d)^2}{2a^2\left(1+it/a^2\right)}\right] + \exp\left[-\frac{(x+d)^2}{2a^2\left(1+it/a^2\right)}\right]\right\} \tag{11.5}$$

and in momentum space, according to (11.4),

---

[2] For an explicit Fourier transformation the integral

$$\frac{1}{\sqrt{2\pi}}\int_{-\infty}^{\infty} dx\, \exp[-ip_x x]\, \exp[-\alpha x^2 + \beta x - \gamma] = \frac{1}{\sqrt{2\alpha}}\exp[-\gamma]\, \exp\left[\frac{(\beta - ip_x)^2}{4\alpha}\right]$$

(for $\mathrm{Re}\,\alpha > 0$) is useful.

$$\langle \mathbf{p} | \psi_{k_z}(t) \rangle = 2Aa^3 \exp\left[-\frac{1}{2}a^2(k_z - p_z)^2 - \frac{1}{2}a^2(p_x^2 + p_y^2) - \frac{1}{2}ip^2t \right] \cos p_x d \; .$$

(11.6)

For the probability density in position space one obtains the lengthy and time-dependent expression

$$|\langle \mathbf{r} | \psi_{k_z}(t) \rangle|^2 = |A|^2 \rho_a(y, t)\rho_a(z - k_z t, t)\{\rho_a(x - d, t) + \rho_a(x + d, t) + I(x, t)\} \; ,$$

(11.7)

with

$$\rho_a(\xi, t) = \left[1 + (t/a^2)^2 \right]^{-1/2} \exp\left[-\frac{\xi^2}{a^2[1 + (t/a^2)^2]} \right]$$

(11.8)

and the interference term

$$I(x, t) = 2\left[1 + (t/a^2)^2 \right]^{-1/2} \exp\left[-\frac{x^2 + d^2}{a^2[1 + (t/a^2)^2]} \right] \cos\left[\frac{2xdt/a^2}{a^2[1 + (t/a^2)^2]} \right] \; .$$

(11.9)

In momentum space the probability density is stationary and simply given by

$$|\langle \mathbf{p} | \psi_{k_z} \rangle|^2 = 4|A|^2 a^6 \exp\left[-a^2(k_z - p_z)^2 - a^2(p_x^2 + p_y^2) \right] \cos^2 p_x d \; , \quad (11.10)$$

from which the position of the interference maxima and minima can be immediately inferred. The function $\cos^2 \varphi$ has maxima (zeros) at $\varphi = n\pi$ ($\varphi = (2n + 1)\pi/2$), $n = 0, 1, 2, \ldots$. Hence we have

$$p_x d = n\pi \qquad \text{(maxima)} \; , \tag{11.11}$$

$$p_x d = (2n + 1)\frac{\pi}{2} \qquad \text{(minima)} \tag{11.12}$$

(note that the slit separation is $2d$ in our case). For $t/a^2 \gg 1$ the interference term (11.9) is proportional to $\cos(2dx/t)$ so that interference maxima (minima) in (11.7) are expected for $2dx/t = 2n\pi$ (for $2dx/t = (2n + 1)\pi$). Identifying x/t with the lateral momentum p_x this result agrees, of course, with (11.11) and (11.12). On the other hand, with $k_z = z/t = 2\pi/\lambda$ one obtains the condition for constructive interference as usually given in textbooks, i.e., in terms of the wavelength λ, the distance between the slits and the screen $L = z$, and the fringe distance from the central maximum x,

$$\frac{n\lambda}{2d} = \frac{x}{L} \qquad \text{(constructive interference)} \; . \tag{11.13}$$

Figure 11.1 illustrates the interference of two Gaussian wave packets in position space.

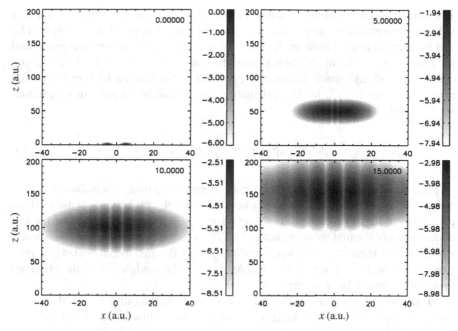

Fig. 11.1 Interference of two Gaussian wave packets in position space. The wave packets of width a and amplitude A start at $x = \pm d$ with equal momentum k_z in z-direction. The actual parameters are $A = a = 1$, $k_z = 10$, and $d = 5$. The contours of the probability density are scaled logarithmically. The time t is given in the *upper right* of each *panel*

11.2.2 Double-Slit in Time

Let us now consider two Gaussian wave packets "born" at the same spatial position $\mathbf{r} = 0$ with vanishing group velocity $\mathbf{k} = 0$ and the same amplitude A and initial width a at times $t = 0$ and $t = t'$. The wave function in position space for times $t > t'$ is then given by

$$\langle \mathbf{r} | \psi(t) \rangle = \frac{A}{\left(1 + \mathrm{i}t/a^2\right)^{3/2}} \exp\left[-\frac{r^2}{2a^2\left(1 + \mathrm{i}t/a^2\right)}\right]$$

$$+ \frac{A}{\left[1 + \mathrm{i}(t - t')/a^2\right]^{3/2}} \exp\left[-\frac{r^2}{2a^2\left[1 + \mathrm{i}(t - t')/a^2\right]}\right] \quad (11.14)$$

and in momentum space by

$$\langle \mathbf{p} | \psi(t) \rangle = Aa^3 \exp\left[-\frac{1}{2}a^2 p^2 - \frac{1}{2}\mathrm{i}p^2 t\right] \left\{1 + \exp\left[\frac{1}{2}\mathrm{i}p^2 t'\right]\right\} . \quad (11.15)$$

The interference term in the position space probability density $|\langle \mathbf{r}|\psi(t)\rangle|^2$ appears to be quite complicated, and in the more general case of nonvanishing momenta $\mathbf{k}_1$ and $\mathbf{k}_2$ even more so. However, it can be shown that far away from the origin and for $t/a^2 \gg t'/a^2$ the modulation is proportional to $\cos[r^2 t'(1 + t'/t)/(2t^2)]$, i.e., chirped in time and space. Identifying r/t with p, the leading term in the cosine argument is just $p^2 t'/2$. In fact, calculating the probability density in momentum space from (11.15) we obtain

$$|\langle \mathbf{p}|\psi\rangle|^2 = 2|A|^2 a^6 \exp\left[-a^2 p^2\right] \left\{ 1 + \cos\left[\frac{1}{2}p^2 t'\right] \right\} . \tag{11.16}$$

Figure 11.2 shows the probability density in space and time. In a gedankenexperiment one may observe at a fixed position in space the probability density which passes by. In Fig. 11.2 the result is shown for $x, y = 0$ and $z = 100$: first the fast components of the first wave packet arrive. Then the fast components of the second wave packets reach the observer, interfering with the first wave packet. At later times the slower components pass by. As a result the modulation in the observed probability density has a negative chirp.

Equations (11.15) and (11.16) can be easily generalized to the case of nonvanishing center-of-mass momenta $\mathbf{k}_1, \mathbf{k}_2$ and different amplitudes A_1, A_2 where

$$\langle \mathbf{p}|\psi_{\mathbf{k}_{1,2}, A_{1,2}}(t)\rangle = A_1 a^3 \exp\left[-\frac{1}{2}a^2(\mathbf{k}_1 - \mathbf{p})^2 - \frac{1}{2}\mathrm{i}p^2 t\right]$$
$$+ A_2 a^3 \exp\left[-\frac{1}{2}a^2(\mathbf{k}_2 - \mathbf{p})^2 - \frac{1}{2}\mathrm{i}p^2(t - t')\right] \tag{11.17}$$

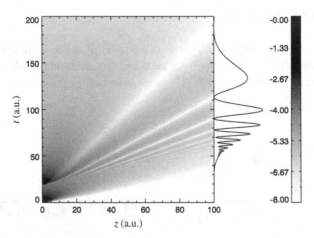

Fig. 11.2 Interference of two Gaussian wave packets born with a time delay t' in the zt-plane. The wave packets of width a and amplitude A start at $\mathbf{r} = 0$ with $k = 0$. The actual parameters are $A = a = 1$ and $t' = 20$. The contours of the probability density are scaled logarithmically. The probability density at the fixed spatial position $z = 100$, $x = y = 0$ as a function of time is indicated at the right border of the contour plot

and

$$|\langle \mathbf{p}|\psi_{\mathbf{k}_{1,2},A_{1,2}}\rangle|^2 = a^6 \left\{ |A_1|^2 \exp\left[-a^2(\mathbf{k}_1 - \mathbf{p})^2\right] + |A_2|^2 \exp\left[-a^2(\mathbf{k}_2 - \mathbf{p})^2\right] \right.$$
$$\left. + \exp\left[-\frac{1}{2}a^2[(\mathbf{k}_1 - \mathbf{p})^2 + (\mathbf{k}_2 - \mathbf{p})^2]\right] 2\mathrm{Re}\, A_1 A_2^* \exp\left[-\frac{1}{2}\mathrm{i}p^2 t'\right] \right\} .$$

(11.18)

The interference is therefore governed by the term

$$\sim 2\mathrm{Re}\, A_1 A_2^* \exp\left[-\frac{1}{2}\mathrm{i}p^2 t'\right] ,$$

(11.19)

while the center-of-mass momenta $\mathbf{k}_1$, $\mathbf{k}_2$ affect the envelope of the momentum spectrum only. If A_1 and A_2 are both real, the condition for interference maxima reads

$$\mathcal{E}_{\mathrm{kin}}t' = 2n\pi, \qquad n = 0, 1, 2, \dots ,$$

(11.20)

where $\mathcal{E}_{\mathrm{kin}} = p^2/2$. The interference term (11.19) is one of the examples for a time–energy uncertainty relation promised in our introductory remarks in Sect. 11.1. It is a relation between the time delay of preparation of the two wave packets and the kinetic energy. The bigger the time delay t' the faster are the oscillations due to interference in the energy or momentum spectra. Suppose we want to measure the time delay by determining the interference minima or maxima. Then expression (11.20) tells us that we need a spectrometer with an energy resolution better than $\simeq \pi/t'$.

Figure 11.3 shows cuts ($p_x = p_y = 0$) through the probability density in momentum space for various center-of-mass momenta and amplitudes. Panel (a) shows basically $1 + \cos(p_z^2 t'/2)$ multiplied by a Gaussian envelope. In case (b) the interference causes only minor modulations of the two momentum wave packets moving with $p_z = k_z = \pm 2$. In case (c) the complex amplitude A_2 changes the "phase of the slit," resulting in a sine-like modulation instead of a cosine-like. Finally, panel (d) corresponds to two wave packets with $k_{2z} = 2k_{1z}$ so that the later-emitted wave packet overtakes the previously emitted one.

11.2.3 Grating in Time

Combinations of single- and double-slits in time and space were thoroughly studied in [11]. We move directly on to the time analogue of a spatial grating. Let us assume that the time delay between two subsequent emissions is constant and given by $t' = T = 2\pi/\omega$, and the total number of time slits is N. For times $t > (N-1)T$, the wave function in momentum space then reads

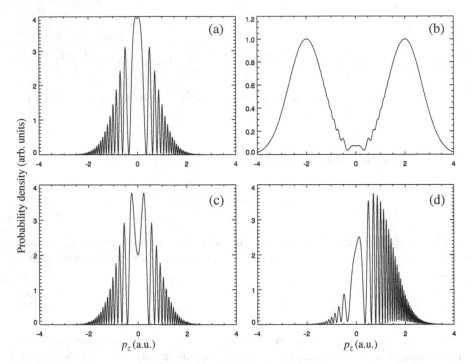

Fig. 11.3 Cuts ($p_x = p_y = 0$) through the probability density in momentum space for the case of two Gaussian wave packets of width $a = 1$, "born" with a delay $t' = 50$, and (a) $\mathbf{k}_1 = \mathbf{k}_2 = \mathbf{0}$, $A_1 = A_2 = 1.0$; (b) $\mathbf{k}_1 = -2.0\mathbf{e}_z = -\mathbf{k}_2$, $A_1 = A_2 = 1.0$; (c) $\mathbf{k}_1 = \mathbf{k}_2 = \mathbf{0}$, $A_1 = 1.0$, $A_2 = -i$; (d) $\mathbf{k}_1 = 0.5\mathbf{e}_z$, $\mathbf{k}_2 = 1.0\mathbf{e}_z$, $A_1 = A_2 = 1.0$

$$\langle \mathbf{p} | \psi_\mathbf{k}(t) \rangle = a^3 \, \exp\left[-\frac{1}{2} a^2 (\mathbf{k} - \mathbf{p})^2 - \frac{1}{2} i p^2 t \right] \Sigma_{N,T}(p) \, , \tag{11.21}$$

with

$$\Sigma_{N,T}(p) = \sum_{n=0}^{N-1} A_n \exp\left[i \mathcal{E}_{\text{kin}} n T \right] \, . \tag{11.22}$$

If the phase-slip between subsequent emissions is constant, i.e., $A_n = A \exp[in\varphi]$, we obtain

$$\Sigma_{N,T}(p) = A \sum_{n=0}^{N-1} \exp\left[i (\mathcal{E}_{\text{kin}} T + \varphi) n \right] = A \sum_{n=0}^{N-1} \exp\left[i \mathcal{E} n T \right] = A \frac{\exp[i \mathcal{E} N T] - 1}{\exp[i \mathcal{E} T] - 1} \, , \tag{11.23}$$

where $\mathcal{E} = \mathcal{E}_{\text{kin}} + \varphi / T$. The probability density in momentum space thus is

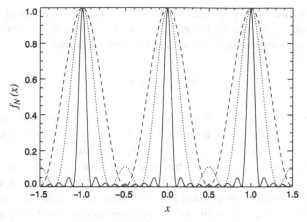

Fig. 11.4 The function $f_N(x) = \sin^2(Nx\pi)/[N^2 \sin^2(x\pi)]$ for $N = 2$ (*dashed*), 3 (*dotted*), and 9 (*solid*). The factor $1/N^2$ was introduced to normalize all the absolute maxima at integer x to 1

$$|\langle \mathbf{p} | \psi_\mathbf{k} \rangle|^2 = |A|^2 a^6 \, \exp\left[-a^2 (\mathbf{k} - \mathbf{p})^2\right] \frac{\sin^2[\mathcal{E}N\pi/\omega]}{\sin^2[\mathcal{E}\pi/\omega]} \, . \tag{11.24}$$

The grating function $f_N(x) = \sin^2(Nx\pi)/[N^2 \sin^2(x\pi)]$ is shown in Fig. 11.4 for $N = 2, 3$, and 9. As for a spatial grating the maxima of $f_N(x)$ at integer x become sharper and sharper with increasing number of slits N, while all other modulations in between are more and more suppressed. $N^2 f_N(x)$ approaches a δ-comb in the limit $N \to \infty$. Integer x corresponds to peaks at $\mathcal{E} = j\omega$, $j \in \mathbb{Z}$. Hence, if we think of the wave packets being generated by a long laser pulse of frequency ω, the photo electron peaks in the energy spectra are expected to be separated by the photon energy $\hbar\omega$ ($= \omega$ in atomic units).

11.2.4 Continuous Slit in Time

If the electronic population in the continuum is generated continuously, e.g., by ionization in strong fields, it is reasonable to replace (11.21) by

$$\langle \mathbf{p} | \psi_\mathbf{k}(t) \rangle = a^3 \, \exp\left[-\frac{1}{2} i p^2 t\right] S(\mathbf{p}) \tag{11.25}$$

with

$$S(\mathbf{p}) = \int_0^{T_p} g(t') \, \exp\left\{-\frac{1}{2} a^2 [\mathbf{k}(t') - \mathbf{p}]^2 + \frac{1}{2} i p^2 t'\right\} \, dt' \, , \tag{11.26}$$

where $g(t')$ is some complex function which weights the electronic source and replaces the coefficients A_n in (11.22). We assume that $g(t')$ is only nonvanishing within the time interval $[0, T_p]$ and allow the center-of-mass momentum $\mathbf{k}$ to vary with the emission time. Having in mind the ionization of atoms in strong laser fields as the source of electronic wave packets in the continuum, we set

$$\mathbf{k}(t) = -\mathbf{A}(t) , \qquad (11.27)$$

with $\mathbf{A}(t)$ the vector potential of the laser pulse because a classical electron, which is instantaneously "born" in a laser field at time t with vanishing initial velocity, will have a drift momentum $\mathbf{k}(t) = -\mathbf{A}(t)$ once the laser pulse is over. If the laser pulse is short (meaning that the electron does not leave the focal region while the laser pulse is on) $\mathbf{k}(t) = -\mathbf{A}(t)$ will also be the drift momentum with which the electron reaches the detector. Since the ionization probability of atoms strongly increases with the absolute value of the applied electric field of the laser pulse $\mathbf{E}(t) = -\partial_t \mathbf{A}(t)$, the absolute value of the weight function $|g(t)|$ should be chosen accordingly. Figure 11.5 shows cuts ($p_x = p_y = 0$) through the probability density in momentum space $|\langle \mathbf{p}|\psi \rangle|^2$ for a vector potential describing a linearly polarized laser pulse in dipole approximation

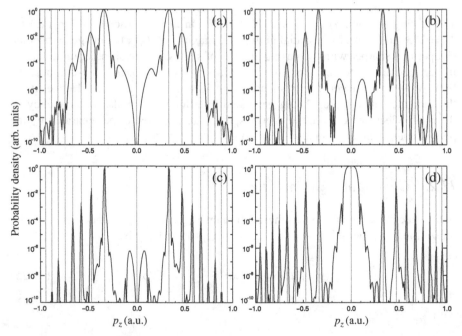

Fig. 11.5 (Color online) Cuts ($p_x = p_y = 0$) through the probability density in momentum space for the continuous slit according to Eqs. (11.25), (11.26), (11.27), and (11.28) with $g(t) = E(t)$ and the laser pulse parameters $\omega = 0.056$, $\varphi_{\mathrm{cep}} = \pi/2$, $\hat{E} = 0.05$. Panel (**a**) shows the result for an $n_{\mathrm{cyc}} = 5$-cycle pulse of the form (11.28), (**b**) for a 10-cycle, (**c**) for a 25-cycle, and (**d**) for a 25-cycle pulse but $g(t) = |E(t)|$. The *vertical lines* indicate momenta p_z fulfilling $p_z^2/2 = j\omega$, $j \in \mathbb{Z}$

$$\mathbf{A}(t) = \begin{cases} \hat{A}\mathbf{e}_z \sin^2\left[\frac{\omega t}{2n_{\text{cyc}}}\right] \sin(\omega t + \varphi_{\text{cep}}) & \text{for } t \in [0, n_{\text{cyc}}2\pi/\omega] \\ 0 & \text{otherwise} \end{cases} . \tag{11.28}$$

Here, $\hat{A} = -\hat{E}/\omega$ is the vector potential amplitude corresponding to an electric field amplitude $\hat{E}$, n_{cyc} is the number of laser cycles in the pulse, and φ_{cep} is the carrier-envelope phase. Figure 11.5 shows that with increasing pulse duration the peaks in the momentum spectra narrow around the positions $\mathcal{E}_{\text{kin}} = p_z^2/2 = j\omega$, $j \in \mathbb{Z}$ – another example for a time–energy uncertainty relation.

11.3 Time-Domain Double-Slit Experiments

The first experiment specifically designed to investigate interference phenomena in a dynamical double-slit setup was carried out by Sillitto and Wykes in 1972 [62]. The motivation for that experiment dates back to a paper by Mandel in 1959. Based on classical coherence theory, Mandel predicted that interference may be observed in a double-slit arrangement where the slits are opened and closed alternately, even if both shutters are never open at the same time [43]. The condition is that the coherence time is longer than the time lag between closing the one and opening the other slit. Interestingly, Mandel's work was inspired by a paper discussing a variant of the Schrödinger-cat paradox [32], which, however, has little to do with the type experiments discussed in the following. In fact, even the Sillitto and Wykes paper differs from the experiments to be discussed, as it still considers a *spatial* double-slit where certain transmission properties of each slit can be controlled in time. In fact, without a spatial separation of the slits, no diffraction or interference effects are to be expected for light [10]! Therefore we will focus on experiments with massive particles where interference of de Broglie waves is observed. The observation of diffraction when using massive particles belongs, of course, to the classics of modern physics. For the first time it was realized with electrons scattered at crystals by Davisson and Germer [18]. In the 1960s, Jönsson succeeded to realize the classic double-slit arrangement and to observe the respective interference pattern [33, 34]. Interference of thermal neutrons in interferometers cut from a single crystal has been the next step for this kind of experiments. They have opened unique possibilities for investigating, e.g., the Aharonov–Bohm effect [21] and dynamical diffraction [58]. With the advent of laser-cooled atoms and the invention of optical elements like mirrors and beam splitters for cold atoms, the number of possible applications of atom interferometry has grown such that it is impossible to summarize them in a few words.

11.3.1 Double-Slit in Time Using Cold Atoms

Virtually all of these experiments are analogues of static optical setups. Time-dependent interferometers have rarely been realized with de Broglie waves. For the Sillitto–Wykes experiment, e.g., a neutron interferometer has been proposed [9] but not yet realized. One notable exception is the experiment by Szriftgiser et al. [63] that uses laser-cooled cesium atoms. These are dropped onto an atom mirror and bounce there three times like on a trampoline. Being based on a laser, this trampoline can be switched on and off at any time ad libitum. Most of the time, the jumping sheet is in fact missing. This allows to use the first bounce for selecting atoms with well-defined velocity. When the atoms return to the mirror at the second bounce, the mirror may in fact be switched on twice, i.e., two wave packets will be reflected. In this way, the double-slit in time is realized. The third bounce together with subsequent optical detection of the atoms that made it through this sequence then is used to analyze the energy distribution (or, what is equivalent, the time of flight) of the atoms upon their second return to the atom mirror. The various places of use of lasers in this experiment should not lead to the misunderstanding that internal states of the atoms being used were altered. This would in fact destroy the effects to be discussed below, could however be used in the future for other which-way experiments.

A schematic of the experimental setup is displayed in Fig. 11.6. It consists in fact of two magneto-optical traps (MOTs), realized by shining diode lasers of

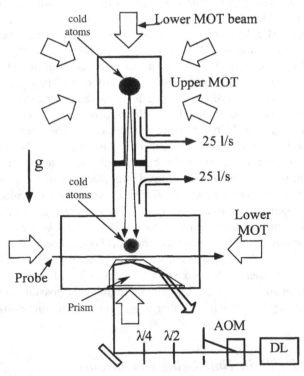

Fig. 11.6 The experimental setup consists of two magneto-optical traps (MOTs). The upper one at relatively high pressure is used to cool a sufficiently large number of atoms. Subsequently, these are dropped and recaptured by the lower MOT some 3 mm above the prism. The evanescent wave created on the prism surface serves as atom mirror (from Szriftgiser et al. [63])

appropriate wavelength from all directions into two glass cubes aligned vertically with 70 cm distance. They are each pumped by a 25-ℓ ion pump and connected such that differential pumping is effective through a 140 mm long thin glass tube. This allows high gas pressures and thus a large number of trapped atoms ($\approx 10^8$) in the upper cube while maintaining good vacuum in the lower cube where the actual experiment is performed. The atoms are released from the upper trap by switching off the respective laser and magnetic field. After falling some 375 ms through the connecting tube they are caught with 20% efficiency in the second MOT, just $h = 3$ mm above the atom mirror. Cooled to 5 μK, in the $F = 4$ hyperfine ground state, every 1.6 s some 10^7 atoms can be dropped onto the atom mirror.

This mirror is formed by an evanescent light field created by total internal reflection at a super-polished fused silica prism using a laser blue-detuned from the cesium D line, see Fig. 11.6 [16]. Accordingly, the atoms are low-field seekers and may be reflected in the light field that decays rapidly in vertical direction like $\exp(-2\kappa z)$, where $\kappa^{-1} = 190$ nm. The spatial extend of the evanescent field in vertical direction requires that it remains switched on for at least $\tau_{\min} = 2\kappa/v$, where $v = \sqrt{2gh}$ is the mean velocity of the atoms at the surface of the atom mirror and $g = 9.81$ m/s^2. This leads to $\tau_{\min} = 1.5$ μs.

The potential energy of the atoms, when released from the MOT, exceeds the thermal energy by about a factor of 10. The temperature of 5 μK would translate into a coherence time $1/\Delta v = 10$ μs. However, a further reduction in energy spread of the atoms is desirable. As already mentioned, this can be achieved by switching the atom mirror active for a brief period τ, of course $\tau > \tau_{\min}$. Obviously, this reduces the energy spread of the reflected atoms to $E_0 \tau/T$ when they return after another $2T$ to the mirror surface for a second time. Here $E_0 = mgz_0$ and $T = 25$ ms is the drop time of the atoms.

In order to observe quantum effects induced on the second reflection of the atoms on the mirror, the (classical) energy spread $\Delta E_{\mathrm{cl}} = E_0 \tau/T$, given by cooling and subsequent selection of atoms as described, has to be smaller than the energy uncertainty ΔE_{qu} due to quantum effects, i.e., due to diffraction of de Broglie waves. This sets an upper limit for the duration τ during which the mirror is active at the second bounce.[3] Otherwise the former would mask the latter. As for any square-shaped mask, the diffraction pattern after the mirror made active by a single laser pulse of duration τ has the familiar sinc-shape form with a width of $\Delta E_{\mathrm{qu}} = h/\tau$. An equivalent statement would be to demand that the coherence time $1/\Delta v = h/\Delta E_{\mathrm{cl}}$ has to be longer than τ. Both lead to $\tau < \sqrt{hT/E_0} = 150$ μs.

Quite elegantly, the energy distribution of the atoms after the second reflection can be analyzed by another reflection taking place around $5T$ after the release of the atoms from the MOT. In order to probe the energy distribution of the atoms, the third pulse is shifted in a time interval corresponding to a few times ΔE_{qu}. What remains to be done is to determine the number of atoms reflected at the third bounce in

---

[3] For simplicity, τ is chosen to be equal for all three reflections.

dependence of the position in time of the third pulse. This can be done by exciting the atoms after the third reflection with yet another laser pulse, but now tuned to the $6s_{1/2}$–$6p_{3/2}$ transition. The fluorescence is detected by a photomultiplier. At this point, the low-target gas pressure made possible by the twin-MOT becomes essential. As one would expect, the transmission of the apparatus is very low. Out of the 10^7 atoms captured in the lower MOT, just a few make it to the detection interaction area. A few improvements in this respect are certainly possible and some are in fact implemented in the work described in more detail in [63]. The sequence of pulses is shown in Fig. 11.7.

Now it is straightforward to perform a double-slit experiment by switching on the atom mirror twice at the second return of the atoms. This means that two interfering de Broglie waves generated with a brief delay will emerge. Of course, the separation of these two pulses should be shorter than the coherence time. Just like for diffraction with a single pulse as discussed above, the interference fringes can be recorded by scanning the third pulse. Thus, the interference pattern is measured as a function of the time of flight (see Fig. 11.8). It is even possible to shift the fringe pattern. In case one of the two pulses that form the double-slit is weaker than the other, the atoms are reflected a little bit closer to the surface of the atom mirror. Therefore, those atoms experience an additional phase shift. With $\Delta\phi = 2\pi \cdot \Delta\ell/\Lambda_{\mathrm{DB}}$, where $\Lambda_{\mathrm{DB}} = h/(mgT)$ is the de Broglie wavelength close to the mirror surface, it can easily be calculated that a difference in path length of $\Delta\ell = h/(2mgT) \approx 12\,\mathrm{nm}$ results in destructive interference ($\Delta\phi = \pi$). This means that a fringe shift by half a period is observed, if the atom for one of the slits approaches the surface by 6 nm closer than the other. A reduction in intensity of 5% for one of the atom mirror pulses is sufficient to realize this phase shift.

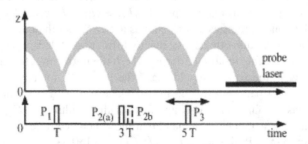

Fig. 11.7 Trajectories of the atoms following their release from the lower trap. The atoms bounce like on a trampoline at the atom mirror which can be switched on and off by pulsing the laser that creates the evanescent wave. The first reflection by applying a laser pulse P_1 is used to narrow down the velocity distribution, thus increasing the coherence time. The second reflection induces diffraction in time. A double-slit in time can be realized by switching the atom mirror on twice with a pair of pulses P_{2a} and P_{2b}, separated by a brief time lag. The atom mirror is switched on for a third time by a laser pulse P_3 in order to probe the diffraction pattern. P_3 is shifted around the nominal return time and the number of atoms that have successfully completed the entire trajectory are detected (from Szriftgiser et al. [63])

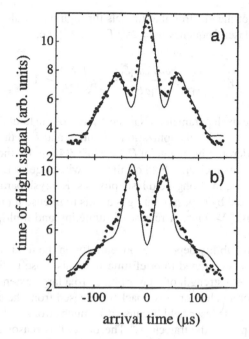

Fig. 11.8 Interference fringes of cold atoms after reflection at a temporal double-slit. For the *upper panel* both atom trajectories have the same length. For the *lower panel*, pulse P_{2b} was given a slightly smaller intensity so that atoms reflected by this pulse would approach the prism surface a few nanometers closer, thus giving rise to a fringe shift of half a period (from Szriftgiser et al. [63])

11.3.2 The Attosecond Double-Slit

Strong-field ionization provides mechanisms for realizing a double-slit in time under quite different conditions. In this case, the temporal slits are brief windows in time within the cycles of the optical field that induce ionization. The decisive fact is that the phase (or time) within any optical cycle, at which ionization takes place, determines the photoelectron momentum and thus its kinetic energy. This means that the instant of ionization is mapped onto the momentum of the photoelectron. Consequently, for more than one optical cycle, there is more than one possibility to create photoelectrons of a given momentum and interference will be observed. Few-cycle laser pulses consisting of two or even less optical cycles within their full width at half maximum (FWHM) are particularly well suited to create double-slit interferences [41].

In the present context, a laser field is considered strong if optical field ionization, i.e., tunneling, is the leading process of ionization. As for regular field ionization, the barrier through which bound electrons may tunnel is formed by the laser field and the atomic potential. For linear ionization, this implies that the tunnel opens and closes twice during one optical cycle. The transient nature of the potential barrier suggests that a description of ionization based on tunneling can only be meaningful if the frequency ω_t of the electron that tunnels through the Coulomb barrier is high as compared to the laser frequency $\omega = 2\pi/T$ [36, 37], i.e.,

$$\gamma := \frac{\omega}{\omega_t} = \frac{\omega\sqrt{2m|\mathcal{E}_0|}}{|e\hat{E}|} = \sqrt{\frac{|\mathcal{E}_0|}{2U_P}} < 1 . \tag{11.29}$$

γ is the so-called Keldysh parameter, $\mathcal{E}_0$ is the energy of the bound state from which the electron starts, i.e., $|\mathcal{E}_0|$ is the ionization potential, and $\hat{E}$ is the amplitude of the laser field. The ponderomotive potential $U_P = e^2\hat{E}^2/(4m\omega^2)$ is the quiver energy of an unbound electron in an electric field oscillating with angular frequency ω and an important scaling factor in strong-field laser physics. Keldysh parameters $\gamma \approx 1$ can fairly easily be realized by exposing rare gas atoms to intense femtosecond lasers at intensities around $10^{14}\,\mathrm{W/cm^2}$. A review of tunneling and multiphoton ionization can be found in [55].

The tunneling probability depends exponentially on the field strength. As a consequence, ionization is confined to brief time intervals close to the maxima of the oscillating electric field strength of a laser pulse. To a large extent, the trajectory of an electron (or rather an electron wave packet) released from the atom by tunneling will be dominated by the laser field because it is much stronger than the field of the ion for the biggest part of the trajectory.[4] Therefore, it is reasonable to model in a first approximation ionization by classical mechanics, i.e., by calculating classical electron trajectories. Assuming, for the time being, a continuous, linearly polarized laser field of the form

$$E(t) = \hat{E}\,\cos(\omega t) \tag{11.30}$$

and tunneling at a phase ωt_0, one immediately obtains in atomic units (see also the remark after Eq. (11.27) in Sect. 11.2.4)

$$v(t) = -\frac{\hat{E}}{\omega}\,[\sin(\omega t) - \sin(\omega t_0)] \tag{11.31}$$

for the velocity of the electron, if $v = 0$ is assumed for the electron at the instant of tunneling. For the drift energy $\mathcal{E}_{\mathrm{kin}}$ of the photoelectron, i.e., the quantity measured in an experiment, this results in

$$\mathcal{E}_{\mathrm{kin}} = 2U_P\,\sin^2(\omega t_0) . \tag{11.32}$$

Equation (11.32) is quite an important result as it shows what has been announced at the beginning: The kinetic energy is determined by the instant at which the electron enters the continuum. In addition, we note that electrons tunneling at $\omega t_0 = 0$ will have zero drift energy, whereas those ejected into the field at $\omega t_0 = \pi/2$ will obtain

---

[4] This is in fact the essence of the strong-field approximation (SFA) to be introduced in Sect. 11.4.

the maximum kinetic energy of $2U_P$. Tunneling is, of course, much more likely at the peak of the field at $\omega t = 0$, which explains that photoelectron spectra show way more electrons at low energies than at high energies. This simple theory of strong-field ionization is known as the "simple man's model" since the late 1980s [65] and has been beautifully confirmed in microwave experiments using Rydberg atoms [25, 26].

The same result can be obtained taking advantage of conservation of the canonical momentum

$$p_{can} = p + eA = \text{const.} , \qquad (11.33)$$

where A is the vector potential and $p = mv$ ($= v$ in atomic units) is the kinetic momentum. Using the same approximations as above, i.e., $p(t_0) = 0$, remembering that $p(t \to \infty) \equiv p_{drift}$, and taking into account $A(t \to \infty) = 0$ for a pulse with no DC component [48], the identity $p_{can}(t_0) = p_{can}(\infty)$ yields

$$p_{drift} = eA(t_0) . \qquad (11.34)$$

Thus, the electron momentum is given by the vector potential at the instant of ionization. Or, reading this statement the other way round: Selecting electrons with a given momentum is equivalent to selecting electrons leaving the atom at a well-defined phase. This result holds for arbitrary pulse forms, in particular also for few-cycle pulses. For an illustration, see Fig. 11.9. Choosing the electron momentum implies choosing the instant of ionization with sub-cycle resolution, i.e., on the attosecond scale. For a given electron energy and emission direction (parallel to the laser

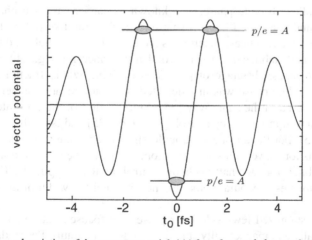

Fig. 11.9 Temporal variation of the vector potential $A(t)$ for a few-cycle laser pulses with a $-$sine-like temporal variation of the field, i.e., $E(t) = \hat{E}(t) \cos(\omega t + \pi/2)$. The temporal slits are given by the condition $p - eA(t_0) = 0$. For a $-$sine-like pulse, this leads to a double-slit in the negative (since $e = -|e|$) direction and a single slit in the opposite direction. Each slit can be resolved into a pair of slits (from Lindner et al. [41])

polarization), the temporal evolution of the field of a few-cycle laser pulse can be chosen such that for one emission direction, just a single optical (half-) cycle may generate electrons of this energy, while for the opposite direction there are two such (half-) cycles. This is in fact the essence of the attosecond double-slit.

The strength of the classical model is the intuitive insight it provides. Hardly more than the number and position of the solutions of $p - eA(t_0) = 0$ for given p has been used in order to explain double-slit behavior for strong-field ionization induced by few-cycle laser pulses. The respective solutions $t_0(p)$ in a quantum model that will be discussed below in Sect. 11.4.3 are complex, thus allowing access to classically forbidden electron energies. However, the symmetry of these solutions stays the same and so do the results qualitatively.

Generation of few-cycle laser pulses is remarkably simple: Pulses of a conventional femtosecond laser system are sent through a hollow fiber filled with a few hundred millibar of Argon or Neon [51, 52]. Due to self-phase modulation the spectrum broadens and the pulses can be compressed to as few as 5 fs (FWHM). The method works best for sufficiently powerful (1 mJ) and short (<25 fs) femtosecond pulses to start with. The duration of the compressed pulse has to be compared with the optical period of $T = 2.5$ fs for the typical wavelength of 800 nm of titanium-sapphire femtosecond lasers. This means the temporal evolution of the field $E(t)$ depends on the phase of the carrier wave with respect to the maximum of the pulse envelope. This can be seen by writing $E(t)$ as a product of envelope $\hat{E}(t)$ and carrier wave:

$$E(t) = \hat{E}(t) \cos(\omega t + \varphi) . \tag{11.35}$$

φ is the so-called absolute phase, also known as carrier-envelope phase. Special cases are cosine-like, sine-like, $-$cosine-like, and $-$sine-like pulses for $\varphi = 0$, $-\pi/2$, π, and $\pi/2$. This phase determines the evolution of the field and therefore its measurement and stabilization is of pivotal importance for, e.g., attosecond laser physics which takes advantage of processes evolving within fractions of optical cycles. The absolute phase was first detected in photoionization experiments [53] and later measured in the same way [54]. Stabilization of the absolute phase has first been achieved in frequency metrology [59, 64, 35] and later also for amplified laser pulses [5]. For the attosecond double-slit, the absolute phase determines the number of slits for a given emission direction. More precisely, it determines the ionization probability for each half-cycle by controlling their amplitude. The analogue to a regular double-slit would be that the number and the width of the slits can be controlled.

The phase-stabilized few-cycle laser pulses are focused onto a gas jet situated in an electrically and magnetically shielded vacuum apparatus. Perpendicular to the laser beam, two electron detectors are installed to the left and to the right of the laser focus, thus forming a twin (or stereo) time-of-flight (TOF) spectrometer. The setup is depicted in Fig. 11.10. The laser polarization is linear and parallel to the axis of the TOF spectrometers.

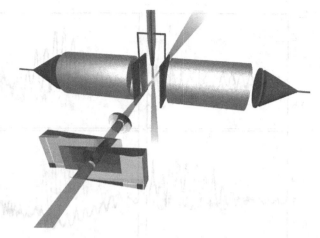

Fig. 11.10 (Color online) Stereo-photoelectron spectrometer. Two opposing electrically and magnetically shielded time-of-flight spectrometers are mounted in an ultrahigh vacuum apparatus. Argon atoms fed in through a nozzle from the top are ionized in the focus of a few-cycle laser beam. The laser polarization is linear and parallel to the flight tubes. Note that the laser field changes sign while propagating through the focus. Slits with a width of 250μm are used to discriminate electrons created outside the laser focus region chosen. The slits can be moved from outside the vacuum system. A photodiode and micro-channel plates (shown in *gray* (*red* online) and *black*) detect the laser pulses and photoelectrons, respectively. A pair of glass wedges is used to optimize dispersion and to change the absolute phase (from Paulus et al. [54])

Figure 11.11 displays measured electron spectra. In Fig. 11.11(a) the spectra recorded at the left and the right detectors are shown for ±cosine-like and ±sine-like pulses. A problem in presenting such spectra is that they quickly roll off with increasing electron energy. This roll-off was eliminated by dividing the spectra by the average of all spectra over the pulse's phase. Clear interference fringes with varying visibility are observed as expected from the discussion above. The highest visibility is observed for −sine-like pulses in the positive ("right") direction. For the same pulses, the visibility is very low in the opposite direction. Changing the phase by π interchanges the role of left and right as expected. The most straightforward explanation is to assume that, for −sine-like pulses, there are two slits and no which-way information for the positive direction and just one slit and (almost) complete which-way information in the negative direction. The fact that the interference pattern does not entirely disappear is caused by the pulse duration, which is still slightly too long to create a perfect single slit.

Under the conditions of this experiment, each argon atom emits at most one electron, whose various options of how to reach a given final state lead to interference. For sine-like pulses, these options correspond to a double-slit in time in one direction and to a single-slit in the other and are created for each atom separately by the few-cycle laser pulse. Therefore, even though there is more than one argon atom in the laser focus, the experiment operates under single-electron conditions. On the scale of the electron's de Broglie wavelength, other atoms are far away and,

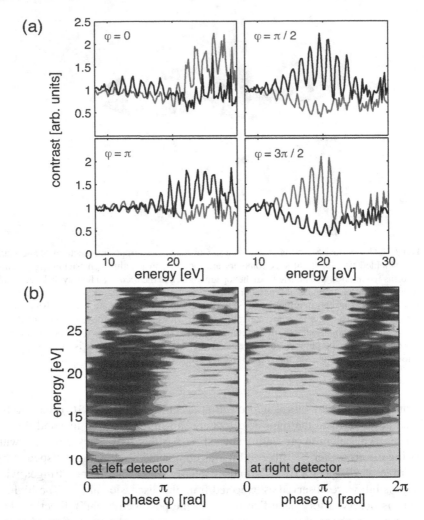

Fig. 11.11 (Color online) Photoelectron spectra of argon measured with 6-fs laser pulses at an intensity of 1×10^{14} W/cm^2 as a function of the absolute phase. Panel (**a**) displays the spectra for $\pm$sine- and $\pm$cosine-like laser fields. The *gray* (*red* online) curves are spectra recorded with the left detector (negative direction), while the black curves relate to the positive direction. For $\varphi = \pi/2$ the fringes exhibit maximum visibility for electron emission to the right, while in the opposite direction minimum fringe visibility is observed. In addition, the fringe positions are shifted. Panel (**b**) displays the entire measurement where the fringe visibility is coded in false colors (online). The fringe positions vary as the phase φ of the pulse is changed. This causes the wave-like bending of the stripes in these figures. Both *panels*, in principle, show the same information because a phase shift of π mirrors the pulse field in space and thus reverses the role of positive and negative direction. However, the data shown were recorded simultaneously but independently where the phase ϕ was varied between 0 and 2π (from Lindner et al. [41])

moreover, randomly distributed. This is in contrast to the double-slit in space where the beam has to be sufficiently dilute to ensure a one-electron measurement.

The fringe pattern exhibits an envelope. From Fig. 11.11 a width of this envelope of about four fringes is inferred. Just as for a double-slit experiment, the width of this envelope can be associated with the width of the slits. However, Fig. 11.9 suggests that what is seen here is not the width of the slit. Rather, each slit can be resolved into a pair of slits whose separation is inversely proportional to the width of the envelope.

It might be worth reflecting that this version of the double-slit experiment has quite some remarkable features. Besides the fact that it is realized in the time–energy domain, there are slits being windows in time of attosecond duration. These "slits" can be opened or closed by changing the temporal evolution of the field of a few-cycle laser pulse by controlling the absolute phase. At any given time there is only a single electron in the double-slit arrangement. In addition, the presence and absence of interference may be observed for the same electron at the same time by choosing the direction of observation.

11.3.3 More Slits

Fringes in photoelectron spectra recorded under strong-field conditions are not new at all. In fact, strong-field photoionization is also known as above-threshold ionization (ATI), and the signature of ATI spectra is a series of peaks separated by the photon energy. The effect has first been observed in 1979 [1]. The conventional interpretation of the effect is an extension of multiphoton ionization. Accordingly, the energy position $\mathcal{E}_s$ of the sth ATI peaks is predicted to be given by a generalized Einstein law $\mathcal{E}_s = (n+s)\hbar\omega - |\mathcal{E}_0| - U_P$. Here, n is the minimum number of photons necessary to subdue the ionization threshold which is raised by U_P due to light shift [15]. However, it is hardly possible to generalize this picture to few-cycle laser pulses. For these the notion of ponderomotive level shifts is questionable. Moreover, the varying contrast of the fringes as well as the fringe shifts in dependence of the absolute phase are difficult to explain.

It is much easier to extend diffraction in the time domain from few-cycle to longer pulses. In place of the double-slit one would then speak of a grating in time, as discussed in Sect. 11.2.3. The respective temporal windows for photoelectron emission in a given direction are separated by the optical period T. Therefore, the separation of the ATI peaks by $\hbar\omega$ is an immediate consequence. With slightly more effort, the ATI peak shift due to the ponderomotive shift of the ionization potential can be explained. For this, the phases of the trajectories given by $S/\hbar$ have to be computed. Hereby $S = \int \mathcal{L}dt$ is the classical action of the trajectory calculated by integrating the Lagrange function $\mathcal{L}$ along the trajectory. For constructive interference, trajectories launched during subsequent optical cycles must have a phase difference of an integer multiple of 2π. This is the case when the generalized Einstein law is fulfilled.

The times t_0 at which the temporal slits are open for a given photoelectron momentum are determined by Eq. (11.34). However, t_0 determines not only the drift momentum of the photoelectrons via Eq. (11.34) but also the probability with which such an electron is ionized in the first place. For tunneling the ionization probability depends exponentially on $-1/|E(t_0)|$.[5] Ionization probability and drift momentum can be disentangled by inducing ionization with attosecond XUV pulses. Attosecond pulses are much shorter than an optical period in the visible (VIS) or near-infrared (NIR) spectral range. If used for ionizing atoms exposed to an additional optical field in the VIS or NIR, they can promote an electron at a defined phase ωt_0 into the optical field. At that instant, the kinetic energy of that photoelectron is $\mathcal{E}(t_0) = \hbar\omega_{\mathrm{XUV}} - |\mathcal{E}_0|$. This is different from the situation discussed in Sect. 11.3.2 where we had $\mathcal{E}(t_0) = 0$. Nevertheless, also in this situation it is possible to calculate the photoelectron's drift momentum using Eq. (11.33):

$$p_{\mathrm{drift}} = \pm\sqrt{2m\mathcal{E}(t_0)} + eA(t_0) . \tag{11.36}$$

If this process is repeated in subsequent optical cycles, interference of the electron wave packets created in each cycle can be expected. Due to the periodicity with T, the spacing of the fringes must be equal to the photon energy $\hbar\omega$.

It might seem to be exceedingly difficult to implement the idea just described experimentally because one needs to synchronize the XUV attosecond pulses with the optical field with attosecond precision, which corresponds to sub-micron spatial precision. Fortunately, the prevalent method for generation of attosecond pulses offers an intrinsic solution to this problem. XUV attosecond pulses can be generated by high-harmonic generation (HHG) in gases. HHG was discovered in the 1980s [24]: An intense NIR femtosecond pulse is focused on rare gases at a density of 10^{16}–10^{17} cm^{-3}. Depending on intensity and other experimental parameters, all odd harmonics up to a few hundred orders may be produced. The mechanism of this process is not very relevant to the present topic. Briefly, it can be explained by a simple extension of the classical model of strong-field ionization discussed in Sect. 11.3.2. Electron wave packets leaving the atom near the peak of the NIR field may be driven back to the ion core. Depending on the phase at which the wave packet was launched, the kinetic energy of the electrons upon return may be as high as $3.17\,U_{\mathrm{P}}$. In case the electron recombines about $3/4\,T$ later, short-wavelength radiation with $\hbar\omega_{\mathrm{XUV}} < 3.17U_{\mathrm{P}} + |\mathcal{E}_0|$ is created [40, 17] and XUV attosecond pulses are emitted collinear to the NIR femtosecond laser pulses in each *half*-cycle. For reviews on HHG, see [61], and for one on re-colliding electrons in general [7]. The

---

[5] The tacit assumption is that tunneling is an instantaneous process. This assumption may be wrong. In that case, one would need to distinguish between the time t_{01} when the electron enters the tunnel and the time t_{02} when the electron leaves the tunnel and the trajectory begins to evolve in the laser field. The field strength relevant for calculating the ionization probability would be $E(t_{01})$, while the drift momentum would be given by $eA(t_{02})$. Obviously, this complication would not alter the mechanisms discussed so far. The only consequence would be a change of the shape of the spectrum.

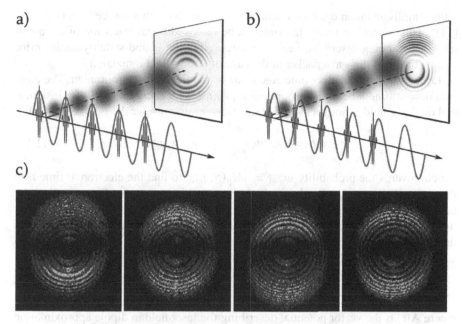

Fig. 11.12 (Color online) Attosecond pulse trains synchronized to an NIR laser field are used to create electron wave packets by ionizing rare gas atoms. Successive electron wave packets have a delay of one optical cycle. The ionization probability can be controlled by shifting the attosecond pulses with respect to the NIR laser field. The electron wave packets interfere and create a diffraction pattern. Panel (**c**) displays experimental results obtained for XUV–NIR delays of 0, $T/4$, $T/2$, and $3T/4$ (from Mauritsson et al. [47])

re-collision process explains why the train of attosecond pulses are synchronized with the optical field by which they are created.

In order to choose t_0, one has to delay the attosecond pulses with respect to the optical field. This is certainly technically demanding, but possible. The remaining problem is that the attosecond pulses are created every half-cycle, i.e., for different signs of the field which results in a π phase shift for consecutive attosecond pulses. Also this problem can be solved by using a two-color femtosecond laser pulse [46]. The idea of realizing a multiple slit diffraction experiment in this way has been realized recently [47], see Fig. 11.12. Admittedly, the purpose of that experiment reaches beyond the present discussion by addressing potential applications outside the scope of this chapter.

11.4 Strong-Field Approximation and Interfering Quantum Trajectories

In our theoretical analysis of time slits in Sect. 11.2, we tacitly assumed that wave packets somehow appear in the continuum and evolve freely afterward. Such a dynamics does not correspond to a unitary time evolution. Instead, the wave function rather fulfills an inhomogeneous Schrödinger equation with a source term (see, e.g., [11]). This source term may, in principle, be removed by an extension of the quantum system being described, i.e., by incorporating the bound state dynamics prior to the "birth" of the wave packet in the case of strong-field ionization.

Let us start with an electronic eigenstate of the field-free Hamiltonian. The electron may at time t_i be in the ground state $|\Psi_0(t_i)\rangle$ with energy $\mathcal{E}_0 < 0$, for instance, and the laser field is not yet switched on. Now let us consider the matrix element

$$M_{\mathbf{p}}(t_f, t_i) = \langle \Psi_{\mathbf{p}}(t_f) | \hat{U}(t_f, t_i) | \Psi_0(t_i) \rangle , \qquad (11.37)$$

which governs the probability $w_{i \to f} = |M_{\mathbf{p}}(t_f, t_i)|^2$ to find the electron at time t_f in the scattering state $|\Psi_{\mathbf{p}}(t_f)\rangle$ where $\mathbf{p}$ is the asymptotic momentum far away from the atom (where the measurement is performed). We assume that at time t_f the laser field is off again. $\hat{U}(t, t') = \hat{U}^{\dagger}(t', t)$ is the time-evolution operator associated with the time-dependent Schrödinger equation

$$i \frac{\partial}{\partial t} |\Psi(t)\rangle = \hat{H}(t) |\Psi(t)\rangle , \qquad \hat{H}(t) = \frac{1}{2} [\hat{\mathbf{p}} + \mathbf{A}(t)]^2 + \hat{V}(\mathbf{r}) , \qquad (11.38)$$

where $\mathbf{A}(t)$ is the vector potential describing the laser field in dipole approximation and $\hat{V}(\mathbf{r})$ is the binding potential. The minimum coupling Hamiltonian $\hat{H}(t)$ can be split in various ways:

$$i \frac{\partial}{\partial t} |\Psi(t)\rangle = [\hat{H}_0 + \hat{W}(t)] |\Psi(t)\rangle = [\hat{H}_V(t) + \hat{V}(\mathbf{r})] |\Psi(t)\rangle , \qquad (11.39)$$

with

$$\hat{H}_0 = \frac{\hat{p}^2}{2} + \hat{V}(\mathbf{r}) , \qquad \hat{H}_V(t) = \frac{\hat{p}^2}{2} + \hat{W}(t) , \qquad (11.40)$$

and $\hat{W}(t)$ the interaction with the laser field,

$$\hat{W}(t) = \hat{\mathbf{p}} \cdot \mathbf{A}(t) + \frac{1}{2} A^2(t) \qquad \text{(velocity gauge)} . \qquad (11.41)$$

The gauge transformation of the potentials (both scalar potential ϕ and vector potential $\mathbf{A}$) and the wave function $|\Psi(t)\rangle$,

$$\mathbf{A}' = \mathbf{A} + \nabla \chi(\mathbf{r}, t) , \qquad \phi' = \phi - \frac{\partial \chi(\mathbf{r}, t)}{\partial t} , \qquad |\Psi'(t)\rangle = e^{-i\chi(\mathbf{r}, t)} |\Psi(t)\rangle ,$$

where $\chi(\mathbf{r}, t)$ is an arbitrary differentiable scalar function, leaves the electric and the magnetic field unchanged:

$$\mathbf{E} = -\partial_t \mathbf{A} - \nabla \phi = \mathbf{E}', \qquad \mathbf{B} = \nabla \times \mathbf{A} = \mathbf{B}' . \qquad (11.42)$$

This gauge invariance offers the possibility to choose a gauge that suits us best, e.g., with respect to computational simplicity. However, this statement only holds true as long as all approximations we make do not destroy the gauge invariance (see Sect. 11.4.6). Transformation to the so-called length gauge is achieved by choosing

$$\chi(\mathbf{r}, t) = -\mathbf{A}(t) \cdot \mathbf{r} . \tag{11.43}$$

Because of $\nabla \chi = -\mathbf{A}$ the vector potential is "transformed away" while $\phi' = -\partial_t \chi = -\mathbf{E} \cdot \mathbf{r}$. The Hamiltonian in length gauge reads

$$\hat{H}'(t) = \frac{\hat{p}^2}{2} + \hat{V}(\mathbf{r}) - \phi'(\mathbf{r}, t) = \frac{\hat{p}^2}{2} + \hat{V}(\mathbf{r}) + \mathbf{E}(t) \cdot \hat{\mathbf{r}} \tag{11.44}$$

(one could also absorb $\hat{V}(\mathbf{r})$ in ϕ and ϕ'). Note that the transformation of the wave function

$$|\Psi'(t)\rangle = e^{i\hat{\mathbf{r}} \cdot \mathbf{A}(t)} |\Psi(t)\rangle \tag{11.45}$$

can be interpreted as a translation in momentum space. In fact, while in velocity gauge the quiver momentum is effectively subtracted from the kinetic momentum, leading to a canonical momentum different from the kinetic momentum, in length gauge kinetic and canonical momenta are equal.

From (11.44) we infer

$$\hat{W}'(t) = \mathbf{E}(t) \cdot \hat{\mathbf{r}} \quad \text{(length gauge)}, \tag{11.46}$$

with $\mathbf{E}(t) = -\partial_t \mathbf{A}(t)$.

11.4.1 Volkov Wave Functions

The Volkov–Hamiltonian $\hat{H}_V(t)$ governs the free motion of the electron in the laser field. It fulfills in velocity gauge

$$i\frac{\partial}{\partial t}|\Psi_{V\mathbf{p}}(t)\rangle = \hat{H}_V(t)|\Psi_{V\mathbf{p}}(t)\rangle = \frac{1}{2}[\hat{\mathbf{p}} + \mathbf{A}(t)]^2 |\Psi_{V\mathbf{p}}(t)\rangle . \tag{11.47}$$

Thanks to the dipole approximation the Volkov–Hamiltonian is diagonal in momentum space. The solution of (11.47) is thus readily written down:

$$|\Psi_{V\mathbf{p}}(t, t_i)\rangle = e^{-iS_{\mathbf{p}}(t, t_i)}|\mathbf{p}\rangle, \qquad S_{\mathbf{p}}(t, t_i) = \frac{1}{2}\int_{t_i}^{t} dt' [\mathbf{p} + \mathbf{A}(t')]^2, \tag{11.48}$$

where $|\mathbf{p}\rangle$ are momentum eigenstates, $\langle \mathbf{r}|\mathbf{p}\rangle = e^{i\mathbf{p}\cdot\mathbf{r}}/(2\pi)^{3/2}$. Note that the lower integration limit t_i affects the overall phase of the Volkov solution only. As mentioned above, the transition to the length gauge corresponds to a translation in momentum space. It is thus easy to check that in length gauge one has

$$|\Psi_{V\mathbf{p}}{}'(t, t_i)\rangle = e^{-iS_\mathbf{p}(t, t_i)}|\mathbf{p} + \mathbf{A}(t)\rangle \qquad \text{(length gauge)}, \qquad (11.49)$$

with the same action $S_\mathbf{p}(t, t_i)$ as in (11.48). In passing we note that in the original paper by Wolkow [66] the solution to the Dirac equation (i.e., relativistic and thus without dipole approximation) was presented.

11.4.2 Strong-Field Approximation (SFA)

The time-evolution operator $\hat{U}(t, t')$ satisfies the time-dependent Schrödinger equation (11.39),

$$i\partial_t \hat{U}(t, t') = [\hat{H}_0 + \hat{W}(t)]\hat{U}(t, t') . \qquad (11.50)$$

Its formal solution is given by the integral equations

$$\hat{U}(t, t') = \hat{U}_0(t, t') - i \int_{t'}^{t} dt'' \, \hat{U}(t, t'')\hat{W}(t'')\hat{U}_0(t'', t')$$

$$= \hat{U}_0(t, t') - i \int_{t'}^{t} dt'' \, \hat{U}_0(t, t'')\hat{W}(t'')\hat{U}(t'', t') , \qquad (11.51)$$

where $\hat{U}_0(t, t')$ is the evolution operator corresponding to the time-dependent Schrödinger equation with $\hat{H}_0$ only. Inserting (11.51) into the matrix element (11.37) leads to

$$M_\mathbf{p}(t_f, t_i) = -i \int_{t_i}^{t_f} dt' \, \langle \Psi_\mathbf{p}(t_f)|\hat{U}(t_f, t')\hat{W}(t')|\Psi_0(t')\rangle , \qquad (11.52)$$

where use of $\langle \Psi_\mathbf{p}(t_f)|\hat{U}_0(t_f, t_i)|\Psi_0(t_i)\rangle = \langle \Psi_\mathbf{p}(t_f)|\Psi_0(t_f)\rangle = 0$ was made because $|\Psi_\mathbf{p}(t_f)\rangle$ is a scattering state orthogonal to $|\Psi_0(t_f)\rangle$ and $\hat{U}_0(t', t_i)|\Psi_0(t_i)\rangle = |\Psi_0(t')\rangle$. Since the propagator $\hat{U}(t, t')$ also satisfies the integral equations

$$\hat{U}(t, t') = \hat{U}_V(t, t') - i \int_{t'}^{t} dt'' \, \hat{U}_V(t, t'')\hat{V}\hat{U}(t'', t')$$

$$= \hat{U}_V(t, t') - i \int_{t'}^{t} dt'' \, \hat{U}(t, t'')\hat{V}\hat{U}_V(t'', t') , \qquad (11.53)$$

where $\hat{U}_V(t, t')$ is the evolution operator corresponding to the time-dependent Schrödinger equation (11.47), one obtains, upon inserting (11.53) in (11.52) [42],

$$
M_{\mathbf{p}}(t_f, t_i) = -i\left[\int_{t_i}^{t_f} dt' \,\langle\Psi_{\mathbf{p}}(t_f)|\hat{U}_V(t_f, t')\hat{W}(t')|\Psi_0(t')\rangle\right.
$$
$$
\left. -i\int_{t_i}^{t_f} dt'' \int_{t''}^{t_f} dt' \,\langle\Psi_{\mathbf{p}}(t_f)|\hat{U}_V(t_f, t')\hat{V}\hat{U}(t', t'')\hat{W}(t'')|\Psi_0(t'')\rangle\right].
$$

$$(11.54)$$

Using $\int_{t_i}^{t_f} dt'' \int_{t''}^{t_f} dt' = \int_{t_i}^{t_f} dt' \int_{t_i}^{t_f} dt'' \,\Theta(t' - t'') = \int_{t_i}^{t_f} dt' \int_{t_i}^{t'} dt''$ expression (11.54) may be recast in the form [42]

$$
M_{\mathbf{p}}(t_f, t_i) = -i\int_{t_i}^{t_f} dt' \,\langle\Psi_{\mathbf{p}}(t_f)|\hat{U}_V(t_f, t')\left[\hat{W}(t')|\Psi_0(t')\rangle\right.
$$
$$
\left. -i\int_{t_i}^{t'} dt'' \,\hat{V}\hat{U}(t', t'')\hat{W}(t'')|\Psi_0(t'')\rangle\right].
$$

$$(11.55)$$

Equation (11.55) is still exact and gauge invariant. Whatever is missed in the first term of (11.55) is included in the second term where the full but unknown time-evolution operator $\hat{U}(t', t'')$ appears. Neglecting the second term, replacing the final state $|\Psi_{\mathbf{p}}(t_f)\rangle$ with a plane wave $|\mathbf{p}\rangle$, and making use of the expansion of the Volkov propagator into Volkov waves

$$
\hat{U}_V(t, t') = \int d^3k\,|\Psi_{V\mathbf{k}}(t, t_i)\rangle\langle\Psi_{V\mathbf{k}}(t', t_i)|
$$

$$(11.56)$$

(t_i is arbitrary since it cancels), we obtain, up to an irrelevant overall phase $\exp[-iS_{\mathbf{p}}(t_f, t_i)]$, the SFA or so-called Keldysh amplitude [36, 37, 22, 60]

$$
M_{K\mathbf{p}}(t_f, t_i) = -i\int_{t_i}^{t_f} dt\,\langle\Psi_{V\mathbf{p}}(t_f, t)|\hat{W}(t)|\Psi_0(t)\rangle.
$$

$$(11.57)$$

The SFA transition amplitude integrates over all ionization times t where the transition from the bound state $|\Psi_0(t)\rangle$ to the Volkov state $|\Psi V_{\mathbf{p}}(t_f, t)\rangle$, mediated by the interaction with the laser field $\hat{W}(t)$, may take place.[6] Expression (11.57) thus may be thought of replacing in a self-consistent manner Eq. (11.25) in the simple ad hoc model of a continuous slit.

In length gauge the matrix element (11.57) reads

$$
M_{K\mathbf{p}}(t_f, 0) = -i\int_0^{T_p} dt'\,e^{-iS_{\mathbf{p}}(t_f, t')}\langle\mathbf{p} + \mathbf{A}(t')|\mathbf{r}\cdot\mathbf{E}(t')|\Psi_0\rangle\,e^{-i\mathcal{E}_0 t'}.
$$

$$(11.58)$$

Here we assumed that the electric field $\mathbf{E}(t)$ is nonvanishing only for $t \in [t_i = 0, T_p]$. Multiplication by the constant phase factor $\exp[iS_{\mathbf{p}}(t_f, 0)]$ finally gives

---

[6] The second term in the matrix element (11.55) may be taken into account in a similar, approximate way, allowing for the description of rescattering processes [48].

$$M_{\mathbf{Kp}} = -i \int_0^{T_p} dt' \langle \mathbf{p} + \mathbf{A}(t') | \mathbf{r} \cdot \mathbf{E}(t') | \Psi_0 \rangle \, e^{i S_{\mathbf{p}, \mathcal{E}_0}(t')} , \qquad (11.59)$$

where

$$S_{\mathbf{p}, \mathcal{E}_0}(t) = \int_0^t dt' \left(\frac{1}{2} [\mathbf{p} + \mathbf{A}(t')]^2 - \mathcal{E}_0 \right) . \qquad (11.60)$$

In the case of a hydrogen-like ion starting from the ground state the matrix element needed in (11.59) reads

$$\langle \mathbf{k} | \mathbf{r} \cdot \mathbf{E}(t) | \Psi_0 \rangle = -i 2^{7/2} (2|\mathcal{E}_0|)^{5/4} \frac{\mathbf{k} \cdot \mathbf{E}(t)}{\pi (k^2 + 2|\mathcal{E}_0|)^3} . \qquad (11.61)$$

In experiments one usually measures the differential ionization probability $w_{\mathbf{p}}$, which is the probability to find an electron of final energy $\mathcal{E}_{\mathbf{p}} = p^2/2$ emitted in a certain direction, i.e., into the solid angle element $d\Omega_{\mathbf{p}}$ that is covered by the measuring device. The probability $w_{\mathbf{p}}$ is related to the transition matrix element $M_{\mathbf{p}}$ through

$$w_{\mathbf{p}} \underbrace{d\mathcal{E}_{\mathbf{p}}}_{p \, dp} d\Omega_{\mathbf{p}} = |M_{\mathbf{p}}|^2 \, d^3 p = |M_{\mathbf{p}}|^2 \, p^2 dp \, d\Omega_{\mathbf{p}}, \qquad (11.62)$$

so that

$$w_{\mathbf{p}} = p \, |M_{\mathbf{p}}|^2 . \qquad (11.63)$$

Hence, in order to calculate the SFA differential ionization probability $w_{\mathbf{Kp}} = p \, |M_{\mathbf{Kp}}|^2$ for a given final momentum $\mathbf{p}$ and hydrogen-like ions only the time integral in (11.59) remains to be evaluated. With nowadays computers, the brute force numerical evaluation of such an integral is so fast that an entire photoelectron momentum or angle-resolved energy spectrum can be obtained in a couple of seconds. However, more insight is gained if we employ the saddle-point approximation to evaluate the time integral in (11.59), as will be shown in the following.

11.4.3 Interfering Quantum Trajectories

The saddle-point times contributing most to the time integral in (11.59) are determined by the stationary phase equation $\partial_t S_{\mathbf{p}, \mathcal{E}_0}(t)|_{t=t_s} = 0$, which leads to

$$\frac{1}{2} [\mathbf{p} + \mathbf{A}(t_s)]^2 = \mathcal{E}_0 . \qquad (11.64)$$

Because $\mathcal{E}_0 < 0$ the saddle-point times t_s are necessarily complex. For the case of a hydrogen-like ion the matrix element (11.59) can be rewritten in terms of the saddle-point solutions as [48]

$$M_{\mathbf{Kp}} \simeq -2^{-1/2}(2|\mathcal{E}_0|)^{5/4} \sum_s \frac{\exp[iS_{ps}]}{S''_{ps}} \, , \tag{11.65}$$

where

$$S_{\mathbf{ps}} = S_{\mathbf{p},\mathcal{E}_0}(t_s) \tag{11.66}$$

and

$$S''_{ps} = \frac{d^2 S_{\mathbf{p},\mathcal{E}_0}(t)}{dt^2}\Bigg|_{t=t_s} = -\mathbf{E}(t_s) \cdot [\mathbf{p} + \mathbf{A}(t_s)] \, . \tag{11.67}$$

In the following we shall restrict ourselves to linearly polarized laser pulses where $\mathbf{A}(t) = A(t)\mathbf{e}_z$. The photoelectron momentum spectra then have azimuthal symmetry about the z-axis, and with $p_\parallel = \mathbf{p} \cdot \mathbf{e}_z$ and $\mathbf{p}_\perp = \mathbf{p} - p_\parallel \mathbf{e}_z$, Eq. (11.64) can be rewritten as

$$A(t_s) = -p_\parallel \pm i\sqrt{2|\mathcal{E}_0| + p_\perp^2} \, . \tag{11.68}$$

For an infinite pulse $A(t)$ is periodic and infinitely many saddle-point times t_s for a given final momentum $\mathbf{p}$ contribute in the sum (11.65), giving rise to discrete peaks in the photoelectron energy spectra, separated by ω. For a finite pulse only a finite number of solutions exist, and only those with the smallest imaginary part of S_{ps} are important. If there are two or more solutions of similar maximum weight, i.e., similar $\mathrm{Im}S_{ps}$, we expect an interference pattern in the photoelectron momentum or energy spectra. If, on the other hand, only one saddle-point solution is dominating the sum in (11.65), no interference pattern is expected.

The action (11.66) is complex but otherwise classical. One may therefore interpret the matrix element (11.65) as a sum over complex classical trajectories [48] or so-called quantum trajectories. Each saddle-point solution corresponds to a quantum trajectory which starts at the complex time t_s (ionization) and which arrives with the desired asymptotic momentum at a detector. The initial conditions can be chosen such that $\mathrm{Re}\,\mathbf{r}(t_s) = \mathbf{0}$, and all entities become real once the electron leaves from the tunnel exit $\mathbf{r}(\mathrm{Re}\,t_s)$ into the classically allowed region. We do not need to calculate these trajectories explicitly for our purposes. However, we would like to mention that they were successfully used as a starting point for incorporating Coulomb corrections into plain SFA (see Sect. 11.4.6).

The connection of (11.65) with the time slits discussed in Sect. 11.2 is now established. In Sect. 11.2 we put the time slits "by hand" upon assuming that at certain time instances wave packets appear in the continuum. Each of these Gaussian wave packets contained all momenta. Instead, in expression (11.65) each saddle-point solution corresponds to a time slit for a given final momentum. The corresponding quantum trajectory connects the time slit with the detector. Several quantum trajectories for a given final momentum may interfere. This gives rise to interference

patterns in the momentum spectra $|M_{Kp}|^2$ or differential energy spectra (11.63).[7]
While in (11.26) we put some weighting function $g(t)$ "by hand," it is automatically
taken care of in (11.65) via the imaginary part of S_{ps}: If the ionization probability
at time $\mathrm{Re}\, t_s$ is too low, the imaginary part of S_{ps} is large, thus suppressing the
contribution of this saddle point to the transition matrix element.

11.4.4 Analysis of Few-Cycle Ionization Dynamics: Attosecond Time Slits

Let us illustrate the saddle-point approach to the SFA for a particular example.
We assume a laser pulse of the form (11.28) with $\omega = 0.056$, $n_{cyc} = 4$, and
$\hat{E} = -\omega \hat{A} = 0.1$. We will first discuss the case $\varphi_{cep} = 0$ and concentrate on
photoelectron spectra along the laser polarization direction, i.e., $\mathbf{p}_\perp = \mathbf{0}$.

Figure 11.13 helps to analyze the dominating saddle-point solutions of (11.68)
for $\mathbf{p}_\perp = \mathbf{0}$ and $p_\| \in [-2.2, 2.2]$. Panel (a) shows the saddle points t_s in the
complex time plane. The vector potential $A(t)$ (solid) and the electric field $E(t)$
(dashed) are indicated in order to show that $\mathrm{Im}\, t_s$ is smallest whenever $|E(t)|$ has a
local maximum, i.e., ionization is probable. The signs $+$ and $-$ refer to the signs in
(11.68), the colors to positive (black) or negative (gray; red online) final momentum
$p_\|$, as is clearly seen in panel (b) where the final momentum $p_\|$ is plotted vs the
ionization time $\mathrm{Re}\, t_s$: The negative of the value of the vector potential at the time of
ionization determines the final momentum at the detector. High final momenta are
achieved if ionization occurs at low $|E(t)|$, i.e., high $|A(t)|$. However, the ionization
probability is expected to be low for small $|E(t)|$, as is confirmed in panel (c) where
$\mathrm{Im}\, S_{ps}$ is plotted, which determines the weight of the saddle-point solution in the
sum in (11.65). Hence, the dashed-dotted saddle-point solutions are the dominating
ones, followed by the dotted, dashed, and solid ones. We identify three dominant
time slits, centered around the three largest local maxima in $|E(t)|$. The central slit
("slit 1") is the most dominant one since it is connected to the absolute maximum of
the electric field, which coincides with the maximum of the pulse envelope for the
carrier-envelope phase chosen. Slits 2a and 2b have already a lower weight since
$\mathrm{Im}\, S_{ps}$ is larger. For our choice of the pulse (even number of cycles and carrier-
envelope phase $\varphi_{cep} = 0$), the saddle-point solutions for $\pm|p_\||$ are symmetric in
the sense that $\mathrm{Im}\, S_{ps}$ is independent of the sign of $p_\|$. As a consequence, the SFA
predicts a perfect "left/right symmetry" of the photoelectron momentum spectrum
so that it is sufficient to analyze the differential photoelectron energy spectrum in,
say, polarization direction $\mathbf{e}_z$, i.e., $p_\| > 0$.

Figure 11.14 shows the result of such an analysis where the number of time slits
in the evaluation of the matrix element (11.65) is increased stepwise. Including only

---

[7] In order to obtain an analytical formula for the spectra one may try to rewrite the discrete sum
(11.65) as an integral, i.e., to introduce a continuous slit [27].

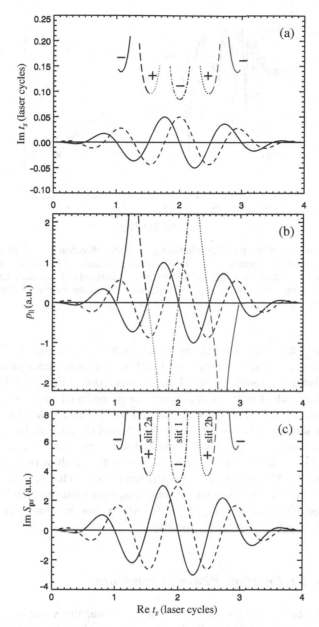

Fig. 11.13 (Color online) Saddle-point analysis for an $n_{cyc} = 4$-cycle linearly polarized $\sin^2$-pulse (11.28) with $\omega = 0.056$ and $\hat{E} = -\omega\hat{A} = 0.1$. In each of the panels, the vector potential $A(t)$ (*solid*) and the electric field $E(t)$ (*dashed*) are indicated (arbitrarily normalized) and the same color and line styles are used. (**a**) Saddle times t_s in the complex time plane; the signs $+$ and $-$ refer to the signs in (11.68), the colors to positive (*black*) or negative (*gray; red* online) final momentum $p_{\parallel}$; only saddle points within plus/minus one cycle around the pulse maximum are shown. (**b**) Final momentum $p_{\parallel}$ vs the ionization time Re t_s. (**c**) Imaginary part of the saddle-point action Im S_{ps} vs the ionization time Re t_s

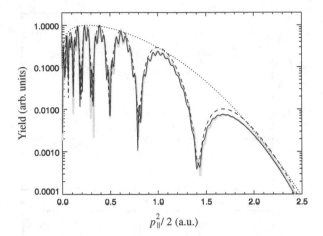

Fig. 11.14 Photoelectron spectra (11.63) in laser polarization direction for the pulse analyzed in Fig. 11.13. The exact SFA spectrum from the numerical integration of (11.59) (*gray* and *bold*) is compared with the saddle-point results (11.65), taking into account only the dominating quantum trajectory (*dotted*), the two most dominating ones (*dashed*), and the three most dominating ones (*solid*). All spectra are shifted vertically such that their maximum is at 1

the dominating saddle-point solutions drawn dash-dotted in Fig. 11.13 gives the dotted spectrum, i.e., just the overall slope without any interference pattern. Including the next branch of saddle-point solutions (drawn dotted in Fig. 11.13) gives the dashed spectrum, which reproduces correctly the pronounced interference pattern of the full, numerical SFA result (plotted gray and bold). These saddle-point solutions constitute the second half of slit 2a and the first half of slit 2b. Finally, incorporating also the second halves of slits 2a and 2b (dashed saddle-point solutions in Fig. 11.13) gives the spectrum drawn solid, which reproduces the small-scale modulations in the full, numerical SFA spectrum. The spectrum does not change anymore if more saddle-point solutions are taken into account (the next most important ones would be the solutions drawn solid in Fig. 11.13 for which, however, Im S_{ps} is already too large to have a sizable effect on the spectrum).

11.4.5 Carrier-Envelope Phase Dependence

Figure 11.13 shows that for the above-chosen pulse, the value $\varphi_{cep} = 0$ corresponds to the case of perfect left/right symmetry: Each time slit has equally weighted parts leading to left ($p_\parallel < 0$) and right ($p_\parallel > 0$) going electrons. Such a pulse is frequently called a "cosine pulse" since the electric field behaves like a cosine with respect to the maximum of the pulse envelope. It is obvious that the situation changes with changing φ_{cep}. The most asymmetric case is expected for $\varphi_{cep} = \pm\pi/2$, i.e., a "sine pulse." In fact, as is seen from Fig. 11.15 there are two groups of two equally weighted slits ("slit 1a," "slit 1b" and "slit 2a," "slit 2b,"

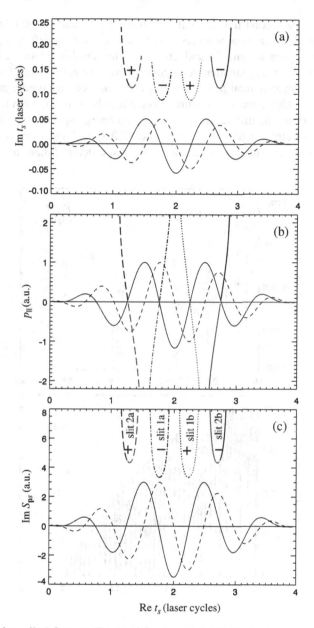

Fig. 11.15 (Color online) Same as Fig. 11.13 but for a "sine-like" pulse ($\varphi_{cep} = \pi/2$). Now there are two most dominant time slits (indicated "slit 1a" and "slit 1b" in panel (**c**)) and two next most important ones ("slit 2a" and "slit 2b"). However, the situation is now asymmetric with respect to the sign of the final momentum, as is seen from panel (**b**)

respectively). The second half of "slit 1a" and the first half of "slit 1b" give rise to right-going electrons now whereas it is vice versa for the left-going electrons. The opposite applies to slits 2a and 2b where the absolute value of the electric field, however, is already smaller. As a consequence, quantum trajectories leading to positive and negative final momentum $p_\parallel$ are weighted differently, and the symmetry in the photoelectron momentum spectrum is broken. This is clearly visible in Fig. 11.16 where the differential photoelectron energy spectra are shown for the "right-going" electrons with $p_\parallel > 0$ and the "left-going" electrons with $p_\parallel < 0$. The spectrum for the right-going electrons is less structured than the spectrum in

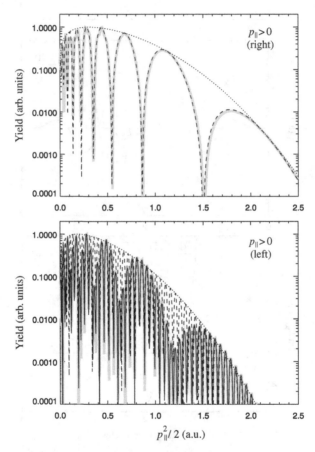

Fig. 11.16 Photoelectron spectra (11.63) in laser polarization direction (*upper panel*, "right-going" electrons) and in the opposite direction (*lower panel*, "left-going" electrons) for the pulse analyzed in Fig. 11.15. The exact SFA spectra from the numerical integration of (11.59) (*gray* and *bold*) are compared with the saddle-point results (11.65), taking into account only one of the dominating quantum trajectories starting from slit 1a or 1b for a given final momentum $p_\parallel$ (*dotted*), both trajectories from slit 1a *and* 1b (*dashed*), and all four trajectories (slits 1a, 1b, 2a, and 2b, *solid*). For the right-going electrons (*upper panel*) slits 2a and 2b are not important so that the interference pattern is less complex than for the left-going electrons (*lower panel*)

Fig. 11.14 for $\varphi_{cep} = 0$: The bigger scale interference pattern (arising from slits 1a and 1b) is similar but the small-scale modulations are absent because slits 2a and 2b are not important for the right-going electrons. However, for the left-going electrons the situation is the opposite: The small-scale modulations are much more pronounced than in Fig. 11.14 because quantum trajectories originating from both slit 2a *and* slit 2b are contributing to the spectrum of the left-going electrons. Hence, by varying the carrier-envelope phase between $+\pi/2$ and $-\pi/2$ the attosecond time slits are shifted and the relative weight of their contributions is affected. As a result the pronounced interference structure visible in Fig. 11.16, lower panel, can be continuously transferred from the left-going electrons to the right-going electrons, with the left/right-symmetric spectrum shown in Fig. 11.16 being the intermediate case $\varphi_{cep} = 0$.

11.4.6 Beyond the SFA

We do not want to give the wrong impression that the plain SFA (i.e., without rescattering and Coulomb corrections) introduced above is sufficient to understand all aspects of strong-field laser atom interaction. Replacing the final state $|\Psi_{\mathbf{p}}(t_f)\rangle$ by a plane wave $|\mathbf{p}\rangle$ and neglecting further interactions with the Coulomb potential in the transition from the (still exact) Eq. (11.55) to the Keldysh amplitude (11.57) is not as innocent as it may seem. First of all, gauge invariance is lost, which can even have consequences on a qualitative level, as was shown in [6]. The formal replacement of the final state and the omission of the second term in (11.55) may be physically interpreted as neglecting the Coulomb interaction between the electron and the ion once the electron is emitted. However, this Coulomb interaction after the emission of the electron influences significantly momentum distributions, including the left/right asymmetry discussed in Sect. 11.4.4 [13, 14], the interference structure of spectra, and the angular distributions of the photoelectrons in, e.g., elliptically polarized laser fields [56]. It is good news that the quantum orbit approach to the SFA introduced in Sect. 11.4.3 also offers the possibility to overcome these (and other) shortcomings [56, 57].

Acknowledgments DB is grateful to Sergei Popruzhenko for illuminating discussions and proofreading. GGP acknowledges support by The Welch Foundation (A-1562) and the Air Force Office of Scientific Research (FA9550-07-1-0069).

References

1. P. Agostini, F. Fabre, G. Mainfray, G. Petite, N. Rahman, Phys. Rev. Lett. **42**, 1127 (1979)
2. Y. Aharonov, D. Bohm, Phys. Rev. **122**, 1649 (1961)
3. M. Arndt, P. Szriftgiser, J. Dalibard, A.M. Steane, Phys. Rev. A **53**, 3369 (1996)
4. G. Badurek, H. Rauch, D. Tuppinger, Phys. Rev. A **34**, 2600 (1986)

5. A. Baltuška, T. Udem, M. Uiberacker, M. Hentschel, E. Goulielmakis, C. Gohle, R. Holzwarth, V.S. Yakovlev, A. Scrinzi, T.W. Hänsch, F. Krausz, Nature **421**, 611 (2003)
6. D. Bauer, D.B. Milošević, W. Becker, Phys. Rev. A **72**, 023415 (2005)
7. W. Becker, F. Grasbon, R. Kopold, D.B. Milošević, G.G. Paulus, H. Walther, Adv. At. Mol. Opt. Phys. **48**, 35 (2002)
8. M. Born, E. Wolf, *Principles of Optics* (Cambridge University Press, Cambridge, 2003)
9. H.R. Brown, J. Summhammer, R.E. Callaghan, P. Kaloyerou, Phys. Lett. A **163**, 21 (1992)
10. Č. Brukner, A. Zeilinger, Phys. Rev. Lett. **79**, 2599 (1997)
11. Č. Brukner, A. Zeilinger, Phys. Rev. A **56**, 3804 (1997)
12. P. Busch, in *Time in Quantum Mechanics*, J.G. Muga, R. Sala Mayato, I.L. Egusquiza (eds.) (Springer, Berlin Heidelberg, 2002), p. 69
13. S. Chelkowski, A.D. Bandrauk, A. Apolonski, Phys. Rev. A **70**, 013815 (2004)
14. S. Chelkowski, A.D. Bandrauk, Phys. Rev. A **71**, 053815 (2005)
15. S.-I. Chu, J. Cooper, Phys. Rev. A **32**, 2769 (1985)
16. R.J. Cook, R. Hill, Opt. Commun. **43**, 258 (1982)
17. P. Corkum, Phys. Rev. Lett. **71**, 1994 (1993)
18. C. Davisson, L.H. Germer, Nature **119**, 558 (1927)
19. A. del Campo, J.G. Muga, M. Moshinsky, J. Phys. B: At. Mol. Opt. Phys. **40**, 975 (2007)
20. A. del Campo, J.G. Muga, M. Kleber, Phys. Rev. A **77**, 013608 (2008)
21. W. Ehrenberg, R. Siday, Proc. Phys. Soc. B **62**, 8 (1949)
22. F.H.M. Faisal, J. Phys. B **6**, L89 (1973)
23. J. Felber, G. Müller, R. Gähler, R. Golub, Physica B **162**, 191 (1990)
24. M. Ferray, A. L'Huillier, X.F. Li, L.A. Lompre, G. Mainfray, C. Manus, J. Phys. B: At. Mol. Opt. Phys. **21**, L31 (1988)
25. T.F. Gallagher, Phys. Rev. Lett. **61**, 2304 (1988)
26. T.F. Gallagher, T. Scholz, Phys. Rev. A **40**, 2762 (1989)
27. S.P. Goreslavskiĭ, S.V. Popruzhenko, JEPT **90**, 778 (2000)
28. A. Góźdź, M. Dębicki, K. Stefańska, Phys. At. Nucl. **71**, 892 (2008)
29. Th. Hils, J. Felber, R. Gähler, W. Gläser, R. Golub, K. Habicht, P. Wille, Phys. Rev. A **58**, 4784 (1998)
30. L.P. Horwitz, Phys. Lett. A **355**, 1 (2006)
31. L.P. Horwitz, Found. Phys. **37**, 734 (2007)
32. K.L. Jánossy, Ann. Phys. **452**, 115 (1956)
33. C. Jönsson, Z. Phys. **161**, 454 (1961)
34. C. Jönsson, Am. J. Phys. **42**, 4 (1974)
35. D. Jones, S. Diddams, J. Ranka, A. Stentz, R. Windeler, J. Hall, S. Cundiff, Science **288**, 635 (2000)
36. L.V. Keldysh, Zh. Eksp. Teor. Fiz. **47**, 1945 (1964)
37. L.V. Keldysh, Sov. Phys. JETP **20**, 1307 (1965)
38. M. Kiffner, J. Evers, C.H. Keitel, Phys. Rev. Lett. **96**, 100403 (2006)
39. M. Kleber, Phys. Rep. **236**, 331 (1994)
40. K.C. Kulander, K.J. Schafer, K. Krause, in *Super-Intense Laser-Atom Physics*, B. Piraux, A. L'Huiller, K. Rzążewski (eds.) (Plenum Press, New York, 1993) p. 95
41. F. Lindner, M.G. Schätzel, H. Walther, A. Baltuška, E. Goulielmakis, F. Krausz, D.B. Milošević, D. Bauer, W. Becker, G.G. Paulus, Phys. Rev. Lett. **95**, 040401 (2005)
42. A. Lohr, M. Kleber, R. Kopold, W. Becker, Phys. Rev. A **55**, R4003 (1997)
43. L. Mandel, JOSA **49**, 931 (1959)
44. L. Mandelstam, I.G. Tamm, J. Phys. (USSR) **9**, 249 (1945)
45. V.I. Man'ko, M. Moshinsky, A. Sharma, Phys. Rev. A **59**, 1809 (1999)
46. J. Mauritsson, P. Johnsson, E. Gustafsson, A. L'Huillier, K.J. Schafer, M.B. Gaarde, Phys. Rev. Lett. **97**, 013001 (2006)
47. J. Mauritsson, P. Johnsson, E. Mansten, M. Swoboda, T. Ruchon, A. L'Huillier, K. Schafer, Phys. Rev. Lett. **100**, 073003 (2008)

48. D.B. Milošević, G.G. Paulus, D. Bauer, W. Becker, J. Phys. B: At. Mol. Opt. Phys. **39**, R203 (2006)
49. M. Moshinsky, Phys. Rev. **88**, 625 (1952)
50. M. Moshinsky, Am. J. Phys. **44**, 1037 (1976)
51. M. Nisoli, S. De Silvestri, O. Svelto, Appl. Phys. Lett. **68**, 2793 (1996)
52. M. Nisoli, S. De Silvestri, O. Svelto, R. Szipöcs, K. Ferencz, Ch. Spielmann, S. Sartania, F. Krausz, Opt. Lett. **22**, 522 (1997)
53. G.G. Paulus, F. Grasbon, H. Walther, P. Villoresi, M. Nisoli, S. Stagira, E. Priori, S. De Silvestri, Nature **414**, 182 (2001)
54. G.G. Paulus, F. Lindner, H. Walther, A. Baltuška, E. Goulielmakis, M. Lezius, F. Krausz, Phys. Rev. Lett. **91**, 253004 (2003)
55. V.S. Popov, Phys. Uspekhi **47**, 855 (2004)
56. S.V. Popruzhenko, G.G. Paulus, D. Bauer, Phys. Rev. A **77**, 053409 (2008)
57. S.V. Popruzhenko, D. Bauer, J. Mod. Opt. **55**, 2573 (2008)
58. H. Rauch, D. Petrascheck, in *Neutron Diffraction*, H. Dachs (ed.) (Springer, Berlin, 1978), p. 303
59. J. Reichert, R. Holzwarth, Th. Udem, T.W. Hänsch, Opt. Commun. **172**, 59 (1999)
60. H.R. Reiss, Phys. Rev. A **22**, 1786 (1980)
61. P. Salières, A. L'Huillier, Ph. Antoine, M. Lewenstein, Adv. At. Mol. Opt. Phys. **41**, 83 (1999)
62. R.M. Sillitto, C. Wykes, Phys. Lett. **39A**, 333 (1972)
63. P. Szriftgiser, D. Guéry-Odelin, M. Arndt, J. Dalibard, Phys. Rev. Lett. **77**, 4 (1996)
64. H. Telle, G. Steinmeyer, A. Dunlop, J. Stenger, D. Sutter, U. Keller, Appl. Phys. B **69**, 327 (1999)
65. H.B. van Linden van den Heuvell, H.G. Muller, in *Multiphoton Processes*, S.J. Smith, P.L. Knight (eds.) (Cambridge University Press, Cambridge, 1988)
66. D.M. Wolkow, Z. Phys. **94**, 250 (1935)

Chapter 12
Optimal Time Evolution for Hermitian and Non-Hermitian Hamiltonians

Carl M. Bender and Dorje C. Brody

> *The shortest path between two truths in the real domain passes through the complex domain.*
>
> Jacques Hadamard, *The Mathematical Intelligencer* **13** (1991)

12.1 Introduction

Interest in optimal time evolution dates back to the end of the seventeenth century, when the famous brachistochrone problem was solved almost simultaneously by Newton, Leibniz, l'Hôpital, and Jacob and Johann Bernoulli. The word *brachistochrone* is derived from Greek and means shortest time (of flight). The classical brachistochrone problem is stated as follows: A bead slides down a frictionless wire from point A to point B in a homogeneous gravitational field. What is the shape of the wire that minimizes the time of flight of the bead? The solution to this problem is that the optimal (fastest) time evolution is achieved when the wire takes the shape of a cycloid, which is the curve that is traced out by a point on a wheel that is rolling on flat ground.

In the past few years there has been much interest in the *quantum brachistochrone* problem, which is formulated in a somewhat similar fashion: Consider two fixed quantum states, an initial state $|\psi_I\rangle$ and a final state $|\psi_F\rangle$ in a Hilbert space. We then consider the set of all Hamiltonians satisfying the energy constraint that the difference between the largest and smallest eigenvalues is a fixed energy ω: $E_{\max} - E_{\min} = \omega$. Some of the Hamiltonians in this set allow the initial state $|\psi_I\rangle$ to evolve into the final state $|\psi_F\rangle$ in time t:

$$|\psi_F\rangle = \mathrm{e}^{-\mathrm{i}Ht/\hbar}|\psi_I\rangle . \qquad (12.1)$$

C.M. Bender (✉)
Department of Physics, Washington University, St. Louis, MO 63130, USA, cmb@wustl.edu

D.C. Brody
Department of Mathematics, Imperial College London, London SW7 2AZ, UK,
d.brody@imperial.ac.uk

Bender, C.M., Brody, D.C.: *Optimal Time Evolution for Hermitian and Non-Hermitian Hamiltonians.* Lect. Notes Phys. **789**, 341–361 (2009)
DOI 10.1007/978-3-642-03174-8_12   © Springer-Verlag Berlin Heidelberg 2009

The quantum brachistochrone problem is to find the *optimal* Hamiltonian, that is, the Hamiltonian that accomplishes this evolution in the shortest possible time, which we denote by τ.

In this chapter we show that for Hermitian Hamiltonians the shortest evolution time τ is a nonzero quantity whose size depends on the Hilbert-space distance between the fixed initial and final state vectors. However, for complex non-Hermitian Hamiltonians, the value of τ can be made arbitrarily small. Thus, non-Hermitian Hamiltonians permit arbitrarily fast time evolution.

Of course, a non-Hermitian Hamiltonian may be physically unrealistic because it may possess complex eigenvalues and/or it may generate nonunitary time evolution, that is, time evolution in which probability is not conserved. However, there is a special class of non-Hermitian Hamiltonians that are PT symmetric, that is, Hamiltonians that are invariant under combined space and time reflection. Although such Hamiltonians are not Hermitian in the Dirac sense, they *do* have entirely real spectra and give rise to unitary time evolution. Thus, such Hamiltonians define consistent and acceptable theories of quantum mechanics. We show in this chapter that if we use Hamiltonians of this type to solve the quantum brachistochrone problem, we can achieve arbitrarily fast time evolution without violating any principles of quantum mechanics. Thus, if it were possible to implement faster-than-Hermitian time evolution, then non-Hermitian Hamiltonians might have important applications in quantum computing.

This chapter is organized as follows: In Sect. 12.2 we introduce and describe PT quantum mechanics and explain how a Hamiltonian that is not Dirac Hermitian can still define a consistent theory of quantum mechanics. Then in Sect. 12.3 we explain why complex classical mechanics allows for faster-than-conventional time evolution. In Sect. 12.4 we discuss the quantum brachistochrone for Hermitian Hamiltonians. Then, in Sect. 12.5 we extend the discussion in Sect. 12.4 to Hamiltonians that are not Dirac Hermitian. In Sect. 12.6 we explain how it might be possible for a complex Hamiltonian to achieve faster-than-Hermitian time evolution.

12.2 *PT* Quantum Mechanics

Based on the training that one receives in a traditional quantum mechanics course, one would expect a theory defined by a non-Hermitian Hamiltonian to be physically unacceptable for a closed system[1] because the energy levels would most likely be complex and the time evolution would most likely be nonunitary (not probability conserving). However, theories defined by a special class of non-Hermitian Hamiltonians called PT-symmetric Hamiltonians can have positive real energy levels and can exhibit unitary time evolution. Such theories are consistent quantum theories. It

---

[1] N. of E.: For *open* systems non-Hermitian Hamiltonians and non unitary evolution may be perfectly physical, see e.g., Chap. 6 by G. Hegerfeldt or Chap. 4 by A. Ruschhaupt et al., this volume.

may be possible to distinguish these theories experimentally from theories defined by Hermitian Hamiltonians because, in principle, non-Hermitian Hamiltonians can be used to generate arbitrarily fast time evolution.

We use the following terminology in this chapter: By *Hermitian*, we mean *Dirac* Hermitian, where the Dirac Hermitian adjoint symbol† represents combined matrix transposition and complex conjugation. The *parity operator* P performs spatial reflection and thus in quantum mechanics it changes the sign of the position operator x and the momentum operator p: $PxP = -x$ and $PpP = -p$. Because the parity operator P is a reflection operator, its square is the unit operator: $P^2 = 1$. The *time-reversal operator* T performs the time reflection $t \rightarrow -t$, and thus it changes the sign of the momentum operator p, $TpT = -p$, but it leaves the position operator invariant: $TxT = x$. The square of T is the unit operator $T^2 = 1$. We require that the operators P and T individually leave invariant the fundamental Heisenberg algebra of quantum mechanics $[x, p] = i$. Thus, while P is a linear operator, we see that T must perform *complex conjugation* $TzT = z^*$, and hence T is an *antilinear* operator.[2]

The first class of PT-symmetric quantum mechanical Hamiltonians was introduced in 1998 [6]. Since then there have been many papers on this subject by a wide range of authors. There have also been three recent review articles [4, 5, 30]. In [6] it was discovered that even if a Hamiltonian is not Hermitian, its energy levels can be all real and positive so long as the eigenfunctions are symmetric under PT reflection.

These new kinds of Hamiltonians are obtained by deforming ordinary Hermitian Hamiltonians into the complex domain. The original class of PT-symmetric Hamiltonians that was proposed in [6] has the form

$$H = p^2 + x^2(ix)^\varepsilon \qquad (\varepsilon > 0) \,, \tag{12.2}$$

where ε is a real deformation parameter. Two particularly interesting special cases are obtained by setting $\varepsilon = 1$ to get $H = p^2 + ix^3$ and by setting $\varepsilon = 2$ to get $H = p^2 - x^4$. Surprisingly, these Hamiltonians have real, positive, discrete energy levels even though the potential for $\varepsilon = 1$ is imaginary and the potential for $\varepsilon = 2$ is upside down. The first complete proof of spectral reality and positivity for H in (12.2) was given by Dorey et al. [28, 29].

The philosophical background of PT quantum mechanics is simply this: One of the axioms of quantum mechanics requires that the Hamiltonian H be Dirac Hermitian. This axiom is distinct from all other quantum mechanical axioms because it is mathematical rather than physical in character. The other axioms of quantum mechanics are stated in physical terms; these other axioms require locality, causality, stability and uniqueness of the vacuum state, conservation of probability, Lorentz

---

[2] Another way to see that T is associated with complex conjugation is to require that the time-dependent Schrödinger equation be invariant under time reversal. This implies that time reflection $t \rightarrow -t$ must be accompanied by complex conjugation i $\rightarrow$ $-$i. See [60] for a discussion of the properties of antilinear operators.

invariance, and so on. The condition of Dirac Hermiticity $H = H^\dagger$ is mathematical, but the condition of PT symmetry $H = H^{PT} = (PT)H(PT)$ (space–time reflection symmetry) is physical because P and T are elements of the Lorentz group.

The spectrum of H in (12.2) is real, which poses the question of whether this Hamiltonian specifies a *quantum mechanical* theory. That is, is the theory specified by H associated with a Hilbert space endowed with a positive inner product and does H specify unitary (norm-preserving) time evolution? The answer to these questions is *yes*. Positivity of the inner product and unitary time evolution was established in [11, 12] for quantum-mechanical systems having an unbroken PT symmetry (an analogous result was obtained by Mostafazadeh [51]) and in [13] for quantum field theory.

To demonstrate that the theory specified by H in (12.2) is a quantum mechanical theory, we construct a linear operator C that satisfies the three simultaneous algebraic equations [11]: $C^2 = \mathbb{1}$, $[C, PT] = 0$, and $[C, H] = 0$. Using C, which in quantum field theory is a Lorentz scalar [8], we can then construct the appropriate inner product for a PT-symmetric Hamiltonian: $\langle a|b \rangle \equiv a^{CPT} \cdot b$. This inner product, which uses the CPT adjoint, has a strictly positive norm: $\langle a|a \rangle > 0$. Because H commutes with both PT and C, H is *self-adjoint* with respect to CPT conjugation. Also, the time-evolution operator $e^{-iHt/\hbar}$ is unitary with respect to CPT conjugation. Note that the Hilbert space and the CPT inner product are *dynamically determined* by the Hamiltonian itself.

We have explained why a PT-symmetric Hamiltonian gives rise to a unitary theory, but in doing so we raise the question of whether PT-symmetric Hamiltonians are useful. The answer to this question is simply that PT-symmetric Hamiltonians have *already* been useful in many areas of physics. For example, in 1959 Wu showed that the ground state of a Bose system of hard spheres is described by a non-Hermitian Hamiltonian [61]. Wu found that the ground-state energy of this system is real and he conjectured that all of the energy levels were real. Hollowood showed that the non-Hermitian Hamiltonian for a complex Toda lattice has real energy levels [42]. Cubic non-Hermitian Hamiltonians of the form $H = p^2 + ix^3$ (and also cubic quantum field theories having an imaginary self-coupling term) arise in studies of the Lee-Yang edge singularity [25, 26, 32, 62] and in various Reggeon field-theory models [23, 39, 40]. In all of these cases a non-Hermitian Hamiltonian having a real spectrum appeared mysterious at the time, but now the explanation is clear: In every case the non-Hermitian Hamiltonian is PT symmetric. Hamiltonians having PT symmetry have also been used to describe magnetohydrodynamic systems [35, 38] and to study nondissipative time-dependent systems interacting with electromagnetic fields [31].

An important application of PT quantum mechanics is in the revitalization of theories that have been thought to be dead because they appear to have ghosts. *Ghosts* are states having negative norm. We have explained above that in order to construct the quantum mechanical theory defined by a PT-symmetric Hamiltonian, we must construct the appropriate adjoint from the C operator. Having constructed the CPT adjoint, one may find that the so-called ghost state is actually not a ghost at all because when its norm is calculated using the appropriate definition of the adjoint, the norm turns out to be positive. This is what happens in the case of the Lee model.

The Lee model was proposed in 1954 as a quantum field theory in which mass, wave-function, and charge renormalization could be performed exactly and in closed form [48]. However, in 1955 Källén and Pauli showed that when the renormalized coupling constant is larger than a critical value, the Hamiltonian becomes non-Hermitian (in the Dirac sense) and a ghost state appears [45]. The importance of the work of Källén and Pauli was emphasized by Salam in his review of their paper [58] and the appearance of the ghost was assumed to be a fundamental defect of the Lee model. However, in 2005 it was shown that the non-Hermitian Lee-model Hamiltonian is PT symmetric and when the norms of the states of this model are determined using the C operator, which can be calculated exactly and in closed form, the ghost state is seen to be an ordinary physical state having positive norm [9]. Thus, the following assertion by Barton [3] is *not correct*: "A non-Hermitian Hamiltonian is unacceptable partly because it may lead to complex energy eigenvalues, but chiefly because it implies a nonunitary S matrix, which fails to conserve probability and makes a hash of the physical interpretation."

Another example of a quantum model that was thought to have ghost states, but in fact does not, is the Pais–Uhlenbeck oscillator model [16–18]. This model has a fourth-order field equation, and for the past several decades it was thought (incorrectly) that all such higher-order field equations lead inevitably to ghosts. It is shown in [16] that when the Pais–Uhlenbeck model is quantized using the methods of PT quantum mechanics, it does not have any ghost states at all.

There are many potential applications for PT quantum mechanics in areas such as particle physics, cosmology, gravitation, quantum field theory, and solid-state physics. These applications are discussed in detail in the recent review article in [5]. Furthermore, there are now indications that theories described by PT-symmetric Hamiltonians can be observed in table-top experiments [49, 55, 56].

Having shown the validity and potential usefulness of PT quantum mechanics, one may ask why PT quantum mechanics works. The reason is that CP is a positive operator, and thus it can be written as the exponential of another operator Q: $CP = e^Q$. The square root of e^Q can then be used to construct a new Hamiltonian $\tilde{H}$ via a similarity transformation on the PT-symmetric Hamiltonian H: $\tilde{H} \equiv e^{-Q/2} H e^{Q/2}$. The new Hamiltonian $\tilde{H}$ has the same energy eigenvalues as the original Hamiltonian H because a similarity transformation is isospectral. Moreover, $\tilde{H}$ is *Dirac Hermitian* [52]; PT quantum mechanics works because there is an isospectral equivalence between a non-Hermitian PT-symmetric Hamiltonian and a conventional Dirac Hermitian Hamiltonian.

There are a number of elementary examples of this equivalence, but a nontrivial illustration is provided by the Hamiltonian H in (12.2) at $\varepsilon = 2$. This Hamiltonian is not Hermitian because boundary conditions that violate the L^2 norm must be imposed in Stokes wedges in the complex plane in order to obtain a real, positive, discrete spectrum. The exact equivalent Hermitian Hamiltonian is $\tilde{H} = p^2 + 4x^4 - 2\hbar x$, where $\hbar$ is Planck's constant [10, 24, 44]. The term proportional to $\hbar$ vanishes in the classical limit and is thus an example of a quantum anomaly.

Since PT symmetry is equivalent by means of a similarity transformation to conventional Dirac Hermiticity, one may wonder whether PT quantum mechanics is actually fundamentally different from ordinary quantum mechanics. The answer is *yes* and, at least in principle, there is an experimentally observable difference between PT-symmetric and ordinary Dirac Hermitian Hamiltonians. The quantum brachistochrone provides a setting for examining this difference and provides a way to discriminate between the class of PT-symmetric Hamiltonians and the class of Dirac Hermitian Hamiltonians.

12.3 Complex Classical Motion

It is implicitly assumed in the derivation of the classical brachistochrone that the path of shortest time of descent is *real*. However, it is interesting that if one allows for the possibility of complex paths of motion, one can achieve an even shorter time of flight. In this section we consider a simple classical-mechanical system. Our purpose is to explain heuristically how extending a dynamical system into the complex domain can result in faster-than-real time evolution.

To demonstrate that a shorter time of flight can be achieved by means of complex paths, let us consider the classical harmonic oscillator, whose Hamiltonian is

$$H = p^2 + x^2 . \tag{12.3}$$

If a particle has energy $E = 1$, then the classical turning points of the motion of the particle are located at $x = \pm 1$. The particle undergoes simple harmonic motion in which it oscillates sinusoidally between these two turning points. This periodic motion is indicated in Fig. 12.1 by a solid line connecting the turning points. However, in addition to this oscillatory motion on the real-x axis, there are an infinite number of other trajectories that a particle of energy E can have [7]. These classical trajectories, which are also shown in Fig. 12.1, are all ellipses whose foci are located at precisely the positions of the turning points. All of the classical orbits are periodic, and all orbits have the same period $T = 2\pi$. Thus, a classical particle travels faster along more distant ellipses.

Now suppose that a classical particle of energy $E = 1$ travels along the real-x axis from some point $x = -a$ to $x = a$, where $a > 1$. If the potential $V(x)$ is everywhere zero along its path, then it will travel at a constant velocity. However, suppose that the particle suddenly finds itself in the parabolic potential $V(x) = x^2$ when it reaches the turning point at $x = -1$ and that it suddenly escapes the influence of this potential at $x = 1$. Then, the time of flight from $x = -a$ to $x = a$ will be changed because the particle does not travel at a constant velocity between the turning points. Next, let us imagine that the potential $V(x) = x^2$ is suddenly turned on *before* the particle reaches the turning point at $x = -1$. In this case, the particle will follow one of the elliptical paths in the complex plane around the positive real axis. Just as the particle reaches the positive real axis the potential is turned off, so the particle proceeds onward along the real axis until it reaches

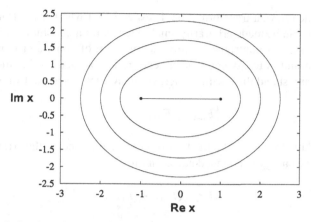

Fig. 12.1 Classical trajectories in the complex-x plane for the harmonic oscillator whose Hamiltonian is $H = p^2 + x^2$. These trajectories represent the possible paths of a particle whose energy is $E = 1$. The trajectories are nested ellipses with foci located at the turning points at $x = \pm 1$. The real *line segment* (degenerate *ellipse*) connecting the turning points is the usual periodic classical solution to the harmonic oscillator. All closed paths have the same period 2π

$x = a$. This trip will take less time because the particle travels faster along the ellipse in the complex plane.

We have arrived at the surprising conclusion that if the classical particle enters the parabolic potential $V(x) = x^2$ immediately after it begins its voyage up the real axis, its time of flight will be exactly half a period, or π. Indeed, by traveling in the complex plane, a particle of energy $E = 1$ can go from the point $x = -a$ to the point $x = a$ in time π, no matter how large a is. Evidently, if a particle is allowed to follow complex classical trajectories, then it is possible to make a drastic reduction in its time of flight between two given real points.

12.4 Hermitian Quantum Brachistochrone

The quantum brachistochrone problem, as described briefly in Sect. 12.1, is similar to the classical counterpart except that the optimization takes place in a Hilbert space. Specifically, we are given a pair of quantum states, an initial state $|\psi_I\rangle$ and the final state $|\psi_F\rangle$, and we would like to find the one-parameter family of unitary operators $\{U_t\}$ that achieves the transformation $|\psi_I\rangle \rightarrow |\psi_F\rangle = U_t|\psi_I\rangle$ in the smallest possible time t. Since a one-parameter family of unitary operators can be formed in terms of a Hermitian operator H as $U_t = \exp(-iHt/\hbar)$, the problem is equivalent to finding the Hermitian operator H that realizes the transformation $|\psi_I\rangle \rightarrow |\psi_F\rangle$ in the shortest possible time.

The Hermitian operator H can be thought of as representing the Hamiltonian, so the quantum brachistochrone problem is equivalent to finding the optimal Hamiltonian H satisfying $\exp(-iHt/\hbar)|\psi_I\rangle = |\psi_F\rangle$. However, it is intuitively clear that if we are allowed to have access to an unbounded energy resource, then the time required for the relevant transformation, irrespective of whether the Hamiltonian is optimal or not, can be made arbitrarily small. Hence, for a quantum brachistochrone problem to possess a nontrivial solution, some form of constraint is needed. The simplest constraint is to assume that the energy is bounded so that the difference between the largest and the smallest energy eigenvalues has a fixed value ω:

$$E_{max} - E_{min} = \omega . \tag{12.4}$$

A short calculation shows that if the Hamiltonian H is bounded, then (i) the standard deviation of the energy is bounded according to

$$\Delta H \leq \tfrac{1}{2}(E_{max} - E_{min}) , \tag{12.5}$$

and (ii) the state with maximum energy uncertainty is $(|E_{max}\rangle + |E_{min}\rangle)/\sqrt{2}$. It follows that the energy constraint (12.4) is equivalent to a constraint on energy uncertainty.

The brachistochrone problem of this type has been analyzed recently and a solution was obtained by means of a variational method [27]. It has also been solved in terms of a more elementary approach making use of the geometry of quantum state space [20, 21]. We shall follow closely the latter approach here.

Let us now state the simplest form of the quantum brachistochrone problem: We have a quantum system represented by an N-dimensional Hilbert space $\mathcal{H}$ and a prescribed pair of states $|\psi_I\rangle$ and $|\psi_F\rangle$ on $\mathcal{H}$. The problem is (a) to find the Hamiltonian H satisfying the constraint (12.4) such that the unitary transformation $\exp(-iHt/\hbar)|\psi_I\rangle = |\psi_F\rangle$ is achieved in the shortest possible time and (b) to find the time required to realize such an operation.

A little geometric intuition allows us to find the solution to this problem with minimum effort. Recall that the time required for transporting a state along a path in $\mathcal{H}$ is given by the *distance* divided by the *speed*. Hence, all we have to do is first identify the shortest path and measure its length and then allow the state to evolve along the path with the greatest possible speed without violating the constraint (12.4).

In quantum mechanics the notion of distance is closely linked to the notion of transition probability [22, 43]. In particular, by looking at the transition probability between neighboring states we can derive the expression for the metric on the space of quantum states. This allows us to measure distances between states. The idea can be sketched as follows: Consider a state $|\psi\rangle$ in $\mathcal{H}$ and a neighboring state $|\psi\rangle + |d\psi\rangle$. The transition probability between these states is

$$\cos^2\left(\tfrac{1}{2}ds\right) = \frac{(\langle\psi| + \langle d\psi|)|\psi\rangle\langle\psi|(|\psi\rangle + |d\psi\rangle)}{\langle\psi|\psi\rangle(\langle\psi| + \langle d\psi|)(|\psi\rangle + |d\psi\rangle)} , \tag{12.6}$$

where ds defines the line element on the space of pure states. By using

$$\cos^2\left(\tfrac{1}{2}ds\right) \approx 1 - \frac{1}{4}ds^2 , \tag{12.7}$$

expanding the right side of (12.6), and retaining terms of quadratic order, we find
that the line element is

$$ds^2 = 4\frac{\langle\psi|\psi\rangle\langle d\psi|d\psi\rangle - \langle\psi|d\psi\rangle\langle d\psi|\psi\rangle}{\langle\psi|\psi\rangle^2} . \tag{12.8}$$

This line element is known in geometry to arise from the *Fubini–Study metric* [47]
and it can be used to measure the distance of the shortest path joining a pair of points
on the space of pure quantum states.

If the Hilbert space is two dimensional, then a generic normalized state vector
$|\psi\rangle$ can be expressed in the form

$$|\psi\rangle = \begin{pmatrix} \cos\frac{1}{2}\theta \\ \sin\frac{1}{2}\theta\, e^{i\phi} \end{pmatrix} . \tag{12.9}$$

A short calculation then shows that the Fubini–Study line element (12.8) reduces in
this case to the expression

$$ds^2 = d\theta^2 + \sin^2\theta\, d\phi^2 , \tag{12.10}$$

which we recognize as the line element on the Bloch sphere $\mathcal{S}$. (The Bloch sphere
is the state space of two-level systems.)

In the case of an N-dimensional Hilbert space $\mathcal{H}$, if we are given a pair of dis-
tinct states $|\psi_I\rangle$ and $|\psi_F\rangle$, then the linear span of these two states forms a two-
dimensional subspace of $\mathcal{H}$. It should be intuitively clear that the shortest path join-
ing $|\psi_I\rangle$ and $|\psi_F\rangle$ should lie on this two-dimensional subspace. Thus, irrespective of
the dimensionality of $\mathcal{H}$ *the solution to our quantum brachistochrone problem can
be obtained by analyzing the two-dimensional subspace spanned by* $|\psi_I\rangle$ *and* $|\psi_F\rangle$.
Even when we restrict our attention to this subspace, there still remain infinitely
many unitary orbits that realize the transformation $|\psi_I\rangle \rightarrow |\psi_F\rangle = U_t|\psi_I\rangle$. How-
ever, since the two-dimensional state space is just the Bloch sphere $\mathcal{S}$ endowed with
the spherical metric (12.10), we see that there is a unique great circle arc that joins
$|\psi_I\rangle$ and $|\psi_F\rangle$ on $\mathcal{S}$. (This assumes, of course, that $|\psi_I\rangle$ and $|\psi_F\rangle$ are not antipodal
points of the sphere. Otherwise, there are infinitely many such paths.) In this way
we have identified the shortest path joining $|\psi_I\rangle$ and $|\psi_F\rangle$. The shortest distance
s_{min} between these two points of $\mathcal{S}$ is thus given by

$$s_{min} = 2\arccos\left(\frac{|\langle\psi_I|\psi_F\rangle|}{\sqrt{\langle\psi_I|\psi_I\rangle\langle\psi_F|\psi_F\rangle}}\right) . \tag{12.11}$$

This result can also be obtained by integrating the line element (12.10) along the
great circle arc on $\mathcal{S}$.

Having obtained the distance of the shortest path we proceed to find the maximum speed at which the state can evolve unitarily. For the evolution of the state

we must consider the general Schrödinger equation, but we also need to express the equation in the correct form. This is the so-called modified Schrödinger equation

$$\frac{d|\psi_t\rangle}{dt} = -\frac{i}{\hbar} \tilde{H}|\psi_t\rangle \, . \tag{12.12}$$

In this equation the mean-adjusted Hamiltonian $\tilde{H}$ is given by

$$\tilde{H} = H - \langle H \rangle \, , \tag{12.13}$$

where

$$\langle H \rangle = \frac{\langle \psi | H | \psi \rangle}{\langle \psi | \psi \rangle} \, . \tag{12.14}$$

Note that $\langle \tilde{H} \rangle = 0$ and that according to (12.12) the tangent vector $\frac{d}{dt}|\psi_t\rangle$ is everywhere orthogonal to the direction of the state [46]. Since the energy expectation $\langle H \rangle$ depends on the state $|\psi\rangle$, the modified Schrödinger equation appears to be nonlinear. However, this is not the case. The point is that the expectation value of the Hamiltonian is a constant of the motion under the Schrödinger dynamics. Thus, given the initial state $|\psi_I\rangle$, we calculate $\langle H \rangle$ in this state and subtract this number from the Hamiltonian. Since the Hamiltonian in quantum mechanics is defined only up to an additive constant, this modification does not alter the physics in any way. It is worthwhile noting that the modified Schrödinger equation (12.12) is canonical and reduces to the standard eigenvalue problem when the state $|\psi_t\rangle$ is time independent without one having to evoke the correspondence principle [22].

If the initial state vector $|\psi_I\rangle$ is normalized, then the evolution (12.12) preserves the norm. It follows that $\langle \psi | d\psi \rangle = 0$. Since the speed v of quantum evolution is given by $v = ds/dt$, we find from (12.8) and (12.12) that

$$v^2 = \frac{4}{\hbar^2} \langle \psi_t | (H - \langle H \rangle)^2 | \psi_t \rangle = \frac{4}{\hbar^2} \langle \psi_I | (H - \langle H \rangle)^2 | \psi_I \rangle \, . \tag{12.15}$$

This shows that the speed of quantum evolution is given by the energy uncertainty. The expression (12.15) for the speed of quantum evolution is known as the *Anandan–Aharonov relation* [1]. Since we know from (12.5) that under the constraint (12.4) the energy uncertainty ΔH is bounded by $\frac{1}{2}\omega$, we find that the maximum speed of quantum evolution is given by

$$v_{\max} = \frac{\omega}{\hbar} \, . \tag{12.16}$$

By using the results in (12.11) and (12.16) we deduce that the minimum time required for realizing the unitary transportation $|\psi_I\rangle \rightarrow |\psi_F\rangle = U_t|\psi_I\rangle$ is given by the ratio $s_{\min}/v_{\max}$. In particular, if $|\psi_I\rangle$ and $|\psi_F\rangle$ are orthogonal, then they correspond to antipodal points on the Bloch sphere $\mathcal{S}$, and we have $s_{\min} = \pi$. In

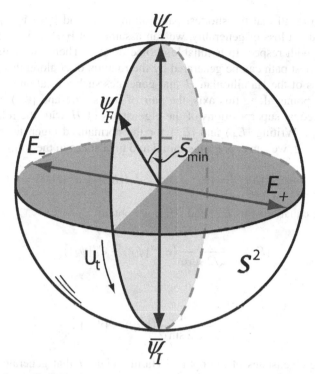

Fig. 12.2 (Color online) Optimal Hamiltonian for quantum state transportation. The two-dimensional complex Hilbert space spanned by the initial state $|\psi_I\rangle$ and the final state $|\psi_F\rangle$ can be visualized in real terms as a Bloch sphere $\mathcal{S}$. The two states $|\psi_I\rangle$ and $|\psi_F\rangle$ can then be identified as a pair of points on $\mathcal{S}$. Assuming that these two points are not antipodal, there exists a unique great circle arc joining these two points, which determines the shortest path joining the two states. The optimal way of unitarily transporting $|\psi_I\rangle$ into $|\psi_F\rangle$ is therefore to rotate the *sphere* along the axis orthogonal to the great *circle*. The axis of rotation then specifies two quantum states: $|E_-\rangle$ and $|E_+\rangle$. These states are the eigenstates of the optimal Hamiltonian H

this case, the minimum time required to orthogonalize the state (that is, for the state to evolve into a new state that is orthogonal to the original state) is known as the *passage time* τ_P [19, 59]. The passage time is explicitly

$$\tau_P = \frac{\pi\hbar}{2\Delta H} = \frac{\pi\hbar}{\omega}. \tag{12.17}$$

The passage time (12.17) provides the bound in Hermitian quantum mechanics for transporting a state into an orthogonal state and is sometimes referred to as the *Fleming bound* [33].

The ratio s_{min}/v_{max} gives the solution to part (b) of our quantum brachistochrone problem. To solve part (a), that is, to find the optimal Hamiltonian, we argue as follows: Since the problem is confined to a two-dimensional subspace of $\mathcal{H}$, the solution can be obtained by elementary trigonometry on the Bloch sphere $\mathcal{S}$. The key idea is to recall that the shortest path joining $|\psi_I\rangle$ and $|\psi_F\rangle$ is a great circle arc on S. Without loss of generality, we can assume that $|\psi_I\rangle$ and $|\psi_F\rangle$ lie on the equator of S with respect to a suitable choice of axis. Then, the unitary motion along the shortest path can be generated by the rotation of S along this axis. Since the eigenstates of the Hamiltonian H that generates such a rotation correspond to the antipodal points along this axis, the pair of states $|\psi_I\rangle$ and $|\psi_F\rangle$ can both be expressed as equal superpositions of the eigenstates of H with the relative phase shifted by $s_{\min}$. Writing $|E_+\rangle$ and $|E_-\rangle$ for the normalized eigenstates of H and using $\alpha = s_{\min}/2$, we can express the initial and final states in the form

$$\tfrac{1}{\sqrt{2}}\left(|E_-\rangle + e^{-i\alpha}|E_+\rangle\right) = |\psi_I\rangle \quad \text{and} \quad \tfrac{1}{\sqrt{2}}\left(|E_-\rangle + e^{i\alpha}|E_+\rangle\right) = |\psi_F\rangle . \quad (12.18)$$

Solving these equations for $|E_+\rangle$ and $|E_-\rangle$, we obtain

$$|E_-\rangle = \frac{i}{\sqrt{2}\sin\alpha}\left(e^{-i\alpha}|\psi_F\rangle - e^{i\alpha}|\psi_I\rangle\right) \quad (12.19)$$

and

$$|E_+\rangle = -\frac{i}{\sqrt{2}\sin\alpha}\left(|\psi_F\rangle - |\psi_I\rangle\right) . \quad (12.20)$$

These are the eigenstates of the optimal Hamiltonian H that generate the unitary motion $|\psi_I\rangle \rightarrow |\psi_F\rangle = U_t|\psi_I\rangle$ along the shortest path. The eigenvalues of the optimal H can be arbitrary as long as condition (12.4) is satisfied. Without loss of generality, we may assume H to be trace free, and we obtain the solution to part (a) of the quantum brachistochrone problem:

$$H = \tfrac{1}{2}\omega|E_+\rangle\langle E_+| - \tfrac{1}{2}\omega|E_-\rangle\langle E_-| . \quad (12.21)$$

This is the "minimal" solution to the problem in the sense that H acts only on the two-dimensional subspace of $\mathcal{H}$ while leaving the rest of $\mathcal{H}$ unperturbed.

As a special example, consider the problem of a spin flip, that is, turning a spin-up state into a spin-down state unitarily. In this case the initial and final states can be written as

$$|\psi_I\rangle = \begin{pmatrix} 1 \\ 0 \end{pmatrix} \quad \text{and} \quad |\psi_F\rangle = \begin{pmatrix} 0 \\ 1 \end{pmatrix} \quad (12.22)$$

in the spin-z basis. Substituting these into (12.19) and (12.20), we find that the eigenstates of the optimal Hamiltonian are

$$|E_-\rangle = \frac{1}{\sqrt{2}}\begin{pmatrix} 1 \\ 1 \end{pmatrix} \quad \text{and} \quad |E_+\rangle = \frac{i}{\sqrt{2}}\begin{pmatrix} 1 \\ -1 \end{pmatrix} \quad (12.23)$$

because in this case we have $\alpha = \pi/2$. Substituting this result into (12.21) yields

$$H = \frac{1}{2} \begin{pmatrix} 0 & -\omega \\ -\omega & 0 \end{pmatrix} \tag{12.24}$$

for the optimal Hamiltonian. By using the relation

$$e^{i\phi\sigma\cdot\mathbf{n}} = \cos\phi\mathbb{1} + i\sin\phi\sigma\cdot\mathbf{n}, \tag{12.25}$$

where $\mathbf{n}$ is a unit vector and

$$\sigma_1 = \begin{pmatrix} 0 & 1 \\ 1 & 0 \end{pmatrix}, \quad \sigma_2 = \begin{pmatrix} 0 & -i \\ i & 0 \end{pmatrix}, \quad \sigma_3 = \begin{pmatrix} 1 & 0 \\ 0 & -1 \end{pmatrix} \tag{12.26}$$

are Pauli matrices, we obtain the following expression for the optimal unitary operator:

$$U_t = \begin{pmatrix} \cos\left(\frac{\omega t}{2\hbar}\right) & -i\sin\left(\frac{\omega t}{2\hbar}\right) \\ -i\sin\left(\frac{\omega t}{2\hbar}\right) & \cos\left(\frac{\omega t}{2\hbar}\right) \end{pmatrix}. \tag{12.27}$$

It follows at once that the optimal unitary orbit $|\psi_t\rangle = U_t|\psi_I\rangle$ is given by

$$|\psi_t\rangle = \begin{pmatrix} \cos\left(\frac{\omega t}{2\hbar}\right) \\ -i\sin\left(\frac{\omega t}{2\hbar}\right) \end{pmatrix}. \tag{12.28}$$

We find that the first time at which $|\psi_t\rangle$ reaches $|\psi_F\rangle$ is given by the condition $\omega t/2\hbar = \pi/2$, that is, when $t = \tau_P$, where τ_P is the passage time given in (12.17).

We have seen how the simplest form of a quantum brachistochrone problem can be solved in Hermitian quantum mechanics by considering a two-dimensional Hilbert subspace combined with elementary geometric constructions on it. In a more general situation the unitary motion may be constrained further so that the optimal Hamiltonian (12.21) may not be implementable. For example, the constraint may enforce the path of the unitary evolution to lie in a three- rather than two-dimensional subspace. To determine what happens let us work out the passage time for this example.

Since in this case the minimal solution H to the brachistochrone problem is a three-dimensional matrix, we can express the initial state $|\psi_I\rangle$ in terms of the three eigenstates of H according to

$$|\psi_I\rangle = \cos\alpha|E_i\rangle + \sin\alpha\cos\beta e^{i\phi}|E_j\rangle + \sin\alpha\sin\beta e^{i\varphi}|E_k\rangle, \tag{12.29}$$

where α, β are angular coordinates, ϕ, φ are phase variables, and we assume that $E_i < E_j < E_k$. If a unitary operator U_T transforms this state into an orthogonal state, then the condition

$$\cos^2\alpha + \sin^2\alpha\cos^2\beta e^{-i\omega_{ji}T/\hbar} + \sin^2\alpha\sin^2\beta e^{-i\omega_{ki}T/\hbar} = 0 \tag{12.30}$$

must be satisfied, where $\omega_{ji} = E_j - E_i$ and $\omega_{ki} = E_k - E_i$. To render the analysis more tractable, we simplify this constraint by assuming that $\alpha = \beta = \pi/4$. Then (12.30) implies that a necessary condition for the state $|\psi_I\rangle$ to evolve into an orthogonal state is given by the relation

$$\frac{\omega_{ki}}{\omega_{ji}} = \frac{2m - 1}{2n - 1} , \qquad (12.31)$$

where m, n are natural numbers such that $m \neq n$.

Thus, the solution to the brachistochrone problem must be such that the eigenvalues of H satisfy condition (12.31) as well as the constraint $E_{\max} - E_{\min} \leq \omega$. Assuming that these constraints are indeed satisfied, the initial state evolves into an orthogonal state $|\psi_F\rangle$. The first time that $|\psi_I\rangle$ evolves into a state orthogonal to $|\psi_I\rangle$, in particular, is given by

$$T = \frac{\pi\hbar}{\omega_{ji}} = \frac{3\pi\hbar}{\omega_{ki}} . \qquad (12.32)$$

However, since in this case $U_t|\psi_I\rangle$ does not describe a geodesic path, T will be larger than Fleming's passage time τ_P given in (12.17). Indeed, without loss of generality we may set $E_i = 0$. Then, it is straightforward to verify that $T = \sqrt{6}\tau_P$. This follows from the fact that under the constraint $\omega_{ki} = 3\omega_{ji}$ that comes from (12.32), the squared energy dispersion in the state (12.29) with $\alpha = \beta = \pi/4$ is given by $\Delta H^2 = \frac{3}{2}\omega_{ji}^2$.

12.5 Non-Hermitian Quantum Brachistochrone

We have seen how the solution to the simple brachistochrone problem can be obtained in the Hermitian quantum theory. What happens if we extend the quantum theory into the complex domain by looking at a PT-symmetric theory? We saw earlier that in classical mechanics if we were to allow for a complex path interpolating a pair of real points of the coordinate space, then it is possible (at least mathematically) to transport a particle across a large distance in virtually no time. It turns out that an analogous situation emerges in the PT-symmetric theory. Here we present a simple algebraic calculation of the optimal evolution time from an initial state to a final state by using a simple 2×2 Hamiltonian. As we have remarked above, the 2×2 model suffices to cover general cases because in the case of our simple brachistochrone problem the solution is found on the two-dimensional subspace of the Hilbert space spanned by the initial state $|\psi_I\rangle$ and the final state $|\psi_F\rangle$. In the case of a PT-symmetric Hamiltonian the variational approach gives a more direct way to handle the brachistochrone problem. Thus, we shall first briefly revisit the Hermitian case but expressed in the variational formalism and then we will compare the result to its PT-symmetric counterpart.

12.5.1 Hermitian 2 × 2 Matrices

We choose a basis so that the initial and final state vectors take the form

$$|\psi_I\rangle = \begin{pmatrix} 1 \\ 0 \end{pmatrix} \quad \text{and} \quad |\psi_F\rangle = \begin{pmatrix} a \\ b \end{pmatrix} , \tag{12.33}$$

where the condition that $|\psi_F\rangle$ be normalized is $|a|^2 + |b|^2 = 1$. The most general 2×2 Hermitian Hamiltonian is

$$H = \begin{pmatrix} s & r\,e^{-i\theta} \\ r\,e^{i\theta} & u \end{pmatrix} \quad (r,\ s,\ u,\ \theta \text{ real}) . \tag{12.34}$$

For this Hamiltonian the eigenvalue constraint (12.4) takes the form

$$\omega^2 = (s - u)^2 + 4r^2 . \tag{12.35}$$

To find the optimal Hamiltonian satisfying this constraint, we rewrite H as a linear combination of Pauli matrices:

$$H = \tfrac{1}{2}(s + u)\mathbb{1} + \tfrac{1}{2}\omega\boldsymbol{\sigma}\cdot\mathbf{n} , \tag{12.36}$$

where

$$\mathbf{n} = \frac{1}{\omega}(2r\cos\theta,\ 2r\sin\theta,\ s - u) \tag{12.37}$$

is a unit vector. Then by the use of identity (12.25) the evolution equation $|\psi_F\rangle = e^{-iHt/\hbar}|\psi_I\rangle$ can be expressed in the form

$$\begin{pmatrix} a \\ b \end{pmatrix} = e^{-\frac{1}{2}i(s+u)t/\hbar} \begin{pmatrix} \cos\frac{\omega t}{2\hbar} - i\frac{s-u}{\omega}\sin\frac{\omega t}{2\hbar} \\ -i\frac{2r}{\omega}e^{i\theta}\sin\frac{\omega t}{2\hbar} \end{pmatrix} . \tag{12.38}$$

The second component of this equation gives $|b| = \frac{2r}{\omega}\sin\frac{\omega t}{2\hbar}$, which allows us to find the required time of evolution:

$$t = \frac{2\hbar}{\omega}\arcsin\frac{\omega|b|}{2r} . \tag{12.39}$$

We must now minimize the time t over all $r > 0$ while maintaining the energy constraint in (12.35). This constraint tells us that the maximum value of r is $\frac{1}{2}\omega$. At this value we have $s = u$. Because H can be made trace free, we can set $s = u = 0$. The variable θ in (12.36) does not affect the eigenvalues, so we may set $\theta = \pi$. Then we recover the optimal Hamiltonian obtained in (12.24). As regards the minimum evolution time τ we have

$$\tau = \frac{2\hbar \arcsin |b|}{\omega} .$$

(12.40)

In the special case for which $a = 0$ and $b = 1$ so that $|\psi_I\rangle$ and $|\psi_F\rangle$ are orthogonal, we recover the passage time $\tau = \tau_P = \pi \hbar/\omega$, the smallest time required for a spin flip .

Although the form of the result in (12.40) resembles the statement of the uncertainty principle, it is merely the statement indicated above that *rate ×time=distance*; the maximum speed of evolution is given by ΔH, and the distance between $|\psi_I\rangle$ and $|\psi_F\rangle$, assuming they are normalized, is given by $2\arccos(|\langle\psi_F|\psi_I\rangle|)$. Since $|\langle\psi_F|\psi_I\rangle| = |a|$ and $|a| = \sqrt{1 - |b|^2}$, we obtain (12.40) from the relation

$$\arccos\sqrt{1 - |b|^2} = \arcsin|b| .$$

(12.41)

12.5.2 Non-Hermitian 2 × 2 Matrices

We now show by direct calculation that for a PT-symmetric Hamiltonian, τ can be arbitrarily small. This is because a PT-symmetric Hamiltonian whose eigenvalues are all real is equivalent to a Hermitian Hamiltonian via $\tilde{H} = e^{-Q/2} H e^{Q/2}$, where Q is Dirac Hermitian. The states in a PT-symmetric theory are mapped by $e^{-Q/2}$ to the corresponding states in the Dirac Hermitian theory. But, the overlap distance between two states does not remain constant under a similarity transformation. We can exploit this property of the similarity transformation to overcome the Hermitian lower limit on the time τ. (The detailed calculation is explained in [14].)

We consider the general class of PT-symmetric 2 × 2 Hamiltonians having the form

$$H = \begin{pmatrix} r\,e^{i\theta} & s \\ s & r\,e^{-i\theta} \end{pmatrix} \qquad (r,\ s,\ \theta \text{ real}) .$$

(12.42)

The time-reversal operator T performs complex conjugation and the parity operator in this case is given by

$$P = \begin{pmatrix} 0 & 1 \\ 1 & 0 \end{pmatrix} .$$

(12.43)

The two eigenvalues

$$E_\pm = r\cos\theta \pm \sqrt{s^2 - r^2\sin^2\theta}$$

(12.44)

are real if $s^2 > r^2\sin^2\theta$. This inequality defines the region of unbroken PT symmetry. The unnormalized eigenstates of H are

$$|E_+\rangle = \begin{pmatrix} e^{i\alpha/2} \\ e^{-i\alpha/2} \end{pmatrix} \quad \text{and} \quad |E_-\rangle = \begin{pmatrix} ie^{-i\alpha/2} \\ -ie^{i\alpha/2} \end{pmatrix}, \tag{12.45}$$

where α is given by $\sin\alpha = (r/s)\sin\theta$. Note that the condition of unbroken PT symmetry of H in (12.42) implies that α is real. The C operator required for defining the Hilbert space inner product is

$$C = \frac{1}{\cos\alpha} \begin{pmatrix} i\sin\alpha & 1 \\ 1 & -i\sin\alpha \end{pmatrix}. \tag{12.46}$$

It is easy to verify that the CPT norms of both eigenstates have the value $\sqrt{2\cos\alpha}$.

To calculate τ we express the Hamiltonian H in (12.42) as

$$H = (r\cos\theta)\mathbb{1} + \tfrac{1}{2}\omega\sigma\cdot\mathbf{n}, \tag{12.47}$$

where

$$\mathbf{n} = \frac{2}{\omega}(s, 0, ir\sin\theta) \tag{12.48}$$

is a unit vector. The energy constraint requires that the squared difference between energy eigenvalues is

$$\omega^2 = 4s^2 - 4r^2\sin^2\theta. \tag{12.49}$$

The positivity of ω^2 is ensured by the condition of unbroken PT symmetry. Notice that (12.49) differs from (12.35) by a sign. We can think of (12.49) as being *hyperbolic* in character, while (12.35) is *elliptic* in character. The technical advantage of the constraint in (12.49) is that because of the minus sign both terms on the right side can become large without violating the condition that ω be fixed. We will see that it is this fact that allows the non-Hermitian Hamiltonian H in (12.42) to achieve faster-than-Hermitian time evolution.

To determine τ we write down the PT-symmetric time-evolution equation in vector form:

$$e^{-iHt/\hbar} \begin{pmatrix} 1 \\ 0 \end{pmatrix} = \frac{e^{-itr\cos\theta/\hbar}}{\cos\alpha} \begin{pmatrix} \cos(\frac{\omega t}{2\hbar} - \alpha) \\ -i\sin\left(\frac{\omega t}{2\hbar}\right) \end{pmatrix}. \tag{12.50}$$

In particular, consider the pair of vectors used in the Hermitian spin-flip case as in (12.22). Observe that the evolution time t needed to reach $|\psi_F\rangle = \begin{pmatrix} 0 \\ 1 \end{pmatrix}$ from $|\psi_I\rangle = \begin{pmatrix} 1 \\ 0 \end{pmatrix}$ is given by

$$t = \frac{(2\alpha - \pi)\hbar}{\omega}. \tag{12.51}$$

Optimizing this result over allowable values for α as α approaches $\frac{1}{2}\pi$, the optimal time τ tends to zero, a dramatic change from the Hermitian result in (12.17)! Note, however, that the two vectors in (12.22) are not orthogonal with respect to the CPT inner product. This is the reason that the Fleming bound in (12.17) is not violated.

12.6 Extension of Non-Hermitian Hamiltonians to Higher-Dimensional Hermitian Hamiltonians

We have seen how a quantum state can be transported unitarily into another state in arbitrarily short time by using a bounded Hamiltonian if we allow for a complex path interpolating them in the space of unitary motions. Can such an operation be implemented in practice? If the answer is affirmative, then the implication is immense in quantum information, computation, cryptography, and other related fields. For example, if a quantum computer were to exist, then solutions to difficult optimization problems can in principle be found in arbitraily short time, and this in turn would have important implications in society as a whole.

A *gedanken* experiment was proposed in [14] to realize this effect in a laboratory. The setup is as follows: we use a Stern–Gerlach filter to create a beam of spin-up electrons. The beam then passes through a *black box* containing a device governed by a PT-symmetric Hamiltonian that flips the spins unitarily in a very short time. The outgoing beam then enters a second Stern–Gerlach device that verifies that the electrons are now in spin-down states. In effect, the black box device is *applying a complex magnetic field* **B**:

$$\mathbf{B} = (s, 0, ir\sin\theta) \, . \tag{12.52}$$

If the field strength has sufficiently large amplitude, then spins can be flipped in virtually no time because the complex path joining these two states is arbitrarily short without violating the energy constraint in (12.49). We emphasize that the fact that the field strength can be made large without violating the energy constraint (12.4) is a consequence of the hyperbolic representation in (12.49).

The arbitrarily short alternative complex pathway from an up state to a down state, as illustrated by this thought experiment, is reminiscent of the short alternative distance between two widely separated space–time points as measured through a wormhole in general relativity. This comparison is of course controversial, and it has subsequently motivated much research and it has generated a lively debate in the literature [2, 15, 34, 36, 37, 50, 53, 54, 57]. We emphasize that the entire package of flipping the spin is not realized by a unitary operation. This follows from the fact that PT-symmetric quantum theory is unitary, and as such it respects the Fleming bound (12.17) applicable to all unitary theories [14]. The point is that the "black box" scheme described above actually consists of three regimes: (i) the preparation of a spin-up state in the Hermitian setup, (ii) the fast unitary motion to flip the spin using a PT-symmetric Hamiltonian, and (iii) the recovery of a spin-down state in the Hermitian setup. Thus, the operation is locally unitary, but the switching between Hermitian and PT-symmetric description is not unitary. This three-step process of switching Hamiltonians is analogous to the classical procedure described in Sect. 12.3 for obtaining faster-than-real time evolution. Recall that in the classical case we were able to transport a finite-energy particle across a large distance in a short time by switching the potential through which it was traveling. Note that in the classical case there is no question of violating unitarity because the particle does not get lost.

The question of unitarity and faster-than-Hermitian time evolution has been reexamined in more detail in [36, 54] by means of a geometric approach and also in [2], where a more general class of non-Hermitian Hamiltonians that are not necessarily PT symmetric are considered. In particular, Mostafazadeh has emphasized the role of quantum observables in such an experiment; the spin operator in Hermitian quantum mechanics cannot be interpreted as a spin operator in the PT-symmetric counterpart, thus leading to ambiguities regarding the physical interpretation of the *gedanken* experiment described above.

An intriguing alternative proposal for an implementation of the fast spin flip has been made more recently by Günther and Samsonov [37]. The idea is to embed the problem into a Hermitian setup represented by a higher-dimensional Hilbert space. Specifically, take our PT-symmetric Hamiltonian H in (12.42). The eigenstates of H are not orthogonal with respect to the Hermitian inner product. Since H is not Hermitian, its Hermitian conjugate defines a new matrix $H^\dagger$. The eigenstates of $H^\dagger$ thus also define a pair of nonorthogonal states in the Hermitian theory. When these four states are suitably normalized, they can be used to form an over-complete basis set in the Hermitian two-dimensional Hilbert space. Such an over-complete set of basis is also known as a positive operator-valued measure (POVM), commonly used in the analysis of quantum information theory. A key idea is that such a basis can be embedded in a higher-dimensional Hilbert space to form an orthogonal basis by using the Naimark dilation [41]. A Hermitian Hamiltonian can then be constructed – in this case a 4×4 matrix – such that its eigenstates are precisely the four states thus obtained. Using this Hamiltonian it is possible to construct a standard unitary motion in such a way that the induced motion obtained by the projection onto the two-dimensional subspace is characterized by the PT-symmetric motion (12.50).

In this way, Günther and Samsonov were able to show that the fast spin flip can in principle be realized in the standard Hermitian quantum mechanics by a combination of a unitary motion and a projection in a larger-dimensional Hilbert space. In practical terms this means that one should couple the spin to an auxiliary particle (this can be done either by a projection or by a unitary operation), apply a unitary evolution in the larger Hilbert space of the combined system, and finally project out the auxiliary particle to recover the spin in the transported state. The net effect of such an operation can then be characterized by (12.50). The passage time is apparently violated due to the general fact that when a unitary motion is projected to a subspace of a Hilbert space, the resulting dynamics need not respect laws of unitarity. It would be of considerable interest to find out whether the Günther and Samsonov scheme can actually be implemented in a laboratory, and if not, what might be the difficulty preventing the violation of the Fleming bound.

Acknowledgments We have benefited greatly from many discussions with Drs. U. Günther and B. Samsonov. We thank Dr. D.W. Hook for his assistance in preparing the figures used in this chapter. CMB is supported by a grant from the US Department of Energy.

References

1. J. Anandan, Y. Aharonov, Phys. Rev. Lett. **65**, 1697 (1990)
2. P. Assis, A. Fring, J. Phys. A: Math. Theor. **41**, 244002 (2008)
3. G. Barton, *Introduction to Advanced Field Theory* (John Wiley & Sons, New York, 1963), Chap. 12
4. C.M. Bender, Contemp. Phys. **46**, 277 (2005)
5. C.M. Bender, Rep. Prog. Phys. **70**, 947 (2007)
6. C.M. Bender, S. Boettcher, Phys. Rev. Lett. **80**, 5243 (1998)
7. C.M. Bender, S. Boettcher, P.N. Meisinger, J. Math. Phys. **40**, 2201 (1999)
8. C.M. Bender, S.F. Brandt, J.-H. Chen, Q. Wang, Phys. Rev. D **71**, 065010 (2005)
9. C.M. Bender, S.F. Brandt, J.-H. Chen, Q. Wang, Phys. Rev. D **71**, 025014 (2005)
10. C.M. Bender, D.C. Brody, J.-H. Chen, H.F. Jones, K.A. Milton, M.C. Ogilvie, Phys. Rev. D **74**, 025016 (2006)
11. C.M. Bender, D.C. Brody, H.F. Jones, Phys. Rev. Lett. **89**, 270401 (2002)
12. C.M. Bender, D.C. Brody, H.F. Jones, Am. J. Phys. **71**, 1095 (2003)
13. C.M. Bender, D.C. Brody, H.F. Jones, Phys. Rev. Lett. **93**, 251601 (2004)
14. C.M. Bender, D.C. Brody, H.F. Jones, B.K. Meister, Phys. Rev. Lett. **98**, 040403 (2007)
15. C.M. Bender, D.C. Brody, H.F. Jones, B.K. Meister, arXiv:0804.3487 [quant-ph]
16. C.M. Bender, P.D. Mannheim, Phys. Rev. Lett. **100**, 110402 (2008)
17. C.M. Bender, P.D. Mannheim, Phys. Rev. D **78**, 025022 (2008)
18. C.M. Bender, P.D. Mannheim, J. Phys. A: Math. Theor. **41**, 304018 (2008)
19. D.C. Brody, J. Phys. A: Math. Gen. **36**, 5587 (2003)
20. D.C. Brody, D.W. Hook, J. Phys. A: Math. Gen. **39**, L167 (2006).
21. D.C. Brody, D.W. Hook, J. Phys. A: Math. Gen. **40**, 10949 (2007).
22. D.C. Brody, L.P. Hughston, J. Geom. Phys. **38**, 19 (2001)
23. R. Brower, M. Furman, M. Moshe, Phys. Lett. B **76**, 213 (1978)
24. V. Buslaev, V. Grecchi, J. Phys. A: Math. Gen. **26**, 5541 (1993)
25. J.L. Cardy, Phys. Rev. Lett. **54**, 1354 (1985)
26. J.L. Cardy, G. Mussardo, Phys. Lett. B **225**, 275 (1989)
27. A. Carlini, A. Hosoya, T. Koike, Y. Okudaira, Phys. Rev. Lett. **96**, 060503 (2006)
28. P. Dorey, C. Dunning, R. Tateo, J. Phys. A: Math. Gen. **34**, L391 (2001)
29. P. Dorey, C. Dunning, R. Tateo, J. Phys. A: Math. Gen. **34**, 5679 (2001)
30. P. Dorey, C. Dunning, R. Tateo, J. Phys. A: Math. Theor. **40**, R205 (2007)
31. C.F. de M. Faria, A. Fring, Laser Phys. **17**, 424 (2007)
32. M.E. Fisher, Phys. Rev. Lett. **40**, 1610 (1978)
33. G.N. Fleming, Nuov. Cim. A **16**, 232 (1973)
34. U. Günther, I. Rotter, B.F. Samsonov, J. Phys. A: Math. Theor. **40**, 8815 (2007)
35. U. Günther, B.F. Samsonov, F. Stefani, J. Phys. A: Math. Theor. **40**, F169 (2007)
36. U. Günther, B.F. Samsonov, arXiv: 0709.0483 [quant-ph]
37. U. Günther, B.F. Samsonov, Phys. Rev. Lett. **101**, 230404 (2008)
38. U. Guenther, F. Stefani, M. Znojil, J. Math. Phys. **46**, 063504 (2005)
39. B. Harms, S. Jones, C.-I. Tan, Phys. Lett. B **91**, 291 (1980)
40. B. Harms, S. Jones, C.-I. Tan, Nucl. Phys. B **171**, 392 (1980)
41. A.S. Holevo, *Probabilistic and Statistical Aspects of Quantum Theory* (North-Holland, Amsterdam, 1982)
42. T. Hollowood, Nucl. Phys. B **384**, 523 (1992)

43. L.P. Hughston, Geometric aspects of quantum mechanics. In *Twistor Theory*, S. Huggett (ed.) (Marcel Dekker, New York, 1995)
44. H.F. Jones, J. Mateo, Phys. Rev. D **73**, 085002 (2006)
45. G. Källén, W. Pauli, Dansk Vid. Selsk. Mat.-Fys. Medd. **30**, 23 (1955)
46. T.W.B. Kibble, Commun. Math. Phys. **65**, 189 (1979)
47. S. Kobayashi, K. Nomizu, *Foundations of Differential Geometry*, vol. 2 (Wiley, New York, 1969)
48. T.D. Lee, Phys. Rev. **95**, 1329 (1954)
49. K.G. Makris, R. El-Ganainy, D.N. Christodoulides, Z.H. Musslimani, Phys. Rev. Lett. **100**, 103904 (2008)
50. D. Martin, arXiv: quant-ph/0701223v2
51. A. Mostafazadeh, J. Math. Phys. **43**, 3944 (2002)
52. A. Mostafazadeh, J. Phys. A: Math. Gen. **36**, 7081 (2003)
53. A. Mostafazadeh, Phys. Rev. Lett. **99**, 130502 (2007)
54. A. Mostafazadeh, arXiv:0709.1756 [quant-ph]
55. Z.H. Musslimani, K.G. Makris, R. El-Ganainy, D.N. Christodoulides, J. Phys. A: Math. Theor. **41**, 244019 (2008)
56. Z.H. Musslimani, K.G. Makris, R. El-Ganainy, D.N. Christodoulides, Phys. Rev. Lett. **100**, 030402 (2008)
57. I. Rotter, J. Phys. A: Math. Theor. **40**, 14515 (2007)
58. A. Salam, Review of *on the Mathematical Structure of T. D. Lee's Model of a Renormalizable Field Theory* by G. Källén and W. Pauli, MathSciNet Mathematical Reviews on the Web MR0076639 (17,927d) (1956)
59. L.S. Schulman, Jump time and passage time. In *Time in Quantum Mechanics*, J.G. Muga, R. Sala Mayato, I.L. Egusquiza (eds.), Lect. Notes Phys. **m72** (Springer-Verlag, Berlin, 2002), pp. 99–121
60. E.P. Wigner, J. Math. Phys. **1**, 414 (1960)
61. T.T. Wu, Phys. Rev. **115**, 1390 (1959)
62. A.B. Zamolodchikov, Nucl. Phys. B **348**, 619 (1991)

Chapter 13
Atomic Clocks

Robert Wynands

13.1 Introduction

Time is a strange thing. On the one hand it is arguably the most inaccessible physical phenomenon of all: both in that it is impossible to manipulate or modify – for all we know – and in that even after thousands of years mankind's philosophers still have not found a fully satisfying way to understand it. On the other hand, no other quantity can be measured with greater precision. Today's atomic clocks allow us to reproduce the length of the second as the SI unit of time with an uncertainty of a few parts in 10^{16} – orders of magnitude better than any other quantity. In a sense, one can say [1]

> Time? We don't know what it is, but we know how to measure it!

This chapter is written in this spirit, from the point of view of an "atomic clockmaker". The goal is to explain why and what for we need man-made clocks at all to tell us the time, how a clock and today's atomic clocks, in particular, work, and what the atomic clocks of the future might look like. In the initial parts of this chapter the material will be presented on an elementary level suitable for a general audience, and in later sections the physical details will be covered with more rigor, including discussion of the influence of quantum aspects of its operation.

There exist a number of books on atomic clocks and atomic frequency standards, of course. The "bible", so to speak, is the two-volume book by Vanier and Audoin [2]. It covers in great detail the most commonly used frequency standards, with the exception of fountain clocks and optical clocks which did not exist at the time this standard reference work was written. Some modern developments are covered in the book by Riehle [3].

R. Wynands (✉)
Physikalisch-Technische Bundesanstalt, Bundesallee 100, Braunschweig 38116, Germany,
robert.wynands@ptb.de

Wynands, R.: *Atomic Clocks*. Lect. Notes Phys. **789**, 363–418 (2009)
DOI 10.1007/978-3-642-03174-8_13

13.2 Why We Need Clocks at All

Why, indeed? Everybody knows a day consists of 24 h of 60 min of 60 s each. So why not just divide the interval between noon today and noon tomorrow, as indicated by the Sun reaching the highest point along its trajectory across the sky, by 86,400, and then we know how long a second is?

13.2.1 Why Not Use Celestial Motion?

The problem with this definition of the length of the second is that the "solar day", defined as the time interval between two successive local noons, is not constant over the course of the year. In temperate latitudes it varies by as much as a minute, with the solar day in the Northern winter half a minute longer than the yearly average and half a minute shorter in September.

But even basing the second on the *average* duration of the day is not a good option. In the 1930s Scheibe and Adelsberger at Physikalisch-Technische Reichs-anstalt were developing highly precise quartz clocks. Not only could they observe the gradual slow down of Earth's rotation due to tidal friction but they were the first to detect the irregularities in the rotation of the Earth due to mass redistributions within the Earth (magma movements, earthquakes) and on the Earth (weather, ocean currents) [4–7]. With today's technology these irregularities in the position and orientation of Earth's axis of rotation and in the length of the day can be monitored on a day-by-day basis. Measured data and predictions over several months can be obtained via the Internet from the web site of the International Earth Rotation and Reference Systems Service [8].

Until 1967 the length of the second in the International System of Units (SI) was defined in a rather complicated fashion based on the length of the year. In that year the astronomy-based definition was abandoned in favour of a definition still valid today:

> The second is the duration of 9 192 631 770 periods of the radiation corresponding to the transition between the two hyperfine levels of the ground state of the ^{133}Cs atom.

In order to realize this definition with highest possible accuracy one therefore has to build a caesium atomic clock.

13.2.2 A Brief History of Clocks

The passing of time has always held a special importance to mankind. Calendar sticks and cave paintings dating back to the Stone Age have been found; solar clocks were used in ancient Egypt, as were water clocks (clepsydras). The first mechanical clocks appeared in central Europe during the thirteen century and became an immediate success: Although they lost or gained a significant part of an hour each day

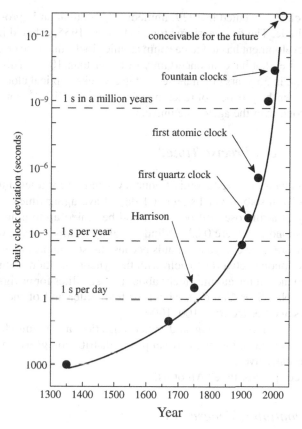

Fig. 13.1 Historical development of mankind's clock-making ability, using the daily clock error as a criterion

they were good enough to fill the magistrates of neighbouring towns with envy, who then hurried to acquire such a symbol of progress, too (Fig. 13.1).

The first practical pendulum clocks were built by Huygens around 1670. Their daily error was of the order of 10 s, or a relative error of only 10^{-4}. This remarkable feat serves to show how special the measurement of time is in metrology. It is not easier to measure the passing of time compared to other metrological tasks, but with proper effort it can be done much better than for any other quantity. It was like this in the seventeen century, and it is still in this way in the twenty-first century.

Another development of note in the time line (Fig. 13.1) is the invention of the chronometer by John Harrison in the eighteen century. It was developed and used in navigation at sea to determine one's longitude. The interesting story surrounding this multi-decade quest by an English carpenter has been told in a famous book [9]. The best mechanical clocks, relying on a mechanical pendulum swinging in a vacuum, could reach inaccuracies of about a second per year; they reached the end of their development around 1900.

The quartz clock, invented in 1929, immediately brought an improvement by an order of magnitude. The first caesium atomic clock, in 1955, ushered in a rapid and accelerating development has led to caesium atomic clocks that lose or gain less than 100 ps per day, i.e. that have an uncertainty of better than 10^{-15}. This corresponds to a microsecond in 35 years or a nanosecond in 2 weeks. Optical clocks envisioned for the future offer the prospect of reaching uncertainties of 10^{-18}, corresponding to only half a second over the age of the universe.

13.2.3 Who Needs Precise Time?

Today one can buy a fist-sized atomic frequency standard for a few hundred euros. It offers a relative instability at 1 s and at 1 day of averaging time of about 10^{-11}. If this stability of a microsecond per day could be scaled up to a human lifetime, it would correspond to a mere 0.02 s. Mind-boggling as this sounds, we all rely on this kind of performance everyday. That is because those compact atomic clocks are employed in cellphone networks to help with the synchronization within a cell [10].

The performance of the atomic clocks aboard the satellites for navigation systems like GPS, GLONASS or Galileo is about 10-fold better and of the clocks in the ground stations another factor of 100–1000.

Other, less demanding timing and synchronization are required for financial transactions, in the managing of electrical power distribution grids, and many other contexts of our daily lives.

So, who needs precise time? All of us!

13.2.4 International Timescales

About 300 atomic clocks from more than 50 institutions worldwide contribute to the generation of the various international timescales. This process is organized by the International Bureau of Weights and Measures (BIPM) near Paris. Most of the clocks are either commercial caesium atomic clocks with an uncertainty down to 2×10^{-13} for the best models or hydrogen masers. Two primary clocks, PTB-CS1 and PTB-CS2, enter at this stage, too; for several decades they have been the only primary clocks actually running continuously as a clock to assist in the generation of international timescales.

Participating laboratories mutually compare their clocks (using a variety of satellite-based techniques) "round the clock", so to speak. Some details can be found in the tutorial by Levine, for instance [11]. The data from all these continuous pair-wise comparisons are used by BIPM scientists to form a weighted mean, called *Echelle Atomic Libre* (EAL), free atomic timescale. Stable clocks obviously get a higher weight than less stable clocks, up to some maximum weight [12]. EAL is a very stable timescale but its scale unit is not necessarily coincident with the SI second.

In a second step, the rate of EAL is adjusted by a correction (6.758×10^{-13} as of June 2008 [13]) to obtain a timescale with a scale interval as close as possible to the SI second. This scale is called *International Atomic Time* (TAI). The rate of TAI is calibrated by comparing any one of the clocks participating in the generation of EAL to a primary clock, nowadays typically a caesium fountain clock.

The rates of EAL and TAI are published in the bulletin *Circular-T* around the middle of each month, based on the measured data for the previous month. This delay is one reason why the scale interval of TAI is not exactly the SI second at any one time. For instance, in June 2008 the scale interval of TAI differed from the SI second by 3.3×10^{-15} [13]. Another reason is that most users of atomic timescales place more importance on a stable scale even if its ticks are not spaced by exactly 1 SI second. Therefore any steering corrections can only be applied at a maximum "slew rate". And finally, there is another variable delay of weeks or months until the laboratories running primary clocks have communicated their results of TAI calibrations to BIPM.

The timescale that we use in everyday life is derived from *Coordinated Universal Time* (UTC), with an adjustment for the respective time zone. UTC in turn is derived from TAI by subtraction of an integer number of seconds, 33 as of this writing (Fig. 13.2).

Unfortunately, after 24 h on a TAI clock Earth has (on average) not quite completed a full rotation around its axis – there is a lag of about 2 ms. This is because at the time when the astronomical definition of the second was fixed the Earth rotated slightly faster around its axis, but since then it has been slowed down by tidal friction (see also Sect. 13.2.1). Over the course of a year and a half on average, the daily lags add up to about 1 s. By international convention, it is then that an additional second, a *leap second*, is introduced into UTC and the times in the various time zones around the Earth. This additional second is usually introduced into UTC as 23:59:60 h on December 31 or June 30. As of this writing, the last time this happened was on 31 December 2005 (Fig. 13.3). At what time another leap second will be inserted depends on the irregular rotation of the Earth, which is hard to predict with sufficient

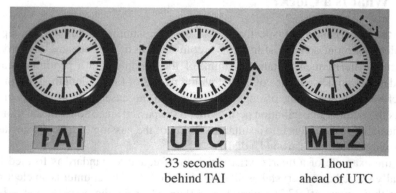

33 seconds behind TAI   1 hour ahead of UTC

Fig. 13.2 The relation between timescales TAI, UTC, and the standard time in a time zone other than that of Greenwich (here Central European Time, "MEZ" in German). Obviously, this photograph was taken before 31 December 2008 leap second increased the lag of UTC with respect to TAI to 34 s

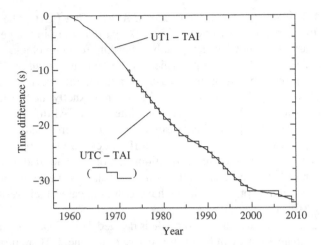

Fig. 13.3 The difference between UTC and TAI over the last decades (*step-like curve*) and the time lag of Earth's rotation phase UT1 with respect to TAI (*smooth curve*). The data beyond July 2008 are an extrapolation calculated by IERS

accuracy over more than a few months. A leap second is announced about half a year in advance by a bulletin of the International Earth Rotation and Reference Systems Service (IERS) [8]. For instance, Issue 36 of Bulletin-C of the IERS published on 04 July 2008 announces another leap second at 23:59:60 h UTC on 31 December 2008 [14]. From then on UTC will lag TAI by 34 s – until the next leap second.

More information on the history and present situation regarding timescales can be found in a review [15].

13.3 What Is a Clock?

From the clock-maker's perspective, which we are assuming in this chapter, a practical description of a clock is this: It is a combination of a frequency standard and a counter. The frequency standard (think of the pendulum in a pendulum clock) provides a periodic sequence of events with a period of known length. The counter (the clockwork consisting of cogs and wheels in the case of a pendulum clock) counts the number of oscillation periods that have passed since some initial moment in time and translates this into a suitable display of the passing of time, for instance, by moving clock hands around a dial.

In the example of a quartz wrist watch the frequency standard is formed by a specially shaped quartz crystal oscillating at 32,768 Hz. The counter is an electronic circuit that counts off 32,768 periods and then advances the display by 1 s. Even the good old hour glass fits the above definition. The period is given by the time the grains of sand or egg shell need to trickle through into the lower compartment, while the counting is done by a human operator reversing the hour glass once all grains have trickled through.

All the clockmaker's expertise goes into ensuring that the period of the frequency standard remains constant, not only when the clock itself ages but also when the conditions around it change. For instance, in an ill-constructed pendulum clock the oscillation period might increase when a temperature rise leads to a lengthening of the pendulum. Issues of air pressure, humidity and also aging properties of materials and greases need to be taken into account and compensated. In an atomic clock, other effects can lead to a change of the period, like changes in magnetic fields or collisions of atoms among themselves or with the walls of a container.

Let us define a few technical terms here for use in the remainder of this chapter (for the full formal definitions of metrological terms, see [16]). When the clock period is different from its nominal value, this is called a *frequency bias*. Such a bias is typically caused by a *systematic effect*, for instance, by the Zeeman effect shifting an atomic energy level in the presence of a magnetic field. The *uncertainty* of a frequency standard describes how sure we can be that the period matches its expected or nominal value, for instance, how well a clock reproduces the length of the SI second. Of course there will always be a certain amount of period-to-period fluctuation (noise) in the length of the period, as characterized by the standard deviation of the noise; this is called the *instability* of the standard. If a clock is switched off and switched on again at a later time the *reproducibility* specifies how well the new clock output frequency matches the previous one; this last parameter is sometimes called *frequency retrace* in technical specifications of commercially available atomic clocks.

The terms *primary* and *secondary* frequency standards both denote a device that has been thoroughly characterized such that all potential frequency biases are either excluded or, if they cannot be avoided, at least understood so well that one can correct the standard's output by a corresponding amount. If such a standard is based on the very transition used to define the length of the SI second, it is called a *primary* standard.

Note that from a pragmatic point of view a clock with a bias is just as usable for precise timing as a clock without bias, provided the bias is known with sufficient precision. All one needs to do is to correct the clock's display by the daily bias multiplied by the number of days that have passed. This only works as well as the value of the bias is known – as quantified by the *uncertainty* of the clock. Determining and minimizing this uncertainty is perhaps the major task of the atomic clock-maker: thinking about effects that could lead to a frequency bias, avoiding them or correcting for them, and ensuring that operating conditions and corrected frequency biases remain constant.

13.4 How an Atomic Clock Works

Simply put: Just like any other clock! The underlying phenomenon of known duration in this case is the period of oscillation of an atomic transition moment between two quantum states or of the electromagnetic wave needed to induce it. The clockwork in general consists of sophisticated electronic circuitry, although lasers are coming more and more into play.

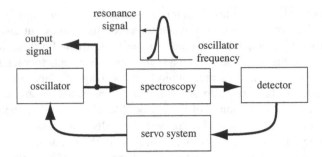

Fig. 13.4 Principle of operation of the atomic frequency standard. The output of an oscillator probes a spectral feature and a detector registers the response of the system. A servo loop uses this signal to retune the oscillator to the desired working point along the spectral feature. The external output of the oscillator is thus referenced to an internal feature of the atom (or molecule or ion)

In an atomic clock the frequency standard is based on a transition between energy levels in a quantum system, for instance, an atom, an ion or a molecule. By stabilizing the frequency of an oscillator to this transition frequency one can counteract technical noise, aging or other sources of drift of the oscillator frequency. Figure 13.4 shows the main principle of operation of such an atomic frequency standard. The oscillator not only provides the output signal of the standard but also is used to interrogate a spectral feature in the quantum system. This spectral line is used as a frequency discriminator. A frequency excursion of the oscillator leads to a change in detected signal. A servo loop then tunes the oscillator back to the desired working point along the profile of the spectral feature, for instance an atomic absorption line. The output of the frequency standard therefore serves as a macroscopic representation of a microscopic property of the atom in question, for instance, the energy difference between two particular energy levels.

In order to characterize and quantify the frequency instability of a clock one could in principle just compute the standard deviation of a sequence of clock readings relative to a perfect (or at least a much better) clock. The smaller the number, the more stable the clock is. However, using this definition a clock with a very stable bias would erroneously be assigned a high instability. Even worse, this standard deviation grows with time!

Instead, the time-and-frequency community uses the two-sample relative Allan standard deviation σ_y, or its square, the Allan variance σ_y^2 [17], which will be described in the following. Consider the sequence of time intervals x_i between the "tick" of the clock under test and the "tick" of a perfect clock, taken at regular time intervals τ. Next compute the normalized instantaneous frequency

$$y_i = (x_{i+1} - x_i)/(2\pi \nu_0 \tau) , \tag{13.1}$$

where ν_0 is the clock's nominal tick frequency. Normalizing the instantaneous frequency by the nominal one has the advantage that the relative stabilities of oscillators running at very different frequencies can be compared in a meaningful way.

The variance of the sequence of y_i estimated from a sample of size N – the N-sample variance – is

$$s_y^2 = \frac{1}{N-1} \left\langle \sum_{i=1}^{N} (y_i - \bar{y})^2 \right\rangle , \tag{13.2}$$

where $\bar{y} = (\sum_{i=1}^{N} y_i)/N$ is the average value of the instantaneous frequency. For the characterization of frequency stability it has turned out [17] that the two-sample variance possesses particularly useful properties. It is obtained by setting $N = 2$ in Eq. (13.2):

$$\sigma_y^2(\tau) = \frac{1}{2} \left\langle (y_2 - y_1)^2 \right\rangle , \tag{13.3}$$

where the τ dependence is implicit in the y_i and the angle brackets indicate the ensemble average.

Equation (13.3) basically represents (except for the factor of 1/2) the average value of the squared relative frequency changes from sample point to sample point. A convenient feature of this definition is that $\sigma_y^2(\tau)$, called the Allan variance in the time-and-frequency community [17], is insensitive to a constant frequency bias between the reference oscillator and the oscillator under test. In that sense the Allan variance is a representative measure of the fluctuations of a frequency around its mean value. The square root of this variance is called the relative Allan standard deviation (or "Allan deviation" for short). Note that in the special case of pure white frequency noise $\sigma_y(\tau)$ coincides with the standard deviation of the Gaussian distribution describing the frequency fluctuations.

In practice, of course, one does not have an infinite series of measurements x_i or y_i at one's disposal, so a finite approximation is used:

$$\sigma_y^2(\tau) = \frac{1}{2(k-1)} \sum_{i=1}^{k-1} (y_{i+1} - y_i)^2 , \tag{13.4}$$

with k sufficiently large to get a meaningful estimate. Lesage and Audoin have published a series of papers dealing with the question of what the uncertainty of that estimate is [18, 19]. For instance, the relative uncertainty in the case of pure white frequency noise is

$$\frac{\Delta(\sigma_y^2)}{\sigma_y^2} = \frac{3k-4}{(k-1)^2} \tag{13.5}$$

$$\frac{\Delta(\sigma_y)}{\sigma_y} = \frac{\sqrt{3k-4}}{2(k-1)} \tag{13.6}$$

for the Allan variance and the Allan deviation, respectively. So typically one should have $k \geq 19$ to reach a 20% relative uncertainty for the Allan deviation, which means a measurement of total duration 20τ.

By looking at the *changes* in instantaneous frequency from data point to data point one removes the influence of a constant bias from the estimate σ_y of the clock's instability. Of course, the Allan variance is still influenced by changes of a bias, for instance, so in general σ_y^2 depends on averaging time τ. For white noise, this dependence takes the form of τ^{-1} for the Allan variance or $\tau^{-1/2}$ for the Allan standard deviation . When $1/f$ noise is present the Allan deviation will level off at some higher τ or even start to increase in the case of drift.

For the case of white frequency noise, which should govern the behaviour of a good primary or secondary frequency standard, the Allan standard deviation can be written as

$$\sigma_y(\tau) = \frac{1}{\pi} \frac{\Delta\nu}{\nu_0} \frac{1}{S/N} \sqrt{\frac{T_{\text{cycle}}}{\tau}} . \tag{13.7}$$

Here $\Delta\nu$ is the width of the atomic transition line of resonance frequency ν_0, S/N is the signal-to-noise ratio of the detection process, T_{cycle} is the period at which any corrections to the clock frequency are applied and τ is the averaging time.

It is clear from Eq. (13.7) that it only makes sense to build an atomic clock when the frequency discriminator has a narrow linewidth $\Delta\nu$. This implies a steep slope at the side of the resonance (or a sharp extremum at the centre of the resonance when a modulation technique is used), so that already a very small frequency excursion of the oscillator gives a detectable signal change.

One also sees that the actual resonance frequency ν_0 should be as high as possible. The choice of caesium is a good one in this respect because not many elements provide a conveniently accessible microwave transition at a frequency as high as 9.2 GHz. The influence of ν_0 is one important reason why the clocks of the future in all probability will be optical clocks. With a ν_0 of hundreds of THz one easily gains several orders of magnitude compared to caesium. Of course, ν_0 is not the only factor determining the instability of a clock, and the other factors can work for or against the various candidate optical clocks. This will be discussed in more detail in Sect. 13.9.

13.5 The "Classic" Caesium Clock

The publication in 1955 by Essen and Parry [20] is regarded as the birth of the caesium atomic clock. Performance has greatly increased since then, but the working principle has remained the same even in today's commercial caesium-beam clocks.

13.5.1 Why Caesium?

Choosing the alkali metal caesium for an atomic clock offers a number of advantages. All radioactive isotopes of caesium have rather short half-lives, so naturally occurring caesium only contains the only stable isotope, ^{133}Cs. As an alkali metal, it has a $^2S_{1/2}$ ground state, which simplifies the hyperfine structure, even though the nuclear spin is characterized by spin quantum number $I = 7/2$ (Fig. 13.5). Since ^{133}Cs is bosonic there exists a pair of ground states with magnetic quantum number $m_F = 0$. Those two states show a Zeeman frequency shift in response to an external quasi-static magnetic field only in second order, which allows an essential relaxation of the experimental requirements for the stability and homogeneity of such an external field. Due to selection rules this "clock transition" can be induced only by an alternating magnetic field polarized along an external static magnetic field.

From Eq. (13.7) we have seen that one wants a transition with a high-resonance frequency. Caesium is well suited in this regard, because its 9-GHz ground state hyperfine splitting is among the largest of all elements and not too large to be conveniently reached by microwave equipment available in the 1950s when the caesium clock was invented. As an alkali atom, caesium easily parts with its outer electron, so a hot-wire ionization detector can provide a high detection efficiency. Other practical advantages are that caesium can be evaporated at a sufficient rate already at rather modest temperatures, that caesium is not poisonous and it is readily available, too.

In retrospect, the choice of caesium was an extremely fortunate one. Caesium arguably is the chemical element that is most easily laser-cooled and laser-manipulated, due to the fact that its level scheme is rather simple and that the 852-nm wavelength of its D_2 line falls right into the range where silicon photodiodes are most efficient and laser diode is available commercially at high powers and with a

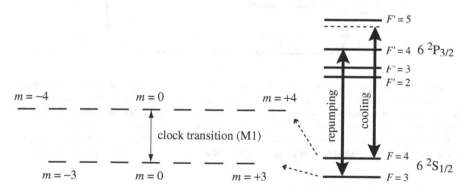

Fig. 13.5 *Right*: Simplified ^{133}Cs energy level diagram (not to scale) showing the hyperfine splitting of the $6\,^2S_{1/2}$ ground state and the $6\,^2P_{3/2}$ excited state. The excitation lines for laser cooling and repumping, used in the caesium fountain clock, are indicated by *arrows*. *Left*: Blow-up of the ground states, indicating the 16 Zeeman sublevels and the clock transition, a magnetic dipole (M1) transition

narrow linewidth. This has made it possible to switch from the thermal-beam clocks to be described next to the fountain clocks described further below. The introduction of fountain clocks reduced the uncertainty in the realization of the length of the SI second by more than an order of magnitude over the very best thermal-beam clocks, and perhaps another order of magnitude in the near future. It is hard to imagine a similar success and a similar improvement in the realization of the unit of time had Essen and Parry chosen another chemical element for their atomic clock.

13.5.2 The Thermal-Beam Caesium Clock

Caesium atoms emanate from an oven at typically 170 °C and cross a first inhomogeneous magnetic field (Fig. 13.6). In PTB-CS1 and PTB-CS2, for instance, this field can be produced by a cylindrical quadrupole or a hexapole or a combination of both. Roughly speaking, atoms in state $|F = 4\rangle$ are collimated by the magnet and those in $|F = 3\rangle$ defocused and lost from the beam. There are some tricky details here which are described in the literature (see [21] and references therein).

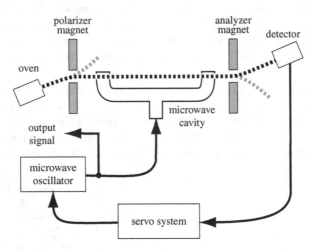

Fig. 13.6 Principle of operation of the "classical" caesium atomic clock making use of a thermal beam of atoms emanating from an oven. The atoms are spin selected by a multipole magnet, then traverse a microwave resonator structure, before a second magnet separates the atoms that did make the transition from those that did not. A servo loop adjusts the microwave frequency such that an optimum number of atoms arrive at a hot-wire detector. In later versions the magnets and detector have been replaced by suitable laser excitation zones, the so-called *optically pumped thermal-beam clocks*

The net result is that after the magnet the caesium beam is spin polarized and then crosses the microwave interaction region. The peculiar two-pronged shape of the microwave cavity is explained in the next section. Note that the microwave interaction has to take place inside a resonant cavity because only in this way the phase of the microwave radiation can stably be controlled over the whole cross section and interaction length of the atomic beam.

A second magnet deflects atoms that made the transition to $|F = 3\rangle$ onto a hot-wire detector ("flop-in configuration"), where the atoms are ionized and a small current ($\approx 10\,\text{pA}$) is registered. So the fact that atoms are arriving at the detector is an indication that the microwave frequency is approximately correct. An electronic servo loop then adjusts the frequency to the optimum value, as registered in the detected signal.

The output signal of the clock consists of a microwave signal of frequency 9, 192, 631, 770 Hz. In other words, 1 s is over once the microwave has completed 9, 192, 631, 770 periods of its oscillation – just as the definition of the SI second says.

The microwave signal feeding the cavity for probing of the atomic transition is typically synthesized from a low-noise voltage-controlled quartz crystal oscillator (VCO). Usually the VCO is weakly phase-locked to a hydrogen maser which serves as a frequency reference for the caesium clock. By proper multiplication and mixing of the VCO frequency with the frequency of a radio frequency synthesizer the 9.2-GHz microwave signal for the interrogation is generated. By square-wave frequency modulation of the synthesizer output frequency with an amplitude of half the linewidth, the atoms are probed alternatively at the left and the right sides of the resonance line. The difference Δp of the two transition probabilities for each side is a measure of the offset of the microwave carrier frequency from the centre of the resonance line. If it is not 0, the synthesizer frequency is corrected accordingly. In this case the series of correction values gives the relative frequency difference between the clock and the frequency reference. Alternatively Δp can be used directly to control the frequency of a VCO in a servo loop. In this case the VCO output directly represents the clock frequency.

The measured clock frequency differs from the unperturbed caesium transition frequency $\nu_0 = 9, 192, 631, 770$ Hz by the sum of all frequency biases. For example, this sum is of the order of 2 Hz in the case of PTB-CS2, a thermal-beam clock in operation since 1985. The correction amounts to about 1 mHz in the case of PTB-CSF1, a fountain primary clock in operation since 1999.

13.6 The Ramsey Technique

The microwave cavity of a caesium atomic beam clock is not a simple rectangular or cylindrical box (Fig. 13.6). Instead, the microwave cavity is split into two interaction zones, so that the *Method of Separated Oscillatory Fields* can be employed. This method is usually called the *Ramsey method*, after Norman Ramsey who invented it in 1949 [22] and received the 1989 Nobel Prize in Physics for it (and for his contributions to the hydrogen maser, to be discussed in Sect. 13.8.2).

A big problem facing the atomic clock-maker is posed by the uncertainty relation between energy (or frequency) and time [23]. If we want to determine the exact transition frequency in an atom with ever higher accuracy, we need to achieve ever smaller linewidths $\Delta \nu$. This means we need to observe the unperturbed transition moment over ever longer time Δt. For the simplest microwave cavity, a cylindrical box roughly one-half wavelength long, the interaction time for caesium atoms with a velocity selected around $\nu \approx 100$ m/s (as in PTB-CS1) would lead to an observed linewidth $\Delta \nu$ of about 10 kHz.

It was found by Ramsey, however, that one could split the interaction zone into two parts separated by some longer distance L and obtain a resonance feature with a width inversely proportional to the transit time L/ν. For a thermal-beam clock the Ramsey method provides a linewidth reduced by l/L, which could amount to one or two orders of magnitude. In the following we will look at the resulting line shape in more detail.

If we assume for simplicity that the microwave, as seen by the atom, is switched on and off instantaneously when the atom enters each arm of the cavity and that its amplitude is constant within each interaction zone, the transition probability is given by the expression [2]

$$
P(\delta) = \frac{4b^2}{\Omega^2} \sin^2(\Omega \tau / 2)
$$
$$
\times \left[\cos(\Omega \tau / 2) \cos[(\delta T + \phi)/2] - \frac{\delta}{\Omega} \sin(\Omega \tau / 2) \sin[(\delta T + \phi)/2] \right]^2 .
$$
$$(13.8)$$

Here τ and T are the transit times of an atom through one interaction region and through the space between the two regions, respectively. The detuning δ is positive when the frequency of the microwave is higher than the resonance frequency of the atom, and $\Omega = \sqrt{b^2 + \delta^2}$. The *Rabi frequency* $b = \mu B_m / \hbar$ is a measure of the strength of the interaction, containing both the magnetic dipole moment μ of the transition and the amplitude B_m of the microwave magnetic field.

Figure 13.7 shows an example of the resulting line shape as a function of δ. The graph of Ramsey transition probability as a function of frequency detuning looks just like the graph of the light intensity as a function of position in the far field of a double-slit diffraction pattern in optics. Of course, this is not a coincidence because in both cases there are two pathways along which the system can reach the same final state from the same initial state. While in the optical diffraction case the photon can have gone through one slit or the other, so to speak, here the atom can undergo the transition in the first or in the second Ramsey zone. Since the end results are indistinguishable, the rules of quantum mechanics state that one has to add the probability amplitudes for the two pathways and square the result to obtain the outcome seen on the detector, leading to interference fringes. The atomic clock with Ramsey excitation therefore constitutes a quantum-mechanical device, an atom interferometer.

One recognizes the dotted line as the line shape when only one zone of twice the length is present, i.e. $T = 0$ in Eq. (13.8) with properly adjusted b. Under this envelope the transition probability oscillates rapidly with detuning. This is

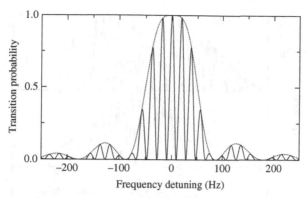

Fig. 13.7 Examples of a calculated Ramsey interference pattern as a function of frequency detuning δ. The parameters used in Eq. (13.8) are $\tau = 11.12$ ms, $T = 40$ ms, $b = 141.21/$s. The *dotted line* is the envelope of the fringe pattern, obtained by averaging over rapidly oscillating terms in (13.8)

analogous to the optical case where the double-slit diffraction pattern consists of high spatial-frequency fringes under an envelope given by the diffraction pattern for a single slit of the same width.

The Ramsey technique has a number of advantages even over a single interaction zone of the same total length $2l + L$, some of which are listed here without further comments:

- The central Ramsey fringe is actually narrower almost by a factor of 2 than if the interaction time would have been $2\tau + T$ in a single long zone.
- For a laterally extended beam of atoms susceptible to Zeeman shift of the resonance frequency, each atom only has to see the same *average* magnetic field along its trajectory. As long as this condition is fulfilled one need not fulfil the much harder condition that the field is constant everywhere on and off axis.
- When the phase difference of the microwave fields in the two zones is constant there is no first-order Doppler shift or broadening.

There are other useful features of the Ramsey technique that are important in other fields of physics. They have been omitted here because they have less relevance for the application of the method in atomic clocks.

Although there is no first-order Doppler broadening of the central Ramsey fringe there is nonetheless a large influence of the velocity distribution of the atoms (Fig. 13.8). In a thermal-beam clock, fast atoms experience a shorter time T between the two parts of the Ramsey interaction, resulting in a larger fringe width than for slower atoms with larger T. As a result, the fringe contrast washes out with increasing detuning from the central fringe. This is once again completely analogous to the optical case, with the role of the atomic velocity distribution played by a limited coherence length of the light illuminating the double slit. The background that the remaining fringes sit on is called the *Rabi pedestal*; it corresponds to the resonance

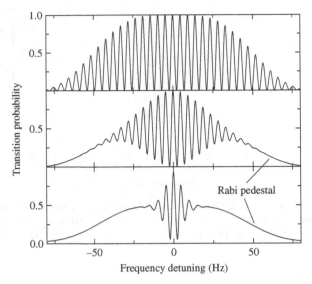

Fig. 13.8 Calculated Ramsey interference fringes as a function of frequency detuning $\delta/2\pi$ for Gaussian velocity distributions of different standard deviations σ_v centred at a mean velocity of $\bar{v} = 2.5$ m/s (other parameters as in Fig. 13.7). *Top*: monokinetic beam. *Centre*: $\sigma_v = 0.1$ m/s. *Bottom*: $\sigma_v = 0.4$ m/s. The fringes wash out to the *Rabi pedestal*, which is one-half of the envelope of the fringe pattern for monokinetic atoms

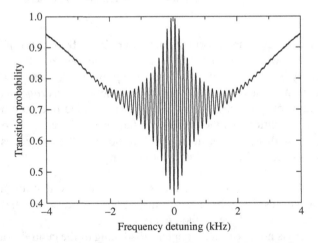

Fig. 13.9 Measured Ramsey fringe pattern for PTB's CS1 thermal-beam clock

linewidth of a single microwave excitation in one of the zones, so it is much wider than a fringe of the Ramsey curve.

A typical example of a measured Ramsey fringe pattern is shown in Fig. 13.9 for the case of PTB-CS1. The curve appears upside down compared to Fig. 13.8 because PTB-CS1 is operated in the so-called flop-out configuration, so that the signal at the detector is at its minimum exactly on resonance. The thermal velocity spread is small enough that more than a dozen fringes can be seen on either side of the maximum. The contrast of the Ramsey fringe is only 55% because the magnetic state selection mostly selects for $|F\rangle$ only, more or less regardless of the magnetic quantum number m_F. Therefore, there is always a background of $|F = 4, m_F \neq 0\rangle$ atoms present at the detector.

The lowest relative uncertainty achieved for thermal-beam clocks with magnetic state selection is 8×10^{-15} in PTB-CS1, using a 73.6-cm separation of the Ramsey zones [21]. Longer zones are not practical, in particular because of the increasing influence of the exact shape of the atomic velocity distribution on potential frequency biases. Some improvement can be obtained by replacing state selection and detection by laser optical pumping and laser-induced fluorescence [24–26]. Two such clocks still are in operation today, SYRTE-JPO [24] and NICT-O1 [26], with relative uncertainties of 6.4 and 6.8 parts in 10^{15}, respectively.

13.7 Atomic Fountain Clocks

New developments in quantum optics, in particular the advent of laser cooling techniques [27–30], opened the door to a radically new approach to the interaction time problem. Using a suitable arrangement of laser beams and magnetic fields one can capture caesium atoms from a thermal vapour and at the same time cool them down to just a few microkelvin above absolute zero temperature. Typically, in a few tenths of a second one can trap 10 million caesium atoms in a cloud a few millimetres in diameter and at a temperature of a microkelvin. At this temperature, the average thermal velocity of the caesium atoms is of the order of 1 cm/s, so the cloud of atoms stays together for a relatively long time. This cloud can be launched against gravity using laser light. Typically, the launch velocity is chosen such that the atoms reach a height of about 1 m above the trapping region before they turn back and fall down the same path they came up. Operation of a fountain clock typically proceeds in a pulsed mode, with the trapping–launching–detecting cycle taking a time $T_{cycle} \sim 1$ s. The motion of the cloud resembles that of the water in a pulsed fountain, hence the name "fountain clock".

On the way up and on the way down the atoms pass through the same microwave cavity. The two spatially separated Ramsey interactions of the thermal-beam clock are thus replaced by two interactions in the same position but with reversed direction of travel. The microwave power is chosen such that on each pass a $\pi/2$ pulse is experienced by the atoms. Ideally, therefore, after the first microwave interaction the atoms are in a quantum-mechanical, coherent superposition of both clock states with equal weight for each state. After the second passage through the cavity the transition from one clock state to the other is "completed" (in the fully resonant case).

Following the second interaction the state of the atom can be probed by laser-induced fluorescence. For a typical launch height around half a metre above the microwave interaction zone it is possible to achieve effective interaction times of

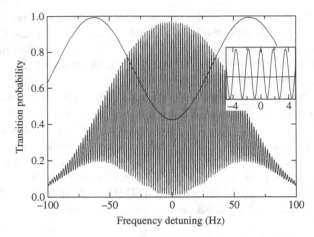

Fig. 13.10 Measured Ramsey fringe pattern for PTB's CSF2 fountain clock. *Inset*: the central part enlarged. The slowly oscillating curve is the Ramsey fringe pattern of the thermal-beam clock PTB-CS1 already shown in Fig. 13.9; its curvature is so small near detuning zero that the curve appears almost flat in the *inset*

more than half a second. The resulting 100-fold reduction in microwave resonance linewidth (Fig. 13.10) is the most obvious advantage of a fountain clock over a thermal-beam clock. Note also that the contrast of the resonance is approaching 100% because the atomic cloud can be prepared such that it consists of $|m_F = 0\rangle$ atoms only.

Equally important is a reduction in the influence of certain systematic effects because the atoms are so slow. Furthermore, the trajectory reversal eliminates the frequency bias that might exist due to a small and all but unavoidable manufacture-related phase difference between the microwave fields in the two interaction zones. It also greatly reduces the influence of any transverse or longitudinal phase gradients of the microwave field inside the cavity.

Employing a laser-cooled sample is essential for a fountain clock because in thermal samples the number of slow atoms is too small to obtain a sufficient signal strength [31] and because one would need a vacuum apparatus of several metres in height in order for some of the slower atoms to come to a stand-still in it and fall back [32]. The first considerations [33] and successful realizations [34] of the fountain principle were followed by the first metrological fountain, FO1, at the Observatoire de Paris/France [35]. Its design [36] became a standard for almost all subsequently constructed fountain clocks, with some variations of course. In the late 1990s the fountains NIST-F1 at the National Institute of Standards and Technology (NIST) in Boulder/USA [37, 38] and CSF1 at Physikalisch-Technische Bundesanstalt (PTB) in Braunschweig/Germany [39–41] became operational as primary standards. Recently, they were joined by the caesium fountain clocks CsF1 at the Istituto Elettrotecnico Nazionale (IEN) in Torino/Italy [42], CsF1 at the National Physical Laboratory (NPL) in Teddington/GB [43], F1 at National Metrology Institute of Japan (NMIJ) [44] and CsF1 at National Institute of Information and

Communication Technology Laboratory (NICT) also in Japan [45]. With FO2 and FOM [46] the Observatoire de Paris (SYRTE) is operating two more primary fountain clocks. A number of other laboratories are currently developing fountain clocks (see [47–54] for an incomplete list of examples). Most of these are employing caesium as the active element.

Because fountain clocks currently constitute mankind's best way to realize the unit of time we will now discuss those clocks a little further. Many more details can be found in a review [55] and the specific references therein.

13.7.1 Operation of a Fountain Clock

Figure 13.11 shows a simplified set-up of the vacuum subsystem of a fountain clock. Six laser beams cross in the centre of the preparation zone, where the cold atom cloud is produced. Above that follows the detection zone which is traversed by laser beams for fluorescence detection of the falling cloud. The microwave interactions take place inside a magnetic shield in the presence of a well-defined internal longitudinal magnetic field (the "C-field"). The clock is operated in a pulsed fashion, with a cycle typically lasting about 1 s (Fig. 13.12).

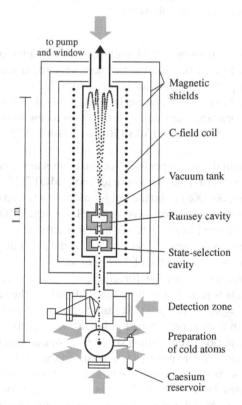

Fig. 13.11 Simplified set-up of the atomic fountain clock PTB-CSF1

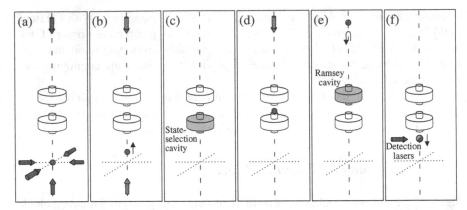

Fig. 13.12 Principle of operation of the atomic fountain clock (simplified). The full cycle (a)–(g) takes about $T_{\text{cycle}} \approx 1$ s. (a) A cloud of cold atoms in states $|F = 4, m_F\rangle$ is trapped in the intersection region of six laser beams. (b) The cloud is launched by frequency detuning of the vertical lasers. (c) Inside the state-selection cavity certain atoms are transferred to $|F = 3, m_F = 0\rangle$. (d) When the cloud exits the state-selection cavity a laser pulse pushes away all atoms not in $|F = 3, m_F = 0\rangle$. (e) The cloud slowly expands during its ballistic flight in the *dark*. On the way up and on the way down it passes through the main microwave cavity, the Ramsey cavity, where the transition to $|F = 4, m_F = 0\rangle$ is driven. (f) Detection lasers are switched on; they probe the population distribution by laser-induced fluorescence

The source of caesium atoms traditionally consists of a temperature-controlled caesium reservoir held at a suitable temperature near room temperature in order to obtain a caesium partial pressure in the order of 10^{-6} Pa in the cooling chamber. Alternatively, a laser-cooled atomic beam can be directed at the main trapping region, thus shortening the trapping time from a few hundred to a few tens of milliseconds; chirp-slowed [46], LVIS [56] and 2D-MOT [57] beams have been used.

To obtain the required low temperatures of the atom samples in an atomic fountain the atoms are cooled in a magneto-optical trap ("MOT") [58] and/or an optical molasses ("OM") [59, 60]. Key to both configurations is a set-up consisting of three mutually orthogonal pairs of counter-propagating laser beams. There are either two vertical and four horizontal beams (the (0,0,1) geometry, Fig. 13.11) or there are three beams, arranged symmetrically around the vertical, coming from below and the other three from above (called the (1,1,1) geometry).

Two laser frequencies are needed for efficient cooling (Fig. 13.5). The main power in the six beams is provided by light tuned slightly to the red (low-frequency) side of the cyclic $|F = 4\rangle \rightarrow |F' = 5\rangle$ caesium transition in order to scatter a large number of photons in a short time. Small polarization imperfections in connection with off-resonant excitation to the excited state $|F' = 4\rangle$ can lead to optical pumping into the $|F = 3\rangle$ state. A repumping laser beam tuned to the caesium transition $|F = 3\rangle \rightarrow |F' = 4\rangle$ is therefore superimposed on at least one of the six cooling laser beams. It depletes the $|F = 3\rangle$ level so that all atoms can continue to participate in the cooling process. Additional laser-cooling steps are typically applied before

and/or after the cloud has been launched towards the cavities. More details on this
are presented in [55].

Typically $10^7, \ldots, 10^8$ atoms are trapped and cooled. This atom number is
mainly limited by the available laser power and laser beam diameters. For reasons
of compactness, low power consumption, reliability and ease of use, most optical
set-ups use exclusively laser diode systems. Their light powers and frequencies need
to be controlled precisely in order to properly cool, launch and detect the atoms.
Furthermore, all laser light has to be blocked completely during the interaction of
the atoms with the microwave field so that the atomic transition frequency is not
shifted by the ac Stark effect [61].

After trapping, the atoms are launched by the "moving molasses" technique
(Fig. 13.12b). The pattern formed by the interference of the vertical trapping beams
can be made to move at a velocity $c\delta v/v_c$ when the upward-directed laser beam is
tuned to a frequency $v_c + \delta v$ and the downward-directed laser beam to $v_c - \delta v$. In this
upward-moving interference pattern the atoms are accelerated within milliseconds
to velocities of several metres per second. At this stage in the fountain cycle the
atomic population is distributed across all ground state sublevels, mostly those with
$|F = 4\rangle$ (Fig. 13.13a).

The last cooling stage consists of a sub-Doppler laser-cooling phase, as described
in [55]. When the lasers are finally switched off altogether the caesium atom

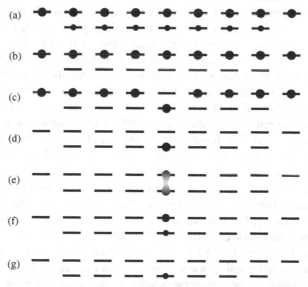

Fig. 13.13 Qualitative population of the Zeeman sublevels in the caesium ground state during key
phases of the fountain cycle. (**a**) After trapping and launching. (**b**) After post-cooling. (**c**) After
passage through the state-selection cavity. (**d**) After the laser blow-away pulse. (**e**) After the first
passage through the Ramsey cavity. (**f**) After the second passage through the Ramsey cavity. Here
is assumed that the microwave frequency is slightly detuned from exact atomic resonance. (**g**) Just
before the atoms in $|F = 3\rangle$ are detected

temperature is about $1\,\mu\mathrm{K}$, corresponding to a thermal velocity of less than 1 cm/s. When the repumping laser is switched off slightly later than the cooling laser all atoms end up in an $|F = 4\rangle$ level (Fig. 13.13b).

As the next step a further state-selection process can be applied to the atoms in order to reduce the background signal and the collisional shift due to atoms in states with $m_F \neq 0$ which do not take part in the "clock transition" between the states with $m_F = 0$. This removal of atoms with $m_F \neq 0$ is a big advantage of fountain clocks over thermal-beam clocks also because it reduces effects like Rabi and Ramsey pulling and frequency shifts due to Majorana transitions [62], apart from improving the contrast of the Ramsey pattern (Fig. 13.10). For the state selection the $|F = 4, m_F = 0\rangle$ atoms are first transferred by a microwave π pulse using the clock transition to the state $|F = 3, m_F = 0\rangle$. This can be done in a microwave state-selection cavity in which the atoms pass before they enter the Ramsey cavity for the first time (Figs. 13.12c and 13.13c). Afterwards all the atoms which remained in the $|F = 4\rangle$ state are pushed away by the strong spontaneous light force exerted by a laser beam tuned to resonance with the $|F = 4\rangle \to |F' = 5\rangle$ transition. This results in a pure atomic sample of atoms all in state $|F = 3, m_F = 0\rangle$ entering the Ramsey cavity (Fig. 13.13d), where subsequently the clock transition $|F = 3, m_F = 0\rangle \to |F = 4, m_F = 0\rangle$ is excited.

Let us note here that this extreme nonequilibrium distribution of atomic population is attractive not only for time keeping. It can be used to perform precise spectroscopic experiments with a state-selected, monokinetic, cold sample of atoms, which can still be manipulated by laser light, microwaves or electric fields for an observation time of about half a second. This single-populated state need not be the $|F = 3, m_F = 0\rangle$ state but can be almost any other of the 16 sublevels when the microwave frequency in the state-selection cavity is tuned to a different transition. Examples for planned or completed experiments include the measurement of the black-body radiation shift of the clock transition when the vacuum tube is heated [63], the investigation of state-dependent Feshbach resonances in the collisional cross sections between cold caesium atoms [64] or the proposed search for parity violation in atoms [65]. The fountain principle can also be used to determine the gravitational constant G [66], the local gravitational acceleration g [67] or the fine structure constant α [68].

One of the most critical parts of a fountain clock is the microwave cavity for the Ramsey interaction. Much work has been done on different realizations of these delicate devices (see [55] for more details and references). Most fountain clocks use a cylindrical microwave cavity with the field oscillating in the TE_{011} mode. This mode (indicated in Fig. 13.14) exhibits particularly low losses which results in a particularly small running-wave component in the cavity. The dependence of the microwave phase on the transverse position of the atomic trajectory is therefore small, as well. The oscillating microwave magnetic field inside the cavity is directed primarily along the vertical, the same direction as the static magnetic C-field of typically 100 nT flux density. Selection rules therefore favour the $\Delta m = 0$ transitions. The static field in addition detunes all other transitions, so that even the small curvature of the microwave field lines near the end caps of the cavity does not lead

Fig. 13.14 Sketch of the microwave field geometry inside a typical TE$_{011}$ cylindrical microwave cavity. *Dotted line*: trajectory of the atoms

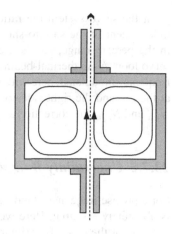

to a substantial amount of parasitic $\Delta m_F = \pm 1$ transitions. As a result, the clock transition $|F = 3, m_F = 0\rangle \rightarrow |F = 4, m_F = 0\rangle$ dominates.

After the first passage through the Ramsey cavity the atomic population looks like as indicated in Fig. 13.13e: a coherent superposition of the two clock states. This atomic coherence precesses at its intrinsic rate, and during the second cavity passage, on the way down, its phase is probed with the microwave again. In the case of exact resonance, for instance, the microwave phase relative to that of the atomic coherence is still the same, so the transition is completed and all atoms end up in the upper state $|F = 4, m_F = 0\rangle$. Figure 13.13f depicts the situation when the microwave frequency is slightly detuned so that the transition probability is not 100%.

In a fountain clock the atoms arriving in $|F = 4\rangle$ and those in $|F = 3\rangle$ are detected separately via laser-induced fluorescence. When the falling atoms pass through a first transverse standing-wave light field tuned to the $|F = 4\rangle \rightarrow |F' = 5\rangle$ transition the fluorescence light emitted by the atoms in $|F = 4\rangle$ is detected with a photodiode. The time-integrated photodetector signal, N_4, is proportional to the number of atoms in the state $|F = 4\rangle$. From the shape of the time-dependent photodetector signal one can infer the axial velocity spread of the atoms in the cloud, which indicates the corresponding kinetic temperature. The $|F = 4\rangle$ atoms are then pushed away by a second beam slightly below, a transverse travelling-wave laser field tuned to the $|F = 4\rangle \rightarrow |F' = 5\rangle$ transition. Only the atoms in state $|F = 3\rangle$ remain (Fig. 13.13g). These are then pumped to the state $|F = 4\rangle$ by a third horizontal detection laser beam, positioned lower again and tuned to the $|F = 3\rangle \rightarrow |F' = 4\rangle$ transition. Superposed or slightly below another laser beam is present, which is again tuned to the $|F = 4\rangle \rightarrow |F' = 5\rangle$ transition. In effect, the atoms in $|F = 3\rangle$ are first pumped to the state $|F = 4\rangle$ and then detected by their fluorescence on the $|F = 4\rangle \rightarrow |F' = 5\rangle$ transition with the help of a second photodetector, giving a measure N_3 for the number of atoms that arrived in the detection zone in the state $|F = 3\rangle$.

In the servo system the ratio $N = N_4/(N_3 + N_4)$ is calculated. This ratio is independent of the shot-to-shot fluctuations in atom number, which typically lies in the percent range, and is used as the input signal to the microwave-frequency servo loop. As in thermal-beam clocks, the interrogation microwave is square-wave frequency modulated, here with a period of $2T_{cycle}$, so that alternate "shots" are taken at the left and the right sides of the central Ramsey fringe. From the difference of N_{left} and N_{right} the correction signal for the microwave carrier frequency is derived.

13.7.2 Uncertainty Budget

For a precise frequency standard a detailed knowledge of all sources of uncertainty is absolutely essential. Here we will present a rather complete list of such contributions for the case of a primary fountain clock. We are choosing this clock as an example because of its great importance today. For the optical clocks of the future the list will be very similar in principle, but of course with some modifications in the details. For example, the role of the cavity-related shifts in a microwave clock, discussed on Page 389, will be played by the wavefront curvature of the interrogation laser beam(s) in an optical clock. Similarly, the Dick effect (explained below) in an optical clock is due to the frequency noise on the interrogation laser rather than on the noise of a microwave oscillator.

Two types of uncertainties have to be considered. *Statistical* uncertainties in general decrease with longer averaging time, whereas *systematic* uncertainties reflect our imperfect knowledge of potential frequency biases and, with the exception of drifts of operating parameters, are independent of averaging time.

13.7.2.1 Statistical (Type A) Uncertainties

In a well-designed fountain clock the following noise contributions to the total instability have to be considered [69]:

(a) *Quantum Projection Noise* [70]: resulting from the fact that a fountain clock is operated alternatingly at the left and the right sides of the central Ramsey fringe where the transition probability is neither 0 nor 1.

(b) *Photon Shot Noise*: resulting from the statistical detection of a large number of photons per atom.

(c) *Electronic Detection Noise*: resulting from the electronic detection process by typically a photodiode followed by a transimpedance amplifier.

(d) *Local Oscillator Noise*: resulting from a downconversion process of local oscillator frequency noise components because of the noncontinuous probing of the atomic transition frequency in a pulsed fountain ("Dick effect"). Basically, phase excursions of the local oscillator while no atoms are in or above the microwave cavity will go undetected and therefore add some of the phase noise of the local oscillator to the output of the clock.

When assuming a white frequency spectrum of the noise sources Eq. (13.7) takes the following form [69]:

$$\sigma_y(\tau) = \frac{\Delta \nu}{\pi \nu_0} \sqrt{\frac{T_{\text{cycle}}}{\tau}} \left(\frac{1}{N_{\text{at}}} + \frac{1}{N_{\text{at}} \varepsilon_c n_{\text{ph}}} + \frac{2\sigma_{\delta N}^2}{N_{\text{at}}^2} + \gamma \right)^{1/2} . \qquad (13.9)$$

The terms in the brackets of (13.9) quantify the four processes (a)–(d) listed above. N_{at} is the number of detected atoms, n_{ph} the average number of photons scattered per atom at the detection and ε_c is the photon collection efficiency. $\sigma_{\delta N}^2$ is the uncorrelated rms fluctuation of the atom number per detection channel.

With sufficiently high numbers of photons detected per atom and using state-of-the-art low-noise electronic components, the noise contributions (a), (b) and (c) can be reduced to such a level that the frequency instability is determined by the Dick effect (d). A quantitative description of the Dick effect is beyond the scope of this chapter; it can be found in [71–74]. With the best currently available voltage-controlled oscillators (VCOs) the relative frequency instability is limited to the order of $10^{-13}(\tau/\text{s})^{-1/2}$.

In order to improve on that one can use a better local oscillator or reduce the dead time of the fountain, i.e. the fraction of the fountain cycle where no atoms are in or above the microwave cavity. However, some dead time is unavoidable in the standard pulsed fountain described so far because one has to wait until the detection process is finished before the next cloud can be launched. A continuous-beam fountain clock, however, would be practically immune from the Dick effect.

Conceptually the easiest way to reduce the influence of the Dick effect on the short-term instability is to use a more stable local oscillator. Using a cryogenic sapphire oscillator developed at the University of Western Australia [75] and a specially designed low-noise microwave synthesis chain the group at SYRTE was able to reach a fractional frequency instability of only $1.6 \times 10^{-14} (\tau/\text{s})^{-1/2}$ [76], which was only limited by quantum projection noise [69].

The additional effort of liquid-helium refrigeration required for low-noise cryogenic sapphire oscillators can be avoided when using a laser frequency comb (see Sect. 13.9.2 below) to transfer the superior short-term stability of a laser stabilized to a high-finesse optical cavity down into the microwave regime and then deriving the 9.2-GHz interrogation signal from that phase-stable microwave. Initial results of the phase stability of such a signal have been reported recently [77].

To speed up the loading and preparation of the cold atomic cloud the atoms can be collected not from the residual background vapour but from a slow atomic beam instead, as was discussed above in Sect. 13.7.1.

Loading from a slow beam is indispensable when implementing a multi-toss scheme. The idea is to have more than one ball of atoms travelling inside the apparatus simultaneously. For instance, the same number of atoms as in a single-ball fountain can be distributed over several balls. In each of them the atomic density is lower, and the collisional shift reduced in proportion. One implementation scheme is to launch several balls (up to 10 or so) in quick succession but with successively decreasing launch height [78]. These balls never meet in the free-flight zone above the cavity where collisions would lead to a frequency shift. But they all come together in the detection zone – where cold collisions do not matter anymore – to produce a strong signal. This idea has been tested at NIST [79] and is envisioned to be implemented in NIST-F2 [80].

Another avenue is the "juggling fountain" [81] where several balls are launched with a time separation smaller than the flight time of an individual cloud. When timing and launch velocities of the individual clouds are chosen such that each time two of them meet their relative energies fulfil the condition for a Ramsauer resonance, they pass through each other basically without scattering, i.e. without additional collisional shifts. It is straightforward to do this with just two "balls", but amazingly it can also be done with more than two balls [82]. The multi-ball scheme requires a precise control of the launch times, velocities and densities of the individual balls. Furthermore it relies on a delicate cancellation of the collisional shifts in successive two-ball collisions, making use of the energy dependence of sign and amplitude of the shift [82].

The Dick effect can be all but eliminated when the fountain clock is operated in a continuous mode rather than pulsed [83, 84]. At the same time, because of the continuous detection a lower density beam can be used, reducing the uncertainty due to cold collisions. However, a continuous fountain poses a number of technical and experimental challenges. First of all, since the preparation and the detection zones have to be spatially distinct the atoms have to fly along a parabolic path. This requires a special geometry for the main cavity. Unfortunately, in this way one loses one of the big advantages of the pulsed fountain design, where the atoms retrace their path through the microwave field and thus mostly cancel any end-to-end phase shifts in the cavity. In the continuous fountain FOCS-1 a special device allows one to rotate the cavity around the vertical axis by precisely $180°$, so that an effective beam reversal occurs [85], in analogy to the procedure in thermal-beam clocks [21].

Furthermore, the suppression of stray light from the preparation zone becomes more complicated. Unlike in the pulsed case the laser light cannot be switched off during the free-flight phase of the atoms, so for the continuous fountain one resorts to mechanical shutters inside the vacuum vessel. A wheel with partially overlapping filters absorbing the laser radiation is rotating rapidly through the atomic beam in such a way that the direct line of sight from the detection zone into the free-flight zone passes through at least one filter at any one time. Not only does this chop thin slices out of the continuous atomic beam but also does one have to have a motor inside the ultra-high vacuum system – which also has to be nonmagnetic!

Details of all design issues can be found in the thesis by Joyet [85]. The first such clock, METAS-FOCS1, is now being commissioned at METAS [57, 47], with an improved version FOCS2 being under development [86]. Design goals are a relative short-term instability of $7 \times 10^{-14} (\tau/\text{s})^{-1/2}$ (using a quartz oscillator as a local oscillator for the 9-GHz synthesis chain) and a relative uncertainty of 10^{-15}.

Obtaining a small frequency instability is indispensable for the evaluation of systematic uncertainty contributions at the level of 10^{-15} or below. Even an instability of only $10^{-13}(\tau/\mathrm{s})^{-1/2}$ still results in a 3.4×10^{-16} statistical uncertainty after one full day of measurement.

13.7.2.2 Systematic (Type B) Uncertainties

To a large degree the sources of frequency bias in a fountain clock are very similar to those in a thermal-beam clock. The latter are described in great quantitative detail in the book by Vanier and Audoin [2]. The fact that now one has a state-selected, monokinetic sample of laser-cold atoms greatly reduces some of these biases but also introduces two new ones: the light shift and the cold-collision shift. Here we will present just an overview and refer the reader to the review [55] and the published formal evaluations of the individual fountain clocks, for example, in [76, 40, 38, 42–45].

(1) *Second-Order Zeeman Effect*
   The applied static magnetic bias field ("C-field") results in a second-order shift of the clock transition by $f_c = 0.0427\,\mathrm{Hz}\times(B_C/\mu\mathrm{T})^2$, where B_C is the magnetic flux density of the C-field, so that almost 5×10^{-14} relative frequency shift is obtained for the typical $B_C \sim 0.1\,\mu\mathrm{T}$. One can use the atoms themselves as probes for the mean C-field strength, its inhomogeneity and its temporal stability by measuring the resonance frequency f_Z of the first-order field-sensitive transition $|F = 4, m_F = 1\rangle \rightarrow |F = 3, m_F = 1\rangle$. In contrast to a conventional beam clock, an atomic fountain provides the advantageous possibility of mapping the C-field by launching the atoms to different heights $h_{\max}$. For a given $h_{\max}$ the measured f_Z is an average over $B_C(z)$ encountered by the atom cloud, weighted with the z- and $h_{\max}$-dependent dwell time of the cloud. The series of $f_Z(h_{\max})$ can be deconvoluted to obtain $B_C(z)$ [87].
(2) *Majorana Transitions*
   Majorana transitions ($\Delta F = 0, \Delta m_F = \pm 1$) between the m_F substates within a hyperfine ground state $|F = 4\rangle$ or $|F = 3\rangle$ can be induced near zero crossings of the magnetic field strength [88]. In real caesium clock cavities it cannot be avoided that due to the field geometry some $\Delta F = \pm 1, \Delta m_F = \pm 1$ transitions occur besides the designated clock transition, albeit with a small probability. In connection with Majorana transitions large frequency shifts can occur [89] because atoms can end up in superposition states with the same F quantum number but different m_F quantum numbers. For these states the hyperfine transition frequency is in general different from the clock transition frequency, with a resulting overall frequency shift. For a well-controlled magnetic field geometry the uncertainty estimate due to Majorana transitions is typically stated as less than 10^{-16}.
(3) *Cavity-Related Shifts: Residual First-Order Doppler Effect and Cavity Pulling*
   A general advantage of an atomic fountain is that the atoms cross the same microwave cavity twice in opposite directions. If the atomic trajectories were

perfectly vertical, frequency shifts due to axial and radial cavity phase variations would be perfectly cancelled as each atom would interact with the field first with velocity v (upwards) and later with $-v$ (downwards). Due to the transverse residual thermal velocity and a possible misalignment of the launching direction atoms in general do not retrace their upward trajectories on the way down, so a first-order Doppler frequency shift can remain. Uncertainty contributions of the order of at most several 10^{-16} are estimated typically [76, 40, 38, 42–45].

The frequency shift due to cavity pulling is usually negligible because atom numbers are still rather low. Furthermore, since cavity pulling is proportional to atom number it is corrected for automatically when the collisional shift correction (see below) is applied.

(4) *Rabi and Ramsey Frequency Pulling*

Frequency shifts due to Rabi and Ramsey frequency pulling [2] can occur in the presence of nonzero and asymmetric (with respect to $m_F = 0$) populations of the $|F, m_F \neq 0\rangle$ substates when the atoms enter the Ramsey cavity. However, the state selection process in a fountain clock strongly reduces the impact of these frequency pulling effects to well below 10^{-16}.

(5) *Microwave Leakage*

The presence of microwave radiation outside the cavities can lead to large frequency shifts despite the partial cancellations due to the near-symmetry of the up-down motion of the cloud. The spread of the cloud, combined with the fundamentally unavoidable radial microwave power variation inside the cavity, makes the atoms see different effective microwave powers during the two cavity passages. This greatly increases the clock's susceptibility to stray field-induced frequency bias [90]. Typical uncertainties evaluated so far fall in the low 10^{-16} range once all microwave devices, cables and connectors have been shielded carefully.

(6) *Electronics and Microwave Spectral Impurities*

The low total systematic uncertainty of a fountain clock requires tight specifications for any electronic subsystems in order to avoid frequency bias caused by servo offsets or microwave spectral impurities [91, 92]. Recent improvements on other systematic uncertainties have prompted several groups to upgrade their microwave electronics so that they no longer limit further progress [93–96].

(7) *Light Shift and Optical Pumping/Heating*

All lasers must be completely blocked while the atoms are in or above the Ramsey cavity. Otherwise, the shift of the atomic energy levels by the AC Stark effect (light shift) [61] will lead to a frequency bias – nanowatts of residual laser power can already be too much. Once the atoms have fallen through the Ramsey cavity they still must be protected from stray light until they have passed the detection zone to avoid heating of the cloud [97] and a reduced contrast of the Ramsey fringe [98]. With combinations of large laser frequency detuning and of acousto-optical and mechanical shutters the corresponding uncertainty can be reduced to below 10^{-16}.

(8) *Blackbody Shift*

AC Stark shift of a different origin is induced by the thermal radiation of the vacuum enclosure. Assuming that this radiation follows the spectral power density distribution of a black body the clock transition frequency is shifted by

$$f_{bb} = -1.579 \times 10^{-4}\,\mathrm{Hz}\, \left(\frac{T}{300\,\mathrm{K}}\right)^4 \times \left[1 + 0.013 \left(\frac{T}{300\,\mathrm{K}}\right)^2\right] \quad (13.10)$$

for a vacuum enclosure at temperature T [99–103, 63]. Hence, for a caesium clock at room temperature the relative frequency shift is of the order of -17×10^{-15}. A corresponding correction must be applied. Its relative uncertainty lies below 10^{-16} when the thermal environment seen by the atoms is known within 0.2 K or so. The NIST-F2 and the NRC-FCs1 fountains under development will all but eliminate this shift when they are operated at liquid-nitrogen temperature [80, 52].

(9) *Collisional Shift*

A major source of uncertainty is the frequency shift due to collisions among the cold atoms in the cloud [104]. The collisional cross section was found to be strongly dependent on energy (i.e. the average temperature of the cloud) [105]. The problem is particularly serious for caesium because it was found that its collisional cross section is unusually large at the low cloud temperatures used in a fountain. Conceptually the simplest solution would be to choose another element. For instance, in rubidium the collisional cross-section is almost two orders of magnitude lower [106, 107]. Indeed, rubidium fountain clocks have been built where the reduction of collisional shift uncertainty was one of the motivations [108, 109].

A number of schemes have been devised to reduce the collisional shift or at least its contribution to the uncertainty budget of the caesium fountain clock, for instance, by lowering the density of the atomic cloud and therefore the collision rate. A reduced number of atoms, however, reduce the detected signal and therefore the signal-to-noise ratio and the short-term stability (Eq. 13.9).

Recent developments are helping to ease this trade-off problem for the case of caesium. The problem of loss of signal for low-density clouds can be circumvented by controlling the preparation of the atoms such that more than one low-density cloud at a time is travelling through the vacuum system, as discussed above in connection with the Dick effect. The continuous fountain also allows one to reduce the influence of collisional shift because it spreads the atoms over the whole trajectory instead of concentrating them in a high-density cloud.

The more traditional approach is to measure the clock's output frequency for two or more effective densities of the atomic cloud and then extrapolate to zero density. Since the frequency shift due to cold collisions is linear in effective atom density, one can extrapolate the measured frequencies obtained with clouds of different densities to zero density using a linear regression. This is the method currently employed by all primary fountain clocks. Since the actual density of the cloud is not readily accessible in the experiment, one substitutes the number N_{at} of detected atoms instead. This, of course, assumes that there is a strict proportionality between N_{at} and effective density. The actual experimental practice in the various laboratories differs in the way that clouds of different densities are prepared. One way is to vary the loading time of the MOT/OM to determine the proportionality factor between atom number and output frequency [55]. In all subsequent frequency measurements one continuously monitors the atom number and scales it with this factor to extrapolate to the zero density value [40]. A disadvantage of this method is that one cannot exclude a difference in the spatial distribution of the atoms within the cloud for the two atom number regimes, which in general could change the proportionality factor between effective density and detected atom number. An additional contribution therefore needs to be included into the uncertainty budget.

As an alternative, the collisional shift can be monitored during the course of a frequency evaluation, by switching between two different atom numbers every few shots [43] or every hour [42]. The switching is done by varying the microwave power or detuning for the state-selection pulse. Once again it cannot be excluded that the density distribution of the state-selected cloud changes between high and low atom number due to inhomogeneities of the microwave excitation.

When a very stable local time reference, for instance, a maser ensemble, is available, one can also run the fountain at extremely low atom number (i.e. very low collisional shift) and just average over a longer time interval [110].

Perhaps the most elegant way is to use the technique of rapid adiabatic passage during the preparation stage (Figs. 13.12c and 13.13c) in the state-selection cavity [111]. This method relies on the fact that the population of state $|F = 4, m_F = 0\rangle$ can be transferred with 100% efficiency into state $|F = 3, m_F = 0\rangle$ when both frequency and amplitude of the microwave radiation inside the selection cavity are ramped with just the right timing [112]. Details of the technique can be found in [112, 111, 113]. Basically, the microwave detuning is ramped from (ideally) minus infinity to infinity, while the amplitude goes from 0 to a maximum (at detuning zero) and back to 0 again. One has to ensure that the rate of change of the microwave frequency has to be much lower at all times than the square of the Rabi frequency (which is proportional to microwave power).

When the pulse is switched off abruptly at detuning $\delta = 0$ the atoms are left in an exactly equal superposition of both states. The important feature of the rapid adiabatic passage is that this happens independently of the actual Rabi frequency an atom sees, i.e. it does not depend on where an atom passes through the field inside the cavity (the field amplitude decreases across the aperture when going away from the centre). The pushing beam therefore removes exactly half of all atoms, without changing the density distribution, temperature or velocity of the cloud – in contrast to the other methods where such changes cannot be excluded.

At SYRTE the ratio of $1:2$ can be prepared and maintained with an accuracy of 10^{-3} [76], allowing for a very precise determination of the collisional shift rate and its correction. The precise control over atomic populations in SYRTE-FO2 has also made it possible to detect Feshbach resonances in the dependence of the collisional shift on magnetic quantum number m_F [64]. Surprisingly, these resonances occur already for flux densities of $2\,\mu T$ or less.

Finally, under certain circumstances one can actually run a caesium fountain clock under operating conditions where the total collisional shift experienced by the cloud is 0 [114]. A typical Cs fountain standard operates with atoms cooled down to 1–$2\,\mu K$ at the time of launch. As the cloud expands during its ballistic flight, correlations build up between atomic position and velocity, an effect particularly pronounced for atoms initially trapped in a small cloud (like in a MOT). Basically, atoms sort themselves spatially into concentric shells of very nearly the same velocity. This means that the effective collisional energy decreases down to a few hundred nanokelvins even before the first Ramsey interaction [115]. At low energies the collision rate coefficients for the clock states $|F = 3, m_F = 0\rangle$ and $|F = 4, m_F = 0\rangle$ differ in sign, which gives rise to a strong variation of the collisional shift if the relative weights of the clock states in the superposition state prepared during the first Ramsey interaction are varied.

This could be confirmed experimentally and theoretically for two independent primary standards [114]. The fraction of the population in a given clock state (e.g. $|F = 4, m_F = 0\rangle$) was changed by adjusting the amplitude

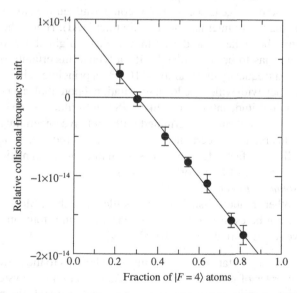

Fig. 13.15 Measurement of the collisional frequency shift in PTB-CSF1 as a function of the population composition during the Ramsey time. The experimental data are fitted by a linear function

of the microwave field in the Ramsey cavity. The collisional frequency shift for atoms captured in a MOT varies linearly with the population fraction (Fig. 13.15). For PTB-CSF1 the collisional frequency shift is cancelled for a 30% fraction of $|F = 4, m_F = 0\rangle$ atoms, which can be achieved experimentally with only a minor reduction of the fringe contrast and short-term stability of the fountain clock.

The practical details of this method have been considered in detail in [116]; under certain circumstances a shortening of the averaging time to reach a certain statistical uncertainty of about an order of magnitude was predicted. So far, this method is not yet applied routinely.

(10) *Background Gas Collisions*

At typical vacuum pressures in the ballistic flight region in the low 10^{-7} Pa range the effect of collisions of cold caesium atoms with residual gas atoms is estimated to be well below 10^{-16}.

(11) *Time Dilatation: Relativistic Doppler Effect*

Relativistic time dilatation reduces the clock frequency observed in the laboratory frame by $f_D \approx \nu_0 \langle v^2 \rangle / (2c^2)$, with $\langle v^2 \rangle$ the mean quadratic velocity of the atoms above the microwave cavity and c the velocity of light. Typically f_D/ν_0 is of the order of 10^{-17}, so that in contrast to thermal-beam clocks time dilatation and the associated uncertainty contribution can be neglected.

(12) *Gravitational Redshift*

Even though the gravitational redshift [117] of about -1.1×10^{-16} per metre of elevation above the geoid is not relevant for the realization of the *proper* second of a clock and for local frequency measurements at the same gravitational potential, its knowledge is necessary for comparing remote clocks, for instance when contributing to international atomic time (TAI). Hence the mean height of the atoms above the geoid during their ballistic flight above the microwave cavity centre has to be determined with a typical uncertainty of 1 m, corresponding to a frequency uncertainty of 10^{-16}. In principle, a limitation is given by the accuracy with which the local gravitational potential can be determined. Even for high-altitude laboratories like NIST in Boulder (≈ 1630 m above sea level) the correction can be determined with a relative uncertainty of 3×10^{-17} [118]. At this relative uncertainty the gravitational redshift will probably not become a limiting factor for clocks based on microwave transitions. This is in contrast to the case of future optical clocks.

(13) *Other Systematic Effects*

There are other frequency-shifting effects (dc Stark shift, Bloch–Siegert shift) [2], which can be estimated to be less than 10^{-17} in a fountain clock, or the microwave recoil shift at around 10^{-16} [119].

In conclusion it can be stated that the main contributions to the systematic uncertainty are of the order of a few 10^{-16} or less. As the individual systematic uncertainty contributions can be assumed to be linearly independent, the resulting total systematic uncertainty is the square root of the sum of squares of the individual contributions.

13.7.3 State of the Art

Worldwide, as of July 2008 there are nine primary fountain clocks in operation that have been used to contribute to the international timescales. The individual clocks have very different characteristics, depending on the design choices made by the teams developing and operating them. Which of the fountains is the "best one" right now? Well, that depends on the criterion chosen.

- *Stability*
  In terms of stability the SYRTE fountains FO1 and FO2 with $\sigma_y(1\,s) = 2 \times 10^{-14}$ are unsurpassed. This low instability is achieved by referencing the 9-GHz microwave to an ultrastable cryogenic sapphire oscillator (CSO) [75, 76, 69]. So far, the other groups have shied back from the enormous technical and financial effort involved in the operation of a CSO at liquid-helium temperature. New developments in the field of femtosecond laser frequency combs have opened up the possibility of using such a comb "in reverse", by locking it to a narrow optical Fabry–Perot resonance and deriving a phase-stable 9-GHz microwave signal from it. Performance similar to that of a CSO has been reported [77].

- *Uncertainty*
  SYRTE-FO1, SYRTE-FO2 and NIST-F1 can now be run under conditions of $3, \ldots, 4 \times 10^{-16}$ relative uncertainty [76, 110, 120], with most other fountain clocks currently in the range of 10^{-15} or slightly below. There appears to be room for substantial reductions in the uncertainties of collisional shift, cavity-related effects and electronics by a more stringent control of operating parameters and a better theoretical understanding of microwave cavities. However, it will be difficult to drive the relative uncertainty much below $1, \ldots, 2 \times 10^{-16}$.

  That is because reducing the uncertainty of the black-body shift appears rather difficult for the existing fountains since it would require a sufficiently detailed knowledge of the effective thermal environment of the atoms during their free-flight phase. This includes not only the actual inner-wall temperature to within better than 0.2 K but also reliable values for the emissivity of the inner surfaces of the vacuum tube (which might change due to physisorption or chemisorption of caesium or residual gases over the years of operation) as well as the influence of radiation coming in through windows and other openings. Any substantial progress on this front might require the cooling of the walls of the vacuum tube. This is exactly what is intended for NIST-F2 and NRC-FCs1, which are currently under construction [80, 52].

- *Robustness*
  Although not as "sexy" as the other two criteria, the operational robustness of a clock is important for its application in the generation of timescales, be it on a local or on an international level, although depending on circumstances some dead time can be tolerated. No fountain can yet run with the same round-the-clock unattended operation as the thermal-beam clocks PTB-CS1 and PTB-CS2 have been doing for decades. The main problem here is the stability of the operating parameters of the laser sources over days and weeks, in particular slow drifts into less favourable operational regimes or even mode hops. Using standard

extended-cavity diode lasers with feedback from an external diffraction grating it is possible today to run a fountain clock with better than 99.8% availability over weeks. This is, for instance, the case with PTB-CSF1, where the small amount of dead time is entirely due to scheduled laser parameter checks [121]; recently, completely unattended and uninterrupted operation for about 3 weeks has been tried successfully.

New laser sources, like the extended-cavity lasers developed for the ACES project [122] or the high-power narrow linewidth DFB laser diodes now available commercially, might simplify long-term operation of fountain clocks. Dead times due to unintended laser dropouts can be minimized by automated laser relock systems, although these cannot always deal with mode hops.

13.8 Other Types of Atomic Clocks

A discussion of atomic clocks would not be complete without at least a brief discussion of some other types of atomic clocks. Out of the many possibilities only two are selected here: the vapour-cell clock, which trades uncertainty for a smaller size, and the hydrogen maser which trades uncertainty for superior short- and medium-term stability.

13.8.1 Vapour-Cell Clocks

Vapour-cell atomic clocks are perhaps the most influential type of atomic clock for everyday life – or if they are not yet, they soon will be. This is because it is rubidium-vapour clocks that work aboard the GPS satellites, helping us navigate unfamiliar terrain, and that are gradually finding their way into cell phone network base stations. With the development of the chip-scale atomic clock even battery-operated, handheld devices like the GPS receivers or the cell phones themselves might soon contain their own atomic clock. And would not an atomic wrist watch be ever so cool?

13.8.1.1 Microwave Vapour-Cell Clock

Caesium and rubidium have two big practical advantages that distinguish them from almost all other chemical elements. They can easily be manipulated with readily available diode laser sources, and the vapour pressure above solid caesium or rubidium is high enough to do useful spectroscopy even at room temperature. It is therefore possible to place a small crumb of Cs or Rb into an evacuated glass cell, shine a light beam through the cell and observe a spectroscopic signal (typically the transmitted power) shaped by the optical properties of the atomic vapour.

In the simplest embodiment this light actually does not come from a laser but from a spectral lamp containing the element of interest, typically rubidium. Natural rubidium consists of a mixture of two isotopes, ^{85}Rb and ^{87}Rb (Fig. 13.16). Most of the light given off by the lamp corresponds to the D_1 and D_2 lines at 794 and 780 nm wavelength, respectively. By a lucky coincidence, the Doppler-broadened $|F = 3\rangle \rightarrow |F'\rangle$ absorption line in ^{85}Rb overlaps the $|F = 2\rangle \rightarrow |F'\rangle$ absorption line in ^{87}Rb; this is true both for the D_1 and the D_2 lines. The $|F = 2\rangle \rightarrow |F'\rangle$ light component emitted by an isotopically pure ^{87}Rb lamp is therefore absorbed in an isotopically pure ^{85}Rb "filter" cell (Fig. 13.17), leaving a transmitted light beam that can only interact with the $|F = 1\rangle \rightarrow |F'\rangle$ absorption line in the main ^{87}Rb cell. Atoms in ground state $|F = 1\rangle$ are optically excited and then decay into either one of the ground states. Atoms in state $|F = 2\rangle$ are trapped because they cannot be excited by the filtered lamp light anymore. State $|F = 2\rangle$ is therefore called a "dark state" for this particular light spectrum.

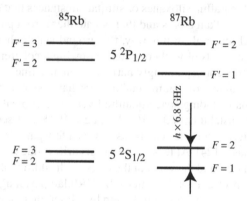

Fig. 13.16 Level scheme for the D_1 excitation line of the two naturally occurring rubidium isotopes

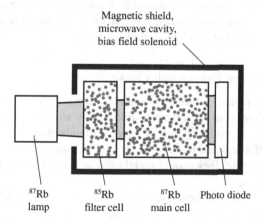

Fig. 13.17 Set-up of a compact optical-pumping rubidium microwave clock. The typical volume of such a clock is 250 cm^3

Sooner or later all of the atomic population will have been optically pumped into state $|F = 2\rangle$, and the main cell becomes transparent for the light beam: one observes maximum signal on the photodetector. When now a 6.8-GHz microwave is present it can induce transitions from $|F = 2\rangle$ to $|F = 1\rangle$. This restores some population to the light-absorbing state, so that the transmission signal decreases. A servo loop adjusts the microwave frequency such that the detector signal is minimized. In the absence of frequency-shifting effects the microwave frequency then is 6, 834, 682, 610. 904, 324 Hz.

However, in order to get a reasonably narrow linewidth one needs to prolong the period of time the atoms spend inside the volume illuminated by the laser beam. One way is to use a chemically inert buffer gas chosen such that the ground state coherence is not destroyed by collisions of buffer gas atoms with the alkali atoms [123]. Typical gases are the noble gases or nitrogen. Alternatively, one can coat the walls of the cell with paraffin, siloxanes or similar substances that exhibit a very low sticking coefficient for alkali atoms and that isolate them from paramagnetic impurities inside the wall material. In this way, the ground state coherence can survive hundreds or even thousands of wall collisions, allowing a coherence lifetime of the order of a second and a correspondingly narrow resonance line.

Over the years a number of commercial designs have been developed for a wide range of applications including space-qualified versions. Typical specifications are a relative short-term instability of 3×10^{-11} and 2×10^{-11} at 1 second and at 1 day, respectively. Long-term drift rates can be as low as a few parts in 10^{14} per day after an initial burn-in phase [124]. At least some of this drift is due to chemical reactions between the alkali atoms and the walls of the cells, with additional contributions due to a changing light spectrum coming out of the ^{87}Rb lamp as it ages.

Easy as it looks to replace the spectral lamp by a laser, there are several problems with this approach [125]. For one thing, the laser frequency noise gets converted by the absorption profile of the vapour cell into amplitude noise [126], thus reducing the short-term stability of the clock, unless special measures are taken to stabilize the laser parameters and to narrow the linewidth. Another issue is the question of long-term stability of the laser characteristics, where until very recently no commercially viable solution of the aging problem could be found.

13.8.1.2 Purely Optical Vapour-Cell Clocks

Small as these clocks are, a further miniaturization is desirable, maybe down to chip-scale. More important would be to find a way to curb the (relatively) large power requirements of optical-pumping clocks, ideally down to a few 10 mW so that it could run off a reasonable-sized battery for an extended period of time. Such small, low-power atomic clocks could find their way into GPS receivers, cell phones or even wrist watches.

The microwave excitation sets a natural size limit to any miniaturization effort, given by the microwave wavelength or some large fraction thereof. This limitation can be overcome by switching to an all-optical excitation scheme (Fig. 13.18). Basically, one uses the same clock transition $|F, m_F = 0\rangle \rightarrow |F + 1, m_F = 0\rangle$ but

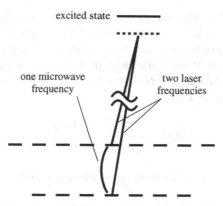

Fig. 13.18 Principle of operation of the dark-state clock. Instead of coupling the levels involved in the "clock transition" with a microwave one uses two phase-stable laser fields with a frequency difference precisely matching the level splitting on the clock transition

instead of coupling those levels with a microwave one uses two phase-stable laser fields with a frequency difference precisely matching the level splitting on the clock transition.

The phenomenon one makes use of here is called *coherent population trapping* (CPT) [127]. CPT in a Λ system (Fig. 13.19) shows interesting properties suitable for a compact atomic clock. Consider first the situation of Fig. 13.19a. A resonant laser field of frequency ω_1 interacts with the transition from state $|1\rangle$ to the excited state $|e\rangle$. Excited atoms can decay back in to either $|1\rangle$ or $|2\rangle$. Since the laser is not resonant with the transition $|2\rangle \rightarrow |e\rangle$, any atom reaching state $|2\rangle$ can no longer be excited by the laser frequency ω_1. State $|2\rangle$ therefore is a *dark state* because after several absorption–emission cycles all population will have accumulated in it; no more absorption can take place, therefore no more spontaneous emission light is given off by the sample.

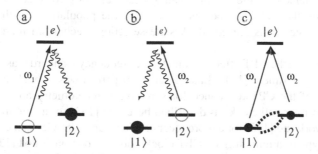

Fig. 13.19 The three-level Λ system. (**a**) A laser resonant with the $|1\rangle \rightarrow |e\rangle$ transition cannot interact with any population in the $|2\rangle$ state, so this state becomes a dark state. (**b**) Likewise, for a laser resonant only with the $|2\rangle \rightarrow |e\rangle$ transition state $|1\rangle$ becomes a dark state. (**c**) For a coherent superposition of the two laser fields the dark state is a coherent superposition of $|1\rangle$ and $|2\rangle$

On the other hand, if the radiation is applied on the other optical transition (Fig. 13.19b) the state $|1\rangle$ becomes a dark state. Surprisingly (or maybe not), when a coherent superposition of the two laser fields is applied, a new dark state consists of a coherent superposition of the two atomic ground states, where the weights of the two components are inversely proportional to the Rabi frequencies Ω_i on the two optical transitions (Fig. 13.19c):

$$|\text{dark}\rangle = \frac{\Omega_2|1\rangle - \Omega_1|2\rangle}{\Omega_1^2 + \Omega_2^2} \ . \tag{13.11}$$

Note that this equation smoothly transforms into the two limiting cases (a) and (b) of Fig. 13.19.

Closer inspection of the mathematics shows that atoms can reach this coherent dark state, the quantum-mechanical superposition of two ground states, only if the laser difference frequency closely matches the frequency splitting between the two ground states. In that case a ground state coherence is generated that prevents any further excitation into the excited state. For a well-designed experimental set-up, where the two laser frequency components are derived from two phase-locked lasers [128] or from two modulation sidebands of a single laser [129], the width of the resonance is only limited by the lifetime of the ground state coherence, for instance by the inverse of the observation time. It is straight-forward to obtain linewidths of just a few 10 Hz as a function of the difference frequency of the two lasers [130] when a buffer gas is used to prevent the atoms in a thermal vapour from leaving the volume illuminated by the laser beams. A servo loop can lock the laser difference frequency to the ground state hyperfine splitting frequency, i.e. 9.2 GHz for caesium vapour or 6.8 GHz for ^{87}Rb vapour.

Note a difference here to the case of a microwave-induced transition in the beam, fountain, or vapour-cell clocks described before. There the signal was connected to the net transfer of population from one clock state to the other, which can only happen when the two levels start with different populations. In contrast, the CPT clock relies on the generation of a ground state coherence, i.e. an off-diagonal element in the density matrix becomes nonzero; this coherence can be induced by the two laser beams even when the populations in the two coupled ground state levels are equal. A state-selection mechanism therefore is not required.

The first use of the CPT effect for an atomic frequency standard was in a sodium atomic-beam experiment [131]. Later a series of papers was published that examined the use of the CPT resonance for sensitive magnetometry and for the construction of vapour-cell clocks as described here (see [132] for a compilation of the pioneering work and [133] for a more recent review). In a US–German collaboration a first prototype of a miniaturized CPT clock could be demonstrated [134]. Shortly thereafter, a truly chip-sized physics package for a CPT clock was constructed and tested [135]. A recent review sums up the current state of the art [136]. A major technological problem there is how to build and fill a millimetre-sized cell with an alkali vapour and a buffer gas at just the right pressure. A first commercial device, although overall not quite at chip-scale yet, will enter the market soon. It will probably offer an instability of 1.6×10^{-11} at 1 second and of below 10^{-12} at several hours, in a package smaller than $50\,cm^3$ [137].

A related device is the CPT maser [138], where the radiation emitted by the oscillating magnetic dipole moment (the ground state coherence) is used. The advantages and disadvantages of this principle are discussed in [138].

13.8.2 The Hydrogen Maser

Another invention by Ramsey, and another reason for why he got the Nobel Prize, is the hydrogen maser (Fig. 13.20) [139, 140, 2]. From a beam of atomic hydrogen, generated from H_2 gas in a radiofrequency discharge, the atoms in the upper ground states $|F = 1, m_F = 0\rangle$ and $|F = 1, m_F = 1\rangle$ are focussed by a polarizer magnet into a storage bulb inside a microwave cavity resonant with the 1.4-GHz frequency of the $|F = 1, m_F = 0\rangle \rightarrow |F = 0, m_F = 0\rangle$ hyperfine transition; atoms in the other ground states ($|F = 1, m_F = -1\rangle$ and $|F = 0, m_F = 0\rangle$) are defocused and do not reach the storage bulb. Therefore, a population inversion exists in the hydrogen atom sample moving around inside the bulb with an average velocity close to 0. When the flux of incoming spin-polarized atoms and their storage time in the bulb is high enough, a self-sustained maser oscillation is possible. A small antenna in the cavity picks up this oscillation, from which the output signal of the frequency standard is derived.

This is the principle of operation of an *active* hydrogen maser, with some variation of course in the individual embodiments. The *passive* hydrogen maser is

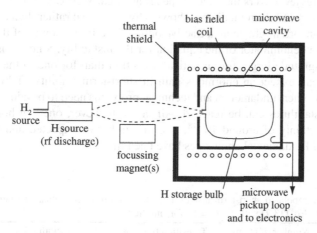

Fig. 13.20 Principle of operation of the hydrogen maser. Molecular hydrogen gas is dissociated in a radiofrequency discharge. Hydrogen atoms in a low-field seeking state are magnetically focussed into a storage bulb where they can emit a microwave photon into the cavity. A pickup loop registers the resulting cavity field and passes it on to the electronics, which in turn generates the maser output signal from it

constructed similarly, except that a 1.4-GHz microwave signal is sent into the cavity and its amplification by the hydrogen atoms inside is monitored. The passive maser in general has a smaller form factor at the expense of larger short-term instability.

A key component is the storage bulb, which is coated with Teflon or similar materials so that an $|F = 1\rangle$ hydrogen atom survives thousands of wall collisions before a spin-depolarization event occurs. Storage times of several seconds are thus possible for the hydrogen atoms.

The main advantage of a hydrogen maser is its low relative short-term instability on the order of $< 10^{-14}$ at 1 s of averaging time. However, its long-term stability suffers from drifts in the parameters of the set-up, in particular the geometrical dimensions of the storage bulb (which can be, for instance, air pressure dependent in simple designs) and chemical changes and againg of its wall coating, which in turn change the frequency shift due to the wall collisions.

The hydrogen maser finds application wherever one needs a stable timescale, for instance in astronomical observatories for pulsar timing or for radio interferometry and, of course, in timing laboratories all over the world.

13.9 Optical Clocks

Perhaps the most dynamic area in the field of atomic clocks is that of optical clocks, where great advances occurred in the last few years. So whatever I write here will be outdated by the time this book appears in print. Therefore, here I will restrict myself to giving only the general ideas and to focussing on some of the recent advances. A fairly recent review covers the state of the art as of early 2005 [141].

A look at Eq. (13.7) immediately shows why an optical rather than a microwave frequency standard might be an attractive option. The frequency ν_0 of the oscillator appears in the denominator of the equation for the instability. Since ν_0 is five to six orders of magnitude larger for an optical transition than for one in the microwave regime one can expect an enormous gain in short-term stability (Table 13.1). Of course, many other fundamental and technical problems need to be addressed before such low instabilities can be reached. All in all, however, one can hope to reach a clock uncertainty of around 10^{-18}, after a number of expected and unexpected technical and fundamental challenges have been met.

Table 13.1 Extrapolated short-term instabilities of some proposed optical frequency standards calculated using Eq. (13.7) with $T_{cycle} = 1$ s for simplicity

Atom / ion	Number of atoms	Transition frequency	Linewidth (Hz)	$\sigma_y(\tau = 1\,\text{s})$
^{133}Cs	6×10^5	9.2 GHz	0.9	4×10^{-14}
^{171}Yb$^+$	1	688 THz	3.1	2×10^{-15}
^{40}Ca	10^7	456 THz	400	9×10^{-17}
^{87}Sr	10^6	430 THz	5	4×10^{-18}

13.9.1 How an Optical Clock Works

The operating principle of an optical clock is once again covered by Fig. 13.4. Here the oscillator is a narrow-band laser source, the spectroscopy is performed either on a single trapped ion or on a cloud of cold atoms (free falling or held in an optical trap). The detection is done by observing the fluorescence light coming from the ion or atoms. One modification, however, needs to be applied to the principle of Fig. 13.4. It is impractical to have an optical signal as the clock output. Instead, one would rather have a microwave signal, which can then be processed by standard electronic techniques. This task, the optical-to-microwave clockwork function, is taken up by a femtosecond laser frequency comb.

The first role of the lasers in an optical clock is to cool the sample. In principle the same cooling techniques are applied as described above for the case of caesium fountain clocks. In the case of a trapped ion clock additional cooling, for instance sideband cooling [142], can be applied to cool the motion of the atom down to its motional ground state. The exact details, like the question of whether and how to apply repumping light, depend on the actual atom/ion in question. As an example, we will consider here the case of a single-ion ^{171}Yb trap (Fig. 13.21) [143]. Two clock transitions are being investigated: the quadrupole transition $S_{1/2} \rightarrow D_{3/2}$ at 435 nm wavelength and the octupole transition $S_{1/2} \rightarrow F_{7/2}$ at 467 nm wavelength, with natural linewidths of 3 Hz and some nanohertz, respectively.

The main laser cooling is done on the 369-nm transition, with some repumping light applied on the other hyperfine transition (Fig. 13.21); a relatively strong magnetic field applied during the cooling phase prevents the population of dark states in the ground state by inducing Larmor precession. In order to excite one of the clock transitions one needs a laser with a linewidth of less than a Hertz, in order to obtain a suitably narrow resonance line. This is achieved by locking the laser to a Fabry–Perot resonance in a high-finesse (finesse > 100,000 is possible) optical resonator

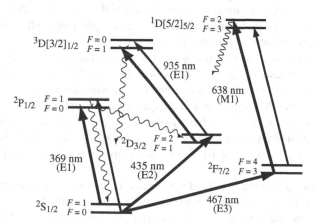

Fig. 13.21 Relevant levels for the ^{171}Yb single ion optical frequency standard, using either the electric-quadrupole resonance line at 435 nm wavelength or the octupole line at 467 nm

with a low-thermal expansion spacer material between the mirrors, typically glass ceramics. This resonator must be well isolated from external perturbations, so it is fixed on a vibration-isolated mount [144]. Impressively small linewidths of about 0.1 Hz are possible in this way [145].

After the ion has been cooled one can switch off the repumping light on the cooling transition, and a little while later the cooling light itself. The atom is then in the lower clock state $^2S_{1/2}$, $F = 0$. Next the probing light (either 435 or 467 nm) is applied for some time (tenths of seconds typically). When the cooling light is switched on again, the amount of fluorescence light observed on the cooling transition immediately after turn-on is an indication of whether the atom has made the transition to the long-lived upper clock state or not: if there still is fluorescence light, it has not! The absence of fluorescence light, in contrast, means that the atom is now in the upper state of the clock transition.

The other laser frequencies are needed to prevent the atom falling into other metastable levels. From the upper level of the cooling transition the atoms can decay spontaneously into both hyperfine levels of the upper clock state $^2D_{3/2}$. The 935-nm light resonantly excites atoms in $F = 1$ to the $^3D[3/2]_{1/2}$ state which decays spontaneously into the ground state. This light beam also nonresonantly excites the transition from the upper clock level, $F = 2$. This weak excitation rate must be taken into account when considering the characteristics of the clock transition but it has the essential and beneficial effect of depopulating the upper clock state much faster than its very long spontaneous lifetime would prescribe. In this way one need not wait for the atom to decay spontaneously via the forbidden clock transition, which could take a very long time indeed! The 638-nm light prevents shelving of the atom in the $^2F_{7/2}$ state, which can happen by spontaneous decay from states omitted from Fig. 13.21 that are populated by collisions with residual gas atoms.

By repeating this cooling and probing cycle for the same detuning one can determine the transition probability for a given detuning of the interrogation laser, and by scanning this laser one can build up a line profile of the clock transition. Its width is Fourier-limited by the duration of the interrogation light pulse when the linewidth of the interrogation laser is sufficiently narrow.

When the servo loop of the optical clock is closed the frequency of the narrow-band interrogation laser is locked to the clock transition. Careful consideration has to be given to the sampling and locking parameters in order to obtain the best stability, as detailed in [146]. A stringent test of the quality of the optical frequency standard realized in this way is to actually build two of them and to compare them. The PTB group has done this for two independent ^{171}Yb$^+$ ion standards and found a frequency difference of only $3.6 \pm 6.1 \times 10^{-16}$, i.e. no difference within the total uncertainty [143].

We will not go into the details of systematic frequency biases of optical clocks here – we have already done that for the example of a fountain clock above. There is once again a long and in principle rather similar list of effects that need to be investigated and controlled, with differences in detail, of course. Of particular importance are questions related to the trapping of the atoms: offsets and motion of an ion due

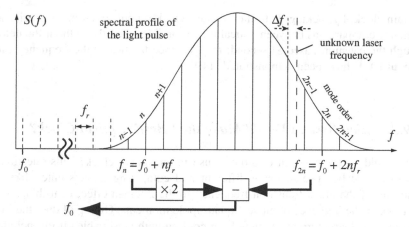

Fig. 13.22 How to measure optical frequencies with a femtosecond laser, or how to transfer an optical frequency into the microwave regime when f_r and f_0 are known and Δf is measured

to static or dynamically generated electrical charges on the materials in the vicinity of the ion, or light shifts (discussed below) in an optical trap for neutral atoms.

13.9.2 Obtaining Time from an Optical Clock

The optical frequency needs to be transferred into the microwave domain so that a time signal can be derived from it. This optical-microwave transfer can be done with the help of a laser frequency comb [147, 148]. The principle, originally devised for the measurement of an unknown optical frequency, is shown in Fig. 13.22. It makes use of the fact that a periodic train of short light pulses with repetition rate f_r, as it comes out of a mode-locked laser, consists of a sequence of equidistant light frequency components separated by f_r in frequency space. Except for a small frequency offset the absolute frequency of one of these modes is a large integer multiple of f_r. The small offset frequency f_0 reflects the phase slip between carrier and envelope of the short pulses as a result of the difference between group and phase velocity in the medium traversed by the pulses. The frequency of an optical lightwave can be determined by measuring its offset (in the radiofrequency range) from the closest mode of the comb.

Using the strong optical nonlinearity of special types of optical fibre one can broaden the frequency spectrum of the pulsed laser beam so much that it spans more than one octave. This allows one to frequency-double one of the low-frequency modes and compare the second harmonic to the closest one of the high-frequency modes. As indicated in Fig. 13.22 this can be used to determine f_0, or even to lock it to some desired value. When the system is run "in reverse", it allows one to transfer the optical frequency of an optical clock into the microwave domain.

Systems like that have been used to determine the frequencies of a number of optical transitions by now, relative to the SI second reproduced by a caesium fountain clock, by locking f_r to the caesium clock output. Even more precise are frequency comparisons between optical transition lines directly, without the detour through the caesium-based SI second: relative uncertainties of the frequency ratio of about 10^{-17} have been demonstrated [149].

13.9.3 Which System Would Make the "Best" Optical Clock?

What would be the best ion or atom to use in an optical clock? This question is hard to decide because so many different and competing aspects enter, like the sensitivity of the clock transition frequency to the Zeeman effect or to blackbody radiation, or the influence of nuclear quadrupole moments. On top of that, the ion's or atom's level structure must allow laser cooling with reasonable effort, including the provision of the required laser wavelengths. It is an interesting idea, therefore, to separate the two functions, that of being laser coolable and that of having a suitable clock transition, into two ions of different chemical species [150]. When the two ions are trapped inside a linear ion trap, for instance, their joint normal modes of motion can be cooled by laser-cooling one ion. The clock transition can be excited and read out by entangling internal states of the ion with the motional states in the trap – the quantum logic clock. This has been demonstrated with a combination of $^9\mathrm{Be}^+$ for cooling and $^{27}\mathrm{Al}^+$ as the clock ion [151].

Whether a single ion or a cloud of cold atoms is the better optical clock depends on whether one is more interested in low instability or in low uncertainty. Certainly, with many atoms one obtains a much higher signal, but on the other hand those atoms could potentially interact with each other. Also, for a single ion the question of inhomogeneity of external magnetic or laser light fields is not an issue. Until recently, the observation time for a cold atom cloud in a clock was limited to the free-fall time in Earth's gravity. In principle, one could trap atoms using the periodic light shift potential they experience in a 1D, 2D, or 3D standing wave. However, in general the trapping light would also shift one of the clock states, thus giving a frequency bias that would be extremely hard to control. Fortunately, it was then pointed out that for strontium, for instance, a specific, "magic" wavelength for the trapping light would cause the same light shift for both trapping states, so that no frequency bias would be caused by the trapping lasers [152]. Similarly, Yb and Ca could be trapped with their respective "magic" wavelengths.

This has certainly helped to push ahead the work on neutral atom clocks, with several groups now working on Sr optical clocks. Measurements of the frequency of the $^1\mathrm{S}_0 \to {}^3\mathrm{P}_0$ transition in $^{87}\mathrm{Sr}$ by three groups agree within 1×10^{-15} with each other [153], making this transition a potential candidate for the redefinition of the second. It is an open question whether a Bose–Einstein condensate of atoms could be useful for an atomic clock; after all, in this dense sample the question of interactions becomes of critical importance.

Today the optical clocks with the lowest estimated frequency bias are ion clocks, in particular the clock transitions $^1\mathrm{S}_0 \to {}^3\mathrm{P}_0$ in $^{27}\mathrm{Al}^+$ and $^2\mathrm{S}_{1/2} \to {}^2\mathrm{D}_{5/2}$ in $^{199}\mathrm{Hg}^+$.

Their current relative systematic uncertainties stand at 2.3×10^{-17} and 1.9×10^{-17}, respectively [149]. This is more than an order of magnitude better than the uncertainty within which the best fountain clocks can reproduce the SI second as it is currently defined.

In 2006 the Meter Convention recommended one microwave and four optical transitions as secondary representations of the second [154]; they are listed in Table 13.2. The elements and transitions in the table are, of course, not the only candidates for a future redefinition of the second – the race is wide open and will still take some time to decide. After all, when the definition of the second is changed one wants to be sure that one really has made the best choice. That is why in all probability at its next official meeting in 2011 the Meter Convention will leave the caesium-based definition of the second untouched for now even though one or more other base units might be redefined [155]. However, at that time we can expect to see recommendations for other secondary representations of the second, as well as reductions in the assigned uncertainties of those in Table 13.2.

Table 13.2 Transitions recommended in 2007 as secondary representations of the second

Atom / ion	Transition	Frequency (Hz)	Relative uncertainty	Group
^{87}Rb	$S_{1/2} \to S_{1/2}$	6,834,682,610.904,324	3×10^{-15}	SYRTE
^{199}Hg$^+$	$S_{1/2} \to D_{5/2}$	1,064,721,609,899,145	3×10^{-15}	NIST
^{171}Yb$^+$	$S_{1/2} \to D_{3/2}$	688,358,979,309,308	9×10^{-15}	PTB
^{88}Sr$^+$	$S_{1/2} \to D_{5/2}$	484,779,044,095,484	7×10^{-15}	NPL
^{87}Sr	$^1S_0 \to {}^3P_0$	429,228,004,229,877	1.5×10^{-14}	U. Tokyo, SYRTE, NIST

13.10 The Future (Maybe)

If we look at clock-making in general, we find a recurring theme. It is always a quest for an oscillating system that is ever better isolated from its surroundings. In that sense a vibrating quartz crystal is better than a mechanical pendulum and an oscillating magnetic transition dipole moment in an atom much better than a macroscopic crystal. If we carry this thought further, one might think about a system even more isolated than an electron in an atom: an oscillation of a nucleus [156].

Certainly, at some point a few years ahead the caesium-based definition of the length of the SI second will be replaced by one based on a transition in the optical frequency range. Until then, caesium fountain clocks will remain the most precise (and still improving) standards of time.

13.10.1 A Nuclear Clock?

Typically, excitations of an atomic nucleus lie in the MeV energy range, but in the case of the thorium isotope ^{229}Th a low-lying energy state a mere 7.6 eV above the ground state has been deduced from the analysis of high-energy decay spectra [157].

It is an intriguing thought that using ultraviolet light one should be able to excite a nuclear transition. A number of interesting physics questions are connected to this issue, like how strong the coupling of that excited state is to the surrounding electron cloud, i.e. how narrow the resonance line is. There is still a lot of work ahead before a single ^{229}Th ion in a Paul trap can provide a new standard of time [156]. The first step must be to find the actual transition frequency in an optical experiment because it is known right now only with an uncertainty of 0.5 eV [157]. With that large uncertainty it is not even clear what laser source would be required.

13.10.2 A Pulsar Clock?

For completeness, one should also mention the idea of going back to an astronomical phenomenon to monitor the passing of time with high precision: pulsars. At some time it was thought that the regular flashes of radio waves emitted by millisecond-period pulsars could form a more stable timescale than Earth-bound clocks could provide [158]. However, there are a number of difficulties that are hard to solve. One is of a principal nature: Because many of the parameters of a pulsar are not known a priori they need to be determined by measurement – which must not involve atomic clocks, or otherwise one would fold their drifts into the parameters of the pulsars. For instance, in general the pulsar must slow down due to energy losses, but this slow-down rate needs to be determined experimentally with very high precision so that it can be corrected for [158, 159].

Already in the mid-1990s it was pointed out that with the atomic clocks available then, such a pulsar timescale would not bring much of an advantage, if at all. Today, after the advent of caesium fountain clocks and with the prospect of optical clocks it has become apparent that man-made clocks will probably remain superior to natural systems like quasars, even for averaging times of several decades. One reason is that one cannot exclude that some violent event (like another body falling into the pulsar) might alter the period of the quasar or shift its beam away from Earth, in effect destroying the clock. But the main reason is that the signal from the quasar on its way to Earth has to traverse whatever is located between the quasar and the Earth. A cloud of interstellar matter will delay the radio signal by an amount connected to its (density-dependent) index of refraction, so cloud movements will in general lead to a jitter or even a long-term drift in the arrival times of the pulses, thus giving frequency noise or even a bias which cannot be controlled. In fact, nowadays the timing jitter of distant pulsars observed on the Earth is analyzed to extract information on a background of gravitational waves traversing the line of sight [160].

13.10.3 Clocks in Space

The same ideas used in fountain clocks can be adapted for clocks in a low-gravity environment. Clouds of cold atoms can be laser-prepared and then gently pushed through a microwave cavity to be detected at its end. One can expect many seconds of interaction time between atoms and microwave radiation, with resulting linewidths around 100 mHz. Three such projects for the International Space Station have been pushed ahead [161]. A NASA project, the caesium clock PARCS [162], recently became a victim of funding redistribution favouring the Mars initiative in the United States. The same fate has befallen the rubidium clock experiment RACE [163], another US-American project. An ESA project, ACES [164], is fully developed but has hit funding problems, too.

Clearly, many design issues must be tackled in order to construct a clock that survives launch and that can work without any in-place user intervention. A major task for all space-based clocks meant to provide timing information for Earth-bound users is to develop ways of transferring the projected stability and accuracy to the ground. It is beyond the scope of this chapter to go into any of the details here.

13.11 Precision Tests of Fundamental Theories

Precise and stable atomic clocks can be employed to learn more about fundamental assumptions about the physical world we live in. For instance, are constants of nature really constant?

13.11.1 Testing Relativity

Local position invariance (LPI) forms a part of the more general Einstein equivalence principle, which in turn is a foundation of Einstein's theory of general relativity. To test its validity, over the course of more than 2 years the frequency difference between PTB-CSF1 and different hydrogen masers was monitored in order to look for variations that are synchronous with the time-varying gravitational potential $\Delta U(t)$ due to the annual elliptical orbital motion of the Earth. No violation of the null result predicting LPI could be detected at the level of 6×10^{-6} of the amplitude of $\Delta U(t)/c^2$ [165, 166]. This result represents an improvement by a factor of about 100 compared to previous similar experiments and is one demonstration that significant improvements in the field of basic research become possible with the help of improved frequency standards like fountain clocks. Recently [167] the analysis was repeated using data of all atomic clocks available at the time, and over the longer data collection interval accumulated since the pioneering work [165]. Another improvement by a factor of 20 has been achieved in this way.

By comparing the frequency values determined for the ^{87}Sr clock transition $^1S_0 \rightarrow {}^3P_0$ at 698-nm wavelength in three different laboratories around the world and at different times, one can derive an even more stringent upper limit for any violation of local position invariance [153].

13.11.2 How Constant Are Fundamental Constants?

The excellent stability and reproducibility of modern atomic clocks can help to detect possible variations in the fundamental constants of nature, by allowing repeated comparisons of different optical and/or microwave transitions over many years [168–175]. This is not the only way to learn about possible time variations of constants. More experimental techniques, applicable on a laboratory and on a cosmological timescale and also covering other fundamental constants, are discussed in a number of recent reviews, some of which can be found under [170, 173, 175]. Here we will mention only a few of those involving atomic clocks in a direct way. But even here, no completeness is claimed or even intended in this very dynamic field.

13.11.2.1 Microwave Comparisons

Currently the SYRTE laboratory in Paris is unique in that they have a metrological-quality rubidium fountain clock at their disposal. In earlier years they could run one of their fountain clocks with ^{87}Rb instead of caesium, leading to a precise determination of the frequency of the Rb clock transition [176]. Combined with a redetermination a few years later one can derive a limit for a temporal variation of the frequency ratio [168]:

$$\left| \frac{d \ln(f_{Rb}/f_{Cs})}{dt} \right| < 7 \times 10^{-16}/\text{year} . \tag{13.12}$$

Since $f_{Cs}/f_{Rb} \propto \alpha^{0.44} \mu_{Cs}/\mu_{Rb}$ [168], one can infer a limit for the temporal variation of the fine structure constant α if one assumes that the ratio of the two nuclear magnetic moments μ_i remains constant. Simply speaking, one therefore has to "decide" whether any observed temporal change should be attributed to a change of the fine structure constant or the magnetic moments of the nuclei, i.e. some properties of the strong nuclear force.

In the meantime fountain SYRTE-FO2 has been modified to run on both atomic species simultaneously. This latter arrangement allows for the precise control and correction of common systematic errors, like drifts of the magnetic holding field above the Ramsey cavity. A recent remeasurement of the rubidium clock frequency has been presented at a conference [109]. Together with the previous measurements an improved limit of

$$\frac{d \ln(f_{Rb}/f_{Cs})}{dt} = (-3.2 \pm 2.3) \times 10^{-16}/\text{year} \tag{13.13}$$

has been obtained.

13.11.2.2 Absolute Frequency Measurements

A model-independent way of determining a possible time variation of α can be found by separating the expression for the frequencies f_1 and f_2 of two electronic transitions into constant and possibly time-dependent terms:

$$f_{1,2} = Ry \times C_{1,2} \times F_{1,2}(\alpha), \tag{13.14}$$

where Ry is the Rydberg constant. The constants $C_{1,2}$ contain only mathematical constants and atomic quantum numbers, while the complete dependence on the fine structure constant α is contained in $F_{1,2}(\alpha)$, which is computed from atomic theory. Taking the time derivative of Eq. (13.14) for the two frequencies f_1 and f_2 one arrives at

$$\frac{d(\ln f_{1,2})}{dt} = \frac{d(\ln Ry)}{dt} + A_{1,2} \times \frac{d(\ln \alpha)}{dt}, \tag{13.15}$$

where $A_{1,2} = d(\ln F_{1,2})/d(\ln \alpha)$. This system of two equations contains two unknowns, the derivatives of the Rydberg and the fine structure constants, and thus can be solved easily and in a model-independent way, i.e. without any assumptions about which quantities do and which do not change.

Using this approach, one can combine absolute frequency measurements (i.e. with respect to a primary caesium fountain clock) for different atomic species and/or transitions taken in different laboratories over some longer time interval. For instance, using the frequency of the clock transition in the ^{199}Hg$^+$ ion, measured at NIST over the last few years, and the frequency of the clock transition in the ^{171}Yb$^+$ ion, measured at PTB since 2000, one arrives at [172, 177]

$$\frac{d \ln \alpha}{dt} = (-2.6 \pm 3.9) \times 10^{-16}/\text{year} \tag{13.16}$$

$$\frac{d \ln Ry}{dt} = (-5.5 \pm 11.1) \times 10^{-16}/\text{year} . \tag{13.17}$$

A number of refinements have been done over the last few years, whenever a new measurement of one of the contributing transitions, or a different transition, has become available. However, for the very best optical frequency standards today, the measurement with respect to the caesium fountains constitutes a limitation because they cannot reproduce the SI second with better than a few parts in 10^{16} relative uncertainty.

13.11.2.3 Optical Frequency Ratios

Even simpler conceptually is the search for a time dependence of α when two optical transition frequencies can be compared directly, without the detour through the Cs clock. Using Eq. (13.14) one obtains

$$\frac{f_1}{f_2} = \frac{C_1}{C_2} \times \frac{F_1(\alpha)}{F_2(\alpha)} \ , \tag{13.18}$$

which results in a simple expression for the time derivative of α only:

$$\frac{d(\ln \alpha)}{dt} = (A_1 - A_2) \times \frac{d[\ln(f_1/f_2)]}{dt} \ . \tag{13.19}$$

Once again, the first term on the right-hand side is known from atomic theory and the second one can be measured.

The most recent test as of this writing is provided by a direct comparison of the "clock" transitions in a trapped Al^+ ion and a trapped Hg^+ ion [149]. From the measured change in the frequency ratio (or lack thereof) over about 2 years one can deduce that the fine structure constant did not change by more than about two parts in 10^{17} per year:

$$\frac{\dot{\alpha}}{\alpha} = (-1.6 \pm 2.3) \times 10^{-17}/\text{year} \ . \tag{13.20}$$

This is just the first example of similar experiments to be expected in the coming years. For instance, Lea has suggested [173] a comparison of an octupole and a quadrupole transition frequency in a single atom, where a number of systematics might be common mode and thus drop out or at least be reduced. A particularly good system is the ytterbium ion [173], where experiments are planned, for instance, at PTB, using the two clock transitions described in Sect. 13.9.

13.12 Conclusion

Atomic clocks and the precise timing and experimentation they allow are continuing to fascinate physicists and a general audience alike. With recent advances and envisioned developments let us hope that this fascination will continue in years to come. Better atomic clocks will allow us to understand the physical world better than today. And perhaps they will make possible technical applications that we cannot even dream of today.

References

1. Used by Physikalisch-Technische Bundesanstalt as a slogan for a clock exhibition in 2005, on the occasion of the 50th birthday of the atomic clock
2. J. Vanier, C. Audoin, *The Quantum Physics of Atomic Frequency Standards* (Adam Hilger, Bristol, 1989)
3. F. Riehle, *Frequency Standards: Basics and Applications* (Wiley-VCH Verlag, Weinheim, 2004)
4. A. Scheibe, U. Adelsberger, Ann. Phys. 5. Folge **18**, 1 (1933)
5. A. Scheibe, U. Adelsberger, Physik. Zeitschr. XXXVII **37**, 185 (1936)

6. A. Scheibe, U. Adelsberger, Physik. Zeitschr. XXXVII **37**, 415 (1936)
7. A. Scheibe, U. Adelsberger, Z. Phys. **127**, 416 (1950)
8. URL http://www.iers.org
9. D. Sobel, W.J.H. Andrewes, *The Illustrated Longitude* (Fourth Estate, London, 1998)
10. J.A. Kusters, C.J. Adams, RF Des. **5**, 28 (1999)
11. J. Levine, Rev. Sci. Instrum. **70**, 2567 (1999)
12. E.F. Arias, Phil. Trans. R. Soc. A **363**, 2289 (2005)
13. BIPM, Circular-T No. 246 (2008)
14. International Earth Rotation and Reference Systems Service, Bulletin-C **36** (2008)
15. B. Guinot, E.F. Arias, Metrologia **42**, S20 (2005)
16. *International Vocabulary of Metrology – Basic and General Concepts and Associated Terms (VIM)* (Bureau International des Poids et Mesures, Paris, 2006)
17. J.A. Barnes, A.R. Chi, L.S. Cutler, D.J. Healy, D.B. Leeson, T.E. McGunigal, J.A. Mullen, Jr., W.L. Smith, R.L. Snydor, R.F.C. Vessot, G.M.R. Winkler, IEEE Trans. Instrum. Meas. **20**, 105 (1971)
18. P. Lesage, C. Audoin, IEEE Trans. Instrum. Meas. **22**, 157 (1973). Erratum in IEEE Trans. Instrum. Meas. **25**, 270 (1976)
19. P. Lesage, C. Audoin, IEEE Trans. Instrum. Meas. **28**, 6 (1979)
20. L. Essen, J. Parry, Nature **176**, 280 (1955)
21. A. Bauch, Metrologia **42**, S43 (2005)
22. N.F. Ramsey, Phys. Rev. **76**, 996 (1949)
23. P. Busch, in *Time in Quantum Mechanics*, vol. 1, G. Muga, R. Sala Mayato, Í. Egusquiza (eds.), Lect. Notes Phys. **734** (Springer, Berlin, Heidelberg, 2008), chap. The time–energy uncertainty relation, pp. 73–105
24. A. Makdissi, E. de Clercq, Metrologia **38**, 409 (2001)
25. J.H. Shirley, W.D. Lee, R.E. Drullinger, Metrologia **38**, 427 (2001). Erratum **39**, 123 (2002)
26. A. Hasegawa, K. Fukuda, M. Kajita, H. Ito, M. Kumagai, M. Hosokawa, N. Kotake, T. Morikawa, Metrologia **41**, 257 (2004)
27. S. Chu, Rev. Mod. Phys. **70**, 685 (1998)
28. C.N. Cohen-Tannoudji, Rev. Mod. Phys. **70**, 707 (1998)
29. W.D. Phillips, Rev. Mod. Phys. **70**, 721 (1998)
30. Special Issue on Laser Cooling, J. Opt. Soc. Am. B **6**(11), 2020 (1989)
31. P. Forman, Proc. IEEE **73**, 1181 (1985)
32. A. De Marchi, Metrologia **18**, 103 (1982)
33. J.L. Hall, M. Zhu, P. Buch, J. Opt. Soc. Am. B **6**, 2194 (1989)
34. M.A. Kasevich, E. Riis, S. Chu, R.G. DeVoe, Phys. Rev. Lett. **63**, 612 (1989)
35. A. Clairon, S. Ghezali, G. Santarelli, Ph. Laurent, S.N. Lea, M. Bahoura, E. Simon, S. Weyers, K. Szymaniec, *Proceedings of the 5th Symposium on Frequency Standards and Metrology* (World Scientific, Singapore, 1996), pp. 45–59
36. A. Clairon, C. Salomon, S. Guellati, W.D. Phillips, Europhys. Lett. **16**, 165 (1991)
37. S.N. Jefferts, D.M. Meekhof, J.H. Shirley, T.E. Parker, F. Levi, *Proceedings of Joint Meeting of the 13th European Frequency and Time Forum – IEEE International Frequency Control Symposium* (Besancon, 1999), pp. 12–15
38. S.R. Jefferts, J. Shirley, T.E. Parker, T.P. Heavner, D.M. Meekhof, C. Nelson, F. Levi, G. Costanzo, A. De Marchi, R. Drullinger, L. Hollberg, W.D. Lee, F.L. Walls, Metrologia **39**, 321 (2002)
39. S. Weyers, U. Hübner, B. Fischer, R. Schröder, Chr. Tamm, A. Bauch, *Proceedings of the 14th European Frequency and Time Forum* (Turin, 2000), pp. 53–57
40. S. Weyers, U. Hübner, B. Fischer, R. Schröder, Chr. Tamm, A. Bauch, Metrologia **38**, 343 (2001)
41. S. Weyers, A. Bauch, R. Schröder, Chr. Tamm, in *Proceedings of the 6th Symposium on Frequency Standards and Metrology*, P. Gill (ed.) (World Sicentific, Singapore, 2002), p. 64
42. F. Levi, L. Lorini, D. Calonico, A. Godone, IEEE Trans. Ultrason. Ferroel. Freq. Control **51**, 1216 (2004)

43. K. Szymaniec, W. Chalupczak, P.B. Whibberley, S.N. Lea, D. Henderson, Metrologia **42**, 49 (2005)
44. T. Kurosu, Y. Fukuyama, Y. Koga, K. Abe, IEEE Trans. Instrum. Meas. **53**, 466 (2004)
45. M. Kumagai, H. Ito, M. Kajita, M. Hosokawa, Metrologia **45**, 139 (2008)
46. C. Vian, P. Rosenbusch, H. Marion, S. Bize, C. Cacciapuoti, S. Zhang, M. Abgrall, D. Chambon, I. Maksimovic, Ph. Laurent, G. Santarelli, A. Clairon, A. Luiten, M. Tobar, C. Salomon, IEEE Trans. Instrum. Meas. **54**, 833 (2005)
47. J. Guéna, G. Dudle, P. Thomann, Eur. Phys. J. **388**, 183 (2007)
48. Yu. S. Domnin, G.A. Elkin, A.V. Novoselov, L.N. Kopylov, V.N. Baryshev, V.G. Pal'chikov, Can. J. Phys. **80**, 1321 (2002)
49. T.Y. Kwon, H.S. Lee, S.H. Yang, S.E. Park, IEEE Trans. Instrum. Meas. **52**, 263 (2003)
50. United States Naval Observatory, Report 2004–01 to the 16th Meeting of the CCTF (2004)
51. National Institute of Metrology, Report 2004–05 to the 16th Meeting of the CCTF (2004)
52. L. Marmet, B. Hoyer, P. Dubé, A.A. Madey, J.E. Bernard, *Proceedings of the IEEE Annual Frequency Control Symposium* (Honolulu, 2008), p. paper 3101
53. D.V. Magalhaẽs, M.S. Santos, A. Bebeachibuli, S.T. Müller, V.S. Bagnato, *Conference on Precision Electromagnetic Measurements* (London, 2004)
54. A. Sen Gupta, personal communication
55. R. Wynands, S. Weyers, Metrologia **42**, S64 (2005)
56. E.A. Donley, T.P. Heavner, S.R. Jefferts, IEEE Trans. Instrum. Meas. **54**, 1905 (2005)
57. P. Thomann, M. Plimmer, G. Di Domenico, N. Castagna, J. Guéna, G. Dudle, F. Füzesi, Appl. Phys. B **84**, 659 (2006)
58. E. Raab, M. Prentiss, A. Cable, S. Chu, D.E. Pritchard, Phys. Rev. Lett. **59**, 2631 (1987)
59. S. Chu, L. Hollberg, J.E. Bjorkholm, A. Cable, A. Ashkin, Phys. Rev. Lett. **55**, 48 (1985)
60. P.D. Lett, W.D. Phillips, S.L. Rolston, C.E. Tanner, R.N. Watts, C.I. Westbrook, J. Opt. Soc. Am. B **6**, 2084 (1989)
61. J.P. Barrat, C.N. Cohen-Tannoudji, J. Phys. Radium **22**, 329 (1961)
62. J. Vanier, C. Audoin, Metrologia **42**, S31 (2005)
63. P. Rosenbusch, S. Zhang, A. Clairon, *Proceedings of the 21st European Frequency and Time Forum* (Geneva, 2007), pp. 1060–1063
64. H. Marion, S. Bize, L. Cacciapuoti, D. Chambon, F. Pereira Dos Santos, G. Santarelli, P. Wolf, A. Clairon, A. Luiten, M. Tobar, S. Kokkelmans, C. Salomon, arXiv:physics/0407064v3 (2004)
65. V. Natarajan, Eur. Phys. J. D **32**, 33 (2005)
66. J.B. Fixler, G.T. Foster, J.M. McGuirk, M.A. Kasevich, Science **315**, 74 (2007)
67. A. Peters, K.Y. Chung, S. Chu, Nature **400**, 849 (1999)
68. A. Wicht, J.M. Hensley, E. Sarajlic, S. Chu, Physica Scripta **T102**, 82 (2002)
69. G. Santarelli, Ph. Laurent, P. Lemonde, A. Clairon, A.G. Mann, S. Chang, A.N. Luiten, C. Salomon, Phys. Rev. Lett. **82**, 4619 (1999)
70. W.M. Itano, J.C. Bergquist, J.J. Bollinger, J.M. Gilligan, D.J. Heinzen, F.L. Moore, M.G. Raizen, D.J. Wineland, Phys. Rev. A **47**, 3454 (1993)
71. G.J. Dick, *Proceedings of the 19th Annual PTTI Systems and Applications Meeting* (Redondo Beach, CA, 1987), pp. 133–147
72. G.J. Dick, J.D. Prestage, C.A. Greenhall, L. Maleki, *Proceedings of the 22th Annual PTTI Systems and Applications Meeting* (Redondo Beach, CA, 1990), pp. 497–508
73. C. Audoin, G. Santarelli, A. Makdissi, A. Clairon, IEEE Trans. Ultrason. Ferroel. Freq. Control **45**, 877 (1998)
74. G. Santarelli, C. Audoin, A. Makdissi, P. Laurent, G.J. Dick, A. Clairon, IEEE Trans. Ultrason. Ferroel. Freq. Control **45**, 887 (1998)
75. A. Luiten, A.G. Mann, D.G. Blair, Electron. Lett. **30**, 417 (1994)
76. S. Bize, Ph. Laurent, M. Abgrall, H. Marion, I. Maksimovic, L. Cacciapuoti, J. Grünert, C. Vian, F. Pereira Dos Santos, P. Rosenbusch, P. Lemonde, G. Santarelli, P. Wolf, A. Clairon, A. Luiten, M. Tobar, C. Salomon, C. R. Physique **5**, 829 (2004)

77. B. Lipphardt, G. Grosche, H. Schnatz, *Proceedings of the 22nd European Frequency and Time Forum*, (Toulouse, 2008)
78. F. Levi, A. Godone, L. Lorini, IEEE Trans. Ultrason. Ferroel. Freq. Control **48**, 847 (2001)
79. S.R. Jefferts, T.P. Heavner, E.A. Donley, J.H. Shirley, T.E. Parker, *Proceedings of the 2003 IEEE International Frequency Control Symposium and PDA Exhibition Jointly with the 17th European Frequency and Time Forum* (Tampa, 2003), pp. 1084–1088
80. M.A. Lombardi, T.P. Heaver, S.R. Jefferts, MEASURE **2**, 74 (2007)
81. C. Fertig, K. Gibble, Phys. Rev. Lett. **81**, 5780 (1998)
82. C. Fertig, J.I. Rees, K. Gibble, *Proceedings of the 2001 International Frequency Control Symposium and PDA Exhibition* (Seattle, 2001), pp. 18–21
83. A. Joyet, G. Mileti, G. Dudle, P. Thomann, IEEE Trans. Instrum. Meas. **50**, 150 (2001)
84. G. Dudle, A. Joyet, P. Berthoud, G. Mileti, P. Thomann, IEEE Trans. Instrum. Meas. **50**, 510 (2001)
85. A. Joyet, Aspects métrologiques d'une fontaine continue à atoms froids. Ph.D. thesis, Université de Neuchâtel (2003)
86. F. Füzesi, M.D. Plimmer, G. Dudle, J. Guéna, P. Thomann, *Proceedings of the 22nd European Frequency and Time Forum* (Toulouse, 2008)
87. S. Weyers, A. Bauch, U. Hübner, R. Schröder, Chr. Tamm, IEEE Trans. Ultrason. Ferroel. Freq. Control **47**, 432 (2000)
88. E. Majorana, Nuovo Cimento **9**, 43 (1932)
89. A. Bauch, R. Schröder, Ann. Phys. **2**, 421 (1993)
90. S. Weyers, R. Schröder, R. Wynands, *Proceedings of the 20th European Frequency and Time Forum* (Braunschweig, 2006), pp. 219–223
91. F. Levi, J.H. Shirley, T.P. Heavner, D.H. Yu, S.R. Jefferts, IEEE Trans. Ultrason. Ferroel. Freq. Control **53**, 1584 (2006)
92. J.H. Shirley, F. Levi, T.P. Heavner, D. Calonico, D.H. Yu, S.R. Jefferts, IEEE Trans. Ultrason. Ferroel. Freq. Control **54**, 2376 (2006)
93. D. Chambon, M. Lours, F. Chapelet, S. Bize, M.E. Tobar, A. Clairon, G. Santarelli, IEEE Trans. Ultrason. Ferroel. Freq. Control **54**, 729 (2007)
94. T.P. Heavner, S.R. Jefferts, E.A. Donley, T.E. Parker, F. Levi, *Proceedings of the IEEE Annual Frequency Control Symposium* (Vancouver, 2005), pp. 308–311
95. A. Sen Gupta, R. Schröder, S. Weyers, R. Wynands, *Proceedings of the 21st European Frequency and Time Forum* (Geneva, 2007), pp. 234–236
96. A. Sen Gupta, J. Metrol. Soc. India **21**, 249 (2006)
97. E.A. Donley, T.P. Heavner, J.W. O'Brien, S.R. Jefferts, *Proceedings of the 2005 International Frequency Control Symposium* (Vancouver, 2005), pp. 292–296
98. P.D. Featonby, C.L. Webb, G.S. Summy, K. Burnett, J. Phys. B: At. Mol. Opt. Phys. **31**, 375 (1998)
99. W.M. Itano, L.L. Lewis, D.J. Wineland, Phys. Rev. A **25**, 1233 (1982)
100. A. Bauch, R. Schröder, Phys. Rev. Lett. **78**, 622 (1997)
101. R. Augustin, A. Bauch, R. Schröder, Kleinheubacher Berichte **41**, 133 (1998)
102. R. Augustin, A. Bauch, R. Schröder, *Proceedings of the 11th European Frequency and Time Forum* (Neuchatel, 1997), pp. 47–52
103. E. Simon, Ph. Laurent, A. Clairon, Phys. Rev. A **57**, 436 (1998)
104. E. Tiesinga, B.J. Verhaar, H.T.C. Stoof, D. van Bragt, Phys. Rev. A **45**, R2671 (1992)
105. P.J. Leo, P.S. Julienne, F.H. Mies, C.J. Williams, Phys. Rev. Lett. **86**, 3743 (2001)
106. S.J.J.M.F. Kokkelmans, B.J. Verhaar, K. Gibble, D.J. Heinzen, Phys. Rev. A **56**, R4389 (1997)
107. Y. Sortais, S. Bize, C. Nicolas, A. Clairon, C. Salomon, C. Williams, Phys. Rev. Lett **85**, 3117 (2000)
108. Y. Sortais, S. Bize, C. Nicolas, G. Santarelli, C. Salomon, A. Clairon, IEEE Trans. Ultrason. Ferroel. Freq. Control **47**, 1093 (2000)

109. J. Guéna, F. Chapelet, P. Rosenbusch, P. Laurent, M. Abgrall, G.D. Rovera, G. Santarelli, M.E. Tobar, S. Bize, A. Clairon, *Proceedings of the 22nd European Frequency and Time Forum* (Toulouse, 2008)
110. T.P. Heavner, S.R. Jefferts, E.A. Donley, J.H. Shirley, T.E. Parker, Metrologia **42**, 411 (2005)
111. F. Pereira Dos Santos, H. Marion, S. Bize, Y. Sortais, A. Clairon, Phys. Rev. Lett. **89**, 233004 (2002)
112. J.C. Camparo, R.P. Frueholz, J. Phys. B **17**, 4169 (1984)
113. H. Marion, Contrôle des collisions froides du 133*Cs*, test de la variation de la constante de structure fine à l'aide d'une fontaine atomique double rubidium–césium. Ph.D. thesis, Université de Paris VI (2005)
114. K. Szymaniec, W. Chalupczak, E. Tiesinga, C.J. Williams, S. Weyers, R. Wynands, Phys. Rev. Lett. **98**, 153002 (2007)
115. W. Chalupczak, K. Szymaniec, J. Phys. B **40**, 343 (2007)
116. K. Szymaniec, W. Chalupczak, S. Weyers, R. Wynands, Appl. Phys. B **89**, 187 (2007)
117. G. Petit, P. Wolf, IEEE Trans. Instrum. Meas. **46**, 201 (1997)
118. N.K. Pavlis, M.A. Weiss, Metrologia **40**, 66 (2003)
119. K. Gibble, Phys. Rev. Lett **97**, 073001 (2006)
120. T.P. Heavner, S.R. Jefferts, E.A. Donley, J.H. Shirley, T.E. Parker, IEEE Trans. Instrum. Meas. **54**, 842 (2005)
121. S. Weyers, R. Wynands, TAI calibration report of PTB-CSF1 to the BIPM (June 2008)
122. X. Baillard, A. Gauget, S. Bize, P. Lemonde, P. Laurent, A. Clairon, P. Rosenbusch, Opt. Commun. **266**, 609 (2006)
123. W. Happer, Rev. Mod. Phys. **44**, 169 (1972)
124. J.C. Camparo, C.M. Klimcak, S.J. Herbulock, IEEE Trans. Instrum. Meas. **54**, 1873 (2005)
125. J. Vanier, C. Mandache, Appl. Phys. B **87**, 565 (2007)
126. J. Kitching, L. Hollberg, S. Knappe, R. Wynands, Opt. Lett. **26**, 1507 (2001)
127. E. Arimondo, Prog. Opt. **XXXV**, 257 (1996)
128. O. Schmidt, R. Wynands, Z. Hussein, D. Meschede, Phys. Rev. A **53**, R27 (1996)
129. C. Affolderbach, A. Nagel, S. Knappe, C. Jung, D. Wiedenmann, R. Wynands, Appl. Phys. B **70**, 407 (2000)
130. S. Brandt, A. Nagel, R. Wynands, D. Meschede, Phys. Rev. A **56**, R1063 (1997)
131. P.R. Hemmer, S. Ezekiel, C.C. Leiby, Jr., Opt. Lett. **8**, 440 (1983)
132. R. Wynands, A. Nagel, Appl. Phys. B **68**, 1 (1999)
133. J. Vanier, Appl. Phys. B **81**, 421 (2005)
134. J. Kitching, L. Hollberg, S. Knappe, R. Wynands, Electron. Lett. **37**, 1449 (2001)
135. S. Knappe, V. Shah, P.D.D. Schwindt, L. Hollberg, J. Kitching, L.-A. Liew, J. Moreland, Appl. Phys. Lett. **85**, 1460 (2004)
136. S. Knappe, Compr. Microsyst. **3**, 571 (2007)
137. J. Deng, P. Vlitas, D. Taylor, L. Perletz, R. Lutwak, *Proceedings of the 22nd European Frequency and Time Forum* (Toulouse, 2008)
138. A. Godone, F. Levi, S. Micalizio, *Coherent Population Trapping Maser* (Edizioni C.L.U.T, Torino, 2002)
139. H.M. Goldenberg, D. Kleppner, N.F. Ramsey, Phys. Rev. Lett. **5**, 361 (1960)
140. R.F.C. Vessot, Metrologia **42**, S80 (2005)
141. P. Gill, Metrologia **42**, S125 (2005)
142. E. Peik, J. Abel, Th. Becker, J. von Zanthier, H. Walther, Phys. Rev. A **60**, 439 (1999)
143. T. Schneider, E. Peik, C. Tamm, Phys. Rev. Lett. **94**, 230801 (2005)
144. T. Nazarova, F. Riehle, U. Sterr, Appl. Phys. B **83**, 531 (2006)
145. W.H. Oskay, W.M. Itano, J.C. Bergquist, Phys. Rev. Lett. **94**, 163001 (2005)
146. E. Peik, T. Schneider, C. Tamm, J. Phys. B: At. Mol. Opt. Phys. **39**, 145 (2006)
147. T.W. Hänsch, Chem. Phys. Chem. **7**, 1170 (2006)
148. L. Hollberg, S. Diddams, A. Bartels, T. Fortier, K. Kim, Metrologia **42**, S105 (2005)

149. T. Rosenband, D.B. Hume, P.O. Schmidt, C.W. Chou, A. Brusch, L. Lorini, W.H. Oskay, R.E. Drullinger, T.M. Fortier, J.E. Stalnaker, S.A. Diddams, W.C. Swann, N.R. Newbury, W.M. Itano, D.J. Wineland, J.C. Bergquist, Science **319**, 1808 (2008)
150. P.O. Schmidt, T. Rosenband, L. Langer, W.M. Itano, J.C. Bergquist, D.J. Wineland, Science **309**, 749 (2005)
151. T. Rosenband, , P.O. Schmidt, D.B. Hume, W.M. Itano, T.M. Fortier, J.E. Stalnaker, K. Kim, S.A. Diddams, J.C.J. Koelemeij, J.C. Bergquist, D.J. Wineland, Phys. Rev. Lett. **98**, 220801 (2007)
152. H. Katori, *Proceedings of the 6th Symposium on Frequency Standards and Metrology*, P. Gill (ed.) (World Sicentific, Singapore, 2002), pp. 323–330
153. S. Blatt, A.D. Ludlow, G.K. Campbell, J.W. Thomsen, T. Zelevinsky, M.M. Boyd, J. Ye, X. Baillard, M. Fouché, R. Le Targat, A. Brusch, P. Lemonde, M. Takamoto, F.-L. Hong, H. Katori, V.V. Flambaum, Phys. Rev. Lett. **100**, 140801 (2008)
154. Procès-Verbeaux des Séances du Comité International des Poids et Mesures **74** (2006) (2007)
155. Comptes Rendus de la 23e Conférence Générale des Poids et Mesures (2007), Resolution 12 (2009). URL http://www.bipm.org/en/CGPM/db/23/12,\citedon2008–08–06
156. E. Peik, Chr. Tamm, Europhys. Lett. **61**, 181 (2003)
157. B.R. Beck, J.A. Becker, P. Beiersdorfer, G.V. Brown, K.J. Moody, J.B. Wilhelmy, F.S. Porter, C.A. Kilbourne, R.L. Kelley, Phys. Rev. Lett. **98**, 142501 (2007)
158. B. Guinot, G. Petit, Astron. Astrophys. **248**, 292 (1991)
159. G. Petit, P. Tavella, Astron. Astrophys. **308**, 290 (1996)
160. D.R. Lorimer, Living Rev. Relativity **8** (2005). [Online Article]: cited [2008–07–24] http://www.livingreviews.org/lrr-2005-7
161. C. Lämmerzahl, G. Ahlers, N. Ashby, M. Barmatz, P.L. Biermann, H. Dittus, V. Dohm, R. Duncan, K. Gibble, J. Lipa, N. Lockerbie, N. Mulders, C. Salomon, Gen. Relativity Gravitation **36**, 615 (2004)
162. S.R. Jefferts, T.P. Heavner, L.W. Hollberg, J. Kitching, D.M. Meekhof, T.E. Parker, W. Phillips, S. Rolston, H.G. Robinson, J.H. Shirley, D.B. Sullivan, F.L. Walls, N. Ashby, W.M. Klipstein, L. Maleki, D. Sidel, R. Thompson, S. Wu, L. Young, R.F.C. Vessot, A. De Marchi, *Proceedings of Joint Meeting of the European Frequency and Time Forum – IEEE International Frequency Control Symposium* (Besancon, 1999), pp. 141–144
163. C. Fertig, K. Gibble, B. Klipstein, L. Maleki, D. Seidel, R. Thompson, *Proceedings of the 2000 International Frequency Control Symposium* (New Haven, 2000), pp. 676–679
164. C. Salomon, N. Dimarcq, M. Abgrall, A. Clairon, Ph. Laurent, P. Lemonde, G. Santarelli, P. Uhrich, L.G. Bernier, G. Busca, A. Jornod, P. Thomann, E. Samain, P. Wolf, F. Gonzalez, P. Guillemot, S. Leon, F. Nouel, C. Sirmain, S. Feltham, C. R. Acad. Sci. Paris, Ser. IV **2**, 1313 (2001)
165. A. Bauch, S. Weyers, Phys. Rev. D **65**, R081101 (2002)
166. A. Bauch, L. Nelson, T. Parker, S. Weyers, *Proceedings of the 2003 International Frequency Control Symposium and 17th European Frequency and Time Forum* (Tampa, 2003)
167. N. Ashby, T.P. Heavner, S.R. Jefferts, T.E. Parker, A.G. Radnaev, Y.O. Dudin, Phys. Rev. Lett. **98**, 070802 (2007)
168. H. Marion, F. Pereira Dos Santos, M. Abgrall, S. Zhang, Y. Sortais, S. Bize, I. Maksimovic, D. Calonico, J. Grünert, P. Lemonde, G. Santarelli, Ph. Laurent, A. Clairon, C. Salomon, Phys. Rev. Lett. **90**, 150801 (2003)
169. S. Bize, S.A. Diddams, U. Tanaka, C.E. Tanner, W.H. Oskay, R.E. Drullinger, T.E. Parker, T.P. Heavner, S.R. Jefferts, L. Hollberg, W.M. Itano, J.C. Bergquist, Phys. Rev. Lett. **90**, 150802 (2003)
170. S. Karshenboim, E. Peik, eds., *Astrophysics, Clocks and Fundamental Constants*, Lect Notes. Phys. **648** (Springer, Berlin, 2004)
171. M. Fischer, N. Kolachevsky, M. Zimmermann, R. Holzwarth, Th. Udem, T.W. Hänsch, M. Abgrall, J. Grünert, I. Maksimovic, S. Bize, H. Marion, F. Pereira Dos Santos, P. Lemonde, G. Santarelli, P. Laurent, A. Clairon, C. Salomon, M. Haas, U.D. Jentschura, C.H. Keitel, Phys. Rev. Lett. **92**, 230802 (2004)

172. E. Peik, B. Lipphardt, H. Schnatz, T. Schneider, Chr. Tamm, S.G. Karshenboim, Phys. Rev. Lett. **93**, 170801 (2004)
173. S.N. Lea, Rep. Prog. Phys. **70**, 1473 (2007)
174. T.M. Fortier, N. Ashby, J.C. Bergquist, M.J. Delaney, S.A. Diddams, T.P. Heavner, L. Hollberg, W.M. Itano, S.R. Jefferts, K. Kim, F. Levi, L. Lorini, W.H. Oskay, T.E. Parker, J. Shirley, J.E. Stalnaker, Phys. Rev. Lett. **98**, 070801 (2007)
175. S.G. Karshenboim, E. Peik, Eur. Phys. J. Special Topics **163**, 1 (2008)
176. S. Bize, Y. Sortais, M.S. Santos, C. Mandache, A. Clairon, C. Salomon, Europhys. Lett. **45**, 558 (1999)
177. E. Peik, B. Lipphardt, H. Schnatz, Chr. Tamm, S. Weyers, R. Wynands, arXiv:physics/0611088v1 (2006)

Index